Tail biting in pigs

Tail biting in pigs

A comprehensive guide to its aetiology, impact and wider significance in pig management

Edited by

Keelin O'Driscoll
Anna Valros

BRILL | WAGENINGEN ACADEMIC

The Library of Congress Cataloging-in-Publication Data is available online at https://catalog.loc.gov
LC record available at https://lccn.loc.gov/2024937795

Typeface for the Latin, Greek, and Cyrillic scripts: "Brill". See and download: brill.com/brill-typeface.

ISBN 978-90-04-69134-6 (hardback)
ISBN 978-90-04-69498-9 (e-book)
DOI 10.3920/9789004694989

This book is printed on acid-free paper and produced in a sustainable manner.

Contents

PART 2
Risk factors associated with the physical environment

PART 3
Methods of addressing the problem

PART 4
The bigger picture

Figures and tables

Figures

Tables

Introduction

Keelin O'Driscoll and Anna Valros

Tail biting is a behavioural disorder in pigs that emerged as a significant problem when pig farming intensified during the middle of the 20th century. At the time, producers recognised that the risk for biting was increasing, so docking of the tail was an obvious and immediately available preventative measure; logically, and in practice, removal of the target of the biting behaviour reduced the risk of it occurring. As a consequence tail docking become so commonplace that today in most countries there is virtually no cultural experience or memory of rearing pigs without docking their tails. In fact it is often perceived to be a necessary procedure to ensure good welfare; if it were not carried out, biting outbreaks are almost certain to occur, and to be so severe that without human intervention pigs would suffer extreme injuries, and in many cases die.

Nevertheless, tail biting still persists at an unacceptably high level. This is because even though the pigs' tails have been removed, the root causes of the motivation for biting have not been. The intensification of the pig industry that occurred during the last century was driven by an overarching objective to maximise food production while lowering its cost to consumers – this objective was overwhelmingly met, enabled by genetic selection, and technological developments that allowed animals to be reared efficiently to slaughter age while reducing inputs that involve costs, such as labour, feed, and the amount of space that animals could access. The importance of accommodating for the behavioural needs of pigs however, was not understood or appreciated at the time. Tail biting was often referred to as a 'vice' – an immoral or wicked behaviour – the implication being that the pigs cognitively engaged in this horrible activity, rather than it evolving as a consequence of the stress induced by the conditions that they were placed in by their human caretakers.

Since then understanding of animal welfare, and specifically welfare associated with an animals mental state, has advanced significantly. Animal welfare emerged as a scientific discipline during the 1960s, and has shifted from being a niche theoretical area, to now being firmly established as a major research field within the animal science community. This, as well as increasing societal concerns regarding how animals are treated, has contributed hugely to the development of national and international policies regarding agricultural animal production. In the EU for instance, it is 30 years since the routine tail docking of pigs has been banned. This is in recognition of the fact that tail docking is merely a remedial measure that permits pigs to be produced economically efficiently in systems that do not meet their needs, and with the aim at the time of moving towards systems that do. Tail biting is not observed in wild or free ranging pigs, and as such can be considered a psychopathological behaviour, a non-adaptive behaviour which in this case is both indicative of

 | DOI:10.3920/9789004694989_002

and causes welfare impairment. The fact that even with the existence of this legislation the vast majority of pigs in the EU are still docked is evidence of the scale of the challenge to 'turn back the clock' to rearing pigs in conditions that at the very least, do not result in a significant risk of such an abnormal behaviour.

Tail biting is often superficially thought of as a single issue in pig production. However, when considered in more depth, it is clear that to resolve the problem a transdisciplinary approach is essential. The reason that tail biting has become entrenched as a problem in pig production is because of the widespread and enduring change in how pigs are managed that has occurred during the past 100 years. Thus the solution to the problem likely necessitates a similar level of systemic change. For this to happen, understanding of the causes and impacts of, and how to ameliorate tail biting is an essential first step. We hope that this book will act as a useful and instructive guide for both academics and all stakeholders in pig production, including farmers, policy makers, and retailers, to help to bring to an end the era of this abnormal behaviour being a 'normal' problem in pig production.

Introduction to the contents of the book

The subject of tail biting can be approached from a wide range of research backgrounds and areas of expertise, so determining a structure to the book that would flow naturally was challenging. In the end, the chapters fell into five general thematic areas, which we ordered so that the reader could move from understanding the historical background and root causes of the disorder, through to the bigger picture of how tail biting interacts with other health, welfare and performance issues, and ethical considerations.

The first section provides the reader with some background to the topic of tail biting, from three different perspectives. In chapter 1, Peter Stevenson provides essential context for the subject of tail biting and docking in the legal sphere, and the challenges in controlling both even when legislation with the aim of doing so has been in place for decades. The chapter explains why the initial EU legislation regarding tail docking and welfare was written, and the various attempts since then to update, clarify, and enforce it, along with the lack of success in doing so. In Chapter 2, Dale Sandercock and his co-authors provide detailed knowledge on the structure of the tail, and how this is affected by both docking and biting. The chapter also describes some of the functions of the tail, namely how tail posture and movement can indicate the pigs' mood. Finally, Rick D'Eath and Marc Bracke round off the first section of the book in Chapter 3, by delving into the evolutionary factors that could underlie the performance of tail biting. The chapter explores the potential ultimate causes of biting (as per Tinbergens four questions), an approach that has been underutilised in much of the tail biting research and literature. The chapter moves through discussion of pig genetics, through to several hypotheses as to the evolutionary origin of the behaviour.

The second and third sections of the book contain three chapters each, all of which tackle a distinct risk factor area for tail biting. Section Two is concerned with the individual pig, and the social environment. In Chapter 4 Laurianne Canario and Catherine Larzul discuss factors including genetic background, sex, individual coping ability, and even the composition of the gut microbiota, and how these can not only directly influence the likelihood that a pig will perform tail biting behaviour itself, but also influence its pen-mates to perform the behaviour. The chapter suggests that improvements in scientific capability to capture individual pig data, including automated recording of behaviour, will hugely aid understanding of these factors and how they influence risk – currently much of the data that is collected is from a sub-sample of animals in affected groups, at limited time-points. As an extension of this, Chapter 5 delves more deeply into how poor health can affect mental state and influence the performance of tail biting. This is an emerging research area even in human medicine, with the complex mechanisms underlying the relationship not fully elucidated. Hege Lund and Janicke Nordgreen hypothesise that the lack of social motivation and depression that animals experience while sick (or in a 'pro-inflammatory state') could act as a trigger for outbreaks of damaging behaviour; thus poor health could have further consequences, beyond the direct experience of illness, when considering animal welfare. Finally, Inonge Reimert and Winanda Ursinus explore in depth the social environment, which is drastically different for commercial pigs to that of their wild living counterparts. A lack of agency, including the ability to exert control over the composition of social groups, is known to be a significant contributor to stress and impaired welfare, which could have knock on impacts on animal behaviour. The authors suggest that it could be worth exploring how groups are typically composed on farm compared to natural distributions with regard to both pig type (e.g. age, size, sex), and individual personalities.

Section three moves on to discussion of physical risk factors, so those that are external to the pig itself. These are broken down into the physical environment the pigs live in, environmental enrichment, and feed and feeding. The first chapter in this section, Chapter 7, describes how both overall farming system type as well as individual aspects of a system can impact welfare in several ways, with many particularly reducing the ability of pigs to express agency. The chapter concludes that by their very nature, indoor intensive systems are more likely to present risks for tail biting. Environmental enrichment, although a component of the physical environment, warrants a chapter of its own; it is considered so essential to ensure good pig welfare that along with the ban on tail docking, appropriate and sufficient enrichment provision has been enshrined in pig legislation in the EU for 30 years. Unfortunately though this is also rarely complied with, as modern conventional pig pens and slurry systems did not evolve to be able to handle large volumes of what the pigs prefer – loose organic material. Stephanie Buijs and her co-authors describe not only the characteristics of the types of enrichment normally used, but also the limitations and challenges associated with provision of optimal types, as well as the importance of not considering it a 'fix-all' solution. Finally, Sandra Edwards provides an overview of how methods of providing feed, and the composition and form of the feed itself, can contribute to the risk of damaging behaviours occurring. As feed is such an important and valuable resource for pigs, any situation that leads to competition will increase this risk. It is also vitally important that nutritionists and producers carefully match the nutritional composition of diets to the pigs' needs, as imbalances can impact the synthesis of brain neurotransmitters, as well as impacting behaviour and mood via the composition and action of the gut microbiome, thus linking back to Chapters 4, 5 and 7.

The next section of the book deals more specifically with how to assess, and potentially reduce, the risk of tail biting occurring. Sabine Dippel has contributed a chapter on risk assessment, which includes detail of not only the level and types of risk that can be assessed (e.g. on a daily basis, for farm planning, and research), but also draws together some of the risk factors that have been highlighted in previous chapters. This chapter also introduces the usefulness of tail lesion scoring both on farm and in the abattoir, which is expanded further in the next chapter, by Anna Valros and Laura Boyle, which deals specifically about where, when, and how tail lesions should be measured. There has been a proliferation of lesion scoring systems both in scientific literature, and practically in abattoirs and on farm. Thus although measuring lesions has many benefits, it is currently difficult to benchmark individual farms, systems, and countries, against each other. The chapter dissects the pros and cons of a range of methods, and proposes that scoring lesions in the abattoir is likely the easiest to standardise, and should thus be the site used for systematic scoring. Indeed the chapter proposes that scores in the abattoir could form an iceberg indicator of pig welfare, a concept that is developed further in chapter 15.

The two other chapters in this section deal not with pigs themselves, but with people, and technology. Ultimately, it is human choices, at a wide range of levels, that will determine how pigs are managed in the future. In Chapter 12 Grace Carroll and her co-authors describe how behaviour change theory can help us to understand the predictors of stakeholder behaviour, rather than depending upon top-down regulatory governance to change it. The chapter also includes an intervention design guide for those working with pigs and stakeholders in the pig industry, with hypothetical and real-world examples to aid people unfamiliar with the psychological foundations of Human Behaviour Change. Finally, the last chapter in this section, by Mona Larsen and colleagues, deals with the subject of precision livestock farming. Ongoing work in this burgeoning scientific area is providing novel and innovative opportunities to reduce the labour associated with identifying tail biting, increasing our ability to identify individual pig factors that could be playing into it, and even potentially improving the likelihood of predicting its occurrence. The ability to automatically isolate and collect behavioural data from individual pigs will contribute hugely to meeting the recommendations in Chapter 4.

The final section of the book has three chapters, each dealing in a way with the consequences of tail biting, and which are more theoretical in nature than the many of the preceding ones. In Chapter 14 Manja Zupan and Antonia Patt describe how tail condition interacts with other aspects of pig health and welfare, and propose a novel theoretical model that considers an intact tail to be an iceberg indicator of pig welfare. Although tail condition had previously been suggested in the literature to be a good indicator of the welfare status of a pig at a specific time, or sometimes during its lifetime, this is the first time that a detailed model has been comprehensively, and so convincingly described. Application of this model has potential to provide a strategy for rapid and reliable welfare assessment, whether on farm, or at the abattoir, and is robust across a range of production systems. Chapter 15, contributed by Jarkko Neimi and Peter Stevenson, addresses the costs associated with tail biting and docking from an economic and legal perspective, and concludes that as preventing tail biting requires constant effort it is likely to be financially prohibitive compared to masking the problem with tail docking. For preventative measures to be adopted, changes to the economic structure of pig industry must thus be implemented so that pig producers are better off after than before the change. Finally, the last chapter in the book by Berenice Bovenkerk and her co-authors delves into ethical issues associated with tail biting and docking. A range of ethical perspectives are explained, and a suggestion that both ideal, and non-ideal perspectives are useful; a combined approach enables a final goal to be aimed for (ideal), yet also includes more potentially achievable steps along the way to be identified (non-ideal). The chapter concludes that overall, there is agreement across theories on a number of issues: the underlying causes of tail biting should be identified, and solutions found, because it is morally problematic to keep pigs in a system where they need to be routinely mutilated, rather than changing the environment to accommodate the pigs needs better.

CHAPTER 1

The legal prohibition on routine tail docking and the long road to achieving compliance

Compassion in World Farming: tail docking of pigs – where are we, and how did we get here

Peter Stevenson

Chief Policy Advisor, Compassion in World Farming, River Court, Mill Lane, Godalming, Surrey GU7 1EZ, UK
Peter@ciwf.org

Abstract

The European Union's prohibition of routine tail docking of pigs was first adopted in 1991. Its wording was strengthened in 2001 and now appears in the codified Pigs Directive of 2008. This provides that before farmers can lawfully tail dock they must have genuinely taken 'other measures' to prevent tail biting and in particular must have changed 'inadequate environmental conditions or management systems'. Over the last twenty years scientific reports, guidance documents and a European Commission Recommendation have examined the steps that can be taken to minimise tail biting and thus to avoid routine tail docking. Over a similar period audit reports by the Commission and investigations by animal welfare organisations have revealed regular breaches of the ban on routine tail docking. Despite this, compliance with the law by the pig sector and enforcement by all Member States except Finland and Sweden is poor. In 2021 the Commission recognised that 'tail-docking of pigs is a routine practice in 26 out of 28 Member States and approximately 150 million pigs annually are subject to this practice'.

Keywords

compliance, legal restriction, docking ban

1 The legislative provisions

On 19 November 1991 the European Communities adopted a prohibition on the routine tail docking of pigs. The prohibition was set out in Council Directive 91/630 laying down minimum standards for the protection of pigs, which came into force on 1 January 1994 (European Communities, 1991). The Directive provided 'Neither tail docking nor tooth clipping must be carried out routinely but only when there is evidence, on the farm, that injuries to sows' teats or to other pigs' ears or tails have occurred as a result of not carrying out these procedures'.

This wording was strengthened in 2001 by Commission Directive 2001/93/EC which came into force on 1 January 2003. The new wording provides: 'Neither tail docking nor reduction of corner teeth must be carried out routinely but only where there is evidence that injuries to sows' teats or to other pigs' ears or tails have occurred. Before carrying out these procedures, other measures shall be taken to prevent tail biting and other vices taking into account environment and stocking densities. For this reason inadequate environmental conditions or management systems must be changed' (European Commission, 2001). The new wording clarifies what is meant by the prohibition on carrying out tail docking 'routinely'; this is considered further below. This is an important legislative development as it requires farmers to improve the conditions in which pigs are kept, rather than resorting to routine tail docking.

The 1991 Directive and the 2001 amendments were, in the interests of clarity, codified by Council Directive 2008/120/EC laying down minimum standards for the protection of pigs. The provision on tail docking in the 2008 Directive is identical to that of the 2001 Commission Directive (European Union, 2008). This means that the ban on tail docking has been in force since 1 January 2003 and, in an earlier form since 1 January 1994. The 1991, 2001 and 2008 Directives are referred to hereafter as 'the Pigs Directive'.

2 Background to the legislation

Scientific research shows that in natural conditions pigs are highly active, spending 75% of their day rooting, foraging and exploring (Stolba and Woodgush, 1989). Such activities are impossible for intensively farmed pigs. The lack of straw or other natural materials prevents pigs from carrying out their innate behaviours. Bored and frustrated, they turn to the only other 'thing' in their bare pens: the tails of other pigs. They begin to chew and then bite those tails. Lack of a stimulating environment is not the only factor causing tail biting, but it is perhaps the most obvious; inspectors can rapidly see if an environment is barren with no enrichment present.

To prevent tail biting, farmers cut off (dock) part of the young piglet's tail. However, docking does not address the underlying issue, namely the unsuitable environment in which pigs are reared. Indeed, scientific research has for many years shown that the correct way to prevent tail biting is not to dock the tails but to keep the pigs in good conditions, in particular to give them straw or other manipulable materials to enable rooting and investigation behaviour. The European Commission's Scientific Veterinary Committee (SVC) in its 1997 report on the welfare of intensively kept pigs stated: 'The problems of injury following tail-biting should be solved by improved management rather than by tail-docking' (European Commission Scientific Veterinary Committee, 1997).

The SVC report concluded that tail-docking is likely to be painful when it is carried out and that in some cases it leads to 'prolonged pain' (European Commission Scientific Veterinary Committee, 1997). EFSA's 2007 Scientific Opinion concluded that since 'tail-docking is likely to be painful, both in the short term and as a result of possible long-term pain from neuroma formation, measures other than tail-docking should be implemented to control tail-biting and its adverse effects for welfare' (EFSA Opinion, 2007).

3 Related legislative provisions on enrichment

The importance of providing enrichment for pigs and its role in minimising tail biting was recognised by the 1991 Pigs Directive. This provided: 'In addition to the measures normally taken to prevent tail-biting and other vices and in order to enable them to satisfy their behavioural needs, all pigs, taking into account environment and stocking density, must be able to obtain straw or any other suitable material or object' (Annex, Chapter I, point 16).

The 2001 Pigs Directive strengthened this requirement. It provides that 'pigs must have permanent access to a sufficient quantity of material to enable proper investigation and manipulation activities, such as straw, hay, wood, sawdust, mushroom compost, peat or a mixture of such, which does not compromise the health of the animals'. This requirement is repeated in the 2008 Pigs Directive (Annex I, Chapter I, point 4).

Farmers who are not providing effective enrichment and who are routinely tail docking are in breach of the 2008 Pigs Directive's provisions on both enrichment and tail docking as the failure to provide enrichment means that they are tail docking without first having tried to prevent tail biting by changing inadequate environmental conditions or management systems.

4 Interpreting the legislation

A Technical Report prepared for EFSA in 2011 has highlighted the importance of the provisions on tail docking and enrichment (Spoolder *et al.*, 2011). It concludes that 'all new data reinforce the importance of providing suitable enrichment materials to allow expression of species relevant behaviours and reduce risk of injurious biting'. It also states that 'an intact curly tail may well be the single most important animal-based welfare indicator for weaned, growing and finishing pigs ... In addition, it stands for high-quality management and respect for the integrity of the pig'.

4.1 *Tail docking*

The 2008 Pigs Directive does not prohibit all tail docking. It provides that before farmers can lawfully tail dock they must have genuinely taken 'other measures' to prevent tail biting and in particular must have changed 'inadequate environmental conditions or management systems'. Scientific research can help interpret which environmental conditions or management systems are 'inadequate' and what are the 'other measures' that must be taken before farmers may lawfully tail dock.

If, for example, pigs are kept in barren conditions with no effective enrichment materials, the farmer has failed to change 'inadequate environmental conditions'. EFSA has concluded that 'the largest risk for being tail bitten is the lack of appropriate enrichment' (EFSA Opinion, 2007). EFSA has also stated that home-made plastic objects (such as 'helicopter' toys) or commercial equivalents (such as plastic 'chew-toys') 'appear to be relatively ineffective at preventing tail biting' (EFSA Report, 2007).

If a competent authority inspector sees that pigs are kept in barren conditions without enrichment materials or with only chains, toys, chewing sticks, plastic objects or tyres, the inspector must – on the basis of EFSA's Opinion and Report on tail docking – conclude that the farmer has failed to change 'inadequate environmental conditions or management systems'. Accordingly, the inspector must conclude that the farmer has not fulfilled the conditions – set out in the 2008 Pigs Directive – that would allow them to lawfully tail dock.

It is not sufficient for farmers to have at some previous time – possibly in the distant past – tried to prevent biting but without success. They must continue to try to prevent biting by measures other than docking. The Pigs Directive requires farmers to make serious and sustained attempts to find other ways of preventing tail biting. This was confirmed by a 2019 presentation by the European Commission (hereafter referred to as 'the Commission') which stated 'As long as tails are being docked, farmers must engage in a continuous cycle of improvement measures to improve on farm conditions' (European Commission, 2019).

The Commission's 2016 Recommendation and accompanying Staff Working Document throw further light on the interpretation of the Directive's provisions on tail docking and enrichment. These documents are considered below.

4.2 *Enrichment*

The 2008 Pigs Directive provides that pigs 'must have permanent access to a sufficient quantity of material to enable proper investigation and manipulation activities, such as straw, hay, wood, sawdust, mushroom compost, peat or a mixture of such ...'.

Farmers may use a material other than one of those specified in the Directive but the term 'such as' means that the material provided must be as effective as those listed in fulfilling the objective of the legislation which is that pigs must be able to engage in 'proper investigation and manipulation activities'. In assessing whether a material that is not specified in the Directive is effective in enabling pigs to engage in 'proper investigation and manipulation activities', competent authorities, veterinarians and farmers should be guided both by their own observations on the farm and by scientific research. A Technical Report submitted to EFSA in 2011 states that an 'appropriate enrichment material can be defined as a material which stimulates exploratory behaviour for an extended length of time' (Spoolder *et al.*, 2011).

An Answer by the Commission (E-5360/09EN) to a Question from a Member of the European Parliament (MEP) states that 'The conclusions and recommendations of the [EFSA] opinions regarding fattening pigs[1] and tail docking[2] are quite clear on enrichment materials. Indeed the scientific opinion on fattening pigs states that since indestructible objects such as chains or tyres are not sufficient to provide for the manipulatory need of pigs, they may be used as ***supplement*** to destructible and rooting materials but not as a ***substitute*** for them' [emphasis added].

5 Attempts to secure compliance with the prohibition on tail docking

The attempts to secure proper compliance with the prohibition of routine tail docking arise not only from a concern about the painful nature of docking, but from a recognition that tail docking plays an important part in propping up intensive pig production. If farmers could not dock, they would have to farm to much higher welfare and health standards.

Gradually over the last 25 years awareness of the importance of avoiding tail docking and how to minimise tail biting have been built up through scientific research, lobbying by animal welfare organisations, and workshops organised by the Commission as well as guidance published, or commissioned, by the Commission.

In 2007 EFSA showed that – despite the ban on routine tail docking – over 90% of EU pigs were still being tail docked (EFSA Report, 2007). In 2008 Compassion in World Farming (Compassion) carried out an 18 month investigation in six key pig producing countries: Germany, Denmark, Hungary, Netherlands, UK and Spain. They visited 74 farms. Frequent breaches of the legislation on tail docking and enrichment were found. Nearly all the pigs seen were tail docked. The vast majority of the farms either provided no enrichment materials or materials that have been found by EFSA to be ineffective. In addition, a 2009 investigation in 13 farms in France found no enrichment materials and routine tail docking.

Formal Complaints were made to the Commission in respect of the failure by the competent authorities of Denmark, Germany, Hungary, Ireland, Belgium, Spain and France to secure compliance with EU legislation that prohibits the routine tail docking of pigs. Although the Commission did not open infringement proceedings against the Member States concerned, a letter dated 7 June 2010 from the

1 Scientific Opinion of the Panel on Animal Health and Welfare on a request from the Commission on animal health and welfare in fattening pigs in relation to housing and husbandry. The EFSA Journal (2007) 564, 1–14.

2 Scientific Opinion of the Panel on Animal Health and Welfare on a request from the Commission on the risks associated with tail biting in pigs and possible means to reduce the need for tail docking considering the different housing and husbandry systems. The EFSA Journal (2007) 611.

Commission to Compassion in World Farming Ireland stated: 'our exchange with the Irish authorities following your complaint has helped raise awareness on the importance of enhancing compliance with the provisions of Directive 2008/120/EC on enrichment material and the prohibition of routine tail-docking'.

Reports by the Commission's Food and Veterinary Office (FVO) of audits carried out in 2008–2012 concerning 19 Member States showed a widespread failure by the pig industry to comply with, and the Member States to enforce (i) the prohibition on routine tail docking and (ii) the requirement to provide enrichment materials (FVO, 2008; FVO, 2009a; FVO, 2009b; FVO, 2009c; FVO, 2009d; FVO, 2009e; FVO, 2009f; FVO, 2009g; FVO, 2010a; FVO, 2010b; FVO, 2010c; FVO, 2010d; FVO, 2010e; FVO, 2010f; FVO, 2010g; FVO, 2010h; FVO, 2011a; FVO, 2011b; FVO, 2011c; FVO, 2011d; FVO, 2011e; FVO, 2011f; FVO, 2012a; FVO, 2012b; FVO, 2012c).

In order to disseminate information on adequate enrichment materials and the avoidance of tail biting, the European Commission organised a workshop on pig welfare which was held on 17 November 2009 in Brussels and where best farming practices were presented (European Commission, 2009). Speaking at the workshop French pig farmer Thierry Schweitzer explained that he did not need to dock his pigs' tails as he provided them with ample straw; the pigs were sufficiently engaged with the straw and so did not turn their attention to the tails of their pen-mates.

In response to comments from two Member State officials that 'it was up to animal welfare organisations to say what constituted effective enrichment', in 2011 Compassion published guidance on *Providing enrichment for pigs* (Compassion, 2011). In 2013 a version of the guidance particularly focussing on countries with hot climates was published.

Some farmers had been saying that their veterinarian had issued a certificate authorising them to tail dock and had told them that, as they had fully slatted floors, they could use metal chains or plastic toys as enrichment. Indeed, a FVO audit of Belgium stated: 'Veterinary certificates are routinely issued by private practitioners on therapeutic grounds authorising farmers to carry out ... tail docking and are accepted by the CA [competent authority] as sufficient grounds for this practice' (FVO, 2009). The Pigs Directive does not indicate that competent authorities can rely on such veterinary certificates. Indeed veterinarians should not issue such certificates on a routine basis but only where they are satisfied that the farmer has first taken genuinely effective measures to change inadequate environmental conditions and management systems as required by the Directive. In January 2012 Compassion wrote to the European Association of Porcine Health Management (EAPHM). The letter said 'Veterinarians clearly have an important role in advising pig farmers as to the steps they need to take to comply with the Directive. In some cases it appears that veterinarians may inadvertently be incorrectly advising their clients that the provision of metal chains or plastic objects meets the requirements of the Directive and that they may regularly tail dock'. The reply from the EAPHM included no recognition of the need to advise veterinarians not to authorise the use of metal chains or plastic objects nor to issue certificates advising their clients that they may tail dock unless they were satisfied that the farmer had first taken measures to change inadequate environmental conditions and management systems (EAPHM, 2012).

Evidence of widespread failures to comply with and enforce the legislation emerged when in 2013 Compassion investigated pig farms in six Member States: Italy, Spain, Poland, the Czech Republic, Ireland and Cyprus. In all 45 farms were visited. All but one of these farms were breaking the legislation on enrichment and tail docking. Details of the investigations and breaches are set out in a report by Compassion entitled *Widespread breaches of pig welfare laws in the EU* (Compassion, 2013).

Despite all the evidence produced over several years by the FVO and animal welfare organisations of failure to implement the Directive's provisions on tail docking and enrichment, little or no improvement in compliance and enforcement was evident. A fresh approach was taken when in 2014 Humane Society International and Compassion presented a petition on tail docking to the European Parliament (European Parliament, 2014a). Another petition on docking was also submitted in 2014 and an earlier petition had been submitted in 2012 (European Parliament, 2012 and 2014b). All three petitions called for proper enforcement of the prohibition on routine docking. These petitions have been considered several times by the Parliament's Petitions Committee. On each occasion the Committee has been sympathetic to the petition and has kept the petition open. The petitions have helped to raise awareness of this issue among MEPs and in the Commission. The petitions were most recently considered by the Petitions Committee on 30 September 2021 when the Committee decided to 'Keep petitions open and ask for an updated reply by the Commission on the implementation of Council Directive 2008/120/EC and Council Directive 98/58/EC, as part of the planning activities for 2020 and beyond' (European Parliament, 2021).

Gradually some Member States were beginning to take enforcement of the prohibition on routine docking seriously. On 29 April 2015 the Agriculture Ministers of Germany, Netherlands, Denmark and Sweden signed a paper requesting a revision of the Pigs Directive including a strengthening of the prohibition of routine tail docking and a reduction in the number of tail docked pigs. The paper stated: 'The occurrence of tail biting has a multi-factorial origin but there is some evidence that the absence of straw and a too high stocking density is more important than many other known factors'.

6 Commission recommendation provides guidance

In 2016 the Commission published Commission Recommendation (EU) 2016/336 in the Official Journal. This sets out 'best practices aimed at reducing the need for tail-docking and optimised solutions for providing enrichment materials' (European Commission, 2016a). Under Article 288 of the Treaty on the Functioning of the EU a recommendation has no binding force.

Paragraph 2 of the Recommendation provides that 'Member States should:

a) ensure that farmers carry out a risk assessment of the incidence of tail-biting based on animal and non-animal based indicators ('the risk assessment'); and
b) establish compliance criteria with the requirements set out in the legislation and make them publicly available on a website'.

Paragraph 3 provides: 'The following parameters should be checked when carrying out the risk assessment:

a) the enrichment materials provided
b) cleanliness
c) thermal comfort and air quality
d) health status
e) competition for food and space
f) diet

These six factors are those that in the Commission's view must be considered when farmers and competent authorities assess whether, before carrying out docking, farmers have met the Directive's requirement to change 'inadequate environmental conditions or management systems'.

Paragraph 3 adds: 'Based on the results of the risk assessment, appropriate management changes in farms should be considered, such as supplying appropriate enrichment materials, comfortable environmental conditions, assuring a good health status and/or providing a balanced diet for pigs'.

6.1 *Enrichment*

The Recommendation then helpfully sets out the Commission's thinking as to what must be done to meet the Directive's requirements on enrichment.

Paragraph 4 provides: 'Enrichment materials should enable pigs to fulfil their essential needs without compromising their health. For that purpose, enrichment materials should be safe and have the following characteristics:

a) Edible – so that pigs can eat or smell them, preferably with some nutritional benefits
b) Chewable – so that pigs can bite them
c) Investigable – so that pigs can investigate them
d) Manipulable – so that pigs can change their location, appearance or structure'.

Paragraph 5 goes into further detail, providing: 'enrichment materials should be provided in such a way that they are:

a) of sustainable interest, that is to say, they should encourage the exploratory behaviour of pigs and be regularly replaced and replenished
b) accessible for oral manipulation

c) given in sufficient quantity
d) clean and hygienic'

Paragraph 6 categories different kind of enrichment materials. It provides: 'In order to fulfil pigs' essential needs enrichments material should meet all the characteristics listed in paragraphs 4 and 5. To that end, enrichments materials should be categorised as:

a) optimal materials – materials possessing all the characteristics listed in paragraphs 4 and 5 and therefore such materials can be used alone
b) suboptimal materials – materials possessing most of the characteristics listed in paragraphs 4 and 5 and therefore such materials should be used in combination with other materials
c) materials of marginal interest – materials providing distraction for pigs which should not be considered as fulfilling their essential needs and therefore optimal or suboptimal materials should also be provided'.

The Recommendation was accompanied by a detailed Staff Working Document on best practices with a view to the prevention of routine tail docking and the provision of enrichment materials to pigs (SWD(2016) 49 final) (European Commission, 2016b). This goes into greater detail on the points covered by the Recommendation.

In addition in 2016 the Commission organised three study visits to Sweden, Finland and Switzerland. In these countries pigs are not tail docked. National experts from other Member States that accompanied the Commission had the opportunity to see and learn from the preventive approach used in these three countries to avoid problems in intensive pig farming.

7 Guidance documents

Ample guidance for farmers and inspectors has been available for some years. In 2014 the EUWelNet published an online training tool for EU inspectors in respect of the provision of enrichment materials, tail biting and tail docking (EUWelNet, 2013). This is entitled 'Understanding environmental enrichment and tail docking requirements for finisher pigs in accordance with EU Directive 2008/120/EC'. It aims to ensure that inspectors are correctly identifying problems and risks on farm and are thus able to support farmers to make management choices so that at least the minimum standards required by the Pigs Directive are achieved. It is available in seven languages.

In 2017 the Commission published a series of factsheets entitled 'Cutting the need for tail-docking' (European Commission, 2017a). These are available in eight languages. It also produced two videos featuring best practice in reducing docking in Finland and Italy (European Commission, 2017b). These are available in nine and eight languages respectively.

8 Commission three-year work programme 2016–2019

In 2016 the Commission embarked on a three-year work programme in which it worked hard to press the Member States and the pig sector to take the ban on routine tail docking and the requirement to provide enrichment seriously. In 2017 the Commission wrote to all Member States asking them to produce National Action Plans with clear objectives and timelines and including compliance criteria for conforming with the legislation on tail docking (European Parliament, 2019). The Action Plans were required to set out what are sufficient and adequate improvement measures by farmers to change inadequate environmental conditions or management systems. The Commission performed a detailed assessment of the content and quality of the 26 Action Plans submitted. The poor quality of the Action Plans was highlighted when in August 2020 Commissioner Kyriakides, in reply to a Parliamentary Question, wrote: 'With the exception of Denmark, all action plans submitted to date lack adequate compliance criteria for one or more of the six tail-docking risk factors' (European Commission, 2020b). (These are the six factors set out in Commission Recommendation 2016/336.) The failure to establish compliance criteria means that farmers and enforcement officials may not be clear about what steps must be taken to comply with the legislation on tail docking.

The failure of Member States to establish compliance criteria for the six tail docking risk factors is inconsistent with:

- Point 2 of the Commission Recommendation which states that Member States should 'establish compliance criteria with the requirements set out in the legislation and make them publicly available on a website'.
- Article 12.1 of the Regulation on official controls (2017/625) which states that 'Competent authorities shall perform official controls in accordance with documented procedures. Those procedures … shall contain instructions for staff performing official controls' (European Parliament and Council, 2017).

In a further answer to a Parliamentary Question, Commissioner Kyriakides wrote in 2020: 'The Commission will continue to closely monitor the action plans of the Member States and will take appropriate measures. In case of persistent failure to apply EC law correctly by a Member State , the Commission may consider the launch of infringement procedure' (European Commission, 2020a). Earlier in a reply to the Parliament's Petitions Committee the Commission stated: 'The Commission is currently considering options for the most effective enforcement of the legislation on the avoidance of tail docking, including the possibility of infringements against certain Member States whose Action Plans do not contain measures to improve compliance with legal requirements' (European Parliament, 2019).

9 Pig Subgroup of the Animal Welfare Platform

The EU Platform on Animal Welfare is an interactive network which aims at promoting dialogue on animal welfare issues among competent authorities, farmers, civil society organisations and scientists. In 2018 the Animal Welfare Platform established a Pig Subgroup. The subgroup aimed to contribute to the objectives of the Platform by providing information, recommendations and proposals for practices which help deliver better pig welfare, in particular on risks for tail biting and avoiding routine tail docking. The subgroup produced a paper focussing on compliance criteria for tail biting risks relating to health, dietary factors, and space and competition (EU Platform on Animal Welfare, 2019a). The subgroup also produced a document entitled 'Measuring tail damage at the slaughterhouse and setting common thresholds'; the Platform acknowledged the principle but decided that dialogue on the technical details was necessary before recommending its dissemination and use (EU Platform on Animal Welfare, 2019b).

10 Still docking after all these years

In 2018 a Court of Auditors report considered how to close the gap between the EU's ambitious goals on animal welfare and practical implementation. The report highlighted the problems encountered in achieving compliance with the ban on routine docking (European Court of Auditors, 2018). The report stated, regarding a number of issues including tail docking, that 'there are still some significant discrepancies between the animal welfare standards established in the EU legislation and the reality on the ground'. The report added: 'In Germany (North-Rhine Westphalia) and in Romania, as in most other Member States, tail docking is routine in intensive pig farms, although this is prohibited by the legislation'.

In 2019 the Federation of Veterinarians of Europe and the European Association of Porcine Health Management adopted a position paper that stressed the need to move away from tail docking (FVE and EAPHM, 2019).

A 2019 presentation by the Commission reported that Commission audits carried out in 2017–18 in Denmark, Germany, Italy and Spain revealed that 98.5–100% of pigs were docked in these Member States. The Commission stated that this constitutes routine docking and stressed that there had been 'No progress / enforcement on avoiding routine tail-docking' (European Commission, 2019). The Commission also referred to audits carried out in 2019 in Austria, France, Hungary and Portugal and stated that there had been 'No change in tail-docking 98.5–100% – i.e. still routine' (European Commission, 2019).

The 2021 Evaluation of the *EU Strategy for the Protection and Welfare of Animals 2012–2015* recognises that: 'tail-docking of pigs is a routine practice in 26 out of 28 Member States and approximately 150 million pigs annually are subject to this practice. … With the exception of a few Member States, such as Finland and Sweden, most EU Member States did not comply with the ban or with providing adequate enrichment materials' (European Commission, 2021a).

11 EU Reference Centre for Animal Welfare-Pigs

The EU Reference Centres for Animal Welfare are established under Article 95 and 96 of the Regulation on official controls (2017/625). They aim at improving the enforcement of the legislation on animal welfare. They provide technical support and coordinated assistance to the Member States in carrying out official controls in the field of animal welfare. In 2018 the Commission designated the first EU Reference Centre for Animal Welfare (EURCAW) focusing on pigs. This has published helpful information on tail biting and docking (EURCAW-Pigs, 2021a). On 9 April 2021 it published a detailed answer to a question about the drivers of tail biting. Table 1.1 sets out sets out some of the key information provided in that answer.

TABLE 1.1 Information provided by EURCAW-Pigs regarding the drivers of tail biting, 2021

Stocking density	EURCAW refers to several studies that 'indicate a clear link between space allowance and tail biting risk'.
Feeding (competition)	Competition for feed is an important risk factor for tail biting. EURCAW states that stress and frustration associated with feeding competition might explain why studies found that as much as 30–50% of tail biting occurred in the vicinity of the feeder.
Pen design; solid flooring	EURCAW states: 'It is very likely that the reduction in tail biting related to the proportion of solid floor is related to the use of manipulable substrates (which are usually offered on solid flooring)'.
Air quality	EURCAW states that ammonia levels should not exceed 20 ppm and ideally should be lower.
Manipulable materials	EURCAW states: 'The relationship between presence of manipulable materials and absence of tail biting has been well documented'.

EURCAW-Pigs has also published a Dossier on tail biting and tail docking (EURCAW-Pigs, 2021b). The dossier aims to help competent authorities and other stakeholders. It includes a section entitled *inspiring examples*. These include one entitled *Active stall fatteners*. Ample straw is provided and straw bales are used to create imaginative structures providing opportunities for exploration and play. In addition, there is access to an outdoor run which also includes straw. EURCAW-Pigs has also produced an Indicator factsheet on enrichment (EURCAW-Pigs, 2021c). This states: 'The provision of sufficient and adequate enrichment material is an essential starting point in preventing tail biting'. It includes a helpful table that analyses the suitability of a range of materials.

12 Failure to enforce the ban on routine docking incompatible with Farm to Fork Strategy

The continuing failure to respect the ban on routine tail docking undermines the Farm to Fork Strategy which has high ambitions regarding animal welfare, stressing that there is an urgent need to improve animal welfare (European Commission, 2020b). Indeed the Strategy states: 'enforcement of existing legislation, notably for animal welfare, pesticide use and protecting the environment, is essential'.

13 Improving compliance with the legislation on docking and enrichment

In some borderline cases it may be difficult for inspectors to assess if a farmer has done enough to comply with the Directive's requirement that, before they may tail dock, they must first have tried to prevent tail biting by changing inadequate environmental conditions and management systems. However, in many cases it will be completely clear that the farmer has made no serious attempt to change inadequate conditions or systems as the inspector will see that the only enrichments provided are chains or plastic toys. In such cases the inspector must take enforcement action.

Of the six compliance criteria established by the Commission 2016 Recommendation, two – health and diet – may be challenging for inspectors to assess other than in extreme cases. Assessment of the other four will often be more straightforward provided of course that the competent authority has established clear compliance criteria e.g. maximum ammonia level and minimum feeder space per pig. It will generally be feasible to determine whether animals and housing are clean and if there is sufficient feeder space to prevent competition at feeding. Devices are available to measure temperature and levels of ammonia, particulate matter and carbon dioxide in pig housing.

Where a breach of the ban on routine docking is established, penalties must, under Article 139 of the Official Controls Regulation, be 'effective, proportionate and dissuasive'. This means that penalties must be sufficient to make farmers recognise that it makes more economic sense to respect the law than to ignore it.

In addition to improving enforcement, other ways of achieving enhanced compliance are available. Retailers, foodservice operators and food manufacturers could be urged not to sell or use meat from pigs that have been docked in breach of the law. The Court of Auditors' report expressed concerns about the use of rural development funds to support farmers who were not in compliance with the legislation on tail docking and enrichment (European Court of Auditors, 2018). Support from the Common Agricultural Policy or other forms of public funding should arguably not be available for producers who breach the ban on routine docking. The Federation of Veterinarians of Europe and the European Association of Porcine Health Management have proposed that 'to ensure a level playing field between EU and third country production, when stopping tail docking in the EU, the EU should also stop the importation of pig meat from countries that allow tail docking' (FVE and EAPHM, 2019).

14 Failure to respect the ban on routine docking could augment disease risks and antimicrobial resistance

Stress is a common element of the risk factors for tail biting (EFSA Report, 2007). Stress is also an important driver of disease in farm animals. A Joint Scientific Opinion by the European Medicines Agency and the European Food Safety Authority states: 'The stress associated with intensive, indoor, large scale production may lead to an increased risk of livestock contracting disease' (EMA and EFSA, 2017). The stress factors that trigger tail biting may also increase susceptibility to disease and hence lead to increased use of antimicrobials. The Commission has pointed out in a study on rearing pigs with intact tails that some of the factors that contribute to low levels of tail biting result in lower use of antimicrobials (European Commission, 2016c). Studies show that use of antimicrobials is much lower in free range and organic pigs than in intensively reared pigs (Alliance to Save our Antibiotics, 2021; Nielsen *et al.*, 2021).

15 World Organisation for Animal Health

At the international level, tail docking is considered by the recommendations on animal welfare produced by the World Organisation for Animal Health (WOAH, often known as the OIE). In 2018 WOAH adopted recommendations on the welfare of farmed pigs (OIE, 2018). These state: 'Tail biting may be reduced by providing an adequate enrichment material and an adequate diet (avoiding deficiencies of minerals or essential amino acids), and avoiding high stocking densities and competition for resources, such as feed and water' (Article 7.13.11).

On enrichment the WOAH recommendations state: 'Animals should be provided with an environment that provides complexity, manipulability and cognitive stimulation to foster normal behaviour (e.g. exploration, foraging such as rooting, biting and chewing materials other than feedstuffs, and social interaction), reduce abnormal behaviour (e.g. tail, ear, leg and flank biting, sham chewing, bar biting and apathetic behaviour) and improve their physical and mental state' (Article 7.13.10). The recommendations add: 'Pigs should be provided with enrichments that aim to improve their welfare through the enhancement of their physical and social environments, such as: – sufficient quantity of suitable materials to enable pigs to fulfil their needs to explore and look for feed (edible materials), bite (chewable materials), root (investigable materials) and manipulate materials. Novelty is another aspect that is important in maintaining interest in the provided materials' (Article 7.13.11).

16 Conclusions

A substantial body of evidence by the European Commission, the European Court of Auditors and animal welfare organisations shows that the prohibition of routine docking is being widely ignored. Detailed guidance on how to minimise tail biting has been published by the European Commission, the EU Reference Centre for Animal Welfare and the EU WelNet. Despite this, the ban on routine docking is regularly breached in 26 Member States.

To achieve improved compliance and enforcement, competent authorities must be willing to place effective penalties on producers who tail dock without first attempting to prevent tail biting by improving inadequate conditions and systems. Although there will be some borderline cases, it will often be clear whether or not a farmer has made a real effort to improve conditions before resorting to docking. For example, if only chains or plastic toys are being provided as enrichment, the farmer has clearly not tried to improve conditions before docking and accordingly effective enforcement action should be taken by the competent authority.

The principle of the legislation is clearly expressed i.e. that farmers may only dock if they have first tried to prevent tail biting by improving inadequate environmental conditions or management systems. However, it would be helpful if the legislation were strengthened to be more specific as to (i) which environmental and management factors must be addressed and (ii) what levels must be achieved for each of these factors before tail docking may be carried out in cases where tail biting continues to be a problem on the farm.

17 Take home messages

- The European Union's prohibition of routine tail docking of pigs – and the related provision requiring pigs to be provided with effective enrichment materials – were first adopted in 1991. Their wording was strengthened in 2001 and now appears in the codified Pigs Directive of 2008.

- The Directive provides that tail docking may only be carried out if the farmer has first tried to prevent tail biting by changing 'inadequate environmental conditions or management systems' but, despite this, still has a tail biting problem.
- Ample guidance for farmers and inspectors has been available for many years. In 2016 the European Commission published a detailed Recommendation. This sets out 'best practices aimed at reducing the need for tail-docking'. It also identifies the characteristics that materials must have in order to provide effective enrichment.
- In 2014 the EUWelNet published an online training tool for EU inspectors in respect of the provision of enrichment materials, tail biting and tail docking. In 2017 the European Commission published a series of factsheets entitled 'Cutting the need for tail-docking'.
- Despite the availability of guidance and advice, routine tail docking is the norm in most Member States, with the exception of Finland and Sweden. Compliance by the pig sector and enforcement by most Member States is very poor. In 2019 the European Commission reported that 98.5–100% of pigs were still being tail docked in many Member States.

References

Alliance to Save our Antibiotics, 2021. Antibiotic use in organic farming. https://www.saveourantibiotics.org/media/1914/20210406_antibiotic_use_in_organic_farming.pdf.

Compassion in World Farming, 2011. Providing enrichment for pigs. https://www.ciwf.org.uk/research/species-pigs/providing-enrichment-for-pigs/.

Compassion in World Farming, 2013. Widespread breaches of pig welfare laws in the EU. http://spolecnostprozvirata.cz/wp-content/uploads/Summary-report_pig_20131.pdf.

EU Platform on Animal Welfare, 2019a. Indicators for tail biting risks relating to compliance criteria on health, dietary factors and space and competition. https://ec.europa.eu/food/system/files/2019-11/aw_platform_plat-conc_indicator-tail-biting.pdf.

EU Platform on Animal Welfare, 2019b. Measuring tail damage at the slaughterhouse and setting common thresholds. Minutes of the meeting of the Platform on 7 October 2019. https://ec.europa.eu/food/animals/animal-welfare/eu-platform-animal-welfare/platform-conclusions_en; https://ec.europa.eu/food/system/files/2019-12/aw_platform_20191007_min.pdf.

EU Reference Centre for Animal Welfare – Pigs, 2021a. Questions to EURCAW Q2E-Pigs-2021-002. 9 April 2021. https://edepot.wur.nl/546256.

EU Reference Centre for Animal Welfare – Pigs, 2021b. Dossier: tail biting and tail docking. https://www.eurcaw.eu/en/eurcaw-pigs/dossiers/tail-biting-and-tail-docking.htm.

EU Reference Centre for Animal Welfare – Pigs, 2021c. Indicator factsheet – risks for tail biting: enrichment. https://edepot.wur.nl/513895.

European Association of Porcine Health Management, 2012. Email dated 27 January 2012 from John Mackinnon, president of the EAPHM to Peter Stevenson of Compassion in World Farming.

European Commission, 2001. Commission directive of 9 November 2001 amending directive 91/630/EEC laying down minimum standards for the protection of pigs. Annex, Chapter I, point 8. Official Journal L316, 01.12.2001 pp. 36–38.

European Commission, 2009. Answer given by Ms A Vassiliou on behalf of the commission. Question reference: E5360/2009. https://www.europarl.europa.eu/doceo/document/E-7-2009-5360-ASW_EN.html.

European Commission, 2016a. Commission Regulation (EU) 2016/336 of 8 March 2016 on the application of Council Directive 2008/120/EC laying down minimum standards for the protection of pigs as regards measures to reduce the need for tail-docking. Official Journal L62, 9.3.2016 pp. 20–22.

European Commission, 2016b. Commission Staff Working Document on best practices with a view to the prevention of routine tail-docking and the provision of enrichment materials to pigs. (SWD(2016) 49 final).

European Commission, 2016c. Overview report on a series of study visits carried out between 25 January and 8 April 2016 in order to share good practice on rearing pigs with intact tails. DG(SANTE) 2016-8987 – MR.

European Commission, 2017a. Cutting the need for tail docking. https://ec.europa.eu/food/animals/animal-welfare/animal-welfare-practice/animal-welfare-farm/pigs/tail-docking_en.

European Commission, 2017b. Reducing tail docking – best practice – videos. https://ec.europa.eu/food/animals/animal-welfare/animal-welfare-practice/animal-welfare-farm/pigs/tail-docking_en.

European Commission, 2019. Outcome of Commission actions to reduce tail biting and prevent routine tail docking of pigs. Presentation to meeting of Civil Dialogue Group, 14 November 2019.

European Commission, 2020a. Answer given by Ms Kyriakides on behalf of the Commission. 12 August 2020. Question reference: E005400/2020. https://www.europarl.europa.eu/doceo/document/E-9-2020-005400-ASW_EN.html.

European Commission, 2020b. A Farm to Fork Strategy for a fair, healthy and environmentally-friendly food system. Brussels, 20.5.2020 COM(2020) 381 final.

European Commission, 2021a. Commission Staff Working Document. Evaluation of the European Union Strategy for the Protection and Welfare of Animals 2012–2015 {SWD(2021) 77 final}. Brussels, 31.3.2021 SWD(2021) 76 final.

European Commission Scientific Veterinary Committee, 1997. European Commission: Scientific Veterinary Committee, Animal Welfare Section. Report on the welfare of intensively kept pigs, 30 September 1997.

European Communities, 1991. Council Directive of 19 November 1991 laying down minimum standards for the protection of pigs (91/630/EEC). Annex, Chapter II, Section III, point 4. Official Journal, L340, 11.12.91 pp. 33–38.

European Court of Auditors, 2018. Special report No 31. Animal welfare in the EU: closing the gap between ambitious goals and practical implementation.

European Food Safety Authority Opinion, 2007. Scientific opinion of the Panel on Animal Health and Welfare on the risks associated with tail biting in pigs and possible means to reduce the need for tail docking considering the different housing and husbandry systems. The EFSA Journal (2007) 611, 1–13.

European Food Safety Authority Report, 2007. Scientific report on the risks associated with tail biting in pigs and possible means to reduce the need for tail docking considering the different housing and husbandry systems. The EFSA Journal (2007) 611, 1–98.

European Medicines Agency and European Food Safety Authority, 2017. Joint scientific opinion on measures to reduce the need to use antimicrobial agents in animal husbandry in the European Union, and the resulting impacts on food safety (RONAFA). The EFSA Journal 2017;15(1):4666. https://efsa.onlinelibrary.wiley.com/doi/abs/10.2903/j.efsa.2017.4666.

European Parliament, 2012. Petition 0336/2012 by C.R. (Danish), concerning the routine tail-docking of piglets in Denmark.

European Parliament, 2014a. Petition No 0721/2014 by Joanna Swabe (Dutch), on behalf of Human Society International, and two signatories, on the routine docking of pigs' tails.

European Parliament, 2014b. Petition No 1141/2014 by Fredrick Federley (Swedish), on the cutting of pigs' tails.

European Parliament, 2019. Notice of 8.3.2019 to Members from Petitions Committee Secretariat including several replies from the European Commission.

European Parliament, 2021. Minutes of the meeting of the Petitions Committee on 30 September 2021. https://www.europarl.europa.eu/doceo/document/PETI-PV-2021-09-30-1_EN.pdf.

European Parliament and Council, 2017. Regulation (EU) 2017/625 of the European Parliament and of the council of 15 March 2017 on official controls and other official activities performed to ensure the application of food and feed law, rules on animal health and welfare, plant health and plant protection products. Official Journal L95, 7.4.2017 pp. 1–142.

European Union, 2008. Council Directive of 18 December 2008 laying down minimum standards for the protection of pigs (2008/630/EEC). Annex I, Chapter I, point 8. Official Journal L 47, 18.02.2009 pp. 5–13.

EUWelNet, 2013. http://pigstraining.welfarequalitynetwork.net/.

Federation of Veterinarians of Europe and European Association of Porcine Health Management, 2019. FVE & EAPHM position on preventing tail docking and tail biting. FVE/018/doc/062.

Food and Veterinary Office, 2008. General audit, Spain – 2008. DG (SANCO)/2008-8347.

Food and Veterinary Office, 2009a. Final report of a specific audit carried out in Belgium from 04 to 13 May 2009 in order to evaluate the implementation of rules on animal welfare in the context of a general audit: DG(SANCO) 2009-8255 – MR FINAL.

Food and Veterinary Office, 2009b. Final report of a general audit carried out in Cyprus during 2009 in order to evaluate official controls put in place by the competent authorities. DG(SANCO)/2009-8783 – MR Final.

Food and Veterinary Office, 2009c. General audit, Estonia – 2008. DG(SANCO)8600/2009.

Food and Veterinary Office, 2009d. Final report of a specific audit carried out in Greece from 16–20 November 2009 in order to evaluate the implementation of controls for animal welfare on farms in the context of a general audit. DG(SANCO) 2009-8243 – MR FINAL.

Food and Veterinary Office, 2009e. Final report of a specific audit carried out in Cyprus from 29 October to 05 November 2009 in order to evaluate the implementation of rules on animal welfare in the context of a general audit. DG(SANCO) 2009-8244 – MR FINAL.

Food and Veterinary Office, 2009f. Final report of a specific audit carried out in Portugal from 02 to 11 December 2009 in order to evaluate the implementation of controls for animal welfare on farms and during transport in the context of a general audit. DG(SANCO) 2009-8242 – MR FINAL.

Food and Veterinary Office, 2009g. Final report of a specific audit carried out in Slovenia from 16 to 24 June 2009 in order to evaluate the implementation of controls for animal welfare on farms, during transport and at the time of slaughter in the context of a general audit. DG(SANCO) 2009-8241 – MR FINAL.

Food and Veterinary Office, 2010a. Final report of a specific audit carried out in Bulgaria from 26 to 30 April 2010 in order to evaluate the implementation of controls for animal welfare on farms and during transport in the context of a general audit. DG(SANCO) 2010-8383 – MR FINAL.

Food and Veterinary Office, 2010b. Final report of a specific audit carried out in Italy from 15 to 26 March 2010 in order to evaluate the implementation of controls for animal welfare on farms and during transport in the context of a general audit. DG(SANCO) 2010-8388 – MR FINAL.

Food and Veterinary Office, 2010c. Final report of a specific audit carried out in the Czech Republic from 01 to 10 June 2010 in order to evaluate the implementation of controls for animal welfare on farms and during transport in the context of a general audit. DG(SANCO) 2010-8384 – MR FINAL.

Food and Veterinary Office, 2010d. Final report of a specific audit carried out in Denmark from 08 to 16 November 2010 in order to evaluate the implementation of controls for animal welfare on farms and during transport in the context of a general audit. DG(SANCO) 2010-8392 – MR FINAL.

Food and Veterinary Office, 2010e. Final report of a specific audit carried out in France from 01 to 12 February 2010 in order to evaluate the implementation of controls for animal welfare on farms and during transport in the context of a general audit. DG(SANCO) 2010-8390 – MR FINAL.

Food and Veterinary Office, 2010f. Final report of a specific audit carried out in Luxembourg from 11 to 15 January 2010 in order to evaluate the implementation of rules on animal welfare on farms and during transport in the context of a general audit. DG(SANCO) 2010-8385 – MR FINAL.

Food and Veterinary Office, 2010g. Final report of a specific audit carried out in Poland from 23 February to04 March 2010 in order to evaluate the implementation of controls on animal welfare on farms and during transport in the context of a general audit. DG(SANCO) 2010-8387 – MR FINAL.

Food and Veterinary Office, 2010h. Final report of a specific audit carried out in Romania from 26 to 30 April 2010 in order to evaluate the implementation of controls on animal welfare on farms and during transport in the context of a general audit. DG(SANCO) 2010-8389 – MR FINAL.

Food and Veterinary Office, 2011a. Final report of a mission carried out in Austria from 20 to 26 January 2011 in order to evaluate the implementation of controls on animal welfare on farms and during transport. DG(SANCO) 2011-6096 – MR FINAL.

Food and Veterinary Office, 2011b. Final report of an audit carried out in Belgium from 27 June to 2011 in order to evaluate the implementation of controls for animal welfare on farms and during transport. DG(SANCO) 2011-6039 – MR FINAL.

Food and Veterinary Office, 2011c. Final report of an audit carried out in Hungary from 05 to 09 2011 in order to evaluate the implementation of controls for animal welfare on farms and during transport. DG(SANCO) 2011-6045 – MR FINAL.

Food and Veterinary Office, 2011d. Final report of an audit carried out in Italy from 09 to 18 November 2011 in order to evaluate the implementation of controls for animal welfare on farms and during transport. DG(SANCO) 2011-6048 – MR FINAL.

Food and Veterinary Office, 2011e. Final report of an audit carried out in Portugal from 17 to 25 May 2011 in order to evaluate the implementation of controls for animal welfare on farms and during transport. DG(SANCO) 2011-6052 – MR FINAL.

Food and Veterinary Office, 2011f. Final report of an audit carried out in Slovakia from 06 to 14 September 2011 in order to evaluate the implementation of controls for animal welfare on farms and during transport. DG(SANCO) 2011-6053 – MR FINAL.

Food and Veterinary Office, 2012a. Final report of an audit carried out in Germany from 06 to 10 February 2012 in order to evaluate the implementation of controls for animal welfare on farms and during transport. DG(SANCO) 2012-6380 – MR FINAL.

Food and Veterinary Office, 2012b. Final report of an audit carried out in Romania from 21 to 29 November 2012 in order to evaluate the implementation of controls for animal welfare on farms and during transport. DG(SANCO) 2012-6374 – MR FINAL.

Food and Veterinary Office, 2012c. Final report of an audit carried out in Slovenia from 16 to 20 April2012 in order to evaluate the implementation of controls for animal welfare on farms and during transport. DG(SANCO) 2012-6375 – MR FINAL.

Nielsen, C., Kongsted, H., Sørensen, J. and Krogh, M. 2021. Antibiotic and medical zinc oxide usage in Danish conventional and welfare-label pig herds in 2016–2018. Preventive Veterinary Medicine. 189:105283 doi: 10.1016/j.prevetmed.2021.105283.

Office International des Epizooties, 2018. Terrestrial Animal Health Code, chapter 7.13, Animal welfare and pig production systems. https://www.oie.int/fileadmin/Home/eng/Health_standards/tahc/current/chapitre_aw_pigs.pdf.

Spoolder, H., Bracke, M., Mueller-Graf, C. and Edwards, E., 2011. Technical report submitted to EFSA. Report 2: preparatory work for the future development of animal based measures for assessing the welfare of weaned, growing and fattening pigs including aspects related to space allowance, floor types, tail biting and need for tail docking.

Stolba, A. and Woodgush, D.G.M., 1989. The behaviour of pigs in a semi-natural environment. Animal Protection 48: 419–425.

CHAPTER 2

Anatomy and physiology of the pig tail: characterisation of tail biting injury and consequences

Dale Sandercock[1,*], *Mette S. Herskin*[2] *and Heli Nordgren*[3]

[1]Animal and Veterinary Sciences, Scotland's Rural College (SRUC), West Mains Road, Edinburgh EH9 3JG, UK; [2]Department of Animal and Veterinary Sciences, AU Foulum, Aarhus University, Blichers Allé 20, 8830 Tjele, Denmark; [3]Department of Veterinary Bioscience, Faculty of Veterinary Medicine, Agnes Sjöbergin katu 2 00790 Helsinki, University of Helsinki, Finland; *Correspondence: v1dsande@ed.ac.uk

Abstract

This chapter describes the function of the porcine tail based on the current available literature and it provides a comprehensive overview of the anatomy and physiology of the pig's tail at both gross and microscopical levels. This chapter also includes details of the somatosensory and motor innervation necessary for the tail to respond to external stimuli such as touch and pain, and to coordinate its movement and posture for communication and expression of emotional state. In addition, this chapter discusses the different types of lesions caused by tail biting and describes the outcomes of lesions of varying severity (e.g. superficial to full thickness tissue injury and tail amputation) and their implications for pig health and welfare.

Keywords

tail anatomy – morphology – myology – osteology – vasculature – innervation – tail lesions – gross pathology – histopathology – immunohistopathology

1 Pig tail anatomy

1.1 *Gross anatomy*

1.1.1 Osteology

The tail bones are a continuation of the sacrum, which in pigs comprises 20–23 caudal (coccygeal) vertebrae (Getty, 1975; Nickel *et al.*, 1986a), which are separated from each other by intervertebral discs. In adult pigs, the first or first few caudal

 | DOI:10.3920/9789004694989_004

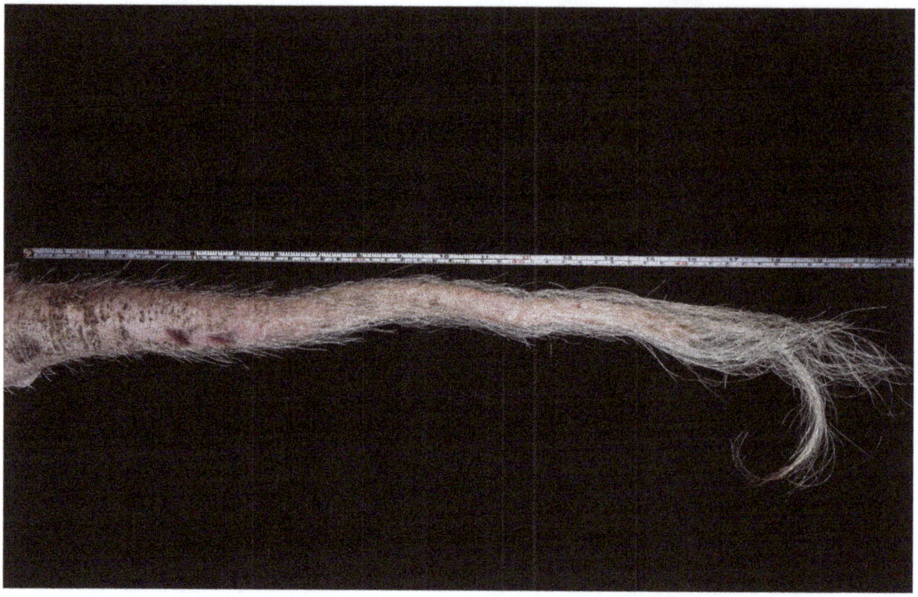

FIGURE 2.1 Post-mortem image of the dorsal surface of a pig tail (fully extended)
PHOTO BY SATU LEPPÄNEN, FACULTY OF VETERINARY MEDICINE, UNIVERSITY OF HELSINKI

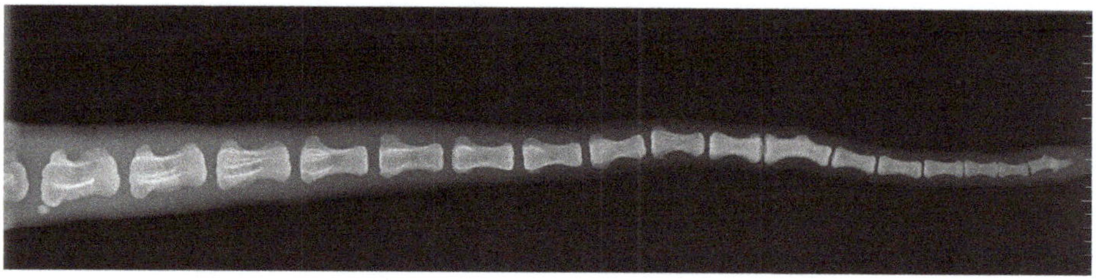

FIGURE 2.2 Post-mortem radiograph (lateral) showing the 18 caudal vertebrae (Ca_4–Ca_{20}) in the pig tail
PHOTO BY DIAGNOSTIC UNIT, FACULTY OF VETERINARY MEDICINE, UNIVERSITY OF HELSINKI

vertebrae may be fused to the sacrum (Singh, 2017a). The first few caudal vertebrae are situated cranial to the base of the tail, with only the 5th vertebrae being level with the anus, and the remaining 15–17 caudal vertebrae make up the external curly part of the pig tail (Figures 2.1 and 2.2). The cylindrical shape and the connection type between the individual caudal vertebrae ensure great flexibility in the pig tail.

The size of the caudal vertebrae diminishes, and their shape simplifies, towards the tail tip, as vertebral arches and processes become less prominent, and the vertebrae achieve a simple form as the tail tapers to the tip (Figure 2.3). The vertebrae are connected to each other by joints, located between the cranial and caudal articular processes of two adjacent vertebrae. In the caudal vertebrae, the cranial articular

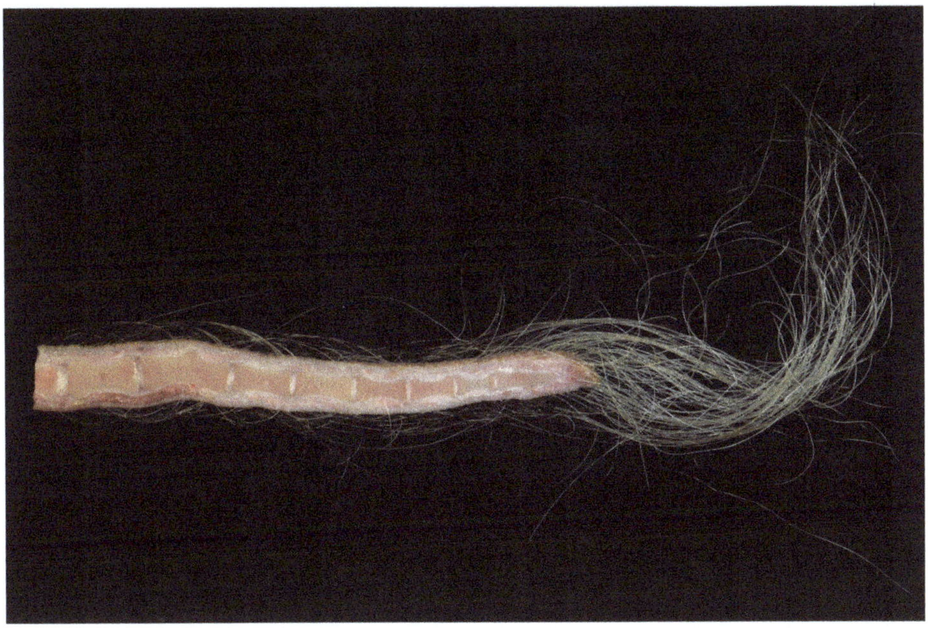

FIGURE 2.3 Dorsoventral transverse section of pig tail showing caudal vertebrae and intervertebral cartilage and articulations. Tail hair plume is apparent at the distal tip of the tail
PHOTO BY SATU LEPPÄNEN, FACULTY OF VETERINARY MEDICINE, UNIVERSITY OF HELSINKI

processes overlap the lateral portion of caudal processes in the eighth most cranial caudal vertebrae. However, only cranial articular processes are left between the 9th and 17th vertebrae, which decrease in size towards the tail tip. Lateral transverse processes, i.e. *Processus transversus*, are fully detectable in only the first three caudal vertebrae, and after the 9th or 10th vertebra, the processes are only visible as small nodules on the lateral side of the vertebral body. Only the first five caudal vertebrae have complete dorsal arches with spinous processes. The vertebral foramen is open in the caudal vertebrae of the first five vertebrae, after which only a groove is detected, until the 8th or 9th vertebra, that eventually disappears completely past the 9th vertebra (Nickel *et al.*, 1986a).

1.1.2 Myology

The pig tail has well-developed muscles, which insert on and enclose the caudal vertebrae (Figures 2.4). These muscles are responsible for the position and movements of the tail. The tail muscles originate from the vertebral column or pelvis (Getty, 1975; Nickel *et al.*, 1986b).

The elevator muscles are continuous with the long trunk muscles, and they are situated dorsally and dorsolaterally in the tail. The muscular actions raise the tail up or, when acting unilaterally, they twist the tail laterally and upwards. Short elevator

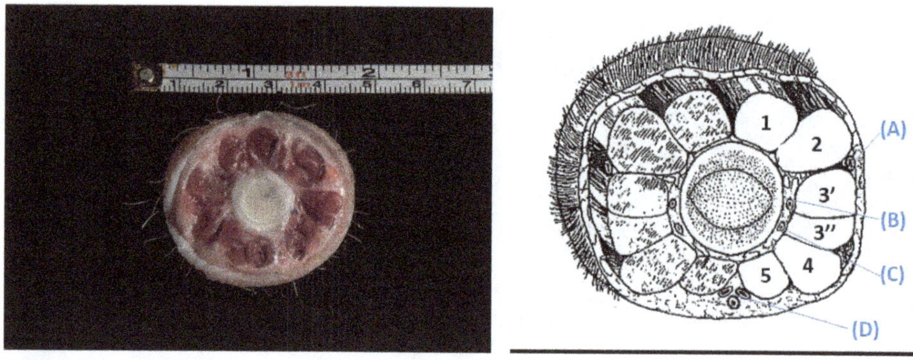

FIGURE 2.4 a) Cross-section of pig tail cut through caudal vertebra. Note, the caudal muscles surrounding the caudal vertebra. Photo by Satu Leppänen, Faculty of the Veterinary Medicine, University of Helsinki; b) Cross-section between 5th and 6th Caudal vertebrae in the pig. 1. *m. sacrocaudalis dorsalis medialis*; 2. *m. sacrocaudalis dorsalis lateralis*, 3′ *m. intertransversarii dorsales caudae*; 3″ *m. intertransversarii ventrales caudae*; 4 *m. sacrocaudalis ventralis lateralis*; 5 *m. sacrocaudalis ventralis medialis*. (A) Cutaneous v.; (B) dorsolateral caudal a.; (C) ventrolateral a. (D) ventromedial a and v; (illustration adapted from Getty, 1975 with permission from Elsevier).

muscles (*m. sacrocaudalis* syn. *sacrococcygeus dorsalis medialis*) are situated most dorsally in the tail. The long elevator muscles (*m. sacrocaudalis dorsalis lateralis*) run dorsolaterally to the tail, on both sides, dorsal to the transverse processes (Nickel *et al.*, 1986b).

The depressor muscles of the tail are located on the ventral and ventrolateral aspect of the tail. Their muscular actions depress the tail or, working unilaterally, move the tail downwards and sideways. The poorly developed bilateral short depressor muscles (*m. sacrocaudalis ventralis medialis*) are located most ventrally. The more developed long depressor muscles (*m. sacrocaudalis ventralis lateralis*) are dorsolateral to the short depressor muscles, ventral to the transverse processes and situated bilaterally (Nickel *et al.*, 1986b).

The flexor muscles (*mm. intertransversarii caudae*) are located laterally on the tail, between the long elevator and the long depressor muscles. In pigs, these muscles are rudimentary (Nickel *et al.*, 1986b).

The long lateral elevator muscle (*m. coccygeus*) runs from the pelvis to the lateral root of the tail attached to the first few caudal vertebrae. When working unilaterally, it flexes the tail from side to side and when acting bilaterally, it pulls the tail down against the anus (Nickel *et al.*, 1986b).

1.1.3 Vascularisation

The unpaired median caudal artery (*a. caudalis mediana*) is the main artery in the tail and is responsible for the arterial blood supply. It runs on the ventral surface of the tail, in the groove between the bilateral short depressor muscles (*mm. sacrocaudales ventrales mediales*), and gives rise to the segmental vessels, i.e. the *rr.*

caudales. Ventrolateral caudal arteries (*a. caudalis ventrolateralis*) run ventrally on both sides to the site of the rudimentary transverse processes of the caudal vertebrae, and supply blood to the skin and subcutaneous tissues of the tail. Above the rudimentary transverse processes run the dorsolateral caudal arteries (*a. caudalis dorsolateralis*), which supply blood to the tail muscles (Nickel *et al.*, 1981a).

The paired median caudal veins (*v. caudalis mediana*) are the main veins in the tail, and they are situated beside the median caudal artery in the vascular groove (Getty, 1975). These vessels give rise to the *rr. caudales*, which supply the musculature of the tail. On the lateral aspect of the tail these vessels join forming paired dorsolateral veins (*vv. caudales dorsolaterales*) above, and ventrolateral veins (*vv. caudales ventrolaterales*) below the transverse processes (Nickel *et al.*, 1981b).

1.1.4 Lymphatic system

Afferent lymph vessels drain from the tail skin and from the subcutaneous tissues to the superficial inguinal lymph nodes (*lnn. inguinales superficiales*), situated in the adipose tissue under the skin of the pubic region. Pigs have 1–6 lymph nodes covering an area of approximately 5–8-cm. The efferent routes run to the iliofemoral, medial, and lateral iliac nodes (*lnn. iliofemorales, ln. iliaci medialis*, and *lateralis*) (Nickel *et al.*, 1981c).

Lymphatic drainage flows from the caudal vertebrae to the ischiatic lymph nodes (*lnn. ischiadici*). These are one or two lymph nodes located on the lateral side of the sacrosciatic ligament in the pelvic cavity. However, in some pigs, lymph nodes may be found on only one side or may be completely absent. Normally, the nodes are 0.4–2.5 cm in size, and their flow is funneled toward the sacral (*ln. sacralis*) and medial iliac (*ln. iliaci medialis*) nodes (Nickel *et al.*, 1981c). In addition, anorectal lymph nodes (*lnn. anorectales*), i.e. 6–10 nodes measuring 0.2–2.2 cm in diameter, may continue as irregular rows between the base of the tail and the dorsolateral border of the rectum (Singh, 2017b; Nickel *et al.*, 1981c). The sacral lymph nodes (*lnn. sacrales*) are situated at the bifurcation of the internal iliac arteries. There are 2–5 of them, measuring 0.7–2 cm in diameter. Their drainage area includes musculature of the tail and caudal vertebrae. The efferent routes of the sacral lymph nodes are to the medial iliac nodes (*ln. iliaci mediales*), 2–6 nodes measuring 0.3–2.5 cm in diameter and situated beside the terminal abdominal aorta (Nickel *et al.*, 1981c).

1.1.5 Innervation

The last sacral nerves (*nn. sacrales*) and caudal nerves (*nn. caudales*) are the peripheral spinal nerves responsible for tail innervation and are found in the cauda equina (Figure 2.5). Pigs have 4–8 (typically 6) pairs of caudal nerves (Gandhi and Getty, 1969). The spinal nerves are mixed nerves composed of a dorsal and a ventral root emerging from the spinal cord. The dorsal root is composed mostly of sensory (afferent) fibers, which form a dorsal root ganglion (DRG) also known as a spinal ganglion. The spinal nerves collect and transmit information from the periphery

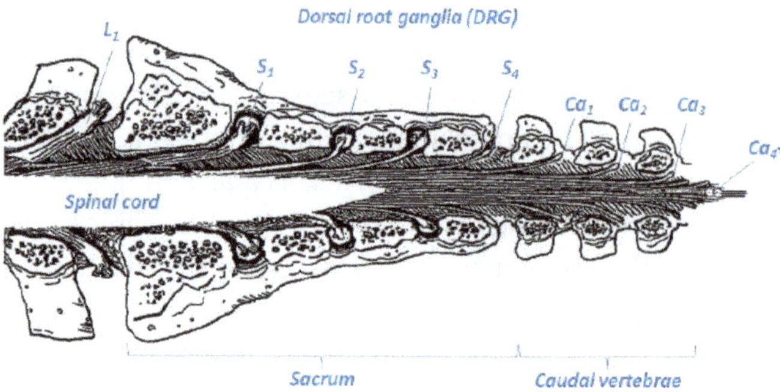

FIGURE 2.5 Diagram of pig spinal cord sacral and caudal innervation (cauda equina)
ILLUSTRATION ADAPTED FROM GETTY, 1975 WITH PERMISSION FROM ELSEVIER

to the brain. The ventral root consists of motor (efferent) neurons, passing information from the central nervous system to the target organs, muscles, and glands (Singh, 2017c).

The dorsal and ventral roots join to create a mixed nerve, leaving the vertebral column through the intervertebral foramen. The nerve then divides into dorsal and ventral neural branches (rami). The dorsal branch of the last sacral nerve merges with the dorsal branch of the first caudal nerve branch. The ventral branch of the last sacral nerve joins with the first ventral caudal branch, creating bilateral caudal dorsal and ventral nerve branches (*n. caudalis dorsalis* and *n. caudalis ventralis*). The caudal dorsal branches (right and left) run dorsal to the rudimentary transverse processes beside the dorsolateral artery all the way to the tip of the tail. Caudal ventral branches run bilaterally below the transverse process beside the ventrolateral artery down to the tail tip. Dorsal and ventral branches supply motor, autonomic and sensory nerve fibers. After the third caudal vertebra, four dorsal root ganglia containing the cell bodies of the distal caudal nerves that innervate the tail to the tip (Ca_4^+) are found aggregated at the distal end of the cauda equina. From this point, the four caudal nerves run exterior to the caudal vertebrae innervating tail tissues all the way to the tail tip (Figure 2.5; Sandercock *et al.*, 2019).

The dorsal elevator muscles of the tail are innervated by dorsal caudal branches and the last sacral nerves and the lateral elevator muscle by ventral branches of the last sacral nerves. The depressor tail muscles are innervated by ventral caudal branches and the last sacral nerves. Small lateral flexor muscles are innervated by the last sacral nerves and by both dorsal and ventral caudal nerves (Nickel *et al.*, 1986b).

The caudal ventral and dorsal branches also supply the skin and subcutaneous tissues of the tail, including the pili erector muscles of the hair, glands, and a wide variety of receptors (e.g. sensory hairs, touch, thermal and nociceptive receptors). Dorsal and ventral branches carry autonomic fibers to the smooth muscles in vessel walls and glands.

1.1.6 Integument

Pig skin shares a significant homology to the skin of most mammals (Nickel *et al.*, 1981d). Most European pig breeds are unpigmented, although exceptions do occur. The skin is thickest on the dorsal parts of the body and thinnest in the ventral areas. The epidermis is thickest in areas that are exposed to frictional forces, where the outermost layer, i.e. the stratum corneum, is considerably thickened. Dermal thickness varies between breeds (Nickel *et al.*, 1981d). The subcutis is rich in adipose tissue and creates a layer of fat, i.e. panniculus adiposus.

Pig hairiness varies in density and type depending on domestication stage, with wild boars having dense hair coats (Nickel *et al.*, 1981). Pig hairs are generally white despite growing from pigmented skin areas. The outer hairs, i.e. bristles, are stiff hairs that divide into strands at their tips. Bristle thickness and stiffness increase with increasing pig age. The length, thickness, degree of bending, and the number of strands at the end of the bristles vary between breeds (Nickel *et al.*, 1981d). The hairs usually grow in groups of one to three, where each hair has its own funnel leading from the hair follicle. Primary hairs enter the skin at a greater incline than the secondary hairs. The fascias (*fascia caudae superficialis* and *fascia caudae profundus*) are beneath the skin. Strong laminae run from the deep fascia to penetrate between the muscles towards the caudal vertebrae (Nickel *et al.*, 1986c).

1.2 *Microscopic anatomy*

1.2.1 Tail histology

A histological cross-section of a pig tail is shown in Figure 2.6. In the image, the caudal vertebra (CV) is readily observable in the centre of the image surrounded by the overlapping longitudinal coccygeal (caudal tail) muscles (CM). The four caudal nerve (CN-blue dashed circles) trunks are located dorso- and ventromedially to the caudal muscles. The muscles are surrounded by the subcutis (S), dermis (D) and epidermis (E). Hair follicle (HF) roots are visible in addition to large lateral and ventral blood vessels (BV).

The caudal vertebra (CV) has a normal vertebral histology. Innermost in the vertebral body lies the bone marrow cavity, where hematopoietic cells (red marrow) and fat cells (yellow bone marrow) occur between cancellus bone trabeculae. The outermost part of the vertebra is comprised of lamellar compact bone. The vertebra itself is surrounded by dense connective tissue. Caudal vertebrae are connected to each other with intervertebral discs, which are integrated to the cartilaginous end plates of adjacent vertebral bodies. A gelatinous nucleus pulposus is in the centre of the disc, surrounded by collagenous annulus fibrosus. The annulus fibrosus is made from concentric rings, and the outermost part of it connects laterally to the ligaments between the vertebrae. The vertebrae are surrounded by the caudal tail muscles (CM). Caudal muscle cell fibres (myocytes) run parallel in the direction of the tail tip and in cross-section are observed as polygonal and have peripherally situated nuclei. Transverse striations of the myocytes, the sarcomeres, may be seen in the longitudinal sections of the striated muscle.

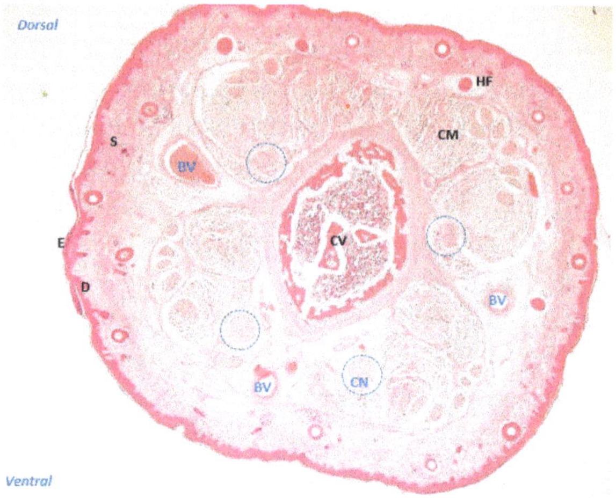

FIGURE 2.6 Histological tissue cross-section of pig tail (ca. 6th caudal vertebra) from a 1-day-old tail docked piglet (H&E ×4 magnification)
PHOTO BY DALE SANDERCOCK, SCOTLAND'S RURAL COLLEGE (SRUC)

Connective tissue surrounds the individual myocytes (endomysium), muscle cell bundles (perimysium), and entire muscles (epimysium). Nerves and blood vessels (arteries, veins, smaller capillaries, and lymphatics) are located in the muscle tissue. Tendons bind muscles to the bones. They consist of regularly arranged dense connective tissue. At the junction between muscle and tendon, the connective tissue of the tendon merges with the connective tissue of the muscle epimysium. (Bacha and Bacha, 2012a).

Peripheral nerve bundles (fascicles) are seen as light, wavy structures in longitudinal sections. A varying number of different-sized myelinated or unmyelinated axons can be seen in the cross sections. The nerves are surrounded by connective tissue (endo-, epi-, and perineurium). Dorsal root ganglia can be recognized as a group of large neuron bodies with centrally located nuclei (Bacha and Bacha, 2012b). Nerves can be identified by using special immunohistochemical stains such as neuron-specific enolase NSE or protein S-100 (Gaitero *et al.*, 2008; Sandercock *et al.*, 2016).

Muscular arteries distribute blood to smaller arterioles, which lead to small capillaries. Arterial and arteriole walls consist of smooth muscle layers and endothelial cell lining the lumen. The capillaries consist of an inner endothelial cell layer and lining connective tissue. The venules and veins are thinner-walled and more irregular in shape compared with arterioles and arteries. The lymphatic vessels are thin-walled, and invisible in the histological samples unless dilated.

The tail is surrounded by skin (integument). The outermost layer is the epidermis, which consists of several layers: the stratum corneum, the stratum granulosum, the stratum spinosum, and the undermost stratum basale. The stratum corneum consist of flat anucleated keratinized cells, which shed from the surface, and are continuously renewed from the mitotically active stratum basale layer (Bacha and Bacha, 2012c). The basal membrane (*lamina basale*), differentiates the epidermis from the dermis, and, in pigs, these two layers are connected tightly together by numerous dermal projections, termed papillae (Nickel *et al.*, 1981d). The dermis consists of fibroblasts producing collagen and elastic fibres.

Nerves, blood vessels, and adnexal structures; hair follicles, erector pili muscles, and sebaceous and sweat gland are situated in the dermis. The form of the sebaceous glands in pigs varies from stretched to semicircular, and they are located as rosettes around the hair follicles of the bristle hairs. Only two opposed sebaceous glands are associated with the wool hair follicles. Pigs have larger sweat glands compared to other domestic species. Their density decreases with age. The sweat glands are situated between the dermis and the subcutis, and their coiled end may reach deep into the subcutis. Their structure varies seasonally on their secretion mode, which is apocrine during warm periods and merocrine during cold months (Nickel *et al.* 1981d).

The subcutis lies beneath the dermis, consists merely of adipose tissue, and attaches the skin to the muscles. It is thickest at the base of the tail and grows thinner towards the end of the tail. Its thickness also varies between individuals depending on their nourishment status.

2 Pig tail physiology

2.1 *Tail morphology*

In the family of pigs, *Suidae*, tail morphology is quite diverse, but a common trait is that the tail in comparison to other domesticated species (e.g. *Bovidae*) is relatively small in proportion to the body. The length of adult wild boar (*Sus scrofa scrofa*) tails has been described to be approximately 20 cm (range: 15–40 cm) (Sprem *et al.*, 2011). In domestic pigs (*Sus scrofa domestica*), Herskin *et al.* (2015) reported a tail length of 9 ± 1 cm at the age of 2–4 days, and 31 ± 1 cm at the time of slaughter (at 22 weeks of age) in a relatively small sample of undocked crossbred (Landrace × Yorkshire × Duroc) finishers.

As evident from Figure 2.7, as well as Figures 2.1, and 2.3, tails of domestic pigs are hairy, the extent of which differs between individuals and breeds. Typically, the hairiness increases towards the tip of the tail, creating a plume (Figure 2.1). The function of the plume is not understood, but may, in part, be to provide protection and insulation of the narrowest and most vunerable part of the tail.

From older literature, the existence of congenital tail-malformations (often called kinks) has been reported (e.g. Nordby, 1934). Recently, Winters *et al.* (2023) examined more than 28,000 tail stumps, collected after tail docking, from 7 Danish

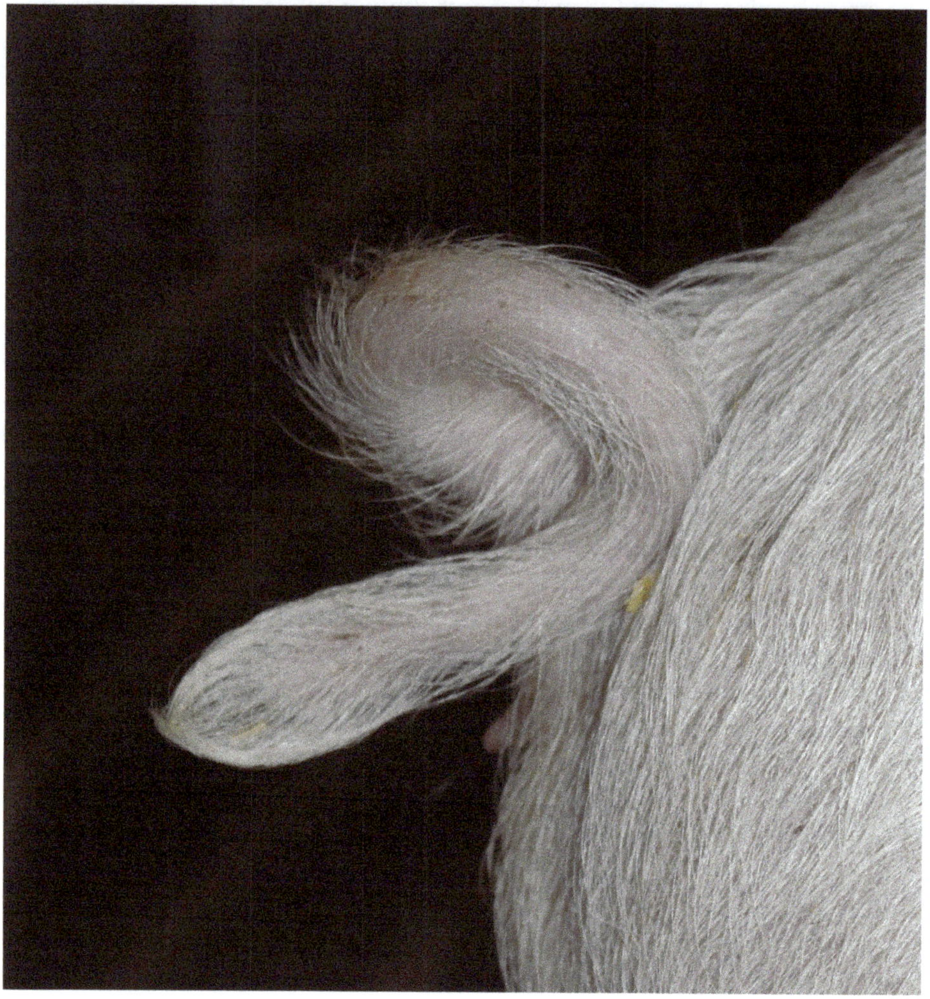

FIGURE 2.7 Photo showing the curled tail posture, the hairiness of pig tails and the tail plume
PHOTO BY MONA LV LARSEN, AARHUS UNIVERSITY

sow herds and found that an average of 1.07% of the tails were malformed, varying considerably between herds (from 0–2.6%). Figure 2.8 shows a selection of examples of kink tails.

Most mammals have tails, flexible appendages to the trunk that are used in various ways (Kiley-Worthington, 1976; Hickman, 1979). Across different mammalian species, tails have multiple functions, such as balance, communication, to protect vulnerable body parts, and to keep insects away. The function and functionality of the tails of modern domestic breeds of pigs have only been studied to a very limited extent (as discussed by Thodberg *et al.* (2018)).

The tail of domesticated pigs can be moved in all directions (omni-directional), and can, as the only member of the *Suidae* family, also be held or maintained in

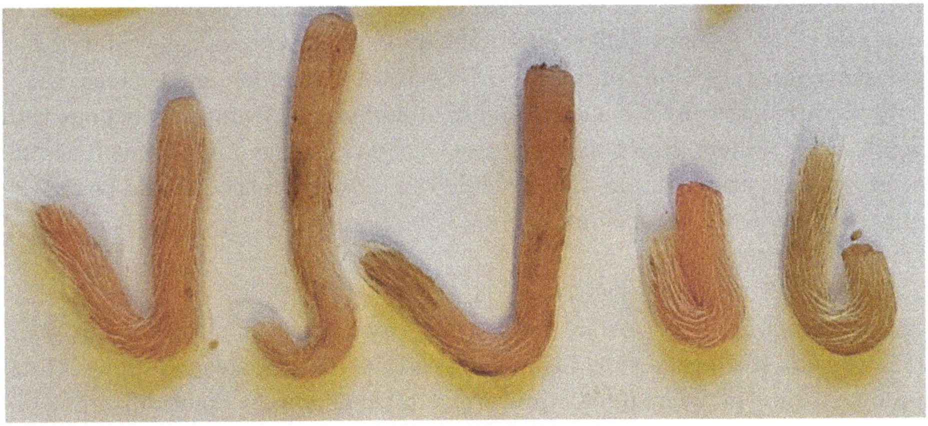

FIGURE 2.8 Piglet tails collected after tail docking showing presence of congenital morphological malformations (often called kinks)
PHOTO JEANET WINTERS, AU

a curled position (Figure 2.1). In recent years, tail posture and motion have been studied to gain information on the emotional state of pigs (e.g., Reimert *et al.*, 2013). In an attempt to standardize the quality of future studies, and their reporting, and to avoid the use of different and/or overlapping definitions, Camerlink and Ursinus (2020) suggested an ethogram including 7 different tail postures (fully curled, half curled, erect, horizontal, active hanging, passive hanging and tucked) and 6 types of movements (motionless, loosely wagging, intensely wagging, flicking, circling and jamming) and their definitions. The authors concluded that, in pigs, tail posture is an important indicator of animal welfare as, at least in part, it reflects the emotional state of the animals (consistent with other mammalian species e.g. *Canidae*) and can also be used as an indicator of on-going damaging behaviour (e.g. tail biting) in pigs reared under intensive conditions. Regarding the underlying emotional state, it is, however, yet not known whether tail posture just reflects the underlying emotional state of the animals, or whether it holds any communicative elements. Below, we introduce the current knowledge on the different types of tail postures and movements studied in pigs, their potential function and underlying motivational states. It is important to remember that the tail is a component of the whole-body expression and probably cannot be looked at independently.

2.2 *Tail function*

2.2.1 Curled tail

As mentioned, domestic pigs differ from other *Suidae* members in having a tail that can be kept curled up. Their closest relative, the wild boar, does not curl its tail. Genotyping studies in pigs (Bertolini *et al.*, 2018) suggest that a curled tail is a phenotypical trait, that can be selected for, or can arise as a consequence of selection of related traits. Camerlink and Ursinus (2020) reviewed the available literature on curly tails, and concluded that, given the relatively short time span of domestication, it seems unlikely that the curled tail has specifically evolved as a communicative signal to conspecifics. In addition, Camerlink and Ursinus (2020) suggested that, despite the societal belief of curled tails being synonymous with good animal welfare, curled tails may also be shown by pigs going through neutral or negative experiences, and that the curled tail position maybe should be considered the neutral default position.

2.2.2 Tail down

Some authors have considered the hanging, relaxed tail as the neutral position (Figure 2.9a). As concluded by Camerlink and Ursinus (2020), across studies of tail posture, the use of different descriptions and different naming of the different types of 'tail down' postures, complicates the interpretation of this posture. The posture of the tail depends also, to some extent, on the posture of the pig. For example, while lying, the tail is also hanging, or lying flat on the surface (Zonderland *et al.*, 2009), but may not necessarily be interpreted in the same way as on an upright animal.

2.2.3 Tail tucked

Several studies at pen group level have linked occurrence of low hanging or even tucked tails with the prevalence of tail injury in tail docked pigs kept under intensive conditions (e.g. Larsen *et al.*, 2018; Chou *et al.*, 2018) as well as in undocked pigs (Lahrman *et al.*, 2018; D'Eath *et al.*, 2018), however, these studies also acknowledge the risk of false positive prediction linked to these tail postures (Larsen *et al.*, 2018). One event, known to elicit tail tucking is situations where the pig is threatened from

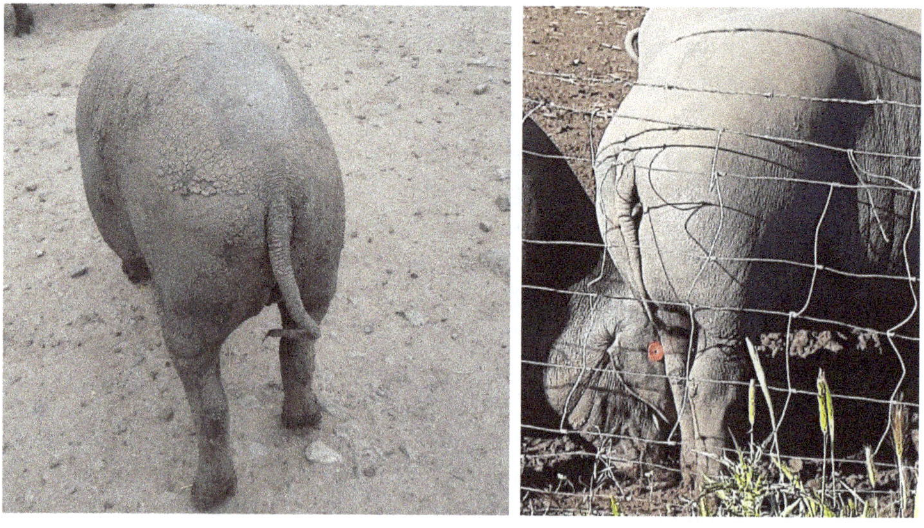

FIGURE 2.9 a) Pig with tail hanging loose (tail down); b) Pig exhibiting tail tucking
PHOTOS BY IRENE CAMERLINK, POLISH ACADEMY OF SCIENCES

behind. Such 'tail jamming' has been described as a more acute tail response and often coinciding with reflexive tucking of the rump and/or lowering of the hindquarters (Camerlink and Ursinus, 2020), but also been discussed in response to tail docking injuries (Rutherford *et al.*, 2009).

2.2.4 Tail up

In ungulates, raised tail has been suggested to be a result of motor reflexes during arousal, such as in situations of fear (Stankowich, 2008). Also in pigs, the tail can be in an upright position during locomotion and during social behaviours (Kiley-Worthington, 1976).

2.2.5 Sideway movements of the tail

Pigs can move their tails sideways in a slow, wagging style or as more abrupt tail flicks (the latter defined by Di Giminiani *et al.* (2017) as rapid horizontal movements of the tail). In the scientific literature, the definition of these two types of tail movements overlap to some extent and is quite variable, which creates challenges for the interpretation of the different movements. Some studies have found relations between tail wagging and play (Newberry *et al.*, 1988; Ocepek *et al.*, 2020). As discussed by Camerlink and Ursinus (2020), slower wagging intensity may be associated with social behaviours, whereas more intense wagging has been linked to tail irritation, frustration or even distress. However, often wagging intensity is not reported.

3 Tail biting in pigs

3.1 *Lesions caused by tail biting*

Tail wounds caused by tail biting are typically classified into groups ranging from mild to severe depending on the type, location and extent of tail injury, i.e. if tail remains attached or is bitten off (Figure 2.10a), representing a full tissue thickness amputation (Penny *et al.*, 1974, Wallgren and Lindahl, 1996). In cases of extreme tail biting the whole tail can be bitten off at the base of the tail (Figure 2.10b). Minor tail wounds tend to cause superficial cutaneous lacerations whereas severe tail wounds, located anywhere along the length of the tail, affect the deeper tissues including the skeletal muscles and the vertebrae and their associated vascular, nervous and lymphatic systems. Traumatic injury to both the skin and deeper tissues caused by tail biting provides a route for systemic infection through direct venous drainage via multiple venous sinuses (Getty and Ghoshal, 1967). Infection can also spread via the sacral lateral lymph node which drain the tail vertebrae (Singh, 2017b) and infection may also spread via the cerebrospinal fluid (Schrøder-Petersen and Simonsen, 2001). The risk of infection is high in bitten pigs and the wounded tail is commonly contaminated leading to tail

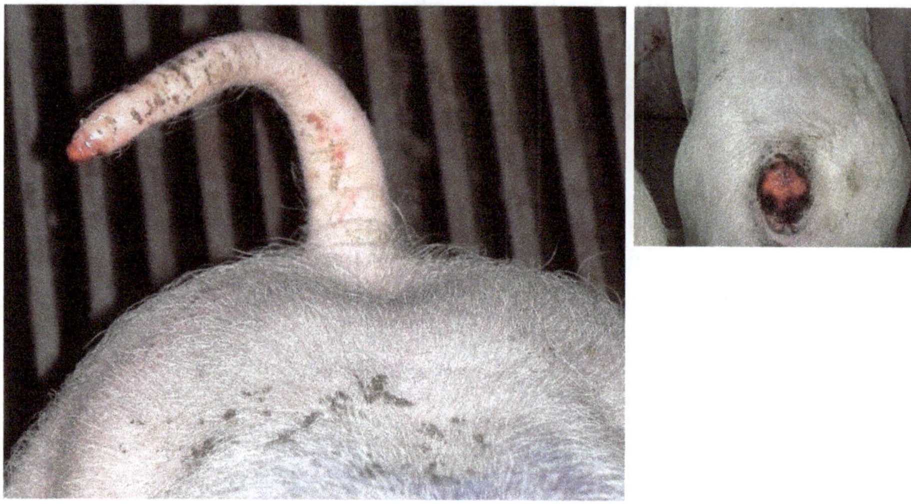

FIGURE 2.10 Examples of bitten tails in standard pig production systems. a) distal part of tail bitten off; b) A pig with a severely bitten tail down to tail root
PHOTOS SANDRA EDWARDS, NEWCASTLE UNIVERSITY

abscesses and necrosis (vom Brocke *et al.* 2019). Secondary infection can occur in the lungs and other tissues (e.g. joints) as a result of pyaemia (Huey, 1996; Kritas and Morrison, 2007).

Histopathological evaluation of bitten tails in pigs is limited to a small number of published studies conducted on tail docked pigs. Superficial tail biting injury is associated with surface erosion, local abscesses, ulceration of the stratified squamous epithelium, and sub-dermal necrosis (Done, 2003, 2011). Cutaneous wound healing after tail biting is associated with hyperkeratosis (ortho- and para-) and epidermal hyperplasia (Yap *et al.*, 2015, unpublished report). In the histological study reported by Yap *et al.* (2015) on pig tails collected at slaughter (5 control, 9 bitten), severe tail biting injury (involving full tail amputation along the whole length of the tail) was further characterised by the presence of a prominent granulation tissue cap formed at the distal tip close to the site of injury (Figure 2.11A), with moderate fibroplasia and dermal neovascularization. Mild dermal neutrophilic inflammation was also evident in six of nine tails with one tail having mild dermal oedema. Mild superficial perivascular inflammation was present in seven pig tails and deep perivascular lymphoplasmacytic inflammation was present in two of the bitten tails. Moderate bone remodelling in four pigs was evident but no osteomyelitis was observed. Traumatic neuromas were present in five of the nine tails showing widespread nerve sheath and neuronal Schwann cell proliferation (Figure 2.11B). Neuromas are known to increase the risk for spontaneous pain and hypersensitivity (Rajput *et al.*, 2012.), and thus may indicate a prolonged pain experience due to tail biting, as well as contributing to increased pain perception if the tail is subsequently injured. Tail bite neuromas tend to follow a pattern of fanning dorsally and

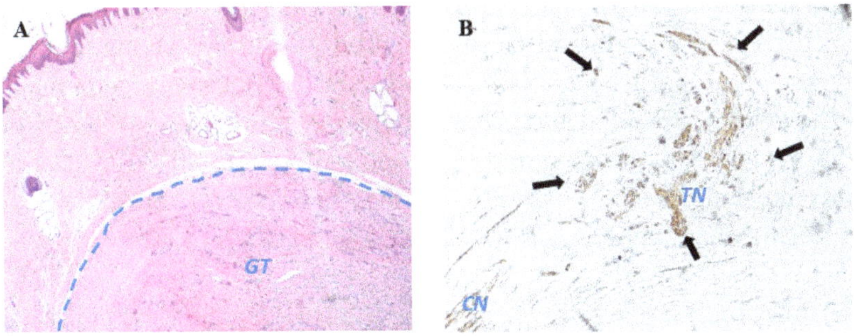

FIGURE 2.11 a) Histopathological image of a bitten pig tail (full amputation) ca. 2 months of age, tissue collected at slaughter age ca. 4 months-old. Image shows presence of granulation tissue (GT) cap (outlined by dashed line) at site of tail resection (H&E × 20); b) Immunohistopathological image showing caudal nerve (CN) and associated traumatic neuroma (TN) in the bitten tail. Regenerating caudal neurons exhibiting a fanning dispersal pattern (arrows) are observed at the site of nerve transection (S100 neural marker immuno-staining × 40)
PHOTOS BY SIONAGH SMITH, RDSVS, EDINBURGH

ventrally blending into the granulation tissue cap near the tip of tail. Differences in neuroma dispersal patterns were observed compared to neuroma formation following hot-iron tail docking. It is speculated that this may be attributable to the effect of tissue cauterization at the time of tail amputation using the hot-iron docking method (Sandercock *et al.*, 2016). Collectively, the presence of tissue infection, inflammation and traumatic neuroma development may affect longer term post-injury tail stump touch and pain sensitivity.

3.2 *Effects of tail lesions on tail function and pig health and welfare*

When pigs are housed under commercial conditions, tail biting behaviour may develop, a behaviour characterized by repeated biting and cannibalizing tails of one or more group members (reviewed by e.g., Schrøder-Pedersen and Simonsen, 2001). The damaging behaviour has considerable economic consequences and is a major animal welfare problem in pig husbandry (as reviewed by Valros, 2018). Tail damage may, however, also be caused by necrosis (Reiner *et al.*, 2019) or injuries such as other pigs accidentally stepping on tails, or when tails are squeezed in doors or other fixtures during transport or other management procedures.

The type of tail damage, subject to far the most scientific attention, is docking of tails – when parts of the tails of new-born pigs are cut off in their first days of life to prevent later outbreaks of tail biting (reviewed by Sutherland & Tucker, 2011). Studies of the consequences of tail docking have shown signs of acute pain during the initial hours, and later formation of neuromas (Herskin *et al.*, 2015; Sandercock *et al.*, 2016), which may have a long-term influence on the sensitivity of the tail stump (Di Giminiani *et al.*, 2017). These studies further are supported by evidence from mechanistic studies demonstrating sustained transcriptomic changes in 185 inflammatory and neuropathic pain associated caudal DRG genes (Sandercock *et al.*, 2019). The existing knowledge on consequences of tail docking in terms of pain have been reviewed by Herskin and Di Giminiani (2024) and Prunier *et al.* (2020). Below, knowledge about consequences of other types of tail damage is introduced, focusing on bitten tails.

3.2.1 Tail function

As mentioned above, increased occurrence of tucked tails in pens of pigs is considered a warning sign of tail biting outbreaks (Wallgren *et al.*, 2019; Larsen *et al.*, 2018), and suggested to be especially relevant in the onset of an outbreak, when tail damage may be difficult to detect from outside the pen (Honeck *et al.*, 2019). Tucked in tails may, though, not only be seen in healthy pen mates, but victims of tail biting also typically keep their tail low and often tucked in between their legs (Torrey *et al.*, 2009; Ursinus *et al.*, 2014; Lahrman *et al.*, 2018). Whether such change in tail posture affects the emotional status or the ability to communicate is not known.

3.2.2 Pig health

The severity of the damage in bitten tails are highly variable, characterized by e.g., local and/or systemic infection associated with immune activation and inflammation potentially leading to sepsis. However, studies of the consequences of bitten tails are challenged by this variability in severity as well as pathological processes. Honeck *et al.* (2019) reviewed the many available studies examining the occurrence of bitten tails with special focus on the methodology used to classify and score the damaged tails. One of the clear conclusions is that the scoring of the bitten tails is done in numerous ways, which makes it difficult to compare findings across studies. As an example, Honeck *et al.* (2019) reviewed how many studies have used, or been inspired by, the Welfare Quality Protocol (2009) – a binary way of scoring the damaged tails, into either 'healthy' or 'bitten'. The score 'bitten' can then be graded as 1 (indication of superficial biting along the length of the tail, but no evidence of fresh blood or of any swelling) or 2 (fresh blood is visible on the tail, there is evidence of some swelling and infection, part of the tissue is missing, and a crust has formed). There are, however, also studies involving more detailed recording of different characteristics such as hair loss, tail loss, presence of blood, blood freshness, redness, swelling, infection and necrosis – and many different combinations of these. Among the latter are several of the cardinal signs of inflammation, of which some can be identified macroscopically (e.g. redness and swelling), whereas others require laboratory diagnostic methodology.

Another challenge associated with studies of bitten tails is the difficulty in obtaining comparable treatment groups, for example in relation to the severity of the damage, the interval since it happened, and the presence of other morbidities which may have increased the underlying risk of the pig to be bitten (Munsterhjelm *et al.*, 2015). In addition, most studies of the consequences of bitten tails are relatively small (e.g., Munsterhjelm *et al.*, 2013; Viitasaari *et al.*, 2015).

It is, however, known that tails damaged by biting may involve full or partial amputation, and is surely painful for the bitten pig (as reviewed by Valros, 2018). In terms of pain, bitten tails involve the acute phase (often repeated several times as victims have been shown to be bitten multiple times (Brunberg *et al.*, 2011)), an inflammatory phase (Heinonen *et al.*, 2010) and potentially also a chronic phase due to the damage to nerves of the tail. In addition, bitten pigs have been reported to show signs of persistent stress (changed pattern of cortisol secretion and adrenal size) (Valros *et al.*, 2013; Munsterhjelm *et al.*, 2013). Valros (2018) reviewed consequences of tail biting on the health of pigs and suggested that pigs with bitten tails have a higher risk of co-morbidities in terms of for example locomotory problems and respiratory lesions. The reason for the increased risk is not fully clear, and the bitten tail may not be the direct cause. However, there is a risk for a direct infectious spread via venous, lymphatic or cerebrospinal routes, and tail biting damage may also increase the risk of opportunistic lesions. Being tail bitten decreases growth at least temporarily (Munsterhjelm *et al.*, 2013) and increases the risk of culling or death (Niemi *et al.*, 2011). At slaughter, tail bitten pigs have a higher prevalence of abscesses, lung lesions and arthritis (Valros *et al.*, 2004; Kritas & Morrison, 2007) due to septicaemia. Also, carcass condemnations increase significantly in tail bitten pigs (Marques *et al.*, 2012; Harley *et al.*, 2014). In a study of 1.7 million pigs slaughtered at the same Danish slaughterhouse, Alban *et al.* (2013) found that 18% of condemned carcasses were recorded as tail bitten at meat inspection. Thus, in addition to being a sign of an underlying welfare problem, tail biting as such is significantly reducing pig welfare, has economic consequences for producers and is contributing to the use of antimicrobials in pig production.

4 Conclusions

This chapter details the gross and microscopical anatomy of the pig tail and clearly shows that the pig tail has a highly-structured multi-tissue composition that is well vascularised and innervated through all tissues present all the way to the tail tip. Due to its high level of somatosensory innervation, the pig tail is highly responsive to touch, and pain caused by injury attributable to tail docking and tail biting, and such stump sensitivity may persist for a considerable time after injury. In addition, it is evident that the tail has a multi-functional use in pigs involving vulval and anal protection, reflection of emotional state and to swipe away insects. The information provided in this chapter represents an up to date review of the anatomy and function of the pig tail and our general understanding of the gross and histopathological lesions caused by tail biting injury and subsequent post-injury tissue reparative responses and consequences. It highlights the necessity to remove the often ineffective option of tail docking as a procedure for managing tail biting and a requirement by producers to optimize pig rearing (e.g. housing, flooring) and husbandry practices (e.g. adequate enrichment provision) to prevent/minimize outbreaks of tail biting for a future, sustainable pig production industry.

5 Take home message

- The tail has a multi-functional use in pigs involving protection of the vulva and anus from conspecifics, and to physically swipe away insects. In addition, the posture of the tail reflects the emotional status of the animals.
- The pig tail has a highly structured multi-tissue composition that is well vascularised and innervated through all tissues present all the way to the tip.
- The tail is highly responsive to touch, and pain caused by injury attributable to tail docking and tail biting, and elevated tail sensitivity may persist for a considerable time after such injury.
- Tail docking can cause long-term pain and is often an ineffective procedure for managing tail biting, and therefore there is a requirement by producers to optimize pig rearing (e.g. housing, flooring) and husbandry practices (e.g. adequate enrichment provision) to prevent/minimize outbreaks of tail biting for a future, sustainable pig production industry.

References

Alban, L., Dahl, J., Andreasen, M., Petersen, J. and Sandberg, M. 2013. Possible impact of the 'yellow card' antimicrobial scheme on meat inspection lesions in Danish finisher pigs. Preventive Veterinary Medicine 108: 334–341. https://doi.org/10.1016/j.prevetmed.2012.11.010.

Bacha, Jr, W.J. and Bacha, L.M. 2012a. *Color atlas of veterinary histology*. 3rd edition. John Wiley & Sons. London, UK, pp. 13–44, 57–51.

Bacha, Jr, W.J. and Bacha, L.M. 2012b. *Color atlas of veterinary histology*. 3rd edition. John Wiley & Sons. London, UK, pp. 45–55.

Bacha, Jr, W.J. and Bacha, L.M. 2012c. *Color atlas of veterinary histology*. 3rd edition. John Wiley & Sons. London, UK, pp. 85–101.

Bertolini, F, Schiavo, G, Tinarelli, S, Santoro, L, Utzeri, V.J., Dall'Olio, S, Fontanesi, L. 2018. Exploiting phenotype diversity in a local animal genetic resource: identification of a single nucleotide polymorphism associated with the tail shape phenotype in the autochthonous Casertana pig breed. Livestock Science 216: 148–152. https://doi.org/10.1016/j.livsci.2018.08.007.

Brunberg, E, Wallenbeck, A, Keeling, L.J. 2011. Tail biting in fattening pigs: associations between frequency of tail biting and other abnormal behaviours. Applied Animal Behaviour Science 133: 18–25. https://doi.org/10.1016/j.applanim.2011.04.019.

Camerlink, I, Ursinus, W.W. 2020. Tail postures and tail motion in pigs: a review. Applied Animal Behaviour Science 230: 105079. https://doi.org/10.1016/j.applanim.2020.105079.

Carr, R.W., Coe, J.E., Forsch, E., Schmelz, M. and Sandercock, D.A. 2015. Structural and functional characterization of peripheral axons in the caudal tail nerve of neonatal pigs: preliminary data. In 9th Congress of the European Pain Federation (EFIC), 2–5 September 2015, Vienna, Austria.

Chou J.Y., D'Eath, R.B., Sandercock, D.A., Waran, N, Haigh, A, O'Driscoll, K. 2018. Use of different wood types as environmental enrichment to manage tail biting in docked pigs in a commercial fully-slatted system. Livestock Science 213: 19–27. https://doi.org/10.1016/j.livsci.2018.04.004Get rights and content.

D'Eath, R.B., Jack, M., Futro, A., Talbot, D., Zhu, Q., Barclay, D., Baxter, E.M. 2018. Automatic early warning of tail biting in pigs: 3D cameras can detect lowered tail posture before an outbreak. PLoS One 13. https://doi.org/10.1371/journal.pone.0194524.

Di Giminiani, P., Sandercock, D.A., Malcolm, E.M., Leach, M.C., Herskin, M.S., Edwards, S.A. 2016. Application of a handheld pressure application measurement device for the characterization of mechanical nociceptive thresholds in intact pig tails. Physiology and Behaviour 165: 119–126. https://doi.org/10.1016/j.physbeh.2016.07.006.

Done, S.H., Guise, J., Chennells, D. (2003). Tail biting and tail docking in pigs. Pig Journal, 51, 136–154.

Done, S.H., Guise, J., Chennells, D. (2011). Tail biting and tail docking in pigs. Pig Journal, 66, 94–102.

Gaitero, L., Añor, S., Fondevila, D., Pumarola, M. 2008. Canine cutaneous spindle cell tumors with features of peripheral nerve sheath tumors: a histopathological and immunohistochemical study. Journal of Comparative Pathology 139:16–33. https://doi.org/10.1016/j.jcpa.2008.03.003.

Gandhi, S.S, and Getty, R. 1969. Cutaneous nerves of the trunk of the domestic pig with special reference to spinal nerves. Part III cutaneous nerves of the sacral and coccygeal regions. Iowa State Journal of Science. 44: 31–43.

Getty, R. 1975. In: Sisson and Grossman's The anatomy of the domestic animals 5th Ed. (Vol. 1). W.B. Saunders & Co. Philadelphia, USA.

Getty, R., Ghoshal, N.G. 1967. Applied anatomy of the sacrococcygeal region of the pig as related to tail bleeding. Veterinary Medicine/Small Animal Clinician 62: 361–367.

Harley, S, Boyle, L.A., O'Connell, N.E., More, S.J., Teixeira, D.L., Hanlon, A. 2014. Docking the value of pig meat? Prevalence and financial implications of welfare lesions in Irish slaughter pigs. Animal Welfare 23: 275–285. https://doi.org/10.7120/09627286.23.3.275.

Herskin M.S., and Di Giminiani P. 2024. Pain in pigs: characterization and indicators. Advances in Pig Welfare, Woodhead Publishing Series in Food Science, Technology and Nutrition, p. 23–48.

Herskin, M.S., Thodberg, K, Jensen, H.E. 2015. Effects of tail docking and docking length on neuroanatomical changes in healed tail tips of pigs. Animal 9: 677–681. https://doi.org/10.1017/S1751731114002857.

Huey, R.J. 1996. Incidence, location and interrelationship between sites of abscess recorded in pigs at a bacon factory in Northern Ireland. Veterinary Record 133:511–414. https://doi.org/10.1136/vr.138.21.511.

Hickman, G.C. 1979. The mammalian tail: a review of functions. Mammal Review 9: 143–159.

Honeck, A., Gertz, M., grosse Beilage, E., Krieter, J. 2019. Comparison of different scoring keys for tail biting in pigs to evaluate the importance of one common scoring key to improve the comparability of studies – a review. Applied Animal Behaviour Science 221:104873. https://doi.org/10.1016/j.applanim.2019.104873.

Kiley-Worthington, M. 1976. Tail movements of ungulates, canids and felids with particular reference to their causation and function as displays. Behaviour 56: 69–115.

Kritas, S.K., Morrison, R.B. 2007. Relationships between tail biting in pigs and disease lesions and condemnations at slaughter. Veterinary Record 160: 149–152. https://doi.org/10.1136/vr.160.5.149.

Lahrman, H.P., Hansen, C.F., D'Eath, R, Busch, M.E., Forkman, B. 2018. Tail posture predicts tail biting outbreaks at pen level in weaner pigs. Applied Animal Behaviour Science 200: 29–35. https://doi.org/10.1016/j.applanim.2017.12.006.

Larsen, M.L.V., Andersen, H.M.L., Pedersen, L.J. 2018. Tail posture as detector of tail damage and an early detector of tail biting in finishing pigs. Applied Animal Behaviour Science 209: 30–35. https://doi.org/10.1016/j.applanim.2018.08.016.

Marques, B.M.F.P.P., Bernardi, M.L, Coelho, C.F., Almeida, M., Morales, O.E., Mores, T.J., Borowski, S.M., Barcellos, D.E.S.N. 2012. Influence of tail biting on weight gain, lesions and condemnations at slaughter of finishing pigs. Pesquisa Veterinária. Brasiliera. 32: 967–974. https://doi.org/10.1590/S0100-736X2012001000003.

Meyer, W., Schwarz, R., Neurand, K. 1978. The skin of domestic mammals as a model for the human skin, with special reference to the domestic pig. Current Problems in Dermatology 7:39–52. https://doi.org/10.1159/000401274.

Munsterhjelm, C., Brunberg, E., Heinonen, M., Keeling, L.J., Valros, A. 2013. Stress measures in tail biters and bitten pigs in a matched case-control. Animal Welfare 22: 331–338. https://doi.org/10.7120/09627286.22.3.331.

Newberry, R.C., Wood-Gush, D.G.M., Hall, J.W. 1988. Playful behaviour of pigs. Behavioural Processes 17: 205–216. https://doi.org/10.1016/0376-6357(88)90004-6.

Nickel, R., Schummer, A., Seiferle, E., Frewein, J., Wilkens, H., & Wille, K.H. 1986a. The locomotor system of the domestic mammals. In: The anatomy of the domestic animals, Vol. 1. Verlag Paul Parey, Berlin and Hamburg, Germany, .p. 42.

Nickel, R., Schummer, A., Seiferle, E., Frewein, J., Wilkens, H., & Wille, K.H. 1986b. The locomotor system of the domestic mammals. In: The anatomy of the domestic animals, Vol. 1. Verlag Paul Parey, Berlin and Hamburg, Germany, pp. 320–232.

Nickel, R., Schummer, A., Seiferle, E., Frewein, J., Wilkens, H., & Wille, K.H. 1986c. The locomotor system of the domestic mammals. In: The anatomy of the domestic animals, Vol. 1. Verlag Paul Parey, Berlin and Hamburg, Germany, .p. 267.

Nickel, R., Schummer, A., Seiferle, E., Frewein, J., Wilkens, H., & Wille, K.H. 1981a. The circulatory system, the skin, and the cutaneous organs of the domestic mammals. In: The anatomy of the domestic animals, Vol. 3. Verlag Paul Parey, Berlin and Hamburg, Germany, p. 159.

Nickel, R., Schummer, A., Seiferle, E., Frewein, J., Wilkens, H., & Wille, K.H. 1981b. The circulatory system, the skin, and the cutaneous organs of the domestic mammals. In: The anatomy of the domestic animals, Vol. 3. Verlag Paul Parey, Berlin and Hamburg, Germany, p. 260.

Nickel, R., Schummer, A., Seiferle, E., Frewein, J., Wilkens, H., & Wille, K.H. 1981c. The circulatory system, the skin, and the cutaneous organs of the domestic mammals. In: The anatomy of the domestic animals, Vol. 3. Verlag Paul Parey, Berlin and Hamburg, Germany, pp. 377–380.

Nickel, R., Schummer, A., Seiferle, E., Frewein, J., Wilkens, H., & Wille, K.H. 1981d. The circulatory system, the skin, and the cutaneous organs of the domestic mammals. In: The anatomy of the domestic animals, Vol. 3. Verlag Paul Parey, Berlin and Hamburg, Germany, pp. 496–498.

Niemi, J.K., Sinisalo, A., Valros, A., Heinonen, M. 2011. The timing and treatment of tail biting in fattening pigs. In: Hultgren, J., Persson, P., Nadeau, E., Fogelberg, F. (Eds): Proceedings of the 24th NJF, Report 7, NJF, Uppsala, Sweden, pp. 50–55.

Nordby, J.E. 1934. Kinky tail in swine. The Journal of Heredity 94: 171–174.

Ocepek, M., Newberry, R.C., Andersen, I.L. 2020. Which types of rooting material give weaner pigs most pleasure? Applied Animal Behaviour Science 231: 105070 https://doi.org/10.1016/j.applanim.2020.105070.

Penny, R.H.C., Hill, F.W.G. 1974. Observations of some conditions in pigs at the abattoir with particular reference to tail biting. Veterinary Record 2: 174–180. DOI: 10.1136/vr.94.9.174.

Rajput, K., Reddy, S., Shankar, H. 2012. Painful neuromas. Clinical Journal of Pain 28:639–645. https://doi.org/10.1097/AJP.0b013e31823d30a2.

Reimert, I, Bolhuis, J.E., Kemp, B., Rodenburg, T.B. 2013. Indicators of positive and negative emotions and emotional contagion in pigs. Physiology and Behaviour 109: 42–50. https://doi.org/10.1016/j.physbeh.2012.11.002.

Reiner, G., Lechner, M., Eisenack, A., Kallenbach, K., Rau, K., Müller, S., Fink-Gremmels, J. 2019. Prevalence of an inflammation and necrosis syndrome in suckling piglets. Animal 13: 2007–2017. https://doi.org/10.1017/S1751731118003403.

Rutherford, K.M.D., Robson, S.K., Donald, R.D., Jarvis, S., Sandercock, D.A., Scott, E.M., Nolan, A.M., Lawrence, A.B. 2009. Pre-natal stress amplifies the immediate behavioural responses to acute pain in piglets. Biology Letters 2009; 5(4):452–454. https://doi.org/10.1098/rsbl.2009.0175.

Sandercock, D.A., Smith, S.H., Di Giminiani, P., Edwards, S.A. 2016. Histopathological characterization of tail injury and traumatic neuroma development after tail docking in piglets. Journal of Comparative Pathology 155: 40–49. https://doi.org/10.1016/j.jcpa.2016.05.003.

Sandercock, D.A., Barnett, M.W., Coe, J.E., Downing, A.C., Nirmal, A.J., Di Giminiani, P., Edwards S.A., Freeman T.C. 2019. Transcriptomic analysis of caudal porcine dorsal root ganglia in tail amputated pigs shows long-term effects on many pain associated genes. Frontiers in Veterinary Science 6:314. https://doi.org/10.3389/fvets.2019.00314.

Schrøder-Petersen, D.I., Simonsen, H.B. 2001. Tail biting in pigs. Veterinary Journal 162: 196–210. https://doi.org/10.1053/tvjl.2001.0605.

Singh, B. 2017a. Dyce, Sack and Wensing's Textbook of veterinary anatomy 5th Edition (E-Book). Elsevier Health Sciences, Amsterdam, Netherlands, p. 34.

Singh, B. 2017b. Dyce, Sack and Wensing's Textbook of veterinary anatomy 5th Edition (E-Book). Elsevier Health Sciences, Amsterdam, Netherlands, p. 776.

Singh, B. 2017c. Dyce, Sack and Wensing's Textbook of veterinary anatomy 5th Edition (E-Book). Elsevier Health Sciences, Amsterdam, Netherlands, pp. 27–29.

Sprem, N., Piria, M., Novosel, H., Florijancic, T., Antunovic, B., Treer, T. 2011. Morphological variability of the Croatian wild boar population. Prethodno Priopcenje 575–583.

Stankowich, T. 2008. Tail-flicking, tail-flagging, and tail position in ungulates with special reference to black-tailed deer. Ethology 114: 875–885. https://doi.org/10.1111/j.1439-0310.2008.01530.x.

Sutherland, M.A., Tucker, C.B. 2011. The long and short of it: a review of tail docking in farm animals. Applied Animal Behaviour Science 135: 179–191. https://doi.org/10.1016/j.applanim.2011.10.015.

Torrey, S., Devillers, N., Lessard, M., Farmer, C., Widowski, T. 2009. Effect of age on the behavioural and physiological responses of piglets to tail docking and ear notching. Journal of Animal Science 87: 1778–1786. https://doi.org/10.2527/jas.2008-1354.

Ursinus, W.W., Van Reenen, C.G., Kemp, B., Bolhuis, J.E. 2014. Tail biting behaviour and tail damage in pigs and their relationship with general behaviour: predicting the inevitable? Applied Animal Behaviour Science 156: 22–36. https://doi.org/10.1016/j.applanim.2014.04.001.

Valros, A., Ahlström, S, Rintala, H.J., Häkkinen, T., Saloniemi, H. 2004. The prevalence of tail damage in slaughter pigs in Finland and associations to carcass condemnations. Acta Agric. Scand. Sec. A: Anim. Sci. 54: 213–219. https://doi.org/10.1080/09064700510009234.

Valros, A., Munsterhjelm, C., Puolanne, E., Ruusunen, M., Heinonen, M., Peltoniemi, O., Pösö, R. 2013. Physiological indicators of stress and meat and carcass characteristics in tail bitten slaughter pigs Acta Veterinaria Scandinavica 55: 75. https://doi.org/10.1186/1751-0147-55-75.

Valros, A. 2018. Tail biting. In: Spinka, M. (Ed.): Advances in pig welfare, Woodhead Publishing Elsevier, Kidlington, UK, pp. 137–166.

vom Brocke, A.L., Karnholz, C., Madey-Rindermann, D. Gauly, M. Leeb, C., Winckler, C. Schrader, L, Dipple, S. 2019. Tail lesions in fattening pigs: relationships with postmortem meat inspection and influence of a tail biting management tool. Animal. 13: 835–844. https://doi.org/10.2527/jas.2008-1354.

Wallgren, P., Lindahl, E. 1996. The influence of tail biting on performance of fattening pigs. Acta Veterinaria Scandinavica 37:453–460. https://doi.org/10.1186/BF03548085.

Wallgren, T., Larsen, A., Gunnarsson, S. 2019. Tail posture as an indicator of tail biting in undocked finishing pigs. Animals 9: 18. https://doi.org/10.3390/ani9010018.

Winters J.F.M., Foldager L., Krogh U., Nørskov N.P., Sørensen M.T. 2023. Impact of gluphosate residues in sow diets on neonatal piglets: tail kinks, stillborn and diarrhoea. Livestock Science 269: 105172.

Yap, M.F., Sandercock, D.A., Smith, S.H. 2015. Histopathological characterisation of tail bite injuries in pigs and comparison with docked tails. BBSRC Research Experience Placement (REP) Final report (B036684). pp. 1–23.

Zonderland, J.J., van Riel, J.W., Bracke, M.B., Kemp, B., den Hartog, L.A., Spoolder, H.A. 2009. Tail posture predicts tail damage among weaned piglets. Applied Animal Behaviour Science 121: 165–170. https://doi.org/10.1016/j.applanim.2009.09.002.

CHAPTER 3

Why does tail biting occur? An evolutionary perspective

Rick D'Eath[1,*] *and Marc Bracke*[2,3]

[1]Animal Behaviour & Welfare Team, Animal and Veterinary Science Research Group, SRUC, Edinburgh, UK; [2]Department of Animal Health and Welfare, Wageningen Livestock Research, Wageningen University & Research, 6708 WD Wageningen, The Netherlands; [3]Agricultural Biosystems Engineering, Wageningen University, 6700 AA, Wageningen, The Netherlands; *Correspondence: Rick.DEath@sruc.ac.uk

Abstract

Tail biting, tail docking and the improper provision of enrichment materials are interrelated and are a major welfare concern in modern pig farming. Many other authors have reviewed the multifactorial causes of tail biting. Here, instead our focus is on the evolutionary origins of the problem. We discuss possible phylogenetic and functional explanations of tail biting in pigs. Tail biting is not a natural behaviour, not being seen in wild or feral pigs. Instead, it is best characterised as abnormal behaviour or a psychopathology, resulting from the thwarting of a motivated source behaviour. The most likely source behaviour is foraging, but there may be a role for other source behaviours which contribute to the expression of tail biting, including resource competition, thermoregulation, sickness behaviour and cannibalism. Since cannibalism is remarkably widespread throughout the animal kingdom, this touches on the question as to whether tail biting may in some sense not be as abnormal as it is generally perceived. Though much of our discussion is rather speculative in nature we believe it is worthwhile exploring hypotheses around the evolutionary origin of tail biting in pigs, and other welfare issues in livestock farming, in more detail in the future. Such an approach can encourage a re-examination of assumptions and generate new hypotheses.

Keywords

proximate and ultimate causation – abnormal behaviour – genetics – evolutionary origins

 | DOI:10.3920/9789004694989_005

1 Introduction

Tail biting is widely recognised as a multifactorial problem (Schrøder-Petersen and Simonsen, 2001; Bracke *et al.*, 2004a, 2004b; EFSA, 2007; D'Eath *et al.*, 2014; Valros, 2018). However, underlying the many potential causes of tail biting, ranging from improper enrichment to thermal stress, gender, tail length and life history (previous biting history), and disease (Boyle *et al.*, 2022), there is also a kind of mono-factorial origin, in that through their management decisions, farmers are affecting many if not all of the risk factors involved in causing tail biting. However, these decisions take place within the constraints of the farm system as designed and built (e.g. with fully-slatted floors), which is in turn constrained by the need for a farm to be profitable in the marketplace for pork. As such farmers, policy makers, citizens and consumers all share some of the (moral) responsibility for tail biting. Improved welfare conditions like high health status, appropriate feeding and the provision of more straw and space allows farmers to raise pigs with intact tails, rather than dock the tails of piglets at an early age (Valros *et al.*, 2016), although to eliminate tail biting entirely would require even less intensive systems. Over a longer time frame, factors related to domestication (genetics) and evolution also contribute to tail biting. The aim of this paper is to explore the origins of this welfare issue from an evolutionary perspective.

Niko Tinbergen put forward a framework for studies of animal behaviour. He proposed that to fully understand any given behaviour we must ask four questions about it (Tinbergen, 1963; Bateson and Laland, 2013). Two are about the immediate or proximate causation of the behaviour: 1) What are mechanisms which cause it to occur in the animal in that moment (Causation or 'Mechanisms of Control' e.g. nervous and endocrine systems, motivation, internal and external triggering stimuli)? And 2) How does the behaviour and its control systems develop during an animal's life? (Ontogeny: the effects of the physical and family/social environment, 'instinct' vs learning). Two further questions concern the ultimate causation of the behaviour: 3) What function does it serve in terms of enhancing the Darwinian fitness of the animal? (Function or 'Current Utility': how the behaviour enhances an animal's chance of surviving and breeding so that it passes its genes into the next generation) and finally, 4) What is the evolutionary history of the behaviour (Phylogeny: how did it evolve, e.g. do similar behaviours occur in closely related species)? It is the last two of these questions that we will discuss in relation to tail biting, informed at times by the other two questions.

In applied ethology and animal welfare science, the focus is often on the causation/mechanism questions. This is for at least two reasons: 1) Animal welfare is the positive or negative affective states experienced by an animal (shorter-term feelings or longer-term mood) (Duncan, 1993; Bracke *et al.*, 1999; Mendl *et al.*, 2010; Dawkins, 2015; Mellor, 2016; Paul *et al.*, 2022). These animal affective states can be thought of as an integral part of motivational systems (avoidance of punishing and seeking of rewarding behaviours) that underlie the control of behaviour (Mendl, Burman and Paul, 2010), so our many indirect indicators of welfare reflect the states of these affective systems (e.g. motivational strength, 'pain faces') or directly measure the physiological state of behavioural control systems (e.g. HPA and SAM axes). 2) To solve a welfare problem, for example to reduce the occurrence of an undesirable behaviour 'in the moment', the most direct solution is often to identify and then alter the external triggering stimuli. For example, in studies aimed at reducing tail biting, researchers have used experiments which alter aspects of the environment (such as the provision of different enrichment regimes/materials) (D'Eath *et al.*, 2014; Buijs and Muns, 2019) and epidemiological studies to identify environmental risk factors on farm (Moinard *et al.*, 2003; Smulders *et al.*, 2008; Scollo, Contiero and Gottardo, 2016; Pandolfi *et al.*, 2017, 2018; Gerster *et al.*, 2022). Studies on the environment during development relate to Tinbergen's second question (Day *et al.*, 2002; van de Weerd *et al.*, 2005; Munsterhjelm *et al.*, 2009; Statham, Green and Mendl, 2011; Telkänranta *et al.*, 2014; Lange *et al.*, 2021), and these ontogenetic studies are also important to fully understand and reduce the behaviour in an animal's lifetime.

In this chapter, we support Tinbergen's view that all four questions are valuable in providing a full explanation of behaviour. While the question of causation/mechanism is often the main focus in applied ethology, there is value in thinking about the ultimate causes too. A description of the full behavioural repertoire of a species in wild conditions can provide a basis for comparison with behaviour in captivity (D'Eath and Turner, 2009), and can highlight behaviours which have received little attention in the applied ethology literature, for example wallowing (Bracke and Spoolder, 2011). Also, theories and explanations for behaviours such as tail biting, while largely about mechanism, often rely on foundations which assume an adaptive origin for the behaviour –or at least an adaptive 'source' behaviour which is thwarted, leading to tail biting. We identify and further explore these foundations in this chapter.

2 The relevance of an animal's evolutionary history in applied ethology and animal welfare

To understand the behaviour of captive domestic animals, insights can be obtained from observing the behaviour of a species' wild ancestors, feral populations, and from experiments where domestic animals are given access to a semi-natural environment, although in the latter case, the developmental history in captivity may affect behaviour. All three kinds of information are available for domestic pigs (D'Eath and Turner, 2009). These studies allow us to identify mismatches between behaviours which are natural for the animals, and the artificial situation in captivity. For example, when kept in a semi-natural environment, preparturient domestic sows seek solitude from their group, gather materials such as grass, leaves and branches, and use them to build nests before giving birth to a litter of piglets (Stolba and Wood-Gush, 1989; Jensen, Stangel and Algers, 1991; Stangel and Jensen, 1991; Jensen, Vestergaard and Algers, 1993). These studies provide a basis for comparison to better understand the restless behaviours and rooting at pen fixtures shown by sows confined in farrowing crates.

Conversely, such studies enable us to identify behaviours which are only seen in captivity, identifying these as unnatural or abnormal. These include wallowing in excreta (Nannoni *et al.*, 2020), stereotypies and tail-biting. Such behaviours often indicate that something is lacking in an animal's past or current physical or social environment, prompting a need for further study to identify the cause and understand the possible welfare impact.

As pigs are domesticated animals, they have been subject to a different set of selection pressures under artificial selection, and even more so in the modern era through quantitative animal breeding. Traits such as reduced fear of humans, and reduced aggression towards them (docility) became advantageous and are selected for. Tinbergen's ultimate questions about the function and evolutionary history of behaviour in the wild pre-domesticated state are still of interest, but we must consider the potential for domestication to lead to unusual or rapid changes in morphology (e.g. lop ears, skin/hair colour patterns, curly tails) and in behaviour, and for those traits sometimes also to acquire a different function. For example, domestic pigs have muscles that are mainly growing passively without being used much by the pigs themselves, serving a main function as human food rather than for moving around and dealing with challenges in the environment.

Pigs also seem to have more pronounced curly tails compared to wild boars. This could conceivably reduce the attractiveness of tails for tail biting as closed loops of material are less frequently bitten than loose ends (Feddes and Fraser, 1994). Tail biting occurs in modern breeds of pigs in captivity but apparently not, as far as we know, in wild pigs. Tail biting is unusual in outdoor farms. There is only one scientific report of it occurring in outdoor pigs. The study took place at five similar farms in the winter in Croatia (Jan–March), when tail biting was known to occur, possibly because pigs spent considerable time in their concrete-floored deep-straw bedded hut for warmth. There was also respiratory disease at these farms (Walker and Bilkei, 2006). For our purposes here, we should consider that tail biting thus might result from putting a 'natural' pig in an unnatural environment, or breeding a pig which has been bred to be no longer 'natural' or a combination of both. On this question, studies of wild boar (or crossbreds) in housing systems which limit their space and resources could be revealing. There have been relatively few studies of this type in wild boar (Gustafsson *et al.*, 1999; Gustafsson, Jensen, de Jonge, Illmann *et al.*, 1999) and none about tail biting, in contrast to the more extensive literature on wild jungle fowl vs domestic fowl and their crosses in captivity (Fallahsharoudi *et al.*, 2017; Katajamaa and Jensen, 2021).

3 'Natural' behaviour

As stated earlier, animal welfare is a concept which assumes that animals have subjective/affective states (feelings) which are important to them (Duncan, 1993; Bracke, Spruijt and Metz, 1999; Dawkins, 2003; Mellor, 2016). Natural behaviour is a quite separate concept, which relates to whether a behaviour is performed in the wild (Yeates, 2018), or perhaps that "animals should lead natural lives through the development and use of their natural adaptations and capabilities" (Fraser *et al.*, 1997). There are numerous examples of behaviours that as seen in the natural environment, and which indicate reduced welfare, or only occur when an animal's welfare is being threatened, for example fleeing from predators. There are also behaviours not seen in the wild that increase welfare, for example positive interactions with humans, and interacting with artificial enrichment devices (Dawkins, 2008; Yeates, 2018; Browning, 2019). This means that the occurrence or absence of a behaviour that is seen in a natural environment, cannot be relied upon as a welfare indicator on its own. Thus, naturalness of behaviour should be understood as a distinct characteristic, separate from welfare (Yeates, 2018).

Tail biting is not recognisable as a natural behaviour, in the sense that it has never been reported in wild boars and other wild pigs, or in domestic pigs in a natural or semi-natural environment (e.g. feral pigs). As such it may be considered as an abnormal behaviour, like stereotypic bar biting or chain chewing, but since tail biting can also be detrimental to production performance it is sometimes also referred to as a 'vice'. From a health and welfare point of view, however, tail biting is perhaps best characterised as a psychopathology, or as it is sometimes referred to in the German literature a technopathy (Reiner *et al.*, 2020).

The biters, especially once an outbreak begins, appear to be highly motivated to bite tails, so it could be argued that it improves their welfare, the excitement and taste of blood providing a novel and rewarding enrichment. However, given that tail biting probably does not occur under (semi-) natural conditions, the elevated motivation to bite pen-mates' tails is probably a sign of poor welfare due to an environment which is inadequate in some way for the normal expression of motivated behaviour. Victim pigs suffer from pain (di Giminiani *et al.*, 2016) and stress (Zupan *et al.*, 2012) during tail biting and may try to avoid being tail bitten, e.g. by tucking the tail down, so their welfare is clearly compromised.

However, rather than being conceived as behaviour occurring in the wild, natural behaviour can also be seen as behaviour which animals tend to perform, often also in a more or less species-specific way, like rooting and wallowing in pigs, or grazing in cattle, as part of their adaptation to functioning well in the natural environment. Generally, these are positively-motivated behaviours. As such, naturalness can be a useful starting point for formulating and testing hypotheses about the performance or prevention of behaviours in captivity, and how these might affect welfare. Under wild conditions, behavioural control systems evolve by natural selection such that animals which survive and reproduce pass on genes for those behaviours to their offspring. So broadly we should expect behavioural control systems to evolve 'decision rules' such that animals are motivated to approach certain types of food, nest sites and opportunities to mate (which enhance biological functioning and Darwinian fitness), while avoiding predators and extremes of environment and so on (which pose potential threats to biological functioning and Darwinian fitness). They may be willing to expend energy and take risks such as fighting with rivals to obtain access to resources which are vital for fitness such as food and mates (Barnard and Hurst, 1996). The decision rules which evolved in one environment – the original wild environment or 'environment of evolutionary adaptation', may no longer have the expected fitness-enhancing results in a captive environment (e.g. a pig using aggression in an attempt to drive away an unfamiliar intruder from her group).

Animal choices and motivation are valuable to assess welfare and help us understand an animal's perspective (Dawkins, 1990, 2003; Mason *et al.*, 1998; Pedersen and Jensen, 2007; Paul *et al.*, 2022). Another way of thinking about studies of wild, feral or free-ranging animals is that they act as endless 'choice tests' where animals continually make choices which indicate their preferences and motivations. We observe that pigs organise themselves into small social groups, roll in mud, and seek to spend substantial amounts of time foraging by sniffing, rooting, digging etc. Therefore, these natural behaviours, behaviours which they obviously have a natural tendency to perform, thus probably contribute positively to their welfare. Similarly, behaviours which they have a natural tendency to avoid, like specific predators and stimuli that evoke fear, are most likely to contribute negatively to their welfare, even when they may be frequent, i.e. normal/prevalent under natural or domesticated conditions. Tail biting is certainly not a natural behaviour in the sense of a behaviour which pigs have a tendency to perform under natural conditions because they are positively motivated to perform that behaviour (so as to enhance their biological functioning). By contrast, even if it were a functional behaviour under more natural conditions (see below the section on cannibalism), it is most probably motivated by frustration or thwarting of motivated behaviour.

4 Genetics of tail biting behaviour

Behaviours have evolved by natural selection or in the case of domesticated animals through a combination of natural and artificial selection. Either way, for change over generations to occur, behaviour must vary between individuals, and some proportion of this variability must be affected by differences between their genes (heritability).

There is good evidence for genetic effects on tail biting (EFSA, 2007; Turner *et al.*, 2018). Tail biting varies between breeds of pig (Breuer *et al.*, 2003; Sinisalo *et al.*, 2012), and anecdotally between different breeding company lines. Heritability of tail-biting is quite low ($h^2 = 0.05 \pm 0.02$ as a 0–1 trait in one of two breeds studied), but at an acceptable level so that breeding against it could be feasible (Breuer *et al.*, 2005).

There are a number of additional challenges to the objective of genetic selection against tail biting (Turner, 2011): the sporadic and unpredictable nature of tail biting means it requires constant vigilance, and unlike physical traits it cannot be measured at the convenience of the breeder. The first victims of tail biting are not easily detected, an untrained/naive observer can fail to notice tail biting, and a tail biting outbreak may rapidly escalate. Once a first pig has been detected with a tail wound, a subsequent inspection reveals many more pigs having tail wounds in the same pen. Identifying biters is difficult and potentially labour intensive, and once many pigs have been bitten, it may be hard to tell the original biter from later 'secondary' biters. Without these challenges, artificial selection pressures against tail biting could be much stronger. Another challenge is that pig farmers generally do not select their own breeding stock, instead using artificial insemination with mixed-sire semen from breeding companies. This means that when growing-fattening pigs are identified as biters or bitten pigs, the problem dam or sire may not be identified easily when parent stock is raised on a different farm and under different rearing conditions. These aspects of modern breeding and farming may contribute to the persistence and future selection of tail biting in pigs raised under modern husbandry conditions.

To get around some of the challenges of phenotyping, researchers have explored the use of tail models as a proxy (Fraser, 1987a; Breuer *et al.*, 2001; Beattie *et al.*, 2005; Statham *et al.*, 2006), with varying success in terms of its relationship with actual tail biting. Others have begun to identify molecular markers which could be used in selection against biters, or perhaps also against victims (Wilson *et al.*, 2012; Brunberg *et al.*, 2013). An alternative approach is to select for social harmony. Genetic selection programmes can be designed to model for 'indirect genetic effects' (Bijma *et al.*, 2007; Bijma *et al.*, 2007; Chen *et al.*, 2007) where the genetic propensity to affect the growth of others in the group is included in the selection process, also called 'group selection'. This has potential to reduce various socially harmful behaviours (aggression, tail biting) and other socially transmitted effects which could affect growth (e.g. disease transmission). When it was applied to pigs, biting at both tails and enrichment materials (jute sacks) was reduced (Camerlink *et al.*, 2015).

A further challenge in breeding against tail biting is that there appears to be a negative genetic relationship between desirable production traits such as rapid efficient lean growth and litter size on the one hand and tail biting on the other hand (Breuer *et al.*, 2005; Grandin and Deesing, 2014; Ursinus *et al.*, 2014). The following explanation for this finding seems plausible: Rapid growth results in increased appetite – feeding motivation, feeding rate, nutrient intake (Carcò *et al.*, 2018). Under resource allocation theory (Knap and Rauw, 2008) we could predict a positive genetic correlation between increased feeding and foraging motivation. As discussed later, foraging motivation is one possible source behaviour for tail biting. Pleiotropic genes affecting more than one aspect of feeding behaviour are known from other species (Anreiter and Sokolowski, 2019). This genetic relationship between production and tail biting, probably through increased foraging motivation, may be part of the origin of the problem in the first place, and certainly could limit genetic progress in correcting it.

Selection for shorter tails has been proposed and could be feasible (D'Eath *et al.*, 2014), but has not to our knowledge been attempted in practice. In theory, it would also be conceivable to select for pigs without tails or even without incisor teeth. Like tail docking itself, however, this may not solve the problem of the pigs perceiving their environment as inadequate and being motivated to perform the behaviour, though perhaps physically prevented from doing so. This practice, like genetically blind chickens or hornless cattle, could be considered 'animal disenhancement' which is a subject of debate among ethicists (Thompson, 2008; Sandoe *et al.*, 2014; Murphy and Kabasenche, 2018). Breeding against pigs' exploratory behaviours such as sniffing, rooting and chewing, in an attempt to reduce tail biting would also fall into this controversial category (D'Eath *et al.*, 2010; Murphy and Kabasenche, 2018).

5 Hypotheses about the evolutionary origins of tail biting

As an abnormal behaviour, tail biting does not have a function that does (or would) improve the Darwinian fitness of pigs. However, abnormal behaviours are derived from source or origin behaviours (Mason, 1991; Mills, 2003) which did originally have a function, but are unable to be fully expressed, i.e. they are 'thwarted' in captive conditions (Clubb *et al.*, 2006). For example, stereotypic bar gnawing and jumping in caged lab mice both originate from different forms of thwarted escape attempts (Würbel *et al.*, 1996; Würbel, 2006). In this section, we explore different hypotheses about what the origin or source behaviours of tail biting could be.

Taylor and co-authors' important literature review characterised three forms of tail biting (two-stage, sudden-forceful and obsessive) (Taylor *et al.*, 2010). In discussing these, different explanations for their underlying mechanistic causes were proposed: Re-directed foraging for two-stage, resource competition for sudden-forceful and some form of abnormal behaviour for obsessive tail biting. A fourth type named 'epidemic tail biting' was suggested by Valros. This is characterised by a rapid onset following a sudden increase in environmental stressors (e.g. rapid temperature change; feed system breakdown/power outage; blocked drinker) in which the combination of chronic and acute stressors quickly reach a tipping point (Valros, 2018).

We will take inspiration from these different manifestations of tail biting, and the source behaviours which they imply as the basis for our discussion of the ultimate causation of tail biting behaviour. We will also consider whether so-called cannibalism, which may result when a tail-biting outbreak is left unchecked/not adequately dealt with, may have an evolutionary basis. We will consider how each explanation could be involved in the original cause/trigger of tail biting or in the maintenance or escalation of tail biting once it has begun. Maintenance and escalation can refer to either the original biter, or to other pigs in the group beginning to perform the behaviour. These factors combine to determine how severe the outcome is in a group.

5.1 *Re-directed foraging*

Two-stage tail biting (Taylor *et al.*, 2010) occurs when pigs begin to investigate tails, showing so-called tail-in-mouth behaviour at an early age (Schrøder-Petersen and Simonsen, 2001) and then this exploratory behaviour escalates into damaging tail biting. An evolutionary origin hypothesis for this is as follows: Pigs have evolved a foraging strategy which involves large periods of time engaging in oral investigatory and foraging behaviours such as sniffing, rooting, chewing and digging in soil where food is mostly hidden. This in turn leads to a motivation for the exploration of novelty and some form of persistent oral behaviour (characterised also as a 'behavioural need' – see later), which is largely unfulfilled in barren environments. Re-direction of motivated exploratory behaviours towards tails can then lead to tail biting.

It is possible that for this to happen a certain motivation build-up is necessary to overcome the pigs' aversion to bite pen-mates' tails. From an evolutionary perspective harming group members is likely to reduce fitness directly as group members generally tend to be related genetically. In a farming context, groups are likely to be made up of both relatives and non-relatives, and once a group is well established, there is no evidence that pigs discriminate between these in aggression, dominance or tail biting. Furthermore, tail investigation may need to overcome the pigs' aversion to be in oral contact with excreta (Bracke, 2007), which may be based on an evolutionary mechanism to avoid e.g. parasite infestation (Nannoni *et al.*, 2020). Compared with their wild counterparts, pigs kept in heated slatted-floor systems, may be exposed to (the smell of and physical contact to) manure to such an extent that it reduces their aversion to investigating pen-mates' tails, which in a more natural environment would more distinctly be associated with an aversion to excreta. From an evolutionary perspective, pigs may be predisposed to a breakdown of the initial aversion to be in contact with excreta since they may perform scent rubbing (a behaviour also seen in carnivores, cattle and deer), and they do not have sweat glands, so they may have evolved the ability to use excreta to cool down when no mud is available for wallowing (Bracke, 2011).

Once the foraging motivation to explore tails has built up sufficiently to trigger tail-in-mouth behaviour the accidental 'discovery' that the biting of tails leads to the taste of blood as well as the avoidance behaviour of the bitten pig, may subsequently enhance the biter pig's motivation to continue tail biting (because it provides a food reward as well as meeting the need for novelty, stimulation and excitement). We will examine the different aspects of this hypothesis.

5.1.1 Pigs' evolved foraging behaviour

Modern molecular-genetic phylogenetic methods show that pigs (Family *Suidae* along with the closely related peccaries *Tayassuidae*) are an outgroup within the even-toed ungulates (artiodactyls) sharing a common ancestor with other more specialised ruminant clades such as giraffe, deer, antelope and cattle (Agnarsson and May-Collado, 2008). Fraser (1984) suggested that pigs can be thought of as an 'early simple and unspecialised member of the cloven-hoof family, with one trotter still firmly in the camp of the carnivores'. Unlike the ruminant artiodactyls, which specialise in grazing and browsing, pigs' feeding habits in the wild are more generalised. They consume a wide variety of foods which include predominantly plants (90%+ of the diet), but also fungi and some animals. Pigs have four main feeding habits – they browse and graze plants, forage on the ground surface for fruit, nuts, fungi and carrion, root/dig underground for roots, fungal rhizomes and invertebrates, and also predate a variety of mostly small vertebrates such as frogs, birds and mice (D'Eath and Turner, 2009; Ballari and Barrios-García, 2014).

Wild boar (Frädrich, 1974) and domestic pigs in a semi-natural environment (Stolba and Wood-Gush, 1989) spend around 50–65% during daylight hours engaged in foraging activities such as rooting and grazing (Jordana Rivero *et al.*, 2019). For pigs in farms, where these activities are directed to substrates, floor, pen fixtures etc, these figures are much lower – closer to 15–20% of the time, but increasing at times when fresh materials are provided (Munsterhjelm *et al.*, 2009; Jensen, Studnitz and Pedersen, 2010; Lahrmann *et al.*, 2015).

An important element of pig foraging behaviour is their rooting disc with its shovel-like tip which has evolved as a tool to perform a considerable amount of physical labour while foraging (Soucek *et al.*, 1999). The pig's nose is also involved in social behaviour e.g. 'social nosing' (Camerlink and Turner, 2013) and it has a well-developed olfactory sense. Pigs rely on odour to locate and remember food sources (Croney *et al.*, 2003) and to detect distant or buried food sources (Oostindjer, 2017). The olfactory sensitivity of pigs is illustrated by the fact that they have 1113 functional genes and 188 pseudogenes encoding their olfactory receptors (Nguyen *et al.*, 2012), which is one of the largest sets known in any mammalian species (Niimura, Ihara and Touhara, 2020). The olfactory areas of the pig brain are large and well-developed (Brunjes, Feldman and Osterberg, 2016). As well as foraging, olfaction is used for social purposes; pigs also have nine glands which produce odours (Pond and Houpt, 1978).

Pigs' evolved foraging behaviour may also offer a potential explanation for the continued performance of tail biting – both in the biter and in conspecifics. The foods that pigs consume in the wild are generally small and patchy in nature (e.g. an area of beech nuts on the forest floor, a bird, mammal or insect nest, or a plant with edible roots or tubers). Pigs may have evolved a mechanism of foraging behaviour control in which the discovery of one food item should facilitate positive feedback to foraging motivation to fully exploit a local patch (Terlouw *et al.*, 1991; Lawrence and Terlouw, 1993; Terlouw *et al.*, 1993), but see Mason and Mendl (1997) for alternative views. Consumption of a small food item appears to increase feed motivation, suggesting there may be positive feedback early in a meal (Terlouw *et al.*, 1993; Haskell *et al.*, 1996). It is plausible that such positive feedback could promote further biting once the 'reward' of the taste and texture of an injured tail is first found. Furthermore, in this environment, pigs foraging in groups should also increase their local search activity in response to a conspecific finding food. Social facilitation and synchrony of feeding in pigs is well known from semi-natural (Hsia and Wood-Gush, 1983, 1984) and farmed environments (Keeling and Hurnik, 1996; Oostindjer *et al.*, 2011). It is plausible that discovery of a tail as a 'novel food source' could then lead to tail biting by other pigs in the group directed towards the same individual or other victim pigs.

5.1.2 Is oral behaviour a behavioural need?

Since behavioural control systems have evolved under natural selection, it follows that under most circumstances which could occur in natural conditions, there should be a reasonably good fit between on the one hand the functional benefits (in terms of Darwinian fitness) that should result from that behaviour and on the other hand the affective states that animals experience and their motivation to perform those fitness-enhancing behaviours (Mendl *et al.*, 2010). For example, animals must have the energy and other nutrients in food to live, grow and reproduce (function) and become hungry (affective state) and consequently be motivated enough to expend energy for the purpose of finding food. But in captivity, if food is provided in abundance, do pigs 'suffer' from not having to spend time searching for food?

The answer to this question depends on the evolved mechanisms underlying motivational control: feeding consists of a food seeking/foraging (appetitive) phase and a food consumption (consummatory) phase. The consumption of food is the goal of this motivational system. So, when it is achieved, negative feedback occurs which both brings an end to food consumption (meal ending – satiation) and delays/postpones/inhibits further food seeking (satiety) allowing an animal to prioritise other behaviours. In captive wild raptors trained for use in falconry, a bird's diet must be carefully controlled to manage its weight, otherwise it will refuse to hunt or return to the lure (Jones, 2001). This suggests that in these birds, motivation to perform the appetitive part of foraging (hunting) is lessened by a successful hunt which ends in food consumption.

This simple conception might suit a species like a raptor which periodically hunts large prey, but pigs appear to persist in performing some level of foraging/

exploratory behaviours even when they are well fed (Munsterhjelm *et al.*, 2009; Jensen and Pedersen, 2010; Jensen *et al.*, 2010; Lahrmann *et al.*, 2015; Read *et al.*, 2020). There has been debate about whether the appetitive phase of feeding (foraging/exploration) is a motivated behaviour in its own right, independent of food consumption, and as such constitutes a so-called 'behavioural need', i.e. a motivation to perform the behaviour where the performance itself is rewarding to the animal (Hughes and Duncan, 1988). The motivational control of feeding will vary between species with different modes of feeding; 'trophic guilds' insectivore, grazer, browser, gleaner etc. (González-Salazar, Martínez-Meyer and López-Santiago, 2014). For an opportunistic omnivorous ground forager/digger of mostly small food items like the pig, 'meal ending' may not be so clear-cut as it would be in a large carnivore, and there may be other benefits to ongoing foraging/exploration including social synchrony, information-gathering and perhaps also 'fattening up' to ensure longer-term survival in case of a subsequent food shortage. Fattening up in autumn (on acorns and beechnuts, for example) to enhance survival during winter appears to be part of wild pigs foraging strategy (Cuevas *et al.*, 2013).

Information-gathering could also form part of the explanation for why pigs continue to explore and forage even when provided with plenty of food. This appears to be an important aspect of the performance and motivation of foraging behaviour across species (Inglis *et al.*, 1997; Bean *et al.*, 1999; Inglis *et al.*, 2001), with animals changing their foraging strategies in response to uncertainty (Anselme and Güntürkün, 2019). Pigs also show conditioned place preference for locations previously associated with foraging for tasty food items (chocolate raisins) hidden in the substrate over a location with the same quantity of tasty food in a trough (de Jonge *et al.*, 2008).

Information gathering, and the related concept of novelty-seeking could be more useful and adaptive for omnivores than it would be for herbivores like sheep or cattle. Grass is often abundant and easier to find, although there can be seasonal shortages, which is a problem often solved by migration. By contrast pigs have to find food in soil and in winter especially they may really have to search for a variety of food items. It seems plausible that given their foraging style, pigs may have evolved an attraction to novelty in addition to an ability to utilise temporarily abundant food availability like those associated with beech or oak 'mast' years.

Since pigs are domesticated, we should consider whether domestication might have influenced the systems which control feeding and foraging behaviour. For most of their domesticated history, pigs' omnivorous habits and foraging behaviour were valuable: pigs were fed on kitchen wastes and were allowed to graze and forage in forests (Clutton-Brock, 1987) so any artificial selection pressure would have reinforced these natural traits. With modern scientific selection, traits like rapid growth and feed conversion efficiency have been selected for. Earlier in the section on genetics we argued that selection for rapid growth may have led to genetic increases in both feeding and foraging motivation.

5.1.3 Is tail biting re-directed foraging?

From the 'environmental enrichment' literature for pigs, we know that materials which are novel (Van de Weerd and Day, 2009), deformable or destructible (Feddes and Fraser, 1994; Bracke, 2007), chewable, odorous and contain ingestible/edible content (Van de Weerd *et al.*, 2003; Studnitz, Jensen and Pedersen, 2007) are more likely to be used by pigs, and they avoid materials soiled with faeces (Bracke, 2007). These elements are consistent with the preferences of a pig foraging in the wild – deformability / destructibility are aspects of foraging by rooting and digging, novelty is consistent with not going over previously-foraged old ground (and with information-gathering); odorous would fit with pigs' very sensitive olfactory sense which they use in foraging; chewing, eating and avoidance of faeces are consistent with food consumption being a motivation for engaging with enrichment.

Provision of substrate and enrichment materials which are preferred by pigs and promote foraging behaviour are known to reduce tail biting risk (see Chapter 8). Other risk factors such as higher stocking density may also be linked with frustrated foraging, since this reduces the floor space and generally also the amount of substrate available for foraging, reduces the access to point-source enrichment items, and means that the bodies of other pigs are a greater proportion of available surfaces. An argument against the 'redirected foraging' hypothesis is that a number of risk factors for tail biting, such as health status, lighting and thermal stress (EFSA, 2007) appear on the face of it to have little to do with foraging behaviour. The availability of manipulable substrates was not ranked very highly as a risk factor by Finnish farmers who emphasised proper ventilation and animal health, although feeding related issues, and the use of straw were also highlighted (Valros *et al.*, 2016). Some of these other risk factors may modulate foraging motivation (for example ill health affects appetite and nutrient requirements) as well as perhaps general restlessness, meaning increased active time, so more opportunity for tail biting. See also Figure 1 in D'Eath *et al.* (2014) for a number of other hypotheses about how different risk factors might relate to the aetiology of tail biting.

Finally, motivation for foraging / feeding may also be part of the explanation for the escalation and maintenance of tail biting once it starts. Biting the tail results in the discovery of a novel tasty food item, so it may be normal for a pig to continue to bite under those circumstances, and for other pigs to be attracted to bitten tails. There are also elements of novelty and excitement that are normally rewarding to the pig.

5.2 *Resource competition*

Sudden-forceful tail biting appears to be an aggressive act motivated by a specific frustration, when pigs cannot access a resource, usually their feeder. Its evolutionary origin might therefore relate to aggressive social behaviours. Pigs have evolved in social groups, where aggression is sometimes necessary to drive intruders away, between males to compete for females and when establishing the social hierarchy. Evolution has equipped them with canines that can cause wounds/scratches, and boars have an area of thick bristles around the shoulders to protect them when fighting (Meese and Ewbank, 1973; Signoret *et al.*, 1975). Since normal agonism is directed to the front part of the pigs, that does not lead to tail biting. However, when pigs block access to a feeder, drinker or other localised resource (enrichment material, preferred lying area), the tail may be the only and main object that can be bitten in frustration. Similar considerations also apply to vulva biting in sows fed sequentially in electronic sow feeding systems (Van Putten and Van de Burgwal, 1990; Rizvi, Nicol and Green, 1998; Bracke, 2007). Such conditions, where the only/main access to a food source is from behind, are less likely to occur in nature where there is usually space for front or side access to resources and to conspecifics.

5.3 *Physical and thermal comfort*

Drafts, seasonal changes in ventilation or temperature, sudden changes in temperature and extremes of temperature have been identified as risk factors for tail biting (Sällvik and Walberg, 1984). These are often said to increase the general level of stress in 'tipping point' models, and perhaps contribute to increased general activity and restlessness, or competition for comfortable lying locations. However these are all mechanistic explanations, and our task here is to consider the pigs' evolved behaviours. Pigs cannot sweat, and in the wild they perform various behaviours for thermoregulation – shade seeking, digging in cool dirt, wallowing, lying further apart to keep cool and huddling or nestbuilding to keep warm (D'Eath and Turner, 2009). Some of these thermoregulatory behaviours have an oral-nasal component. So, it seems at least plausible that pigs' attempts to perform thermoregulatory behaviours in the captive environment could increase oral-nasal activity, increasing the likelihood of tail in mouth and tail biting.

5.4 *Disease, gut microbiota, cytokines and sickness behaviour*

Recent papers have put forward the hypothesis that infectious disease and other stressors which influence the immune system may be a causal factor in tail biting (Nordgreen *et al.*, 2020; Boyle *et al.*, 2022). This is discussed further in chapter 5 'Poor Health and the Brain'. Another recent hypothesis suggests a role of the gut microbiota in tail biting (Brunberg *et al.*, 2016). A recent paper showed that differences in gut microbiota exist between different tail biting phenotypes. Although this is interesting, it does not prove causation (Rabhi *et al.*, 2020).

From an evolutionary perspective, sickness behaviour (including reduced activity, appetite and changed mood) has evolved as an adaptive response to infectious disease, by promoting rest and recovery. The mechanism involves immune system signals such as pro-inflammatory cytokines which signal to the brain (Millman, 2007; Dantzer *et al.*, 2008). The nutrient requirements of pigs are likely to change when mounting an immune response. This could also have an impact on foraging-related behaviours (Boyle *et al.*, 2022). There is a lot more to do to better understand cause and effect, and test these hypotheses, but it at least seems plausible that evolved sickness behaviour could affect biters through alterations to pigs' nutrient requirements or their mood (Munsterhjelm *et al.*, 2019). Inactivity as a result of sickness behaviour could result in victims of tail biting being less reactive to the attentions of biters.

5.5 *Abnormal behavioural control? 'Obsessive' tail biting and passive victims*

Mills (2003) made a helpful distinction between adaptive behaviours which are adaptive for the animal but inconvenient for the keeper, maladaptive behaviours, which are attempts to behave in an adaptive way in an environment that does not accommodate that behaviour, and malfunctional behaviours which result from a disruption of the nervous system. The redirected oral behaviour and resource competition explanations for tail biting fall in the 'maladaptive' category. It is possible that (some forms of) tail biting could also be described as malfunctional.

The third form of tail biting put forward by Taylor *et al.* (2010) was 'obsessive', characterised by certain individuals who perform a lot of tail biting. It is unclear whether this is really a distinct category, or simply an observation about the existence of individual differences in the other two kinds 'two-stage' and 'sudden-forceful' (D'Eath *et al.*, 2014).

It is plausible that the specific sub-category of tail biters known as 'obsessive' biters which go from victim to victim biting tails are showing a form of malfunctional behaviour that could be classified as an abnormal repetitive behaviour (ARB). ARBs may be divided into two subcategories – stereotypic behaviours which involve repeated and unvarying motor patterns (e.g. stereotypic bar mouthing in sows and jumping in laboratory mice), and impulsive/compulsive behaviours which involve the repetition of an inappropriate goal, but with more flexible motor output (e.g. barbering in laboratory mice) (Garner, 2005). Obsessive tail biting might then be a form of impulsive/compulsive (i.e. abnormal) behaviour with a more flexible motor output. This implies a malfunctional behavioural/psychological control system, and would mean that the performance of tail biting, and/or its result (injured tails) are for some reason much more rewarding than normal to that pig. This may resemble the development of emancipated stereotypic behaviour, where at first an animal tries to perform a behaviour in a normal adaptive way, but repeated thwarting by the captive environment results in an ARB developing (Mason and Rushen, 2006).

However, this is an explanation that falls under Tinbergen's questions of mechanism and development, which still relies on some form of normal (evolved) motivated 'source' behaviour being thwarted (like redirected foraging or competition for a resource). For example, thwarted escape attempts lead to stereotypic behaviours in mice (Würbel, Stauffacher and von Holst, 1996; Würbel, 2006) and human obsessive-compulsive disorder can be understood as an extreme form of evolved harm-avoidance strategies (Brüne, 2006). Obsessive tail biting might also have a motivational explanation such as a rebound effect, i.e. the tendency to perform behaviours like play and locomotion in an exaggerated way following periods of deprivation (Hutson *et al.*, 1993; Held and Spinka, 2011). From an evolutionary perspective, it may make sense for certain motivations to have the capacity/tendency to build up over time (Hinde, 1956). On balance, we are not convinced that we need an 'abnormal psychological mechanism' to maintain ongoing tail biting. It may be normal for a pig, once they have learnt that tails are a novel, tasty and exciting food source, to continue biting other tails. It would be fascinating (albeit highly challenging in practice) to identify and study 'obsessive' biters to investigate these hypotheses.

Victims of tail biting are not random: certain individual pigs seem to possess a distinctive phenotype and genotype which is different to that of both biters and uninvolved neutral pigs (Brunberg *et al.*, 2013). For our purposes this raises the interesting hypothesis that some kind of abnormality of behavioural control, such as 'learnt helplessness' results in unusual passivity and lack of response to the painful attentions of biters. However, this concerns the mechanisms of behavioural control. Our proper question here should perhaps be 'Why was it adaptive for the mechanisms which underlie the control of behaviour to evolve in this way?' From an evolutionary perspective, being injured may relate to being under predatory attack and in such circumstances it seems to be adaptive to show learned helplessness or a kind of freezing behaviour (Gilbert, 2001). It may be instructive to contrast tail biting with ear biting here too. It may be that since pain and damage to the ears can occur as part of a social aggression (fighting or bullying), so ear bitten pigs may be more likely to react by avoiding or shaking their head.

5.6 *Epidemic tail biting*

Various authors have proposed 'tipping point' models for tail biting, where an accumulation of different chronic and acute risk factors combines with the propensity of individuals and groups to begin tail biting (Bracke *et al.*, 2012, 2018; Valros, 2018). Such a 'tipping point' could occur when only one or a few pigs reach the threshold to begin biting, or many pigs could become biters. Although this is an explanation about causation / mechanism, the evolved psychological mechanisms of positive feedback and positive contrast (Wright and Panksepp, 2014) may be involved in the escalation of a tail biting outbreak as part of (an excessive) foraging motivation as discussed above.

5.7 *Cannibalism*

This hypothesis could be framed as follows: A wide variety of species engage in various forms of cannibalism. Cannibalism becomes more common under conditions of high population density, food scarcity and stress (Pfennig *et al.*, 1998). The conditions on farms can trigger an evolved cannibalistic response in pigs, as can also occur in poultry (Yngvesson and Keeling, 2001). A different form of this hypothesis is that an evolved cannibalistic response is not the initial trigger for tail biting, but is part of the explanation for the maintenance of tail biting behaviour: its continuation once it begins in an individual biter, and its spread to other pigs in the group.

Cannibalism in various forms is rather widespread throughout the animal kingdom (Fox, 1975; Polis, Myers and Hess, 1984). Examples abound from small invertebrates and particularly in r-selected species where large numbers of larvae or juveniles are present.

In mammals and birds, there are many examples of 'cannibalism' in the form of brood reduction through siblicide and infanticide (Polis *et al.*, 1984). In mice, infanticide occurs in the wild in response to high population density, and may occur also in captivity as a result of the mouse's perception of overcrowding (Latham and Mason, 2004). The few cases of mammals killing (and then sometimes consuming) other adults appear to involve males competing for mates or females defending their young against infanticide (Gómez *et al.*, 2021).

In pigs and wild boar, pre-weaning competition between piglets, which includes fighting with the needle teeth often results in the deaths of the weakest piglets (Fraser and Thompson, 1991). However, piglets do not cannibalise each other in this process, i.e. they do not consume the dead individuals as it occurs pre-weaning when milk is their food source. Dead piglets may be eaten by the sow, and cannibalism of live piglets can also occur (Chen *et al.*, 2008), but not usually in wild or free ranging pigs, suggesting this is an abnormal behaviour.

Wild boar are known to eat carrion. Many scavenging mammalian species avoid or delay the consumption of conspecific carcasses, which can reduce disease or parasite risk (Pfennig *et al.*, 1998; Gonzálvez *et al.*, 2021). Wild boar however do not discriminate in the sense that feeding from conspecific carcasses does occur (Ballari and Barrios-García, 2014; Probst *et al.*, 2017; Cukor *et al.*, 2020).

In line with this explanation, we may report here a personal observation that domestic pigs witnessing the slaughter (electrical stunning followed by bleeding) of a conspecific pig, did not show overt signs of distress, but instead seemed primarily ingestion-motivated as indicated by the fact that witnessing pigs were seen to lick the blood of their slaughtered conspecifics (MB, pers. obs.). The absence of clear signs of distress may be relevant in relation to cannibalism in that pigs and other domesticated animals like farmed deer, laboratory mice and laying hens do not seem to respond aversively to seeing conspecifics getting killed, as people might expect them to do, given our human aversion to the killing of conspecifics (Bracke, 1992).

Despite the instances in which siblicide, infanticide and scavenging from conspecific carcasses occurs in pigs, it may be a bit of a stretch to go from these examples to an evolutionary origin for tail biting, even though it is commonly known in pig farming that tail biting can escalate into pigs eating the back parts of the bitten animal leading to death and consumption of the pig once it has died (Jericho and Church, 1972; McGlone *et al.*, 1990). We can perhaps say that wild boars' failure to avoid conspecific carcasses at least shows that pigs do not have an aversion to the taste and smell of conspecific blood and flesh. Once tail biting begins, other pigs appear to be drawn to the injured tails and also begin biting them. Some studies have suggested that blood at the injured site is attractive and increases biting (Fraser, 1987a, 1987b; Fraser *et al.*, 1991) while others found only a weak preference for bloody tail models over water-soaked tail models (McIntyre and Edwards, 2002), although pigs which grew poorly had a more marked preference for the bloody tail model. This raises the possibility that certain pigs' motivation to satisfy unmet nutritional needs might be a driver for tail biting escalation, spread and persistence.

6 Discussion and conclusion

We hope that our speculative journey through this topic of the evolutionary origins of tail biting has been interesting and instructive to the reader. Hypotheses about the evolutionary origins of behaviour are worth stating and then being examined in detail, as well as alternative hypotheses (including mechanistic explanations) to explain the same observations (Mason and Mendl, 1997).

Abnormal behaviours such at tail biting are often a result of the thwarting or re-direction of motivated behaviours, especially foraging, which have evolved under natural selection and been modified through domestication. Our examination of redirected foraging, social competition, thermoregulation, sickness behaviour and cannibalism as potential origin behaviours suggests that any or all of these may play some role in tail biting, and we hope that our speculations on these aspects will inspire hypothesis testing in further research. For example, the following could be worth investigating further: 1) The genetic relationships between lean carcass growth rate, feed intake, motivation to forage and tail biting. 2) Whether 'obsessive' biters are normal or show signs of abnormal behavioural regulation and whether the 'victim phenotype' is behaviourally unresponsive or passive. 3) The nutritional needs and appetites of biters and how these might be affected by disease, growth rate or stress. 4) Whether cannibalism has been documented in wild or feral pigs, what are the predisposing factors involved and what are the triggers and inhibitions involved in the consumption of conspecific remains.

7 Take home message

- Tail biting is an 'unnatural' behaviour, not normally seen in wild boar, feral pigs or domestic pigs in semi-natural environments.

- Tail biting thus seems to originate from the pigs' natural evolved and motivated behaviour being thwarted in a farm environment.
- Domestication, genetic selection and the pigs' perception of overcrowding seems to have increased pigs' propensity to tail bite.
- Thwarting of pigs' evolved foraging and exploratory behaviours is the best supported hypothesis for a source behaviour
- Alternative/complementary source behaviours include social competition and cannibalism related to a perception of overcrowding.

References

Agnarsson, I. and May-Collado, L.J. (2008) The phylogeny of Cetartiodactyla: the importance of dense taxon sampling, missing data, and the remarkable promise of cytochrome b to provide reliable species-level phylogenies *Molecular Phylogenetics and Evolution* 48:964–985. https://doi.org/10.1016/j.ympev.2008.05.046.

Anreiter, I. and Sokolowski, M.B. (2019) The foraging gene and its behavioral effects: pleiotropy and plasticity, 53:373–392: https://doi.org/10.1146/annurev-genet-112618-043536.

Anselme, P. and Güntürkün, O. (2019) How foraging works: uncertainty magnifies food-seeking motivation *Behavioral and Brain Sciences*, 42 e35, 1–59. https://doi.org/10.1017/S0140525X18000948.

Ballari, S.A. and Barrios-García, M.N. (2014) A review of wild boar Sus scrofa diet and factors affecting food selection in native and introduced ranges *Mammal Review*, 44(2):124–134. https://doi.org/10.1111/mam.12015.

Barnard, C.J. and Hurst, J.L. (1996) Welfare by design: the natural selection of welfare criteria *Animal Welfare*, 5(4):405–433. https://doi.org/10.1017/S0962728600019151.

Bateson, P. and Laland, K.N. (2013) Tinbergen's four questions: an appreciation and an update *Trends in Ecology & Evolution*, 28(12):712–718. https://doi.org/10.1016/j.tree.2013.09.013.

Bean, D., Mason, G.J. and Bateson, M. (1999) Contrafreeloading in starlings: testing the information hypothesis *Behaviour*, 136:1267–1282. https://doi.org/10.1163/156853999500712.

Beattie, V.E., Breuer, K., O'Connell, N.E., Sneddon, I.A., Mercer, T.J., Rance, K.A. Sutcliffe, M.E.M. and Edwards, S.A. (2005) Factors identifying pigs predisposed to tail biting *Animal Science*, 80(3):307–312. https://doi.org/10.1079/ASC40040307.

Bijma, P. *et al.* (2007) Multilevel selection 2: estimating the genetic parameters determining inheritance and response to selection *Genetics*, 175(1):289–299. https://doi.org/10.1534/genetics.106.062729.

Bijma, P., Muir, W.A. and van Arendonk, J.A.M. 2007. Multilevel selection 1: quantitative genetics of inheritance and response to selection *Genetics*, 175(1):277–288. https://doi.org/10.1534/genetics.106.062711.

Boyle, L.A., Edwards, S.A., Bolhuis, J.E., Pol, F., Semrov, M.Z., Schutze, S., Nordgreen, J., Bozakova, N., Sossidou, E.N. and Valros, A. (2022) The evidence for a causal link between disease and damaging behavior in pigs *Frontiers in Veterinary Science*, 8:771682 https://doi.org/10.3389/fvets.2021.771682.

Bracke, M.B.M. (1992) *Can animals have a preference not to be killed? Observations on red deer, laying hens and mice which were exposed to conspecifics being killed. A study in cognition*. University of Edinburgh, UK.

Bracke, M.B.M. *et al.* (2004a) Decision support system with semantic model to assess the risk of tail biting in pigs 1. Modelling *Applied Animal Behaviour Science*, 87:31–44. https://doi.org/10.1016/j.applanim.2003.12.005.

Bracke, M.B.M., Hulsegge, B., Keeling, L. and Blokhuis, H.J. (2004b) Decision support system with semantic model to assess the risk of tail biting in pigs 2 Validation *Applied Animal Behaviour Science*, 87(1–2):45–54. https://doi.org/10.1016/j.applanim.2003.12.006.

Bracke, M.B.M. (2007) Multifactorial testing of enrichment criteria: pigs "demand" hygiene and destructibility more than sound *Applied Animal Behaviour Science*, 107(3–4):218–232. https://doi.org/10.1016/j.applanim.2006.10.001.

Bracke, M.B.M. (2007) 'Vulva biting', in A. Velarde and R. Geers (eds) *On farm monitoring of pig welfare*. Wageningen, Netherlands: Wageningen Academic Publishers, 111(1–2):65–69. https://doi.org/10.1016/j.applanim.2007.10.007.

Bracke, M.B.M. (2011) Review of wallowing in pigs: description of the behaviour and its motivational basis *Applied Animal Behaviour Science*, 132(1–2):1–13. https://doi.org/10.1016/j.applanim.2011.01.002.

Bracke, M.B.M., Vermeer, H.M., Bokma-Bakker, M.H., van der Peet-Schwering, C.M.C., Bolhuis, L., Leeijen, J. (2012) Checklist aanpak staartbijten bij (biologische) varkens (Checklist for tail biting in organic pigs)'. Lelystad, Netherlands: Wageningen Livestock Research. https://research.wur.nl/en/publications/checklist-aanpak-staartbijten-bij-biologische-varkens.

Bracke, M.B.M., Rodenburg, T.B., Vermeer, H.M. and van Niekerk, T.G.C.M. (2018) *Towards a common conceptual framework and illustrative model for feather pecking in poultry and tail biting in pigs-connecting science to solutions*. http://www.henhub.eu/framework-and-model-fp-tb-1-introduction/ (Accessed: 15 June 2022).

Bracke, M.B.M. and Spoolder, H.A.M. (2011) Review of wallowing in pigs: implications for animal welfare *Animal Welfare*, 20(3): 347–363. https://doi.org/10.1017/S0962728600002918.

Bracke, M.B.M., Spruijt, B.M. and Metz, J.H.M. (1999) Overall animal welfare assessment reviewed. Part 1: is it possible? *Netherlands Journal of Agricultural Science*, 47:279–291. https://doi.org/10.18174/njas.v47i3.466.

Breuer, K., Sutcliffe, M.E.M., Mercer, J.T., Rance, K.A., Beattie, V.E., Sneddon, I.A. and Edwards, S.A. (2003) The effect of breed on the development of adverse social behaviours in pigs *Applied Animal Behaviour Science*, 84(1–3):59–74. https://doi.org/10.1016/S0168-1591(03)00147-3.

Breuer, K., Sutcliffe, M.E.M., Mercer, J.T., Rance, K.A., O'Connell, N.E., Sneddon, I.A. and Edwards, S.A. (2005) Heritability of clinical tail-biting and its relation to performance traits *Livestock Production Science*, 93(1): 87–94. https://doi.org/10.1016/j.livprodsci.2004.11.009.

Breuer, K.B., Beattie, V.E., Dunne, L.M., Slade, E.C., Davies, Z., Mercer, J.T., Rance, K.A., Sneddon, I.A., Sutcliffe, M.E.M. and Edwards, S.A. (2001) Validation and development of a behavioural test to predict the predisposition of growing pigs to perform harmful social behaviour such as tail biting *Proceedings of the British Society of Animal Science*, 2001:50. https://doi.org/10.1017/S1752756200004324.

Browning, H. (2019) The natural behavior debate: two conceptions of animal welfare *Journal of Applied Animal Welfare Science*, 23(3):325–337. https://doi.org/10.1080/10888705.2019.1672552.

Brunberg, E., Jensen, P., Isaksson, A. and Keeling, L.J. (2013) Behavioural and brain gene expression profiling in pigs during tail biting outbreaks – evidence of a tail biting resistant phenotype *PLoS ONE*, 8(6): e66513. https://doi.org/10.1371/journal.pone.0066513.

Brunberg, E.I., Rodenburg, T.B., Rydhmer, L., Kjaer, J.B., Jensen, P. and Keeling, L.J. (2016) Omnivores going astray: a review and new synthesis of abnormal behavior in pigs and laying hens *Frontiers in Veterinary Science*, 3:57. https://doi.org/10.3389/fvets.2016.00057.

Brüne, M. (2006) The evolutionary psychology of obsessive-compulsive disorder: the role of cognitive metarepresentation *Perspectives in biology and medicine*, 49(3):317–329. https://doi.org/10.1353/pbm.2006.0037.

Brunjes, P.C., Feldman, S. and Osterberg, S.K. (2016) The pig olfactory brain: a primer *Chemical Senses*, 41(5):415–425. https://doi.org/10.1093/chemse/bjw016.

Buijs, S. and Muns, R. (2019) A review of the effects of non-straw enrichment on tail biting in pigs *Animals*, 9(10):824. https://doi.org/10.3390/ani9100824.

Camerlink, I., Ursinus, W.W., Bijma, P., Kemp, B. and Bolhuis, J.E. (2015) Indirect genetic effects for growth rate in domestic pigs alter aggressive and manipulative biting behaviour *Behavior Genetics*, 45(1):117–126. https://doi.org/10.1007/s10519-014-9671-9.

Camerlink, I. and Turner, S.P. (2013) The pig's nose and its role in dominance relationships and harmful behaviour *Applied Animal Behaviour Science*, 145(3–4): 84–91. https://doi.org/10.1016/j.applanim.2013.02.008.

Carcò, G., Gallo, L., Bona, M.D., Latorre, M.A., Fondevila, M. and Schiavon, S. (2018) The influence of feeding behaviour on growth performance, carcass and meat characteristics of growing pigs, *PLOS ONE*, 13(10): e0205572. https://doi.org/10.1371/journal.pone.0205572.

Chen, C.Y., Johnson, R.K., Newman, S. and Van Vleck, L.D. (2007) A general review of competition genetic effects with an emphasis on swine breeding. *Genetics and Molecular Research*, 6(3):594–606.

Chen, C., Gilbert, C.L., Yang, G., Guo, Y., Segonds-Pichon, A., Ma, J., Evans, G., Brenig, B., Sargent, C., Affara, N. and Huang, L. (2008) Maternal infanticide in sows: incidence and behavioural comparisons between savaging and non-savaging sows at parturition *Applied Animal Behaviour Science*, 109(2–4):238–248. https://doi.org/10.1016/j.applanim.2007.02.008.

Clubb, R., Vickery, S. and Latham, N. (2006) Motivation and Motivational Explanations for Stereotypy in G. Mason and J. Rushen (eds) *Stereotypic Animal Behaviour*. 2nd edn, 12.

Clutton-Brock, J. (1987) A natural history of domesticated mammals London/Cambridge: British Museum and Cambridge University Press.

Croney, C.C., Adams, K.M., Washington, C.G. and Stricklin, W.R. (2003) A note on visual, olfactory and spatial cue use in foraging behavior of pigs: indirectly assessing cognitive abilities *Applied Animal Behaviour Science*, 83(4):303–308. https://doi.org/10.1016/S0168-1591(03)00128-X.

Cuevas, M.F., Ojeda, R.A., Dacar, M.A. and Jaksic, F.M. (2013) Seasonal variation in feeding habits and diet selection by wild boars in a semi-arid environment of Argentina *Acta Theriologica*, 58(1):63–72. https://doi.org/10.1007/s13364-012-0105-x.

Cukor, J., Linda, R., Václavek, P., Mahlerová, K., Šatrán, P. and Havránek, F. (2020) Confirmed cannibalism in wild boar and its possible role in African swine fever transmission *Transboundary and Emerging Diseases*, 67(3):1068–1073. https://doi.org/10.1111/tbed.13468.

Dantzer, R., O'Connor, J.C., Freund, G.G., Johnson, R.W. and Kelley, K.W. (2008) From inflammation to sickness and depression: when the immune system subjugates the brain *Nature Reviews Neuroscience*, 9(1):46–56. https://doi.org/10.1038/nrn2297.

Dawkins, M. (2015) Chapter two – animal welfare and the paradox of animal consciousness edited by H.J.B. Marc Naguib *Advances in the Study of Behavior*, 47:5–38. https://doi.org/10.1016/bs.asb.2014.11.001.

Dawkins, M.S. (1990) From an animal's point of view: motivation, fitness and animal welfare *Behavioral and Brain Sciences*, 13(1):1–61. https://doi.org/10.1017/S0140525X00077104.

Dawkins, M.S. (2003) Behaviour as a tool in the assessment of animal welfare *Zoology*, 106(4):383–387. https://doi.org/10.1078/0944-2006-00122.

Dawkins, M.S. (2008) The science of animal suffering *Ethology*, 114(10):937–945. https://doi.org/10.1111/j.1439-0310.2008.01557.x.

Day, J.E.L., Burfoot, A., Docking, C.M., Whittaker, X., Spoolder, H.A.M. and Edwards, S.A. (2002) The effects of prior experience of straw and the level of straw provision on the behaviour of growing pigs *Applied Animal Behaviour Science*, 76(3):189–202. https://doi.org/10.1016/S0168-1591(02)00017-5.

D'Eath, R., Conington, J., Lawrence, A., Olsson, I. and Sandøe, P. (2010) Breeding for behavioural change in farm animals: Practical, economic and ethical considerations *Animal Welfare*, 19(S1):17–27. https://doi.org/10.1017/S0962728600002207.

D'Eath, R.B., Arnott, G., Turner, S.P., Jensen, T., Lahrmann, H.P., Busch, M.E., Niemi, J.K., Lawrence, A.B. and Sandøe, P. (2014) Injurious tail biting in pigs: How can it be controlled in existing systems without tail docking? *Animal*, 8(9):1479–1497. https://doi.org/10.1017/S1751731114001359.

D'Eath, R.B. and Turner, S.P. (2009) The natural behaviour of the pig animal welfare, 7:13–45. https://doi.org/10.1007/978-1-4020-8909-1_2.

Duncan, I.J.H. (1993) Welfare is to do with what animals feel. *Journal of Agricultural & Environmental Ethics*, 6(S2):8–14.

EFSA (2007) The risks associated with tail biting in pigs and possible means to reduce the need for tail docking considering the different housing and husbandry systems – scientific opinion of the Panel on Animal Health and Welfare *EFSA Journal*, 5(12):1–13. https://doi.org/10.2903/j.efsa.2007.611.

Fallahsharoudi, A., de Kock, N., Johnsson, M., Bektic, L., Ubhayasekera, S.J., Bergquist, J., Wright, D. and Jensen, P. (2017) QTL mapping of stress related gene expression in a cross between domesticated chickens and ancestral red junglefowl *Molecular and Cellular Endocrinology*, 446:52–58. https://doi.org/10.1016/j.mce.2017.02.010.

Feddes, J.J.R. and Fraser, D. (1994) Non-nutritive chewing by pigs: implications for tail-biting and behavioural enrichment *Transactions of the ASAE*, 37(3):947–950. https://doi.org/10.13031/2013.28163.

Fox, L.R. (1975) Cannibalism in natural populations *Annual Review of Ecology and Systematics*, 6(1):87–106. https://doi.org/10.1146/annurev.es.06.110175.000511.

Frädrich, H. (1974) A comparison of behaviour in the Suidae edited by V. Geist and F. Walther. N.S.: IUCN Publishers, pp. 133–143.

Fraser, D. (1984) The role of behaviour in swine production: a review of research *Applied Animal Ethology*, 11(4): 317–339. https://doi.org/10.1016/0304-3762(84)90041-5.

Fraser, D. (1987a) Attraction to blood as a factor in tail-biting by pigs *Applied Animal Behaviour Science*, 17(1–2): 61–86. https://doi.org/10.1016/0168-1591(87)90008-6.

Fraser, D. (1987b) Mineral-deficient diets and the pig's attraction to blood: implications for tail-biting *Canadian Journal of Animal Science*, 67:909–918. https://doi.org/10.4141/CJAS87-096.

Fraser, D., Weary, D.M., Pajor, E.A. and Milligan, B.N. (1997) A scientific conception of animal welfare that reflects ethical concerns *Animal Welfare*, 65(3):187–205. https://doi.org/10.1017/S0962728600019795.

Fraser, D., Bernon, D.E. and Ball, R.O. (1991) Enhanced attraction to blood by pigs with inadequate dietary-protein supplementation *Canadian Journal of Animal Science*, 71:611–619. https://doi.org/10.4141/CJAS91-076.

Fraser, D. and Thompson, B.K. (1991) Armed sibling rivalry among suckling piglets *Behavioral Ecology and Sociobiology*, 29:9–15. https://doi.org/10.1007/BF00164289.

Garner, J.P. (2005) Stereotypies and other abnormal repetitive behaviors: potential impact on validity, reliability, and replicability of scientific outcomes *Ilar Journal*, 46(2):106–117. https://doi.org/10.1093/ilar.46.2.106.

Gerster, U., Sidler, X., Wechsler, B. and Nathues, C. (2022) Prevalence of tail lesions in Swiss finishing pigs *Schweizer Archiv fur Tierheilkunde*, 164(4):339–349. https://doi.org/10.17236/sat00352.

Gilbert, P. (2001) Evolutionary approaches to psychopathology: the role of natural defences *Australian and New Zealand Journal of Psychiatry*, 35(1):17–27. https://doi.org/10.1046/j.1440-1614.2001.00856.

Di Giminiani, P., Brierley, V.L.M.H., Annalisa Scollo, A., Gottardo, F., Malcolm, E.M., Edwards, S.A. and Leach, M.C. (2016) The assessment of facial expressions in piglets undergoing tail docking and castration: Toward the development of the piglet grimace scale *Frontiers in Veterinary Science*, 3(100). https://doi.org/10.3389/fvets.2016.00100.

Gómez, J.M., Verdú, M. and González-Megías, A. (2021) Killing conspecific adults in mammals *Proceedings. Biological sciences*, 288. https://doi.org/10.1098/rspb.2021.1080.

González-Salazar, C., Martínez-Meyer, E. and López-Santiago, G. (2014) A hierarchical classification of trophic guilds for North American birds and mammals *Revista Mexicana de Biodiversidad*, 85(3): 931–941. https://doi.org/10.7550/rmb.38023.

Gonzálvez, M., Martínez-Carrasco, C., Sánchez-Zapata, J.A. and Moleón, M. (2021) Smart carnivores think twice: red fox delays scavenging on conspecific carcasses to reduce parasite risk *Applied Animal Behaviour Science*, 243:105462. https://doi.org/10.1016/j.applanim.2021.105462.

Grandin, T. and Deesing, M.J. (2014) Genetics and animal welfare *Genetics and the Behavior of Domestic Animals*, 435–472. https://doi.org/10.1016/B978-0-12-394586-0.00012-3.

Gustafsson, M., Jensen, P., de Jonge, F.H. and Schuurman, T. (1999) Domestication effects on foraging strategies in pigs (Sus scrofa) *Applied Animal Behaviour Science*, 62(4): 305–317. https://doi.org/10.1016/S0168-1591(98)00236-6.

Gustafsson, M., Jensen, P., de Jonge, F.H., Illmann, G. and Spinka, M. (1999) Maternal behaviour of domestic sows and crosses between domestic sows and wild boar *Applied Animal Behaviour Science*, 65(1): 29–42. https://doi.org/10.1016/S0168-1591(99)00048-9.

Haskell, M.J., Terlouw, E.M.C., Lawrence, A.B. and Erhard, H.W. (1996) The relationship between food consumption and persistence of post-feeding foraging behaviour in sows *Applied Animal Behaviour Science*, 48(3–4): 249–262. https://doi.org/10.1016/0168-1591(95)00644-3.

Held, S.D.E. and Spinka, M. (2011) Animal play and animal welfare *Animal Behaviour*, 81(5): 891–899. https://doi.org/10.1016/j.anbehav.2011.01.007.

Hinde, R.A. (1956) Ethological models and the concept of "drive" *The British Journal for the Philosophy of Science*, 6(24): 321–331. https://doi.org/10.1093/bjps/VI.24.321.

Hsia, L.C. and Wood-Gush, D.G.M. (1983) A note on social facilitation and competition in the feeding behaviour of pigs *Animal Science*, 37(1): 149–152. https://doi.org/10.1017/S0003356100001677.

Hsia, L.C. and Wood-Gush, D.G.M. (1984) Social facilitation in the feeding behaviour of pigs and the effect of rank *Applied Animal Ethology*, 11(3): 265–270. https://doi.org/10.1016/0304-3762(84)90033-6.

Hughes, B.O. and Duncan, I.J.H. (1988) The notion of ethological "need", models of motivation and animal welfare, *Animal Behaviour*, 36(6): 1696–1707. https://doi.org/10.1016/S0003-3472(88)80110-6.

Hutson, G.D., Haskell, M.J., Dickenson, L.G. and Slinger, D.E. (1993) Preferences of pregnant sows for wet and dry concrete floors *Applied Animal Behaviour Science*, 37(2): 91–99. https://doi.org/10.1016/0168-1591(93)90102-U.

Inglis, I.R., Langton, S., Forkman, B. and Lazarus. J. (2001) An information primacy model of exploratory and foraging behaviour, *Animal Behaviour*, 62(3): 543–557. https://doi.org/10.1006/anbe.2001.1780.

Inglis, I.R., Forkman, N. and Lazarus, J. (1997) Free food or earned food? A review and a fuzzy model of contrafreeloading *Animal Behaviour*, 53(6): 1171–1191. https://doi.org/10.1006/anbe.1996.0320.

Jensen, M.B. and Pedersen, L.J. (2010) Effects of feeding level and access to rooting material on behaviour of growing pigs in situations with reduced feeding space and delayed feeding *Applied Animal Behaviour Science*, 123(1–2): 1–6. https://doi.org/10.1016/j.applanim.2009.12.015.

Jensen, M.B., Studnitz, M. and Pedersen, L.J. (2010) The effect of type of rooting material and space allowance on exploration and abnormal behaviour in growing pigs *Applied Animal Behaviour Science*, 123(3–4): 87–92. https://doi.org/10.1016/j.applanim.2010.01.002.

Jensen, P., Stangel, G. and Algers, B. (1991) Nursing and suckling behaviour of semi-naturally kept pigs during the first 10 days postpartum *Applied Animal Behaviour Science*, 31(3–4): 195–209. https://doi.org/10.1016/0168-1591(91)90005-I.

Jensen, P., Vestergaard, K. and Algers, B. (1993) Nestbuilding in free-ranging domestic sows *Applied Animal Behaviour Science*, 38(3–4): 245–255. https://doi.org/10.1016/0168-1591(93)90023-I.

Jericho, K.W. and Church, T.L. (1972) Cannibalism in pigs *The Canadian Veterinary Journal*, 13(7):156–159. https://www.ncbi.nlm.nih.gov/pmc/articles/PMC1695826/ (Accessed: 15 June 2022).

Jones, M.P. (2001) Behavioral aspects of captive birds of prey *Veterinary Clinics of North America: Exotic Animal Practice*, 4(3): 613–632. https://doi.org/10.1016/S1094-9194(17)30026-9.

de Jonge, F.H., Tilly, S.L., Baars, A.M. and Spruijt, B.M. (2008) On the rewarding nature of appetitive feeding behaviour in pigs (Sus scrofa): do domesticated pigs contra-freeload *Applied Animal Behaviour Science*, 114(3–4): 359–372. https://doi.org/10.1016/j.applanim.2008.03.006.

Katajamaa, R. and Jensen, P. (2021) Tameness correlates with domestication related traits in a Red Junglefowl intercross *Genes, Brain and Behavior*, 20(3): e12704. https://doi.org/10.1111/gbb.12704.

Keeling, L.J. and Hurnik, J.F. (1996) Social facilitation and synchronization of eating between familiar and unfamiliar newly weaned piglets *Acta Agriculturae Scandinavica Section A – Animal Science*, 46:54–60. https://www.doi.org/10.1080/09064709609410924.

Knap, P.W. and Rauw, W.M. (2008) Selection for high production in pigs in Wendy M. Rauw (ed.) *Resource allocation theory applied to farm animal production*. Wallingford: CABI, 210–229. http://dx.doi.org/10.1079/9781845933944.0210.

Lahrmann, H.P., Oxholm, L.C., Steinmetz, H., Nielsen, M.B. and D'Eath, R.B. (2015) The effect of long or chopped straw on pig behaviour *Animal*, 9(5):862–870. https://doi.org/10.1017/s1751731114003024.

Lange, A., Hahne, M., Lambertz, C., Gauly, M., Wendt, M., Janssen, H. and Traulsen, I. (2021) Effects of different housing systems during suckling and rearing period on skin and tail lesions, tail losses and performance of growing and finishing pigs *Animals 2021*, 11(8):2184. https://doi.org/10.3390/ani11082184.

Latham, N. and Mason, G. (2004) From house mouse to mouse house: the behavioural biology of free-living Mus musculus and its implications in the laboratory *Applied Animal Behaviour Science*, 86(3–4):261–289. https://doi.org/10.1016/j.applanim.2004.02.006.

Lawrence, A.B. and Terlouw, E.M.C. (1993) A review of the behavioural factors involved in the development and continued performance of stereotypic behaviors in pigs *Journal of Animal Science*, 71(10):2815–2825. https://doi.org/10.2527/1993.71102815x.

Mason, G. (1991) Stereotypies: a critical review *Animal Behaviour*, 41(6):1015–1037. https://doi.org/10.1016/S0003-3472(05)80640-2.

Mason, G., McFarland, D. and Garner, J. (1998) A demanding task: using economic techniques to assess animal priorities *Animal Behaviour*, 55(4):1071–1075. https://doi.org/10.1006/anbe.1997.0692.

Mason, G. and Mendl, M. (1997) Do the stereotypies of pigs, chickens and mink reflect adaptive species differences in the control of foraging? *Applied Animal Behaviour Science*, 53(1–2):45–58. https://doi.org/10.1016/S0168-1591(96)01150-1 (Accessed: 31 May 2022).

Mason, G. and Rushen, J. (2006) Stereotypic animal behaviour: fundamentals and applications to welfare. Wallingford, UK: CABI.

McGlone, J.J., Sells, J.M., Harris, S.C. and Hurst, R.J. (1990) *Cannibalism in growing pigs: effects of tail docking and housing system on behaviour, performance and immune function*. Texas tech Univ. Sci. Tech. Rep. No. T-5-283. http://www.depts.ttu.edu/liru_afs/pdf/CANNIBALISMINGROWINGPIGS.pdf.

McIntyre, J. and Edwards, S.A. (2002) Preference for blood and behavioural measurements of known tailbiting pigs compared to control penmates. Egmond an Zee, Netherlands: Proceedings of the 36th International Congress of the International Society for Applied Ethology, p. 93. http://www.applied-ethology.org/isae_meetings.html.

Meese, G.B. and Ewbank, R. (1973) The establishment and nature of the dominance hierarchy in the domesticated pig *Animal Behaviour*, 21(2):326–334. https://doi.org/10.1016/S0003-3472(73)80074-0.

Mellor, D.J. (2016) Updating animal welfare thinking: moving beyond the "Five Freedoms" towards "A Life Worth Living" *Animals*, 6(3):21. https://doi.org/10.3390/ani6030021.

Mendl, M., Burman, O.H.P. and Paul, E.S. (2010) An integrative and functional framework for the study of animal emotion and mood *Proceedings of the Royal Society B-Biological Sciences*, 277(1696):2895–2904. https://doi.org/10.1098/rspb.2010.0303.

Millman, S.T. (2007) Sickness behaviour and its relevance to animal welfare assessment at the group level *Animal Welfare*, 16(2):123–125. http://dx.doi.org/10.1017/S0962728600031146.

Mills, D.S. (2003) Medical paradigms for the study of problem behaviour: a critical review *Applied Animal Behaviour Science*, 81(3):265–277. https://doi.org/10.1016/S0168-1591(02)00286-1.

Moinard, C., Mendl, M., Nicol, C.J. and Green, L.E. (2003) A case control study of on-farm risk factors for tail biting in pigs *Applied Animal Behaviour Science*, 81(4):333–355. https://doi.org/10.1016/S0168-1591(02)00276-9.

Munsterhjelm, C., Peltoniemi, O.A.T., Heinonen, M., Hälli, O., Karhapää, M. and Valros, A. (2009) Experience of moderate bedding affects behaviour of growing pigs *Applied Animal Behaviour Science*, 118(1–2):42–53. https://doi.org/10.1016/j.applanim.2009.01.007.

Munsterhjelm, C., Nordgreen, J., Aae, F., Heinonen, M., Valros, A. and Janczak, A.M. (2019) Sick and grumpy: changes in social behaviour after a controlled immune stimulation in group-housed gilts *Physiology & Behavior*, 198:76–83. https://doi.org/10.1016/j.physbeh.2018.09.018.

Murphy, K.N. and Kabasenche, W.P. (2018) Animal disenhancement in moral context *NanoEthics 2018*, 12:225–236. https://doi.org/10.1007/s11569-018-0323-z.

Nannoni, E., Aarnink, A.J.A., Vermeer, H.M., Reimert, I., Fels, M. and Bracke, M.B.M. (2020) Soiling of pig pens: a review of eliminative behaviour *Animals 2020*, 10(11):2025. https://doi.org/10.3390/ani10112025.

Nguyen, D.T., Lee, K., Choi, H., Choi, M-K., Le, M.T., Song, N., Kim, J-H., Seo, H.G., Oh, J-W., Lee, K., Kim, T-H. and Park, C. (2012) The complete swine olfactory subgenome: expansion of the olfactory gene repertoire in the pig genome *BMC Genomics*, 13(1):584. http://dx.doi.org/10.1186/1471-2164-13-584.

Niimura, Y., Ihara, S. and Touhara, K. (2020) Mammalian olfactory and vomeronasal receptor families in *The Senses: A Comprehensive Reference*. Elsevier, 3:516–535. http://dx.doi.org/10.1016/B978-0-12-809324-5.24175-8.

Nordgreen, J., Edwards, S.A., Boyle, L.A., Bolhuis, J.E., Veit, C., Sayyari, A., Marin, D.E., Dimitrov, I., Janczak, A.M. and Valros, A. (2020) A proposed role for pro-inflammatory cytokines in damaging behavior in pigs *Frontiers in Veterinary Science*, 7:646. https://doi.org/10.3389/fvets.2020.00646.

Oostindjer, M., Bolhuis, J.E., Mendl, M., Held, S., van den Brand, H. and Kemp, B. (2011) Learning how to eat like a pig: effectiveness of mechanisms for vertical social learning in piglets *Animal Behaviour*, 82(3):503–511. https://doi.org/10.1016/j.anbehav.2011.05.031.

Oostindjer, M. (2017) The role of olfaction in feeding and foraging in B.L. Nielsen (ed.) *Olfaction in animal behaviour and welfare*. CABI, 73–84. http://dx.doi.org/10.1079/9781786391599.0073.

Pandolfi, F., Stoddart, K., Wainwright, N., Kyriazakis, I. and Edwards, S.A. (2017) The "Real Welfare" scheme: benchmarking welfare outcomes for commercially farmed pigs *Animal*, 11(10):1816–1824. https://doi.org/10.1017/s1751731117000246.

Pandolfi, F., Edwards, S.A., Maes, D. and Kyriazakis, I. (2018) Connecting different data sources to assess the interconnections between biosecurity, health, welfare, and performance in commercial pig farms in Great Britain *Frontiers in Veterinary Science*, 5:41. https://doi.org/10.3389/fvets.2018.00041.

Paul, E.S., Browne, W., Mendl, M.T., Caplen, G., Trevarthen, A., Held, S. and Nicol, C.J. (2022) Assessing animal welfare: a triangulation of preference, judgement bias and other candidate welfare indicators *Animal Behaviour*, 186:151–177. https://doi.org/10.1016/j.anbehav.2022.02.003.

Pedersen, L.J. and Jensen, M.B. (2007) Concurrent schedules of reinforcement as a method to quantify the relative attractiveness of two rooting materials *Applied Animal Behaviour Science*, 107(1–2):147–151. https://doi.org/10.1016/j.applanim.2006.07.007.

Pfennig, D.W., Ho, S.G. and Hoffman, E.A. (1998) Pathogen transmission as a selective force against cannibalism *Animal Behaviour*, 55(5):1255–1261. https://doi.org/10.1006/anbe.1997.9996.

Polis, G.A., Myers, C.A. and Hess, W.R. (1984) A survey of intraspecific predation within the class Mammalia *Mammal Review*, 14(4):187–198. http://dx.doi.org/10.1111/j.1365-2907.1984.tb00345.x.

Pond, W.G. and Houpt, K.A. (1978) The biology of the pig. Ithaca, N.Y.: Comstock Pub. Associates.

Probst, C., Globig, A., Knoll, B., Conraths, F.J. and Depner, K. (2017) Behaviour of free ranging wild boar towards their dead fellows: potential implications for the transmission of African swine fever *Royal Society Open Science*, 4(5):170054. https://doi.org/10.1098/rsos.170054.

van Putten, G. and van de Burgwal, J.A. (1990) Vulva biting in group-housed sows – preliminary report *Applied Animal Behaviour Science*, 26(1–2):181–186. https://doi.org/10.1016/0168-1591(90)90097-W.

Rabhi, N., Thibodeau, A., Côté, J-C, Devillers, N., Laplante, B., Fravalo, P., Larivière-Gauthier, G., Thériault, W.P., Faucitano, L., Beauchamp, G. and Quessy, S. (2020) Association between tail-biting and intestinal microbiota composition in pigs *Frontiers in Veterinary Science*, 7:563762. https://doi.org/10.3389/fvets.2020.563762.

Read, E., Baxter, E.M., Farish, M. and D'Eath, R.B. (2020) Trough half empty: pregnant sows are fed under half of their ad libitum intake *Animal Welfare*, 29(2):151–162. https://doi.org/10.7120/09627286.29.2.151.

Reiner, G., Kühling, J., Lechner, M., Schrade, H., Saltzmann, J., Muelling, C., Dänicke, S. and Loewenstein, F. (2020) Swine inflammation and necrosis syndrome is influenced by husbandry and quality of sow in suckling piglets, weaners and fattening pigs *Porcine Health Management*, 6(1):32. https://doi.org/10.1186/s40813-020-00170-2.

Rivero, M.J., Rodríguez-Estévez, V., Pietrosemoli, S., Carballo, C., Cooke, A.S. and Kongsted, A.G. (2019) Forage consumption and its effects on the performance of growing swine – discussed in relation to European wild boar (Sus scrofa L.) in semi-extensive systems: a review *Animals: an Open Access Journal from MDPI*, 9(7): 457. https://doi.org/10.3390/ani9070457.

Rizvi, S., Nicol, C.J. and Green, L.E. (1998) Risk factors for vulva biting in breeding sows in south-west England *The Veterinary Record*, 143(24):654–658. https://doi.org/10.1136/vr.143.24.654.

Sällvik, K. and Walberg, K. (1984) The effects of air velocity and temperature on the behaviour and growth of pigs *Journal of Agricultural Engineering Research*, 30:305–312. https://doi.org/10.1016/S0021-8634(84)80031-1.

Sandøe, P., Hocking, P.M., Förkman, B., Haldane, K., Kristensen, H.H. and Palmer, C. (2014) The blind hens' challenge: does it undermine the view that only welfare matters in our dealings with animals? *Environmental Values*, 23(6):727–742. https://doi.org/10.3197/096327114X13947900181950.

Schrøder-Petersen, D.L. and Simonsen, H.B. (2001) Tail biting in pigs *Veterinary Journal*, 162(3):196–210. https://doi.org/10.1053/tvjl.2001.0605.

Scollo, A., Contiero, B. and Gottardo, E. (2016) Frequency of tail lesions and risk factors for tail biting in heavy pig production from weaning to 170 kg live weight *Veterinary Journal*, 207:92–98. https://doi.org/10.1016/j.tvjl.2015.10.056.

Signoret, J.P., Baldwin, B.A., Fraser, D. and Hafez, E.S.E. (1975) The behaviour of swine in E.S.E. Hafez (ed.) *Behaviour of domestic animals*. Bailliere Tindall, London, 295–329.

Sinisalo, A., Niemi, J.K., Heinonen, M. and Valros, A. (2012) Tail biting and production performance in fattening pigs *Livestock Science*, 143(2–3):220–225. https://doi.org/10.1016/j.livsci.2011.09.019 (Accessed: 31 May 2022).

Smulders, D., Hautekiet, V., Verbeke, G. and Geers, R. (2008) Tail and ear biting lesions in pigs: an epidemiological study *Animal Welfare*, 17(1):61–69. https://doi.org/10.1017/S0962728600031997.

Soucek, G., Breit, S., König, H.E. and Liebich, H.G. (1999) Functional aspects of musles of the external nose in pigs (Sus scrofa f. domestica) *Anatomia, Histologia, Embryologia*, 28(5–6):307–314. https://doi.org/10.1046/j.1439-0264.1999.00211.x.

Stanged, G. and Jensen, P. (1991) Behaviour of semi-naturally kept sows and piglets (except suckling) during 10 days post-partum *Applied Animal Behaviour Science*, 31(3–4):211–227. https://doi.org/10.1016/0168-1591(91)90006-J.

Statham, P., Green, L. and Mendl, M. (2011) A longitudinal study of the effects of providing straw at different stages of life on tail-biting and other behaviour in commercially housed pigs *Applied Animal Behaviour Science*, 134(3–4):100–108. https://doi.org/10.1016/j.applanim.2011.08.009.

Statham, P.T.E., Green, L.E. and Mendl, M. (2006) Development of a propensity to chew test as a predictor of tail-biting in pigs in M. Mendl (ed.) *Proceedings of the 40th International Congress of the ISAE, Bristol, UK*, 184–184.

Stolba, A. and Wood-Gush, D.G.M. (1989) The behaviour of pigs in a semi-natural environment *Animal Production*, 48(2):419–425. https://doi.org/10.1017/S0003356100040411.

Studnitz, M., Jensen, M.B. and Pedersen, L.J. (2007) Why do pigs root and in what will they root? A review on the exploratory behaviour of pigs in relation to environmental enrichment *Applied Animal Behaviour Science*, 107(3–4):183–197. https://doi.org/10.1016/j.applanim.2006.11.013.

Taylor, N.R., Main, D.C.J., Mendl, M. and Edwards, S.A. (2010) Tail-biting: A new perspective *The Veterinary Journal*, 186(2):137–147. https://doi.org/10.1016/j.tvjl.2009.08.028.

Telkänranta, H., Swan, K., Hirvonen, H. and Valros, A. (2014) Chewable materials before weaning reduce tail biting in growing pigs *Applied Animal Behaviour Science*, 157:14–22. https://doi.org/10.1016/j.applanim.2014.01.004.

Terlouw, E.M.C., Wiersma, A., Lawrence, A.B. and Macleod, H.A. (1993) Ingestion of food facilitates the performance of stereotypies in sows *Animal Behaviour*, 46(5):939–950. https://doi.org/10.1006/anbe.1993.1275.

Terlouw, E.M.C., Lawrence, A.B. and Illius, A.W. (1991) Influences of feeding level and physical restriction on development of stereotypies in sows *Animal Behaviour*, 42(6):981–991. https://doi.org/10.1016/S0003-3472(05)80151-4.

Thompson, P.B. (2008) The opposite of human enhancement: nanotechnology and the blind chicken problem *NanoEthics 2008*, 2(3):305–316. http://dx.doi.org/10.1007/978-94-007-1787-9_15.

Tinbergen, N. (1963) On aims and methods of ethology *Zeitschrift für Tierpsychologie*, 20(4):410–433. https://doi.org/10.1111/j.1439-0310.1963.tb01161.x.

Turner, S.P. (2011) Breeding against harmful social behaviours in pigs and chickens: state of the art and the way forward *Applied Animal Behaviour Science*, 134(1–2):1–9. https://doi.org/10.1016/j.applanim.2011.06.001 (Accessed: 31 May 2022).

Turner, S.P., Camerlink, I., Baxter, E.M., D'Eath, R.B., Desire, S. and Roehe, R. (2018) Breeding for pig welfare: opportunities and challenges in M. Špinka (ed.) *Advances in Pig Welfare*. Woodhead Publishing, 399–414. https://doi.org/10.1016/B978-0-08-101012-9.00012-5.

Ursinus, W.W., Wijnen, H.J., Bartels, A.C., Dijvesteijn, N., van Reenen, C.G. and Bolhuis, J.E. (2014) Damaging biting behaviors in intensively kept rearing gilts: The effect of jute sacks and relations with production characteristics *Journal of Animal Science*, 92(11):5193–5202. https://doi.org/10.2527/jas.2014-7918.

Valros, A., Munsterhjelm, C., Hänninen, L., Kauppinen, T. and Heinonen, M. (2016) Managing undocked pigs – on-farm prevention of tail biting and attitudes towards tail biting and docking *Porcine Health Management*, 2(1):1–11. http://dx.doi.org/10.1186/s40813-016-0020-7.

Valros, A. (2018) Tail biting in M. Špinka (ed.) *Advances in Pig Welfare*. Woodhead Publishing, 137–166. https://doi.org/10.1016/B978-0-08-101012-9.00004-6.

Walker, P.K. and Bilkei, G. (2006) Tail-biting in outdoor pig production *Veterinary Journal*, 171(2):367–369. https://doi.org/https://doi.org/10.1016/j.tvjl.2004.10.011.

van de Weerd, H.A., Docking, C.M., Day, J.E.L., Avery, P.J. and Edwards, S.A. (2003) A systematic approach towards developing environmental enrichment for pigs *Applied Animal Behaviour Science*, 84(2):101–118. https://doi.org/10.1016/S0168-1591(03)00150-3.

van de Weerd, H.A., Docking, C.M., Day, J.E.L. and Edwards, S.A. (2005) The development of harmful social behaviour in pigs with intact tails and different enrichment backgrounds in two housing systems *Animal Science*, 80(3):289–298. https://doi.org/10.1079/ASC40450289.

van de Weerd, H.A. and Day, J.E.L. (2009) A review of environmental enrichment for pigs housed in intensive housing systems *Applied Animal Behaviour Science*, 116(1):1–20. https://doi.org/10.1016/j.applanim.2008.08.001.

Wilson, K., Zanella, R., Ventura, C., Johansen, H.L., Framstad, T., Janczak, A., Zanella, A.J. and Neibergs, H.L. (2012) Identification of chromosomal locations associated with tail biting and being a victim of tail-biting behaviour in the domestic pig (Sus scrofa domesticus) *Journal of Applied Genetics*, 53:449–456. https://doi.org/10.1007/s13353-012-0112-2.

Wright, J.S. and Panksepp, J. (2014) An evolutionary framework to understand foraging, wanting, and desire: the neuropsychology of the SEEKING system *An Interdisciplinary*

Journal for Psychoanalysis and the Neurosciences, 14(1):5–39. https://doi.org/10.1080/15294145.2012.10773683.

Würbel, H. (2006) The motivational basis of caged rodents' stereotypies in G. Mason and J. Rushen (eds.) *Stereotypic animal behaviour: fundamentals and applications to welfare: second edition*. CABI Publishing, 86–120. https://doi.org/10.1079/9780851990040.0086.

Würbel, H., Stauffacher, M. and von Holst, D. (1996) Stereotypies in Laboratory Mice – Quantitative and Qualitative Description of the Ontogeny of "Wire-gnawing" and "Jumping" in Zur:ICR and Zur:ICR nu *Ethology*, 102(3):371–385. http://dx.doi.org/10.1111/j.1439-0310.1996.tb01133.x.

Yeates, J. (2018) Naturalness and animal welfare *Animals*, 8(4):53. https://doi.org/10.3390/ani8040053.

Yngvesson, J. and Keeling, L.J. (2001) Body size and fluctuating asymmetry in relation to cannibalistic behaviour in laying hens *Animal Behaviour*, 61(3):609–615. https://doi.org/10.1006/anbe.2000.1616.

Zupan, M., Janczak, A.M., Framstad, T. and Zanella, A.J. (2012) The effect of biting tails and having tails bitten in pigs *Physiology & Behavior*, 106(5):638–644. https://doi.org/10.1016/j.physbeh.2012.04.025.

PART 1

Risk factors associated with individual pigs and the social environment

CHAPTER 4

Individual pig-related factors

Laurianne Canario and Catherine Larzul*

INRAE, 24 chemin de Borde Rouge, 31326 Cedex, Castanet-Tolosan, France;
*Correspondence: laurianne.canario@inrae.fr

Abstract

The suffering of tail damage and the expression of behaviours associated with tail biting are influenced by factors inherent to the pig itself. When the influence of environmental, nutritional and pig-related factors is high, manipulative behaviour on group mates may evolve into an outbreak of detrimental behaviour. Individual pig-related factors, including sex type, growth, susceptibility to stress, microbiota composition and health, are often studied – indeed, increasingly so. Some are expected to be more significant than others, and studies accounting for several of these factors at a time should be undertaken. In this chapter, we will describe the internal factors that explain why pigs move from normal behaviour to this abnormal behaviour towards conspecific group members. We emphasise genetic factors in tail biting associated behaviours. Both direct and indirect effects determine an animals potential to tail bite or to be tail bitten, and they are expressed by each pig. Indirect effects refer to a pig's influence on the tail biting associated behaviours of other pigs it is housed with. The performance of tail biting associated behaviours also relates to previous experience within the group, and to previous developmental stages, with strong effects suspected to be present in the early developmental phase, notably exposure to stress. Epigenetics therefore undoubtedly has a role to play here. Progress needs to be made on the characterisation of animals, including measuring their behaviour over the long term so that sporadic events of tail biting are captured. Relevant analytical models for this are available. The continuous recording of animal and environmental data (incl. social context), and the development of integrated analyses of large populations accounting for internal factors are essential if we are to find solutions that reduce the prevalence of tail biting in pigs.

Keywords

animal-based factors – monitoring – stress – genetics

 | DOI:10.3920/9789004694989_006

1 Introduction

As pigs are intelligent animals, many of their behavioural responses are based on decision-making and motivational processes. With well-developed senses and sophisticated cognitive abilities, pigs have a very good perception of their environment. They can make contextual associations and remember them for a long time. As very social and exploratory animals, pigs like to interact with their fellows. However, at some point, the normal behaviour of sniffing and softly chewing others tails turns into biting which is an abnormal behaviour (Schrøder-Petersen *et al.*, 2003). Tail biting can be accidental, or aggressive, with multiple bites amounting to injury if the targeted fellow pig does not react. We therefore assume that in most cases, caudophagy, or as we shall say, tail biting, is performed in the context of an accurate assessment by the pig of the situation it is in. Tail biting is dictated by a pigs' inner susceptibility to bite others tails and to be tail bitten. Predisposing factors such as stress-sensitivity and exploration-propensity are modulated by its genetic make-up. Many pig behaviours are reproducible and some are inherited (Rydhmer and Canario, 2022). They differ from many other traits in that they are modified by experience and learning. These two compounds affect a pig in its decision to interact physically in a normal manner or not with other pigs. The pig is able to modify its behaviour, i.e. adapt, with acquired knowledge (Gieling *et al.*, 2011). Thus, the response of the pig depends on its genes and internal state at a given time, and the mechanisms are complex to decipher.

Although most reported risk factors for tail biting associated behaviours relate to enrichment, the type of housing, or flooring (Henry *et al.*, 2021), the problem cannot be solved by environmental modifications alone. Indeed tail biting has been observed in all production systems, including organic (Walker and Belkei 2001; See chapter 7). Consequently, a growing body of literature is examining the impact of pig-related factors on tail biting. This chapter describes these risk factors focusing on pig features.

We begin by describing the various factors, in an individual pig, that are expected to play a role in the genesis of tail biting: the contributions of non-genetic factors (developmental status, experience), genetic factors (transmitted by the DNA sequence) and epigenetic factors (DNA regulation). The three different types of tail biting activity, with rather different motivations and potentially different risk factors, complicates the analyses (Taylor *et al.*, 2010). We shall need to discuss the issues using three profiles: being a victim, being a biter and remaining neutral (Brunberg *et al.* 2011), and will describe individual differences associated with these different profiles. It is important to characterise each individual animal in detail, as some pigs change from one profile to another over time or have the dual profile of victim and biter (Brunberg *et al.*, 2011; Zonderland *et al.*, 2011a; Ursinus *et al.*, 2014a). In the second part of the chapter, we will explain what further work is required to improve our knowledge of the ways in which different forms of tail biting arise.

2 Individual features of the pig influencing tail biting behaviour

A large majority of pigs (90%) never come to bite other pigs (Brunberg *et al.*, 2011). So, what characteristics of the animal are decisive for the expression (or not) of biting behaviour? In addressing this question, we need to distinguish three profiles: biting, bitten and neutral. A few studies characterise the biter into one of the two biting profiles: non-aggressive/occasional and aggressive/obsessive (Taylor *et al.*, 2010). Equally, different kinds of victim exist. The status of the victim is characterised using the analysis of the severity of tail lesions. Some pigs react by retaliating or avoiding further bites, others are persecuted, and yet others are placid and allow themselves to be bitten on the tail and hindquarters. By contrast to biters and victims, neutral pigs are not involved in tail biting even though tail biting is going on in the group of pigs around them (Brunberg *et al.*, 2011). In the following subsections, we focus on the physiology underlying tail biting associated behaviour.

D'Eath *et al.* (2014) underline that although outbreaks often start with noticeable changes in the environment, sometimes they do not. Outbreaks of the latter kind could depend on a change in the individual pig's response threshold or fluctuations in its response to stress. This highlights the importance of approaching the problem of tail biting from the point of view of the individual pig. Two well-known general pig characteristics that have been shown to be implicated in tail biting, though not always consistently across studies, are age and sex. Morphological characteristics are also involved in its causation. Increasingly, it is being recognised that animal growth plays a key role. Tail length, rarely measured, is a facilitating factor by default. The individual pig's personality is the focus of new work, as is its ability and motivation to explore. Until recently, tail biting was observed and analysed chiefly in growing and finishing pigs, but it appears that in some populations pre-weaned pigs also practice it. Little is known about the behaviour in piglets. Finally, tail biting behaviour can be induced by a combination of individual factors and disease. The animal health aspects that may be involved (reviewed in Boyle *et al.*, 2022) will not be discussed in detail here (see Chapter 5). Neutral pigs are poorly characterised beyond their behavioural and genetic profiles. It seems safe to assume that their characteristics may be diverse.

2.1 *Physical characteristics*

2.1.1 Sex

Sex is identified as a risk factor but it should be considered in combination with single-sex or multi-sex group composition. Several on-farm studies have shown that groups of females suffer more tail damage than groups of entire males (Zonderland *et al.*, 2010; Canario and Flatres-Grall, 2018). Steinmetz and Pedersen (2009) found that groups of barrows have more tail damage than groups of females. Interestingly, Sinisalo *et al.* (2012) who did not identify an effect of sex on the risk of being bitten in a comparison of groups of barrows plus entire pigs or females, argued on the weakness of the design for analysing the within-group sex effect and suspected an impact of the pig weight. Lesions are likely to be more severe in mixed groups (Keeling *et al.*, 2012), although Steinmetz and Pedersen (2009) did not confirm this when comparing single-sex and multi-sex groups of females and barrows. Studies conducted on farms and in slaughterhouses frequently report that males are bitten more than females, with the findings relating to both barrows (e.g., Schrøder-Petersen *et al.* 2003, Valros *et al.*, 2004; Kritas and Morrison, 2004) and entire males (e.g., Harley *et al.*, 2012; Zonderland *et al.*, 2010, 2011) according to the frequency and severity of lesions. Persistence of tail damage was higher in males in mixed groups (Zonderland *et al.*, 2010). Furthermore, the risk in males increases with the proportion of females in the pen (Kritas and Morrison, 2004). As mixed groups are the norm in pig husbandry, this could explain why more males show tail damage at slaughter.

Schrøder-Petersen *et al.* (2003) showed that females were more frequent biters than males in mixed groups, and that they tended to bite more severely than males (Brunberg *et al.*, 2011). By jointly analysing lesions and behaviour (bites and biting) in the same animals, Li *et al.* (2017) confirmed that females were more frequently biters and males more often victims in mixed groups. However, Sinisalo *et al.* (2012) found no difference between the sexes. Comparison of weaned piglets in single-sex groups revealed that females bit each other more than males do (Zonderland *et al.*, 2010) and later in finishers (Haigh and O'Driscoll, 2019). No sex effect was found for tail biting in early life by Hakanson and Bolhuis (2021). The effect of sex is not yet explained, but it is probably related to the way the mixing of the animals is managed. An interaction involving group composition and other conspecific characteristics is suspected. Taking into account more environmental parameters in future studies of this factor shall facilitate its understanding. From observations of gilts attacking large castrated males when entering the growing stage, Taylor *et al.* (2010) concluded that in later growth stages tail biting is related to sexual maturity. The physical characteristics of pigs are therefore part of the equation that explain tail biting behaviour.

2.1.2 Weight and (lean) growth

Recent studies have pointed out that victims have higher growth rates (Hakansson and Houe 2020; Valros *et al.*, 2021). The fast-growing pigs that become victims may eventually end up with a reduced growth rate when severely bitten (Boyle *et al.*, 2022), bridging the gap with the studies from Camerlink *et al.* (2012) and Sinisalo *et al.* (2012) who reported lower growth rates in bitten pigs. Hakansson and Houe (2020) recently showed that the risk of lesions during lactation, though not severe, is higher in piglets that are heavy at weaning. The risk of becoming a victim increases with higher growth rate in post weaning piglets (Hakansson *et al.* 2020). Moreover, exploratory and playful piglets may be more likely to bite tails of their littermates. Another important point to that could help understand the discrepancies between studies might be the severity of the injury, as a discriminating factor between aggressive and non-aggressive tail biting. More information on whether tail biting is performed in relation to feeding is also needed. Canario and Flatres-Grall (2018) observed no growth difference in bitten and non-bitten sows. Interestingly, however, the bitten sows were leaner than others, in line with previous findings by Breuer *et al.* (2005).

A few decades ago, Sambraus (1985) observed that pigs engaging in tail biting behaviour are the lightest in the group, although this was not confirmed by Breuer *et al.*, (2005). However Van de Weerd *et al.* (2005) reported that although there was no difference in weight between pigs that occasionally bite and non-biting controls, persistently biting pigs were indeed lighter, regardless of their age. Zupan *et al.*, (2012) observed the same pattern. These obsessive biters with lower than average weight have been described as hyperactive pigs that switch from one tail to another during a biting outbreak. They suffer from a growth defect that occurs close to the expression of the behaviour rather than early after birth (Van de Weerd *et al.*, 2005). However, Munsterhjelm *et al.* (2016) observed that in mixed groups of gilts and barrows, pigs that performed tail-in-mouth behaviour were lighter at birth than their littermates, even though no difference was visible at 4 weeks of age. Undernutrition during lactation has been suggested as a predisposing factor for biting behaviour in post-weaning piglets, as reviewed by Prunier *et al.* (2019). Zonderland *et al.* (2011b) also found that post-weaning, biting pigs were lighter than their victims and controls (by an average of 500g). Hakansson and Bolhuis (2021), who studied a population at risk in Denmark (with 25% observed tail biting), found that piglets practising tail biting grow slower and tend to be lighter at weaning than other (neutral and victim) pigs. Indeed tail biting occasionally observed during lactation can be related to limited access to sow's udder.

There are several possible reasons hypothesised for the lower weight of biters. They may expend more energy in their increased activity (e.g. Van de Weerd *et al.*, 2005), or they may have reduced growth as a result of internal causes such as health condition (Valros and Heinonen, 2015). Tail biting can also be deployed as a tactic to keep larger pigs away from the feed trough when access is difficult (D'Eath *et al.*, 2014). In contrast to all this findings, Ursinus *et al.* (2014b) reported that biting gilts had faster growth. Valros *et al.* (2021) further observed that pigs with a high growth rate exerted more directed manipulation towards other pigs. Fast growing pigs that are more likely to engage in tail biting, are certainly under greater stress. The disparities in relationship may be explained by different determining factors between populations, and different types of tail biting.

2.1.3 Tail posture and length

Interestingly, tail posture appears to be correlated with tail biting associated behaviour (Schrøder-Petersen and Simonsen, 2001), and it can therefore indicate tail biting risk in a group. The number of pigs with hanging tails increases in pens around the start of an outbreak (Wedin *et al.*, 2018). Zonderland *et al.* (2009) found that piglets with tails between their legs had more tail damage 2–3 days later, even though at the time of the first observation they had no tail damage. Using 3D cameras, D'Eath *et al.* (2018) found pigs with hanging tails up to two weeks before, and in some cases even two weeks after, an outbreak. Hanging tails are positively but insufficiently correlated with the presence of lesions (Larsen *et al.*, 2018; D'Eath *et al.*, 2021; Wallgren *et al.*, 2019) to use it to detect victims of tail biting. We need to consider the role of tail posture change in various activities and emotions expressed by pigs to confirm it as an early predictor of tail biting (Wallgren *et al.*, 2019). Even if we put the predictor dimension aside, a number of questions relating to variations in tail posture and wagging (Camerlink and Ursinus, 2020) linked with various internal factors (health, genetics) remain. Wilder *et al.* (2021) analysed group interactions after mixing and evidenced an effect of group structure on the tail posture of each pig, and interpreted their finding in connection with stress contagion in a group (Reimert *et al.*, 2013). Health problems could result in a reduction in high or curly tails, which are less easy to catch than hanging tails. In the context of tail docking, in general, the shorter the tail, the lower the risk to be bitten (Harley *et al.*, 2012), as the biter's grip on a short tail is less firm. Unfortunately, the influence of natural tail length has been rarely studied (Kritas and Morrison, 2004).

2.2 *Coping ability*

When a pig does not satisfy its foraging and exploratory needs, frustration grows, and this can lead to a redirection of foraging and exploration towards group mates (Fraser *et al.*, 1987; Taylor *et al.*, 2010). Exploratory behaviour, which is very pronounced in this species can be used as an indicator of the pig's adaptation to its environment. Pigs not affected by environment perturbation maintain this behaviour as part of their normal activity budget. In an impoverished environment, pigs direct their manipulative behaviours towards their conspecifics and this may escalate into biting. Ursinus *et al.* (2014a) did not find a change in the frequency of exploratory behaviour prior to tail biting events. Conversely, chewing enrichment material correlated to tail damage. The response can depend on exposure to enrichment with materials earlier in life; a loss of enrichment post weaning may increase redirected behaviour including tail biting (Munsterhjelm *et al.*, 2009). However, Day *et al.*, (2002) found that when the pigs receive enrichment during the growth phase, previous enrichment has little effect on conspecific tail chewing.

Individuals vary considerably in their social interactions with group mates. This variation is linked with factors including the pigs' level of satisfaction with its environment, experience, and personality. Personality is synonymous with stability of reactions over different repetitions of a test, or of the same situation, with differences in personality reflected in differences in behaviour. Biters and victims alike may fail to adapt to an environment which they perceive as unsatisfactory, poor or insecure. Coping style is a personality trait that describes how pigs react when challenged (Koolhaas *et al.* 2007). The effect of coping style on tail biting activity is mediated by the degree of enrichment of the environment (Bolhuis *et al.*, 2005) at different stages of life. Pigs with a more reactive coping style show more frustration in a poor environment (Jansen *et al.*, 2009), so it could be assumed that they also perform more tail biting when there is a lack, or a loss, of enrichment material in their environment. However, Ursinus *et al.* (2014a) did not find a link between coping style and tail biting.

Non-curious pigs and also those that fail to maintain social contact with other pigs could be considered to be performing unnatural behaviour. The degree of sociality that pigs express is thus important to consider. In general, sociability is variable among pigs, with some seeking more contact and being more promiscuous than others, and social bonds intensified by licking of the body of other pigs. It is thus possible that the non-aggressive form of tail biting can be facilitated in groups with greater social contact and chewing activity. Hakansson and Bolhuis (2021) compared the activity profiles of neutral pigs with those of victims and biters in the same group. Neutral piglets rested more than the other pigs and tended to suckle more, which means they were more likely to satisfy their needs. They also sought social contact less than biters. Brunberg *et al.* (2013a, b) have pointed out that when they are exposed to risk, neutral pigs express fewer behaviours directed towards other pigs (e.g. nose to belly and tail to mouth) than pigs in pens without tail biting. It can be speculated that there is an environment-attention deficit in neutral pigs that makes them less social and less affected by what is happening in their surroundings. Fear is another personality trait related to tail biting but so far, conclusions are inconsistent between studies. In a novel environment, Ursinus *et al.* (2014c) found that tail biters avoided staying close to a novel object, whereas Zupan *et al.* (2012) that they were faster to touch it than non-biters.

Coping ability is modulated by learning mechanisms and experience. Tail biting may be related to the propensity of the animal to copy the behaviour of other (familiar, preferred) mates (Gieling *et al.*, 2011), and then depends less on inner motivation and needs. Pigs develop preferential associations and share activities with certain other individuals, including feeding activity for instance (Docking *et al.*, 2018). One could consider that chewing other pigs could be promoted by synchronisation of activity. Different individuals in a group of young pigs may enter the growing-finishing facility with very different experiences. Tail-in-mouth behaviour, known to be a precursor to tail biting, develops at a young age (Schrøder-Petersen *et al.*, 2003; Taylor *et al.*, 2010) and in some populations, tail biting occurs in piglets before weaning (Hakanson and Bolhuis, 2021). Therefore, tail biting in the growing stage may be connected with unmet nutritional needs during lactation, making the pre-weaning phase important in its aetiology. Mild forms of tail damage were in a few cases prevalent before weaning (Ursinus *et al.*, 2014a; Hakansson and Bolhuis, 2021) and such early life experiences may have an impact on later stress response. Canario and Flatres-Grall (2018) found substantial litter effects on levels of tail biting victimisation in gilts and Ursinus *et al.* (2014b) observed that tail biters were more likely to come from litters with relatively high levels of tail biting behaviour.

Beside the effects of personality and experience, it should be kept in mind that tail biting can be a contextual response. The threshold at which the pig reacts can drop when it feels hunger, illness or fatigue. In other words, a diminished pig can "simply" lose control. Ursinus *et al.* (2014a) found that tail biting was not a repeated behaviour, except in persistent biters. Hakansson and Bolhuis (2021) distinguished two categories of biting piglets, based on the frequency of biting, and suggest that different causalities are in play in the non-obsessive and obsessive type of biters. Low-biting piglets were more engaged in normal social interactions (e.g. play) with their litter-mates, while high-biting piglets performed more oral-anal activities directed both towards their litter-mates and their mothers (Ladewig *et al.*, 1984; Ursinus *et al.*, 2014b).

2.3 *Susceptibility to stress*

Tail biting outbreaks can lead to psychological dysfunction, even if only in some pigs (Zupan *et al.*, 2012), which experience a level of stress that they are unable to handle successfully. Jensen *et al.* (2004) highlighted that animals more vulnerable to stress are at greater risk of developing detrimental behaviours. The acute response of the hypothalamic pituitary adrenocortical (HPA) axis can be studied by monitoring ACTH release, and cortisol levels are often used to measure chronic stress when animal welfare is being assessed (Wiepkema and Koolhaas, 1993). Valros *et al.* (2013) showed that chronically bitten pigs had a lower cortisol reaction to acute stress than non-bitten pigs, indicating hypocortisolism. This may have been induced by the chronic stress caused of being bitten, or by any persistent infections caused by chronic tail biting. Individual differences in ability to adjust to environmental challenges may lead to suboptimal development of stress axis activity in stressful situations. Several physiological systems are implicated in susceptibility to stress, but in pigs the principal system appears to be the HPA axis. The main compounds involved in the HPA acute stress response are cortisol and adrenocorticotropic hormone (Martinez-Miro *et al.*, 2016). The dopaminergic and the serotonergic systems might also be implicated in tail biting in pigs. The former has an important role in the reward system and influences exploration behaviour and emotional reactions such as fear and anxiety, whereas the latter is involved in many physiological functions, including body temperature, appetite, sleep, pain, motricity and cognition (emotion, memory, attention, decision-making). Both systems have an important role in the animal's ability to cope with stressful environmental conditions and frustration. Peripheral blood serotonin is lower in biters than it was in victims (Ursinus *et al.*, 2014c), but it is not known whether these differences in the serotonin metabolism are temporary conditions or stable differences between pigs that are prone to tail biting and pigs with a lower propensity to do so. In any case, differences in blood are difficult to interpret because they do not reflect the metabolism of the whole system. Moreover, associations between tail biting and disease or illness could result from shared risk factors, even in the absence of any causal relationship (Boyle *et al.*, 2022). Interestingly, disease also causes pigs to be less responsive to general oral manipulation by their pen mates, putting them at increased risk of injury (Munsterhjelm *et al.*, 2017).

2.4 *Microbiota*

Brunberg *et al.* (2016) wrote about new avenues of research into the implications of gut bacteria, their metabolites, and the polyamine system on brain and behaviour. The same authors suggested how another mechanism, namely increased saliva production, could play a role in tail biting. It is a possibility that tail chewing increases saliva production in the chewer, which increases stomach pH, and thus changes the environment for gut bacteria. In recent years, research into microbiota, especially in the gut, has attracted growing interest, as the gut microbiota can interact with host physiology and have an impact on animal health and welfare (Kraimi *et al.*, 2019). Relationships between gut microbiota, HPA axis reactivity and the animal's ability to cope with challenges may all play an important role in the development of detrimental behaviour in pigs (Brunberg *et al.* 2016). Reviewing studies on microbiota-gut-brain axis, Kobek-Kjeldager *et al.* (2022) asked whether their results provide insights into the aetiology of tail biting because microbiota can impact emotion, memory, as well as social and feeding behaviours. Recently, both Rabhi *et al.* (2020) and Verbeek *et al.* (2021) found differences in the gut microbiota of biting, bitten and neutral pigs. Rabhi *et al.* (2020) found no differences in gut microbiota composition or diversity between biters and bitten pigs. However, they observed a consistently higher level of the relative abundance of *Lactobacillus* in neutral pigs as compared with the two other groups. Verbeek *et al.* (2021) observed higher relative abundance of *Prevotella_7* in neutral pigs and a tendency for lower relative abundances of *Butyrivibrio* and *Familiy_XIII_AD3011*, again compared with biters and bitten pigs.

As well as considering the composition of the microbial community, Kobek-Kjeldager *et al.* (2022) underlined the beneficial effects of the microbial fermentation products, short chain fatty acids (SCFAs) and especially butyrate, on health, mental state and stress sensitivity, although the role of SFCAs in tail biting is not yet specifically demonstrated (Verbeek *et al.* 2021). Gut microbiota profiles are controlled by environmental factors, but the genetics of the host is known to modulate them. In pigs, in a selection experiment involving relative abundances of *Prevotella/Mitsuokella* and *Ruminococcus/Treponema* genera (Larzul *et al.*, 2021), a higher frequency of tail biting events, combined with more aggressiveness, was observed during the post-weaning period in the high *Ruminococcus* genetic line (C Larzul, personal communication). Indeed Mudd *et al.* (2017) pointed out that microbiota may shape brain development with a particular focus on high gut microbiota *Ruminococcus* abundance related to low cortisol levels. As yet we have limited evidence of the impact of gut microbiota on the development of tail biting, but results from the early studies are promising. The interactions of gut microbiota, detrimental behaviours, the HPA-axis and the serotonergic system remain to be explored in experimental studies of biting, bitten and neutral pigs.

3 Genetic factors

The sensitivity of animals depends on proven genetic bases in addition to past experience, and it is increasingly recognised that genetics is an important determining factor for tail biting. Research in this area is currently limited in quantity. The possibility of reducing tail biting by applying the knowledge of which breeds, or genetic lines, or pigs with specific genes are more inclined to bite pen mates need to be explored in detail.

3.1 *Breed differences*

In the analysis of any behaviour, the fact that breeds differ in that behaviour demonstrates that there is a genetic origin to its occurrence. We looked for information on conventional and local breeds, as well as on crossbreeds. There is little documentation of breed or lineage differences in tail biting. A few studies have experimentally compared breeds in the same environment (e.g., Lund and Simonsen, 2000; Breuer *et al.*, 2003; Sinisalo *et al.*, 2012), to erase the effects of environmental variability. Unfortunately, the lack of an agreed standard for scoring tail bites makes the comparison of breeds from different studies difficult (Honeck *et al.*, 2019) and impairs the possibility to realise a meta-analysis on breed effects.

Lund and Simonsen (2000) reported no clear effect of breed. However in other studies, differences between breeds were found both in tendency to bite and in susceptibility to be bitten. Guy *et al.* (2002) compared three crossbred genotypes (indoor type: Landrace × Large white sows mated to Large White boars; Outdoor type, either 25% Duroc/25% Landrace/50% Large White or 25% Meishan/25% Landrace/50% Large White sows mated to Large White boars), and found no difference in tail biting. Gentz *et al.* (2020) used two crossbreeds (Large White × Landrace) and (Pietrain × (Large White × Landrace)), in their study of tail lesions and losses in different farrowing and rearing systems, but unfortunately although the numbers in each population in their study were large, they did not report on breed differences. This lack of reporting, here and elsewhere, represents a missed opportunity in teasing out the effect of pig breed on biting behaviour and tail damage. Early on in the history of tail biting studies, Landrace was believed to be a breed more prone to tail biting than other pure breeds (EFSA, 2007). However, using the motivation of pigs to chew a rope as an indirect measure of their tendency to tail bite, Breuer *et al.* (2003) later concluded that Duroc pigs were more likely to bite tails than Large White or Landrace pigs. The Duroc differ from other breeds in its level of sociality, with pigs being bolder than pigs from other breeds. A possible interaction with sex type should also not be discounted.

We mentioned above the importance of foraging and exploratory behaviour in pigs, and its possible causal link with tail biting. It is of interest, therefore that breed differences in foraging behaviour have been identified. In a comparison of young Large White, Landrace and Duroc pigs, the Durocs were more exploratory (Breuer *et al.*, 2003), which could tie in with the hypothesis that they have a higher propensity to perform tail biting. In Breuer *et al.* (2005), the prevalence of being bitten was higher in Landrace pigs than it was in Large White, whereas Sinisalo *et al.*, (2012) observed the opposite. Jensen *et al.* (2004) reported that Danish Landrace boars were bitten no more than Yorkshire boars, but that in both breeds tails were more often observed bitten than they were in Duroc. In a Swedish population of male pigs with a low level of tail biting, the pigs found to be bitten were: Yorkshire (3.5%), Landrace (1.8%) and Hampshire (0.1%) (Westin, 2003). As in the study from Breuer *et al.* (2003), the mixed breed composition of the groups might have had an influence on breed effect. Therefore, experimental studies may lead to different conclusions depending on whether mixed or pure groups are compared. Following a survey of 58 Irish pig producers, Haigh and O'Driscoll (2019) reported that the majority (76%) of pig farmers had not noticed any impact of breed on tail biting. The remaining 24% of pig farmers reported previous tail biting problems with the Hampshire breed. However, this Hampshire susceptibility to tail biting behaviour was not confirmed experimentally, as Hampshire breed appears to be more neutral as compared with other breeds (Jensen *et al.*, 2004). Overall, there is only weak and inconsistent evidence for a breed effect on the predisposition to bite or be bitten. In other words, breed comparison provides little evidence of genetic influence on the expression of tail biting behaviours.

3.2 *Molecular polymorphisms*

Only a few studies have been published on the molecular genetic determinism of tail biting (Wilson *et al.*, 2012). Several chromosomal regions of the pig's genome are associated with the perpetration and reception of tail biting behaviours. To date, only one significant association with being a biter has been found, located on chromosome 16. By contrast, several such associations with being bitten have been identified on chromosomes 1, 9, 18 and an unassigned region. Thus being a victim of tail biting and being a tail biter seem to be behaviours that are not genetically linked. Genome-wide linkage or association analyses evidenced several trait-associated loci (QTL) for HPA axis activity. Two main QTL regions include the distal end of chromosome 7, harbouring the SERPINA6 locus coding for corticosteroid binding globulin, and the distal end of chromosome 2, harbouring the NR3C1 locus coding for glucocorticoid receptor. These two genes are involved in the production of HPA axis hormones, and their variation might be beneficial to pig populations by inducing greater flexibility in their response (Kasper *et al.*, 2020). Notwithstanding the importance of HPA axis in stress susceptibility, no relation between these genes and damaging behaviours has yet been evidenced. As far as tail shape can influence tail biting, Bertolini *et al.* (2018) found a single genetic marker, on chromosome 12, associated with the curly tail in the Casertana breed. The near SOX9 gene, known to be involved in skeletogenesis, could be further studied in relation with tail shape and tail biting in other breeds.

Brain (hypothalamus and prefrontal cortex) gene expression studies have been carried out to understand the differences between biting, bitten and neutral pigs, and whether the biters were able to resist tail biting, as reported by Brunberg *et al.* (2016). The biters and bitten pigs were similar, and both had gene expression patterns differing from that of neutral pigs. The identified genes had a known influence on leanness, social behaviour, and exploration. The authors mentioned that a key goal would be to understand why specific individuals stay neutral and do not get involved in tail biting. Identifying neutral behaviour, and in particular resistance to damaging behaviour, could be easier, since more genes are differentially expressed between neutral pigs versus biters and victims than between biters and victims. Some of the genes that were found to be differentially expressed were associated with behaviour-related traits in other species. PDK4 (pyruvate dehydrogenase kinase, isozyme 4) is related to hypersociability in mice and has been identified as a possible biomarker of heat stress. A polymorphism in the epidermal growth factor (EGF) gene, more highly expressed in biters and bitten pigs than in neutral pigs, has been implicated in novelty seeking in humans. Two genes over-expressed in neutral pigs – the GHRL gene coding for ghrelin and the Cartilage Oligomeric Matrix Protein, or COMP, gene – are related to eating behaviour and feather pecking in hens, respectively. Additionally, the functions of the identified genes all correspond to production traits (growth, feed intake, leaner carcass) and behavioural traits. This is consistent with the idea that selective breeding for production traits may have also increased the risk of damaging behaviour as an undesirable side-effect.

As stated by Kasper *et al.* (2020), the combination of low heritability for tail biting related traits, the shortage of association studies investigating genetic markers, and in general the difficulty of recording damaging behaviours, suggest that more holistic studies will be needed if we are to better understand the architecture of this abnormal behaviour. One step would be to collect information on different levels of post-genomic regulation, by conducting transcriptomic, proteomic and metabolomics analyses. Another would be to consider translational research, and to explore candidate genes related to obsessive-compulsive disorders identified in other species.

3.3 *Individual differences*

3.3.1 Direct genetic effects

Direct genetic effects describe how an animal's genotype influences its behaviour. In quantitative genetic studies, which are individual-centred, the proportion of trait variation explained by variation in the individual's genotype takes into account e.g. the number of tail bites of the pig being investigated, which is compared with that of related and unrelated conspecifics through pedigree information. Genetic determinism is estimated at a population level and may differ from one breed to another. Breuer *et al.* (2005) found that tail biting activity was weakly heritable in the Landrace breed ($h^2 = 0.27$) while being not at all heritable in Large White ($h^2 = 0$). The heritability of the trait 'presence of tail damage' was estimated at 0.23 by Aikins-Wilson *et al.* (2021) using repeated measurements and various crosses, and that of 'presence of tail bites' was 0.06 in a Sino-European maternal line (Canario and Flatres-Grall, 2018). Using medication and removal information from a Large White population, Hermesch (2018) found that being tail bitten was heritable at 0.12. Historical differences in selective breeding may explain why lines differ in tail biting in accordance with increased leanness and/or growth rate. This happens through causal (genetic) relationships between traits. Apart from the study of Hermesch (2018), several studies highlight that tail biting and other abnormal behaviours are unfavourably correlated with leanness on the genetic scale (Breuer *et al.*, 2005; Silano *et al.*, 2012; Brunberg *et al.*, 2013; Canario and Flatres-Grall 2018). As a result, it is assumed that a common, or shared, pool of genes underlies both trait categories.

However, there is no clear evidence that tail biting is more frequent in modern highly-selected populations than it is in their ancient counterparts. In the particular case of 'obsessive' tail biting, Taylor *et al.* (2010) hypothesised that genetic determinism may be responsible for these repeated grasping and pulling of the tails of other pigs, and that it would relate to a genetic dysfunction causing increased attraction to blood. The extent to which lack of environmental adaptation triggers such a response is not documented. Where other factors intrinsic to pigs are concerned, exploratory behaviour varies between individuals and should be partly under genetic control (Breuer *et al.*, 2003). The stress response, as measured by the ACTH response, is also heritable. The direct heritability for tail length at birth is moderate ($h^2 = 0.42$), indicating the potential to select for short tails but due to a moderate positive genetic correlation (0.40), pig growth may be reduced as a consequence (Aikins-Wilson *et al.*, 2021).

An important gap in the literature on the study of the genetic determinism of tail biting, and the possibility of reducing it through selective breeding does exist. The low direct genetic variance evidenced raises questions on the variations in the expression and regulation of the individual's genes underlying such abnormal behaviours. Also, one should keep in mind that, not every individual that is genetically predisposed to perform tail biting or be bitten will actually bite or be bitten over the course of its lifetime. The environmental situation has a significant impact on outcomes, and past experience (stress, contentedness) also plays a role in the animal's evaluation of the situation.

3.3.2 Indirect genetic effects

The individuals with which a pig socially interacts are an important part of its environment. Whether a pig expresses biting behaviour, or becomes a victim, at any given time depends both on that animal itself and on its group mates. Each individual makes decisions, partly in response to the group around, and in turn it is able to influence the behaviour of other group members. Animals can influence the traits of those around them either coercively by taking resources that others would have used, or consensually by sharing resources (e.g. food, enrichment materials). When the ability to influence others socially is itself inherited, an individual's genotype affects the traits expressed in its interaction fellows, in other words its group mates' phenotypes (e.g., Moore *et al.*, 1997). The focal pig can therefore impact on the development, growth and behaviour of its conspecifics in stable or event-driven situations. Genetic influences of a pig on the traits of its conspecifics are called 'indirect genetic effects'.

If the trait of interest is biting behaviour, in contrast with direct genetic effect, indirect genetic effects (IGEs) are the effects of the focal pig's genotype on the biting activity of its group mates. In studies describing the presence of tail lesions, the social effect describes causing tail lesions to group mates, which in some case will reflect biting others. Models incorporating both direct and indirect sources of genetic variation have been shown to capture genetic variation more accurately (Ellen *et al.*, 2014; Canario *et al.*, 2017). For instance the comparison of classical heritability with the ratio of total heritable variance to the total phenotypic variance, that respectively do not and do account for social variation, allowed identifying large social genetic effects on bitten gilts (Canario and Flatres-Grall, 2018). These effects explained far more of the observed variation of the trait than direct effects.

Therefore, the interactions with group mates play a major role in the reception of tail biting, at both non-genetic and genetic levels. As a key point, Rydhmer and Canario (2022) explained in detail why the genetic covariance between direct and indirect effects is important in advances in our understanding of the functioning of populations. The correlation is negative (unfavourable) if pigs compete for limited resources. If the genetic correlation is positive, there is cooperation. Null direct-indirect genetic correlations for tail biting reception were found in the Sino-European line mentioned above (Canario and Flatres-Grall, 2018). Accordingly, gilts with high genetic predisposition for being tail bitten did not have larger IGEs. Thus, the probability of other gilts in a group becoming tail bitten would not be increased by selective breeding against tail bitten gilts.

Similar to IGEs for tail biting, IGEs for growth refer to the influence of the focal pig's genotype on the growth of its group mates. In a one-generation selection experiment, Camerlink *et al.* (2015) analysed the biting behaviours of groups selected for low or high IGEs for pig growth in two contrasting environments:

housed in conventional barren pens or in pens enriched with straw and wood shavings. Tail damage and biting, which were much lower in the enriched pens, were almost equally affected by housing conditions in both genetic groups. Groups of pigs with greater IGEs i.e. promoting the growth of their conspecifics, displayed the lowest levels of tail biting behaviour in enriched pens, and pigs with lower IGEs in barren pens had the highest level of tail biting. Overall, selection for greater IGEg reduced biting behaviours in an addition to the greater reductive effects of straw bedding, demonstrating the scope for accounting for IGE to provide opportunities to reduce harmful behaviour in pigs.

3.3.3 Interactions between genetics and the environment

Management practices related to feeding, grouping strategy, available space, climate, and enrichment are some of the many risk factors for tail biting that interact with individual pig related-factors. Differences between pigs in their propensity to bite others tails and/or to be bitten could be related to a greater sensitivity of some individuals to an environmental risk factor (see Figure 4.1). Therefore, as well as being influenced by specific internal (including genetic) factors, tail biting behaviours are influenced by interactions between these and environmental factors, termed genotype-environment interactions (GxE). GxEs refer to the variation in the relative performance of genotypes between environments, or in response to a change in the environment. Thus two genotypes exhibiting the same level of tail biting in one environment may have different levels of tail biting in a second environment. This could result in different genetic parameters being estimated either between populations, or within a population reared under different environmental conditions (Canario *et al.*, 2020). GxEs have been observed in connection with a wide range of behaviours, including indicators of stress management and cognitive ability in human. However, early studies on finishing pigs showed no interaction between genotype and environment for aggressive and non-aggressive tail biting (Hill *et al.*, 1998; Guy *et al.*, 2002). Describing the environment as a gradient, Bodin *et al.* (2015) examined the time spent by pigs handling straw and found the limit point at which further provision of straw does not affect tail biting for one genotype. In the case of tail biting, environmental gradients may be drawn from factors influencing the social context such as group size, and available space. To account for GxE in the analyses, data on a genetic line or breed must be collected in at least two environments or under different conditions.

At a higher level of complexity, social interactions frequently change from one environment to another, and therefore as environments change we can expect the way in which pigs influence each other to change as well. Very few studies have examined how IGEs change between environments. No effect of GxE was found in the selection experiment of Camerlink *et al.* (2015). The result in this case was an additive, rather than interactive, effect of genetics and environment, but it would

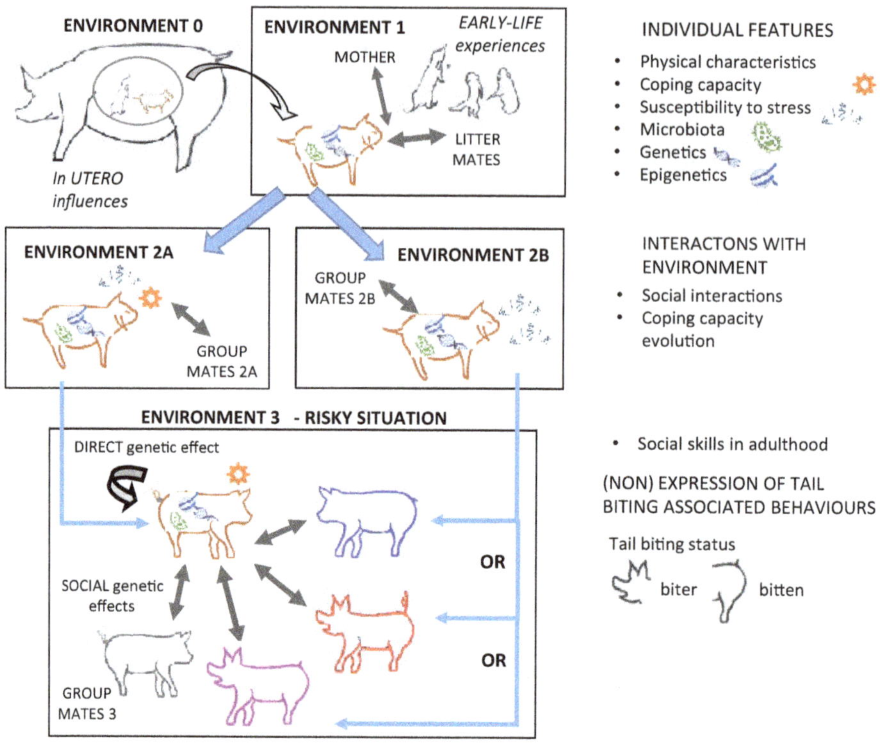

FIGURE 4.1 Schematic describing how experiences acquired in life across various environments shape a pig's personality and susceptibility to be involved in tail biting. Its behaviour depends on interactions between its features and environments. The focal pig going from environment 2A to environment 3 is neutral in its social interactions with group mates. If it is put in environment 2B that generate more stress instead, this pig has a greater probability to become a victim, a biter, or both.

need to be confirmed with a larger study. In the analysis of tail biting lesions of Canario and Flatres-Grall (2018), the correlation of rankings of frequently used boars in two herds according to both their direct and indirect genetic effects indicated a strong GxE effect for bite reception. Thus progeny of boars with stable and advantageous direct and indirect genetic effects in all environments should be selected. Through this method the proportion of gilts with a genetic predisposition to allow other members of the group to bite their tails would be reduced. Physiological characterisation of boars and their offspring, such as sensitivity to stress and leanness, should be studied to confirm if the same determinism is involved in different herds that differ in management. For instance, larger group sizes can favour tail biting, as evidenced by e.g. Holmgren and Lundeheim (2004). Thus mechanisms leading to a change in tail biting behaviour depending upon group size could act differently to those leading to changes in aggressive interactions, where there is often a decrease in interactions in large groups (Canario *et al.*, 2017). Analyses indicate that IGEs for tail biting lesions evolve through a contagion – zero dilution – mechanism instead (L Canario, pers. comm.). The presumed GxE of tail biting means that the genetic predisposition for biting or being bitten is expressed or not depending on the environment. Thus, the choice of the genotype or breed to be used in any environment must be made with caution, to ensure that it is suited for it (Bowman, 1972).

4 Novel areas of research

Controlling outbreaks of tail biting remains very difficult, because we often do not understand how each of the different risk factors contributes to the behaviour, or how they interact with each other. New approaches require the implementation of measures on the individual pig and recourse to Big Data analysis.

4.1 *Automated monitoring of individual pig's behaviour associated with tail biting*

Recording the behaviour of each individual within a group is complicated if the target behaviours are infrequent and of short duration, particularly when many pigs are likely to be interacting with each other. The capacity for continuous human observation is obviously limited, and the consistent and timely recording of behaviour and relevant indirect measures such as tail damage is time consuming, especially as injuries can be subtle. Indeed, the prevalence of severe tail bites is often extremely low, at 3%, in contrast to the frequency of mild tail bites (can be as high as 72%; e.g., van Staaveren *et al.*, 2017). Detailed recording of tail lesions will be eased by image analysis (Brünger *et al.*, 2019). Identifying and characterising the neutral individual pigs that steer clear of interactions that lead to tail biting would also be a great aid to understanding the mechanisms of tail biting. This is the challenge that new technological developments in artificial intelligence (AI) will undoubtedly allow us to address in the near future, as these will offer the opportunity for automatic monitoring of the activity of individual pigs. Advances in automated recording offering concurrent tracking of individuals and the identification of social pairwise interactions, which will be crucial in identifying tail biting (Liu *et al.*, 2020). Projects investigating the potential of automated detection of pig activity and tail biting make use of 2D video analysis to establish individuals' proximity with a multiple object tracking algorithm (Liu *et al.*, 2020) and 3D analysis to observe tail posture and movements (D'Eath *et al.*, 2018, 2021). Algorithms that are capable to detect medium-speed aggression (Chen *et al.*, 2017) could help detecting aggressive tail biting (Liu *et al.*, 2020).

For the time being, video analysis remains unlikely to be able to detect subtle chewing and tail biting behaviours, even though such monitoring may be decisive in understanding why a pig bites a conspecific. Identification of the biter is of considerable interest when we are selecting against this behaviour, so it is unfortunate that this kind of data is as yet only recorded with a degree of uncertainty. However a method based on the imputation of missing data in the analysis of tail biting traits is proposed in Canario *et al.* (2020). Moreover, algorithms are being developed for the monitoring of affective states and pain from facial expression (Neethirajan *et al.* 2021), as they are related to the pig perception of stress, and influence tail biting (Reimert *et al.*, 2013; Lawrence *et al.*, 2018).

Linking behavioural records to individually identified animals in an automated manner remains a challenge. Reliable, long-term methods of identification will be essential for selective breeding at the individual level as individualised physical characteristics (e.g. growth, feeding activity) can be explored to predict the predisposition of pigs to bite or be bitten (Wallenbeck and Keeling, 2012). With data at the level of the individual animal, we will also be able to account for both direct and indirect (genetic) effects. Longitudinal records covering the entirety of the animal's life are required if we are to take into account previous rearing conditions, and past social experiences, including early-life effects. The study of detrimental behaviour at a specific stage of life, with multiple records obtained before and after the event, will allows us to improve the classification of pigs into different categories of e.g., whether a one-time biter, or an obsessive biter (Canario *et al.*, 2020). Finally, multiple observations of an individual would allow us to study changes of state in a dynamic way – e.g. the transition from being unbitten to being bitten (Ursinus *et al.*, 2014a).

4.2 *Analysing derived pathways that lead to tail biting*

Much work on the deciphering of the mechanisms of tail biting remains to be done. The first step will be to finely characterise and record the various forms of tail biting behaviour in large populations on an individual basis, with a view to distinguishing compulsive from occasional behaviour. Behavioural data will be more meaningful if it is aggregated and related to the intrinsic and environmental characteristics of the individual at different times in its life. By understanding the physiological basis of tail biting, there will be the potential to identify biomarkers that can be used on a large scale. Samples from each individual can also provide meta-genomic information (offering a more complete characterisation of the individual and its microbiota) to understand interactions between a pig and its environment. Tail biting occurs due to the physiology (stress, motivation), and cognitive, coping and social abilities of a pig that evolve throughout its life. In particular, the early life environment and experiences of a pig modulates its entire later life. For instance Canario *et al.* (2017) showed that piglets sharing the same environment early in life develop common social skills, generating early social effects which can have a strong influence later on conspecifics. It is also in the early stages of development that epigenetic marks, that regulate gene expression, are fixed.

Epigenetic effects refer to changes caused by modification in gene expression, rather than changes to the genetic code itself. These effects can alter the physiology and metabolism, and ultimately the social ability of young and adult pigs. For instance Seebacher and Krause (2019) illustrated that epigenetic effects on neuroendocrine factors and metabolic functions can modify animal social behaviour in response to environmental stimuli. We also speculate that epigenetic factors mediate the reciprocal feedbacks between the individual pig (e.g., HPA axis response) and its group mates in response to exposure to environmental stress. Even though this is an area where there is as yet little knowledge, the impact of environmental influences on tail biting are more than likely mediated by epigenetics. Hyde *et al.* (2019) suggested epigenetics functions as an interface between the individual and its environment, through which the pig body interprets and responds to stressful experiences and adapts its coping capacity.

An equivalent hypothesis could be formulated with regard to gut microbiota. Psychosocial stressors with an epigenetic impact are encountered in a variety of circumstances and sometimes over long periods. These can range from early life adversity caused by a stressful social situation (e.g. hostile litter mates limiting access to the udder), or management practices such as weaning and mixing of litters leading to physical and/or mental suffering. During both negative and positive experiences, the individual pig learns from interactions with conspecifics (Held *et al.*, 2000; Gieling *et al.*, 2011), and this modulates its coping capacity and presumably its engagement in tail biting.

If epigenetic effects on the genes involved in tail biting or susceptibility to stress are transmitted from generation to generation, the positive treatment of piglets by their mother and litter mates early in life has potential to influence the DNA methylation of their subsequent offspring, so that tail biting associated alleles are inactivated. It will be interesting to decipher and quantify the modes and types of epigenetic transmission of the susceptibility of pigs to be involved in tail biting-related activities. The base hypothesis for genetic determinism is that progeny are the average of their parents' genetic values. But epigenetic effects mean that differences can be transmitted to progeny in a non-genetic way. The transmissibility model (David and Ricard, 2019) allows us to estimate transmission coefficients that differ from 0.5 to take into account non-genetic transmission or inheritance. The refinement of the model by integrating information on epigenetic marks and microbiota composition provides an opportunity to better understand genetic and non-genetic similarities between individuals, and this paves the way to improve complex traits such as those associated with tail biting by selective breeding.

It is through these multiple, interacting internal factors that susceptibility to tail biting is established. Thus when a pig finds itself at experiencing stress, and according to the social situation (with other pigs reacting before it or not doing so), the information it has integrated at the level of its organism (genetic and non-genetic) combines to trigger the expression (or not) of a tail biting-related behaviour.

5 Conclusion

Several risk factors for tail biting that are associated with individual features of the pig have been identified over the past two decades. These include behaviour, physiology and stress regulation, microbiota and genetics. These factors interact to shape the susceptibility, or predisposition, of each animal to tail biting. The differences between biter, bitten and neutral pigs already underlines that the individual pig's mechanisms of adaptation to variations in its living environment are involved in the expression of abnormal behaviours. At present, due to limited tools for monitoring tail biting, conclusions have been established using few individuals from affected populations, which is likely a contributing factor to inconclusive or inconsistent results across studies. Few studies have explored genetics as a precursor of tail biting associated behaviours. Capturing the genetic variance of such traits when they are recorded with a degree of uncertainty is known to be difficult, and is all the more complex since strong cases of GxE are anticipated. Advances in automated recording that could provide repeated measurements of multiple behaviours in a range of environments, and analyses of tail biting traits, will offer opportunities to fill in these gaps. Preliminary results from a small number of studies relating to genetics, coping ability, stress response and host × microbiota interactions demonstrate a need for further research to deepen our understanding of the internal factors at play. The ultimate step is the possibility of selective breeding against tail biting, with the objective to improve the welfare of all group-housed pigs.

6 Short summary

- Some pigs are more likely to bite conspecifics' tails than others and some are more likely to receive tail bites than others. These differences are due to variation in cognition, behaviour, physiology and stress regulation, microbiota and genetics. All these pig-related risk factors were identified, but not so many were accounted for in a same study.
- So far, only a few genetic parameters have been estimated for the reception and performing of tail biting. Strong GxE is suspected, and there is evidence that indirect

genetic effects are involved in the expression of tail biting. Fine monitoring with advanced technologies and individual identification at a population level will be crucial to analyse the genetic determinism of tail biting associated behaviours.
– Two new research questions that particularly merit further studies concern the role of the microbiome in, and epigenetic origins of, tail biting associated behaviours.

References

Aikins-Wilson, S., Bohlouli, M. and König, S., 2021. Maternal and direct genetic parameters for tail length, tail lesions, and growth traits in pigs. Journal of Animal Science 99:1. https://doi.org/10.1093/jas/skaa398.

Bertolini, F., Schiavo, G., Tinarelli, S., Santoro, L., Utzeri, V.J., Dall'Olio, S., Costa, L.N., Gallo, M. and Fontanesi, L., 2018. Exploiting phenotype diversity in a local animal genetic resource: identification of a single nucleotide polymorphism associated with the tail shape phenotype in the autochthonous Casertana pig breed. Livestock Science 216:148–152. https://doi.org/10.1016/j.livsci.2018.08.007.

Bodin, L., Algers, B., Andersson, M., Olsson, A.C., and Botermans, J., 2015. The amount of straw for growing-finishing pigs considering the reduction of time spent in manipulative behavior. Journal of Veterinary Science 1:105. https://doi.org/10.15226/2381-2907/1/1/00105.

Bolhuis, J.E., Schouten, W.G., Schrama, J.W. and Wiegant, V.M., 2005. Behavioural development of pigs with different coping characteristics in barren and substrate-enriched housing conditions. Applied Animal Behaviour Science 93:213–28. https://doi.org/10.1016/j.applanim.

Bowman, J., 1972. Genotype × environment interactions. Les Annales de Génétique et Sélection Animimale 4:117–123. https://doi.org/10.1186/1297-9686-4-1-117.

Boyle, L.A., Edwards, S.A., Bolhuis, J.E., Pol, F., Šemrov, M.Z., Schütze, S., Nordgreen, J., Bozakova, N., Sossidou, E.N., and Valros, A., 2022. The evidence for a causal link between disease and damaging behavior in pigs. Frontiers in Veterinary Science 8:771682. https://doi.org/10.3389/fvets.2021.771682.

Breuer, K., Sutcliffe, M.E.M., Mercer, J.T., Rance, K.A., Beattie, V.E., Sneddon, I.A. and Edwards, S.A., 2003. The effect of breed on the development of adverse social behaviours in pigs. Applied Animal Behaviour Science 84:59–74. https://doi.org/10.1016/S0168-1591(03)00147-3.

Breuer, K., Sutcliffe, M.E.M., Mercer, J.T., Rance, K.A., O'Connell, N.E., Sneddon, I.A., and Edwards, S.A., 2005. Heritability of clinical tail-biting and its relation to performance traits. Livestock Production Science 93:87–94. https://doi.org/10.1016/j.livprodsci.2004.11.009.

Brunberg, E., Wallenbeck, A. and Keeling, L.J., 2011. Tail biting in fattening pigs: associations between frequency of tail biting and other abnormal behaviours. Applied Animal Behaviour Science 133:18–25. https://doi.org/10.1016/j.applanim.2011.04.019.

Brunberg, E., Jensen, P., Isaksson, A. and Keeling, L.J., 2013a. Behavioural and brain gene expression profiling in pigs during tail biting outbreaks – evidence of a tail biting resistant phenotype. PLoS One 8(6), e66513. https://doi.org/10.1371/journal.pone.0066513.

Brunberg, E., Jensen, P., Isaksson, A. and Keeling, L.J., 2013b. Brain gene expression differences are associated with abnormal tail biting behavior in pigs. Genes Brain and Behaviour 12(2):275–281. https://doi.org/10.1111/gbb.12002.

Brunberg, E.I., Bas Rodenburg, T., Rydhmer, L., Kjaer, J.B., Jensen, P. and Keeling, L.J., 2016. Omnivores going astray: A review and new synthesis of abnormal behavior in pigs and laying hens. Frontiers in Veterinary Science 3:1–15. https://doi.org/10.3389/fvets.2016.00057.

Brünger, J., Dippel, S., Koch, R. and Veit, C., 2019. "Tailception": using neural networks for assessing tail lesions on pictures of pig carcasses. Animal 13:1030–1036. https://doi.org/10.1017/S1751731118003038.

Camerlink, I., Bijma, P., Kemp, B. and Bolhuis, J.E., 2012. Relationship between growth rate and oral manipulation, social nosing, and aggression in finishing pigs. Applied Animal Behaviour Science 142:11–17. https://doi.org/10.1016/j.applanim.2012.09.004.

Camerlink, I., Ursinus, W.W., Bijma, P., Kemp, B. and Bolhuis, J.E., 2015. Indirect genetic effects for growth rate in domestic pigs alter aggressive and manipulative biting behaviour. Behaviour Genetics 45(1):117–126. https://doi.org/10.1007/s10519-014-9671-9.

Camerlink, I. and Ursinus, W.W., 2020. Tail postures and tail motion in pigs: a review. Applied Animal Behaviour Science 230. https://doi.org/10.1016/j.applanim.2020.105079.

Canario, L., Bijma, P., David, I., Camerlink, I., Martin, A., Rauw, W.M., Flatres-Grall, L., van der Zande, L., Turner, S.P., Larzul, C. and Rydhmer, L., 2020. Prospects for the analysis and reduction of damaging behaviour in group-housed livestock, with application to pig breeding. Frontiers in Genetics. 11:611073. https://doi.org/10.3389/fgene.2020.611073.

Canario, L. and Flatres-Grall, L., 2018. Genetics of tail-biting receipt in gilts from the Tai Zumu line. Proceedings of the World Congress on Genetics Applied to Livestock Production. 11, 572. http://www.wcgalp.org/system/files/proceedings/2018/genetics-tail-biting-receipt-gilts-tai-zumu-line.pdf.

Canario, L., Lundeheim, N. and Bijma, P., 2017. The early-life environment of a pig shapes the phenotypes of its social partners in adulthood. Heredity 118(6):534–541 https://doi.org/10.1038/hdy.2017.3.

Chen, C., Zhu, W., Ma, C., Guo, Y., Huang, W., Ruan, C., 2017. Image motion feature extraction for recognition of aggressive behaviors among group-housed pigs. Computers and Electronics in Agriculture. 142:380–387. https://doi.org/10.1016/j.compag.2017.09.013.

Day, J.E.L., Spoolder, H.A.M., Burfoot, A., Chamberlain, H.L. and Edwards, S.A., 2002. The separate and interactive effects of handling and environmental enrichment on the behaviour and welfare of growing pigs. Applied Animal Behaviour Science 75(3):177–192. https://doi.org/10.1016/S0168-1591(01)00199-X.

D'Eath, R.B., Arnott, G., Turner, S.P., Jensen, T., Lahrmann, H.P., Busch, M.E., Niemi, J.K., Lawrence, A.B. and Sandøe, P. 2014. Injurious tail biting in pigs: how can it be controlled in existing systems without tail docking? Animal 8(9):1479–1497. https://doi.org/10.1017/S1751731114001359.

D'Eath, R.B., Jack, M., Futro, A., Talbot, D., Zhu, Q., Barclay, D., and Baxter, E.M., 2018. Automatic early warning of tail biting in pigs: 3D cameras can detect lowered tail posture before an outbreak. PloS one, 13(4):e0194524. https://doi.org/10.1371/journal.pone.0194524.

D'Eath, R.B., Foister, S., Jack, M., Bowers, N., Zhu, Q., Barclay, D., and Baxter, E.M., 2021. Changes in tail posture detected by a 3D machine vision system are associated with injury from damaging behaviours and ill health on commercial pig farms. PLoS ONE 16(10): e0258895. https://doi.org/ 10.1371/journal.pone.0258895.

David, I., Canario, L., Combes, S. and Demars, J., 2019. Intergenerational transmission of characters through genetics, epigenetics, microbiota, and learning in livestock. Frontiers in Genetics 10:1058. https://doi.org/10.3389/fgene.2019.01058.

Diana, A., Manzanilla, E.G., Calderón Díaz, J.A., Leonard, F.C., and Boyle, L.A., 2017. Do weaner pigs need in-feed antibiotics to ensure good health and welfare? PLoS ONE. 12:e0185622. https://doi.org/10.1371/journal.pone.0185622.

Dirven, B.C.J., Homberg, J.R., Kozicz, T., and Henckens, M.J.A.G., 2017. Epigenetic programming of the neuroendocrine stress response by adult life stress. Journal of Molecular Endocrinology 59(1):R11–R31. https://doi.org/10.1530/JME-17-0019.

Docking, C.M., van de Weerd, H.A., Day, J.E.L. and Edwards, S.A., 2008. The influence of age on the use of potential enrichment objects and synchronisation of behaviour of pigs. Applied Animal Behaviour Science 110:244–257. https://doi.org/10.1016/j.applanim.2007.05.004.

Ellen, E.D., Rodenburg, T.B., Albers, G.A., Bolhuis, J.E., Camerlink, I., Duijvesteijn, N., Knol, E.F., Muir, W.M., Peeters, K., Reimert, I., Sell-Kubiak, E., van Arendonk, J.A., Visscher, J. and Bijma, P., 2014. The prospects of selection for social genetic effects to improve welfare and productivity in livestock. Frontiers in genetics 5:377. https://doi.org/10.3389/fgene.2014.00377.

EFSA 2007. The risks associated with tail biting in pigs and possible means to reduce the need for tail docking considering the different housing and husbandry systems (question no. EFSA-Q-2006-013). Annex to the EFSA Journal 611, 1–13.

Fraser, D. 1987. Attraction to blood as a factor in tail-biting by pigs. Applied Animal Behaviour Science 17(1):61–68. https://doi.org/10.1016/0168-1591(87)90008-6.

Gentz, M., Lange, A., Zeidler, S., Lambertz, C., Gauly, M., Burfeind, O. and Traulsen, I., 2020. Tail lesions and losses of docked and undocked pigs in different farrowing and rearing systems. Agriculture 10(4):130. https://doi.org/10.3390/agriculture10040130.

Gieling, E.T., Nordquist, R.E. and van der Staay, F.J., 2011. Assessing learning and memory in pigs. Animal Cognition 14(2):151–173. https://doi.org/10.1007/s10071-010-0364-3.

Guy, J.H., Rowlinson, P., Chadwick, J.P. and Ellis, M., 2002. Behaviour of two genotypes of growing–finishing pig in three different housing systems. Applied Animal Behaviour Science 75(3):193–206. https://doi.org/10.1016/S0168-1591(01)00197-6.

Haigh, A. and O'Driscoll, K., 2019. Irish pig farmer's perceptions and experiences of tail and ear biting. Porcine Health Management 5:30. https://doi.org/10.1186/s40813-019-0135-8.

Hakansson, F. and Houe, H., 2020. Risk factors associated with tail damage in conventional non-docked pigs throughout the lactation and rearing period. Preventive Veterinary Medecine 184:105160. https://doi.org/10.1016/j.prevetmed.2020.105160.

Hakansson, F. and Bolhuis, J.E., 2021. Tail-biting behaviour pre-weaning: association between other pig-directed and general behaviour in piglets. Applied Animal Behaviour Science 241:105385 https://doi.org/10.1016/j.applanim.2021.105385.

Hakansson, F., Lahrmann, H.P. and Forkman, P. 2020. A longitudinal study of pre- and post-weaning tail damage in non-docked pigs. Animal 14:2159–2166. https://doi.org/10.1017/S1751731120001081.

Harley, S., More, S., O'Connell, N., Hanlon, A., Teixeira, D. and Boyle, L., 2012. Evaluating the prevalence of tail biting and carcase condemnations in slaughter pigs in the Republic and Northern Ireland, and the potential of abattoir meat inspection as a welfare surveillance tool. Veterinary Record 171:621–662. https://doi.org/10.1136/vr.100986.

Held, S., Mendl, M., Devereux, C. and Byrne, R.W., 2000. Social tactics of pigs in a competitive foraging task: the 'informed forager' paradigm. Animal Behaviour. 59:569–576. https://doi.org/10.1006/anbe.1999.1322.

Hermesch, S., 2018. Identification of genetic factors affecting tail biting in pigs. Final report. APL project 2016/094. Australian Government Department of Agriculture and Water Resources.

Hill, J.D., McGlone, J.J., Fullwood, S.D., and Miller, M.F., 1998. Environmental enrichment influences on pig behavior, performance and meat quality. Applied Animal Behaviour Science 57, 51–68. https://doi.org/10.1016/S0168-1591(97)00116-0.

Honeck, A., Gertz, M., Beilage, E.G., Krieter, J., 2019. Comparison of different scoring keys for tail-biting in pigs to evaluate the importance of one common scoring key to improve the comparability of studies – a review. Applied Animal Behaviour Science 221. https://doi.org/10.1016/j.applanim.2019.104873.

Henry, M.H., Jansen, M., del Rocio Amezcua, M., O'Sullivan, T.L., Niel, L., Shoveller, A.K. and Friendship, R.M., 2021. Tail-biting in pigs: a scoping review. Animals 11(7):2002. https://doi.org/10.3390/ani11072002.

Holmgren, N. and Lundeheim, N., 2004. Risk factors for tail biting. Proceedings of the 18th IPVS congress. Hamburg, Germany, pg 786.

Hyde, L.K., Friso, S. and Choi, S.W., 2019. Introduction to epigenetics; Elsevier Inc.: Amsterdam, The Netherlands, ISBN 9780128045725.

Jansen, J., Bolhuis, J.E., Schouten, W.G., Spruijt, B.M. and Wiegant, V.M., 2009. Spatial learning in pigs: effects of environmental enrichment and individual characteristics on behaviour and performance. Animal Cognition. 12:303–15. https://doi.org/10.1007/s10071-008-0191-y.

Jensen, T., Busch, M., Riis, A., Maribo, H. and Vernersen, A., 2004. Halebid: Manual om forebyggelse og håndtering. Landsudvalget for Svin, Danske Slagterier.

Kasper, C., Ribeiro, D., Almeida, A.M., Larzul, C., Liaubet, L. and Murani, E., 2020. Omics application in animal science – a special emphasis on stress response and damaging behaviour in pigs. Genes 11:920. https://doi.org/10.3390/genes11080920.

Keeling, L.J., Wallenbeck, A., Larsen, A. and Holmgren, N. 2012. Scoring tail damage in pigs: an evaluation based on recordings at Swedish slaughterhouses. Acta Veterinaria Scandinavica 54:32. https://doi.org/10.1186/1751-0147-54-32.

Kobek-Kjeldager, C., Schönherz, A.A., Lene, C. and Pedersen, J., 2022. Diet and microbiota-gut-brain axis in relation to tail biting in pigs: a review. Applied Animal Behaviour Science 246:105514. https://doi.org/10.1016/j.applanim.2021.105514.

Kraimi, N., Dawkins, M., Gebhardt-Henrich, S.G., Velge, P., Rychlik, I., Volf, J., Creach, P., Smith, A., Colles, F., Leterrier, C., 2019. Influence of the microbiota-gut-brain axis on behavior and welfare in farm animals: A review. Physiology and Behavior 210:112658. https://doi.org/10.1016/j.physbeh.2019.112658.

Kritas, S.K. and Morrison, R.B. 2004. An observational study on tail biting in commercial grower-finisher barns. Journal of Swine Health and Production 12:17–22.

Koolhaas, J.M., De Boer, S.F., Buwalda, B. and Van Reenen, K., 2007. Individual variation in coping with stress: a multidimensional approach of ultimate and proximate mechanisms. Brain Behaviour Evolution 70:218–226. https://doi.org/10.1159/000105485.

Larsen, M.L.V., Andersen, H.M.L. and Pedersen, L.J., 2018. Tail posture as a detector of tail damage and an early detector of tail biting in finishing pigs. Applied Animal Behaviour Science 209:30–35. https://doi.org/10.1016/j.applanim.2018.08.016.

Larzul, C., Borey, M., Billon, Y., Rossignol, M.N., Lemonnier, G., Estellé, J., and Rogel-Gaillard, C., 2021. Response to selection on fecal microbiota composition in large white piglets. 38th International Society for Animal Genetics Conference, Virtual, July 26–30, 2021.

Lawrence, A.B., Newberry, R.C. and Spinka, M., 2018. Positive welfare: what does it add to the debate over pig welfare? In: Spinka, M. editor. Advances in pig welfare. Cambridge, UK: Woodhead Publishing Duxford, pp. 415–444.

Li, Y.Z., Zhang, H.F., Johnston, L.J., Martin, W., Peterson, J.D. and Coetzee, J.F., 2017. Effects of tail docking and tail biting on performance and welfare of growing-finishing pigs in a confinement housing system. Journal Animal Science 95(11):4835–4845. https://doi.org/10.2527/jas2017.1571.

Liu, D., Oczak, M., Maschat, K., Baumgartner. J., Pletzer, B., He, D. and Nortin, T., 2020. A computer vision-based method for spatial-temporal action recognition of tail-biting behaviour in group-housed pigs. Biosystems Engineering 195:27–41. https://doi.org/10.1016/j.biosystemseng.2020.04.007.

Lund, A. and Simonsen, H.B., 2000. Aggression and stimulus-directed activities in two breeds of finishing pig. The Pig Journal 45, 123–130.

Martínez-Miró, S., Tecles, F., Ramón, M., Escribano, D., Hernández, F., Madrid, J., Orengo, J., Martínez-Subiela, S., Manteca, X., and Cerón, J.J., 2016. Causes, consequences and biomarkers of stress in swine: an update. BMC veterinary research, 12(1):171. https://doi.org/10.1186/s12917-016-0791-8.

Moore, A.J., Brodie, E.D., Jr, and Wolf, J.B., 1997. Interacting phenotypes and the evolutionary process. I. Direct and indirect genetic effects of social interactions. Evolution. 51:1352–1362. https://doi.org/10.2307/2411187.

Mudd, A.T., Berding, K., Wang, M., Donovan, S.M., and Dilger, R.N., 2017. Serum cortisol mediates the relationship between fecal Ruminococcus and brain N-acetylaspartate in the young pig. Gut Microbes. 8(6):589–600. https://doi.org/10.1080/19490976.2017.1353849.

Munsterhjelm, C., Heinonen, M., and Valros, A. 2016. Can tail-in-mouth behaviour in weaned piglets be predicted by behaviour and performance? Applied Animal Behaviour Science 184:16–24. https://doi.org/10.1016/j.applanim.2016.08.013.

Munsterhjelm, C., Nordgreen, J., Aae, F., Heinonen, M., Olstad, K., Aasmundstad, T., Janczak, A.M. and Valros, A. 2017. To be blamed or pitied? The effect of illness on social behavior, cytokine levels and feed intake in undocked boars. Physiology and Behaviour 179:298–307. https://doi.org/10.1016/j.physbeh.2017.06.024.

Munsterhjelm, C., Peltoniemi, O.A., Heinonen, M., Hälli, O., Karhapää, M. and Valros, A. 2009. Experience of moderate bedding affects behaviour of growing pigs. Applied Animal Behaviour Science 118:42–53. https://doi.org/10.1016/j.applanim.2009.01.007.

Neethirajan, S., .Reimert, I., and Kemp, B., 2021. "Measuring farm animal emotions – sensor-based approaches" Sensors 21, 2:553. https://doi.org/10.3390/s21020553.

Prunier, A., Averos, X., Dimitrov, I., Edwards, S., Hillmann, E., Holinger, M., Ilieski, V., Leming, R., Tallet, C., Turner, S., Zupan, M., and Camerlink, I., 2019. Early life predisposing factors for biting in pigs. Animal 14(3):570–587. https://doi.org/10.1017/S1751731119001940.

Rabhi, N., Thibodeau, A., Côté, J.C., Devillers, N., Laplante, B., Fravalo, P., Larivière-Gauthier, G., Thériault, W.P., Faucitano, L., Beauchamp, G. and Quessy, S., 2020. Association between tail-biting and intestinal microbiota composition in pigs. Frontiers in Veterinary Science 7. https://doi.org/10.3389/fvets.2020.563762.

Reimert, I., Bolhuis, J.E., Kemp, B., Rodenburg, T.B., 2013. Indicators of positive and negative emotions and emotional contagion in pigs. Physiology and Behaviour 109:42–50. https://doi.org/10.1016/j.physbeh.2012.11.002.

Rydhmer, L. and Canario, L., 2022. "Chapter 11 – Behavioral Genetics in Pigs and Relations to Welfare," in Genetics and the Behavior of Domestic Animals (Second Edition), eds. T. Grandin & M.J. Deesing. (San Diego: Academic Press), 397–434. https://doi.org/10.1016/B978-0-323-85752-9.00008-1

Sambraus, H.H., 1985. Mouth-based anomalous syndromes. In World animal science, A5, ethology of farm animals. A comprehensive study of the behavioural features of common farm animals (ed. A.F. Fraser), pp. 391–422. Elsevier, Amsterdam, The Nertherlands.

Schrøder-Petersen, D.L., and Simonsen, H.B., 2001. Tail biting in pigs. The Veterinary Journal 162(3):196–210. https://doi.org/10.1053/tvjl.2001.0605.

Schrøder-Petersen, D.L., Simonsen, H.B. and Lawson, L.G. 2003. Tail-in-mouth behaviour among weaner pigs in relation to age, gender and group composition regarding gender. Acta Agricultura Scandinavica 53:29–34. https://doi.org/10.1080/09064700310002017.

Seebacher, F., and Krause, J., 2019. Epigenetics of social behaviour. Trends in Ecology and Evolution 34:818–830. https://doi.org/10.1016/j.tree.2019.04.017.

Sinisalo, A., Niemi, J.K., Heinonen, M. and Valros, A., 2012. Tail biting and production performance in fattening pigs. Livestock Science 143:220–225. https://doi.org/10.1016/j.livsci.2011.09.019.

Steinmetz, H.V. and Pedersen, M.L., 2009. Kønsvis opstaldnings betydning for forekomsten af halebid (single sex housing – its importance for tail biting). Report Meddelelse nr. 845 Videncenter for Svineproduktion, Danish Pig Research Centre.

Taylor, N.R., Main, D.C.J., Mendl, M. and Edwards, S.A., 2010. Tail-biting: a new perspective. The Veterinary Journal 186(2):137–147. https://doi.org/10.1016/j.tvjl.2009.08.028.

Ursinus, W.W., Van Reenen, C.G., Kemp, B. and Bolhuis, J.E., 2014a. Tail biting behaviour and tail damage in pigs and the relationship with general behaviour: predicting the inevitable? Applied Animal Behaviour Science 156:22–36. https://doi.org/10.1016/j.applanim.2014.04.001.

Ursinus, W.W., Wijnen, H.J., Bartels, A.C., Dijvesteijn, N., van Reenen, C.G., and Bolhuis, J.E., 2014b. Damaging biting behaviors in intensively kept rearing gilts: the effect of jute sacks and relations with production characteristics. Journal of Animal Science 92, 5193–5202. https://doi.org/10.2527/jas.2014-7918.

Ursinus, W.W., Van Reenen, C.G., Reimert, I. and Bolhuis, J.E., 2014c. Tail biting in pigs: blood serotonin and fearfulness as pieces of the puzzle? PLoS ONE. 9:e107040. https://doi.org/10.1371/journal.pone.0107040.

Valros, A., Munsterhjelm, C., Puolanne, E., Ruusunen, M., Heinonen, M., Peltoniemi, O.A.T. and Reeta Pösö, A., 2013. Physiological indicators of stress and meat and carcass characteristics in tail bitten slaughter pigs. Acta Veterinaria Scandinavica 55:75. https://doi.org/10.1186/1751-0147-55-75.

Valros, A., Ahlstrom, S., Rintala, H., Hakkinen, T. and Saloniemi, H., 2004. The prevalence of tail damage in slaughter pigs in Finland and associations to carcass condemnations. Acta Agriculturae Scandinavica Section A-Animal Science 54:213–219. https://doi.org/10.1080/09064700510009234.

Valros, A. and Heinonen, M., 2015. Save the pig tail. Porcine Health Management 1, 2. https://doi.org/10.1186/2055-5660-1-2.

Valros, A., Salib, V., Hällib, O., Saarib, S. and Heinonen, M., 2021. Does weight matter? Exploring links between birth weight, growth and pig-directed manipulative behaviour in growing-finishing pigs. Applied Animal Behaviour Science 245. https://doi.org/10.1016/j.applanim.2021.105506.

van de Weerd, H., Docking, C., Day, J. and Edwards, S., 2005. The development of harmful social behaviour in pigs with intact tails and different enrichment backgrounds in two housing systems. Animal Science 80:289298. https://doi.org/10.1079/ASC40450289.

van Staaveren, N., Vale, A.P., Manzanilla, E.G., Teixeira, D.L., Leonard, F.C., Hanlon, A., and Boyle, L.A., 2016. Relationship between tail lesions and lung health in slaughter pigs. Preventive Veterinary Medecine 127:21–6. https://doi.org/10.1016/j.prevetmed.2016.03.004.

van Staaveren, N., Teixeira, D.L., Hanlon, A. and Boyle, L.A., 2017. Pig carcass tail lesions: the influence of record keeping through an advisory service and the relationship with farm performance parameters. Animal 11:140–146. https://doi.org/10.1017/S1751731116001117.

Verbeek, E., Keeling, L., Lanberg, L., Lindberg, J.E., Dicksved, J., 2021. The gut microbiota and microbial metabolites are associated with tail biting in pigs. Scientific reports 11(1):20547. https://doi.org/10.1038/s41598-021-99741-8.

Wallenbeck, A., and Keeling, L.J., 2013. Using data from electronic feeders on visit frequency and feed consumption to indicate tail biting outbreaks in commercial pig production. J Animal Science 91(6):2879–84. https://doi.org/10.2527/jas.2012-5848.

Walker, P.K. and Bilkei, G., 2006. Tail-biting in outdoor pig production. Veterinary Journal 171:367–369. https://doi.org/10.1016/j.tvjl.2004.10.011.

Wallgren, T., Larsen, A. and Gunnarsson, S. 2019. Tail posture as an indicator of tail biting in undocked finishing pigs. Animals (Basel). 9(1):18. https://doi.org/10.3390/ani9010018.

Wedin, M., Baxter, E.M., Jack, M., Futro, A. and D'Eath, R.B., 2018. Early indicators of tail biting outbreaks in pigs. Applied Animal Behaviour Science 208:7–13. https://doi.org / 10.1016/j.applanim.2018.08.008.

Westin, R., 2003. Svansbitning hos gris relaterat till individuell tillväxt och ras (Tail biting in pigs related to individual growth and breed). Department of Animal Environment and Health, Sveriges Lantbruksuniversitet (SLU), Veterinärmedicinska fakulteten, Veterinärprogrammet.

Wiepkema, P.R., Koolhaas, J.M. 1993. Stress and animal welfare. Animal Welfare 2:195–218.

Wilder, T., Krieter, J., Kemper, N. and Büttner, K., 2021. Network analysis of tail-biting in pigs – indirect effect on tail posture. Livestock Science 251. https://doi.org/10.1016/j.livsci.2021.104668.

Wilson, K., Zanella, R., Ventura, C., Johansen, H.L., Framstad, T., Janczak, A., Zanella, A.J. and Neibergs, H.L., 2012. Identification of chromosomal locations associated with tail biting and being a victim of tail-biting behaviour in the domestic pig (Sus scrofa domesticus). Journal of Applied Genetics 53:449–56. https://doi.org/10.1007/s13353-012-0112-2.

Zonderland, J.J., Bracke, M.B.M., den Hartog, L.A., Kemp, B., Spoolder, H.A.M., 2010. Gender effects on tail damage development in single- or mixed-sex groups of weaned piglets. Livestock Science 2010, 129,151–158. https://doi.org/10.1016/j.livsci.2010.01.018.

Zonderland, J.J., Kemp, B., Bracke, M.B.M., den Hartog, L.A. and Spoolder, H.A.M., 2011a. Individual piglets' contribution to the development of tail biting. Animal 5, 601–607. https://doi.org/10.1017/S1751731110002132.

Zonderland, J.J., Schepers, F., Bracke, M., Den Hartog, L., Kemp, B. and Spoolder, H., 2011b. Characteristics of biter and victim piglets apparent before a tail-biting outbreak. Animal. 5:767–75. https://doi.org/10.1017/S1751731110002326.

Zupan, M., Janczak, A.M., Framstad, T. and Zanella, A.J. 2012. The effect of biting tails and having tails bitten in pigs. Physiology and Behaviour 106:638– 44. https://doi.org/10.1016/j.physbeh.2012.04.025.

CHAPTER 5

Poor health and the brain: how the immune system can influence neurophysiology and behaviour, and how this relates to damaging behaviour in pigs

Hege Lund[1] *and Janicke Nordgreen*[2,*]

[1]Department of Preclinical Sciences and Pathology, Faculty of Veterinary Medicine, Norwegian University of Life Sciences, Aas, Norway; [2]Department of Paraclinical Sciences, Faculty of Veterinary Medicine, Norwegian University of Life Sciences, Aas, Norway; *Correspondence: Janicke.nordgreen@nmbu.no

Abstract

The immune system is essential for protecting the body, and it also influences mental state. Similarities between sickness behaviour and depression have led scientists to look for mechanisms underlying mental illness in the effects of immune activation on the brain. One feature of sickness behaviour is altered social motivation. Many of us have experienced feeling less socially inclined when sick. Unlike us, pigs cannot control their social exposure, and altered social motivation due to a pro-inflammatory state could potentially contribute to outbreaks of damaging behaviour. From studies on humans and on rodents we know that cytokines, immune-active proteins, can have profound effects on how our brain functions, and induce depression. Immune cells in the brain can even 'remember' previous bouts of inflammation and react stronger with time, maybe explaining why some individuals are more sensitive to the negative effects of poor health than others. The literature on the effects of immune-activation on pig social behaviour is sparse but indicates a link between a pro-inflammatory state and changes in interactions with pen-mates. In addition, several risk factors for tail biting can activate the immune system, supporting the hypothesis that cytokines could be the mechanistic link between risk factors and altered behaviour. Poor health has long been accepted as one of the most important animal welfare challenges, and the knowledge presented here gives an additional dimension to the importance of keeping livestock healthy.

Keywords

tail biting – inflammation – depression – immune system – social behaviour

 | DOI:10.3920/9789004694989_007

1 Introduction

Our physical health influences our mood, our emotions and behaviour through substances produced by the immune system. Immune mediators such as cytokines can influence neurotransmitter levels, and as a result, our social motivation and our overall outlook on the world can be negatively affected (Dantzer *et al.*, 2008). A drop in the mood, akin to depression, in relation to the common flu is well known, and in the wake of the covid-19 pandemic, reports are coming in about long-term neuropsychiatric effects (Lerner *et al.* 2021; Nalbandian *et al.*, 2021). This link between immune activation and neurophysiology and mood is also likely to be important for farm animals. Not least because they are kept under conditions that can activate their immune system both through exposure to pathogens, but also through physiological and psychological stress. *Sus scrofa*, just like *Homo sapiens*, is a gregarious species. But unlike us, a pig cannot escape from the throng of its social group when ill, and the effects on neurophysiology wrought by infection or other immune activating events could thus profoundly influence social interactions with damaging behaviour, including tail biting, as one of the results (Nordgreen *et al.*, 2020). The overarching aim of this chapter is to present the link between health and CNS functioning and discuss whether it can be related to tail biting in pigs. Many of the risk factors for tail biting can activate the immune system, supporting the hypothesis that cytokines may be the link between risk factors and damaging behaviour. Although an acutely sick and lethargic pig might be targeted as a victim due to its change in behaviour, the discussion in this chapter focuses on why biters bite. In that perspective, irritability and increased emotional lability that are sometimes reported as depressive symptoms are important. There are other ways that the immune system can influence behaviour and mental health aside from depression. Cytokines have been studied in the context of severe psychiatric disease such as obsessive compulsive disorder (Marazziti *et al.*, 2015) and schizophrenia (Momtazmanesh *et al.*, 2019). However, this review focuses on changes in cognition, behaviour, motivation and emotion either defined as sickness behaviour, core symptoms of depression or as depression-related symptoms. We are mostly concerned with the 'sudden-forceful' and 'two-stage' types of tail biting as described by Taylor *et al.* (2010), and not with the 'obsessive' type.

This chapter consists of four sections. Immune function is central for the subject we are to discuss, and we will **start** by (1) describing how the mammalian immune system functions and outline the immune response following infectious and non-infectious damage. (2) In the **second** part, we will introduce the scientific background on the effects of sickness on behaviour, both from clinical reports from human patients, and from controlled studies in human and animal models. (3) The **third** part is concerned with how the immune system can influence neurophysiology and how our knowledge about these mechanisms fit with hypothesised mechanisms of depression and anxiety. (4) In **the last** section we will review what we know about the effect of immune stimulation on pig neurophysiology and behaviour and discuss the relevance of the theoretical framework from the previous sections for understanding tail biting. The chapter will end with a discussion of the **application** of this knowledge for those working with pig production, health, and welfare.

2 A brief overview of the mammalian immune system

In this section we describe how the mammalian immune system functions and outline the immune response following infectious and non-infectious damage.

The main function of the immune system is to defend the body against infectious microbes such as bacteria, viruses, parasites, and fungi. Also, non-infectious external substances recognized as foreign as well as products of own damaged cells can elicit immune responses. Cells of the immune system also play a crucial role in tissue restoration and in maintaining the integrity of own tissue. The immune system of vertebrates consists of an innate and an adaptive branch that elicit coordinated responses. Whereas the innate immune system responds rapidly and largely in a similar way to each pathogen encounter, the onset of the response of the adaptive immune system is slower but improved and strengthened with each pathogen exposure due to its ability to remember previous encounters. Both branches of the immune system influence neurophysiology largely through signalling molecules called *cytokines*. Cytokines are produced by immunocompetent cells such as macrophages, dendritic cells, microglia, natural killer (NK)-cells and lymphocytes, as well as endothelial cells, and are important in orchestrating the immune-response, balancing pro-and anti-inflammatory activity and ensuring communication between both branches of the immune system.

The innate immune branch consists of outer physical and chemical barriers, such as the epithelial layers of the skin, the gut wall and respiratory tract, and of an inner defence constituted by innate immune cells and soluble proteins. The innate immune system responds both to pathogen associated molecular patterns (PAMPs) and to endogenous damage associated molecular patterns (DAMPs). Unlike the adaptive immune system, which responds with highly specialized cellular receptors and antibodies, the innate immune receptors recognize molecules that are shared by many types of microorganisms, such as lipopolysaccharide from the cell-wall of gram-negative bacteria (PAMPs), and DAMPs that can be found in tissues following many types of sterile injury, as well as cell damage caused by infections. DAMPs and microorganisms both initiate an inflammatory response. Acute inflammation is the initial innate immune reaction to infection or injury, ensuring the recruitment of cells and various molecules in the plasma to the inflamed area with the aim to stop the infection, limit tissue damage and restore tissue. Whereas acute inflammation develops rapidly and lasts for days, chronic inflammation can follow if the infection is not cleared, or the tissue injury is extensive or prolonged. DAMPs and

PAMPs are recognized by pattern recognition receptors found on the surface, in the cytosol and in endosomal membranes of immune cells. An important group of pattern recognition receptors are the Toll-like receptors (TLRs). There are several different TLRs responding to different PAMPs and DAMPs, but their downstream effects center on two sets of transcription factors: Nuclear factor kappa B (NF-κB) (Lawrence, 2009) and interferon response factor (IRF) 3 and 7 (Fitzgerald *et al.*, 2003). NF-κB activation leads to transcription of genes for the pro-inflammatory cytokines interleukin **(IL)-1, IL-6** and tumor necrosis factor alpha **(TNF-α)**, and IRF activation to the transcription of the antiviral type 1 interferons **IFNα** and **IFNβ**.

The most extensively studied cytokines in the context of immune activation and mental illness are IL-1, IL-6 and TNF-α. Tissue macrophages, dendritic cells and microglia, the resident tissue macrophages in the brain, are central in producing these cytokines, which exert their effects through cytokine receptors. They initiate fever (all), activate neutrophils (TNF-α), change tryptophan metabolism (TNF-α) (Lestage *et al.*, 2002; O'Connor *et al.*, 2009; Walker *et al.*, 2013), and stimulate liver cells to produce acute phase proteins (APP) (all) such as **C-reactive protein (CRP)** and **haptoglobin.** Acute phase proteins are used in the emergency room as unspecific markers of inflammation. They (CRP and haptoglobin) have various functions aiming to prevent microbial invasion such as activating the complement system, or exerting antimicrobial activities through iron segregation or adhesion to the surface of microbes and infected cells, leading to the destruction of bacteria and of infected cells. Acute phase proteins are also used as markers of chronic stress (Dhama *et al.*, 2019; Gouin *et al.*, 2012). As mentioned above, IL-1, IL-6 and TNF-α have been most extensively researched for their putatively pro-depressive effect (Alzarea and Rahman, 2019; O'Connor *et al.*, 2009; Remus and Dantzer, 2016). In addition, IFN-α and IL-2 have documented psychological side effects when used as treatment against hepatitis C (Kraus *et al.*, 2003). IFN-α will activate NK cells, which in turn produce IFN-γ. IFN-γ influences tryptophan metabolism by the same mechanism as TNF-α (Weiss *et al.*, 1999), and the effects of this on behaviour and neurophysiology will be adressed in section two and three below. Many risk factors for tail biting, such as an unhygienic barn environment (Jagielo *et al.*, 1996; Sahlander *et al.*, 2012; Roque 2018), lack of enrichment (D'eath *et al.*, 2014), mycotoxins in food (Mikami *et al.*, 2011; Girardet *et al.*, 2011), psychological stress (Maes *et al.*, 1998) and disease (Wyns *et al.*, 2022) can influence the immune system (reviewed in Nordgreen *et al.*, 2020). This is part of the background for the hypothesis that cytokines are the effectors connecting different risk factors to a change in behaviour.

Besides inflammatory cytokines, which are mainly produced by cells from the innate branch, many other cytokines are also produced by the cells of the adaptive immune system. T and B lymphocytes, and especially CD4+ helper T cells, proliferate and differentiate into effector cells whose functions are mediated partly by secreted cytokines, which activate other immune cells. Memory B and T cells also have the ability to remember a previous pathogen and respond rapidly and stronger upon pathogen reencounter. This immunological memory is a hallmark of acquired immunity. A form of "memory" has also been suggested for cells of the innate branch. The concept of "trained immunity" is described as an enhanced level of responsiveness of innate immune cells after an initial stimulus (Netea *et al.*, 2020). However, in contrast to the long-term adaptive immunological memory due to gene recombination, trained immunity is mediated by epigenetic reprogramming and is of shorter duration. A form of "trained immunity" has also been described for the microglia in the brain (Bilbo *et al.*, 2007; Bilbo and Schwarz, 2009; Neher and Cunningham 2019), and this will be addressed later.

The brain is protected from many of the physiological changes and pathogen challenges that the rest of the body is exposed to by the blood-brain-barrier. But as outlined in the introduction, our brains are sometimes strongly influenced by sickness. The immune system can influence the brain through five pathways: by PAMP binding receptors on immunocompetent cells in the circumventricular organs and choroid plexus, through cytokine receptors on peripheral ends of the vagus nerve, by circulating cytokines binding receptors on brain endothelial cells, by immune cell to brain signalling through cytokines that can be transported across the blood brain barrier by carrier proteins (reviewed by Dantzer *et al.*, 2008), and by circulating immunocompetent cells entering brain tissue (D'Mello, 2009). The circumventricular organs and the choroid plexus are sites where the blood-brain barrier is permeable, and where cytokines and immune cells can diffuse into brain tissue (Weiss *et al.*, 1999). The vagus nerve can carry a signal about peripheral inflammation when peripheral nerve endings are stimulated by e.g. TNF-α. When pro-inflammatory cytokines bind to receptors on brain endothelial cells, the endothelium is stimulated to produce messenger molecules like prostaglandins and nitric oxide that are released into the brain parenchyma (Rivest *et al.*, 2000; Eskilsson 2020). Cytokines, prostaglandins and nitric oxide all contribute to activating microglia, which also produce pro-inflammatory cytokines.In paragraph three, we will look closer into the effect this has on neurotransmitter balance and brain function, but first, we will discuss behavioural effects of inflammation and immune activation.

3 The effect of immune activation on behaviour

In this section we introduce the scientific background for what we know about the effects of sickness on behaviour, both from clinical reports from human patients, and from controlled studies in human and animal models.

We often prefer to be alone, and might feel irritable and sad, when we are under the weather. Subdued mood, lethargy, loss of appetite and thirst, decreased interest in the surroundings and disturbed sleep patterns are other components of what we call sickness behaviour (Hart, 1988). While this motivational shift aids in recovery, it also shares features with depression (DSM-5). This overlap (Dantzer *et al.*, 2008; Maes *et al.*, 1995; Smith, 1991), alongside the observation that immune therapy can lead to psychological side effects (ie: unwanted effects of a treatment that manifest as altered cognition, motivation, emotional state and/or behaviour) (Capuron *et al.*, 2000; Constant *et al.*, 2005; Denicoff *et al.*, 1987), and that chronic illness and depression often correlate with increased concentrations of proteins from the immune system as measured in blood (Lasselin and Capuron, 2014; Milrad *et al.*, 2018; Nemeroff and Goldschmidt-Clermont, 2012), led to the hypothesis that immune activation may be a causal factor in some forms of mental illness (Dantzer *et al.*, 2008), and that understanding the neurophysiological changes induced by sickness could help us understand some of the mechanisms behind mental unhealth. In the hypothesis paper called 'the macrophage theory of depression' (1991), Smith described how cytokines released by macrophages could produce symptoms of depression in humans. He argued that macrophage activation could explain the high incidence of depression seen with diagnoses such as rheumatoid arthritis and coronary heart disease. In a sample of patients, the acute phase protein haptoglobin was higher in subjects with major depression and correlated with measures of T-cell activation (Maes *et al.* 1993). The macrophage theory of depression was extended to 'The monocyte-T-lymphocyte theory of major depression' in 1995 (Maes *et al.*, 1995) with a focus on IL-1β, IL-6 and interferons and their effect on CNS. More recently, patients with major depressive disorder (MDD) have been reported to have higher plasma levels of TNF-α (Felger *et al.* 2013; Postal *et al.* 2016), and in a large cross-sectional study, levels of the acute phase protein CRP were positively associated with depression (Cepeda *et al.* 2016). Other examples linking chronic inflammation to mood deterioration in humans can be found in cardiac disease, metabolic syndrome and chronic fatigue syndrome (Lasselin and Capuron, 2014; Milrad *et al.*, 2018; Nemeroff and Goldschmidt-Clermont, 2012).

In addition to being part of the innate and adaptive immune responses, cytokines may also be administered as part of a treatment regime for hepatitis C and some types of cancer. The psychological side effects experienced by the patients have been important empirical evidence supporting the argument that cytokines influence mental state (Denicoff *et al.*, 1987; Capuron *et al.*, 2000; Constant *et al.*, 2005). Fortyfour patients with metastatic cancer were treated with IL-2 and activated NK-cells (Denicoff *et al.*, 1987). Of these, 15 showed a severe time and dose-related change in behaviour, and 22 showed cognitive impairment. In all patients, the changes resolved when treatment was stopped. In the treatment of hepatitis C with IFN-α, the insidence of depression ranged from 0–44 % (Bonaccorso *et al.*, 2002; Castera *et al.*, 2002; Zdilar *et al.*, 2000). IFN-α may act by increasing the concentration of IL-1 and IL-6 (Myint 2009 – references therein), and/or to stimulate NK cells to produce IFN-γ.

In these clinical studies, tracking symptoms as treatments are initiated and stopped makes it possible to determine causality: the psychological side effects reported by the patients appear to be caused by the cytokine treatment. To relate inflammation-induced behavioural changes to neurophysiological mechanisms in more depth and in controlled experiments, several models of inflammation have been developed. Most of these models aim to activate the innate immune system and focus on the effect of immune activation on behaviour and neurophysiology during the first day(s) after stimulation. Lipopolysaccharide, a component of the cell wall of gram-negative bacteria, is often used for acute immune stimulation. It binds toll-like receptor 4, and activates the transcription factor NF-κB leading to production of the pro-inflammatory cytokines IL-1, IL-6 and TNF-α. Mice are injected with LPS at doses that induce transient sickness which typically resolves within the first 12 hours. Then, behaviour is measured in e.g. the tail-hanging test, the forced swim test, and the sucrose consumption test, with results usually (but not always: see Casaril *et al.*, 2021) pointing towards a depressed phenotype after LPS injection (reviewed by Dantzer *et al.*, 2008). This model has been extensively used to describe the systemic and neurophysiological changes that follow inflammation, and this is the focus of the next section.

4 The effect of immune activation on neurophysiology

In this section we describe how the immune system can influence neurophysiology and how our knowledge about these mechanisms fit with hypothesised mechanisms of depression and anxiety.

At the end of section one, we described how the brain can be influenced by the immune system during inflammation. Levels of cytokines, nitric oxide and prostaglandins in brain parenchyma are increased, and microglia are activated producing more pro-inflammatory cytokines which serve to recruit even more microglial cells to the inflammatory response (Bilbo and Schwarz, 2009). Microglia activated by cytokines, DAMPs or PAMPs alter their morphology and biochemistry. They go from highly ramified cells with compressed cell bodies to amoeboid-like cells which eventually lose their ramification and can move within the CNS and phagocytose infectious agents or damaged cells. Another important aspect of microglial cells is that they are long lived (Reu *et al.*, 2017) and have a cellular 'memory' of earlier adverse events. This phenomenon of 'trained immunity' is called 'microglia priming' and means that microglial cells respond stronger to stimulation in an adult if they already experienced activation earlier in life (Bilbo *et al.*, 2007; Bilbo and Schwarz, 2009; Neher and Cunningham 2019). Thus, microglial priming might be a mechanism by which the body stores the impact of earlier negative events, leading to cumulative effects on neurophysiology and behaviour throughout life. The pro-inflammatory activity of microglia can influence several aspects of CNS functioning negatively, including neurotransmitter systems and adult hippocampal neurogenesis. Importantly though, microglia are essential both for normal brain

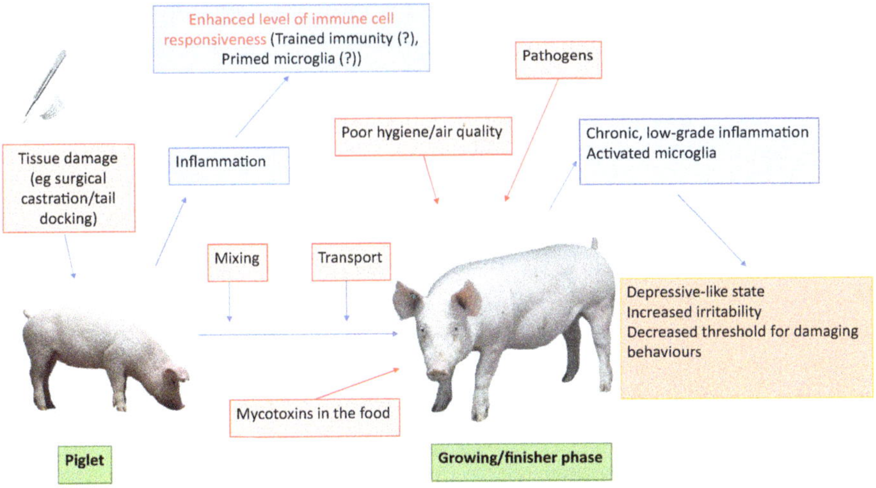

FIGURE 5.1 A visualisation of the hypothesised role of the immune system in the impact of different stressors occuring during the lifetime of a pig in a commercial production system: procedures such as tail docking and castration cause tissue damage and inflammation. This could contribute to an enhanced level of responsiveness of immune cells including priming of microglia and maybe a strengthened response to later stressors such as mixing, transport and suboptimal housing conditions.

development and for protecting the brain from damage during disease, and activation of microglia will not always have detrimental results. The understanding of when microglial priming will occur and have negative consequences and when it merely constitutes a healthy part of the body's protective response is not complete. Below we will take a closer look at what can happen when neurotransmitter systems and neurogenesis are influenced by pro-inflammatory cytokines. This is summarized in Figure 5.1.

Serotonin, dopamine, noradrenaline and glutamate are neurotransmitters. Serotonergic and noradrenergic neurons reside in the brain stem in the raphe nuclei and locus coeruleus, and the projections from these nuclei reach most areas of the brain. They influence important functions such as sleep/wake rhythm, mood, alertness and appetite through different receptor subtypes. Decreased levels of noradrenaline and serotonin were early on considered a possible cause of depression (Schildkraut, 1965), and reuptake inhibitors of both transmitters are used as antidepressants also today. Dopaminergic neurons are found in the ventral tegmental area, substantia nigra and hypothalamus and influence locomotor activity, anticipation and motivation and prolactin secretion. Dopamine is therefore essential for functional, goal-directed behaviour, and a decrease in dopamine is thought to underlie fatigue, anhedonia (inability to feel pleasure) and psychomotor retardation seen in depression (reviewed by Felger, 2016; Haaron *et al.*, 2015). Serotonin, noradrenaline and dopamine act mainly through G-protein coupled receptors. This means that their influence on other neurons is mostly modulatory and happens through molecular cascades elicited by receptor binding. Glutamate, the fourth transmitter we will discuss here, is slightly different in that several of its effects are mediated through ion channels. Glutamate is the most important excitatory neurotransmitter in the CNS. Excess extracellular glutamate can harm neurons through excitotoxicity.

The transmitter systems are interconnected, complicating an understanding of how inflammation influences the brain. It is helpful to keep in mind that each transmitter system can potentially influence mood in its own right (directly), but also indirectly through its effect on other transmitter systems. We will start with tryptophan, the precursor of serotonin, and see how several tryptophan metabolites can influence the glutamatergic system which again impacts the availability of dopamine. The net effect is that of a brain geared towards excitotoxicity, away from neural plasticity and neurogenesis, with depression and anxiety as potential results.

Tryptophan is the amino acid precursor of serotonin. It is catalysed to 5-hydroxytryptophan by tryptophan hydroxylase (TPH), and then further to 5-hydroxytryptamine; serotonin. The TPH enzyme is not saturated at physiological levels of tryptophan. This means that changes in tryptophan concentration could potentially influence serotonin levels too, and that food levels of tryptophan and other amino acids that compete with tryptophan for brain access are important in influencing serotonin synthesis. Therefore, effects of external events on serotonin levels should be viewed and discussed in light of tryptophan availability. In the healthy state, most free tryptophan (ie the proportion not being incorporated into proteins) is used not for serotonin, but for kynurenine synthesis. The enzyme tryptophan 2,3-dioxygenase (TDO) is constitutively expressed in the liver and catalyses the first reaction in the pathway from tryptophan to kynurenine (reviewed in Roth *et al.*, 2021). During inflammation, the proportion of tryptophan being transformed to kynurenine increases as the body steps up tryptophan catabolism to limit the availability of this amino acid for pathogens, and to regulate the immune response (Takikawa, 2005). This increase in kynurenine synthesis is initiated by upregulation of the enzyme indoleamine 2,3 deoxygenase (IDO) by the two pro-inflammatory cytokines TNF-α and IFN-γ (Figure 5.2). IDO catalyses the break-down of tryptophan to kynurenine (Fujigaki *et al.*, 2006). Kynurenine is further metabolised along two pathways, one producing neuroprotective substances and one producing mainly neurotoxic substances (reviewed by Myint and Kim, 2003). Kynurenic acid is produced in the neuroprotective pathway and is a glutamate receptor antagonist. Quinolinic acid (QA) and 3-hydroxy kynurenine (3OHK) are produced in the neurotoxic pathway (Myint and Kim, 2003). As kynurenine synthesis is not unique to inflammation, it seems counterintuitive that kynurenine metabolites should have a role in depression. However, part of the explanation may lie in the balance between the neuroprotective and the neurotoxic pathway (Dantzer, 2016; Müller *et al.*, 2009). Astrocytes (a type of glia cell in the CNS) produce enzymes for the neuroprotective

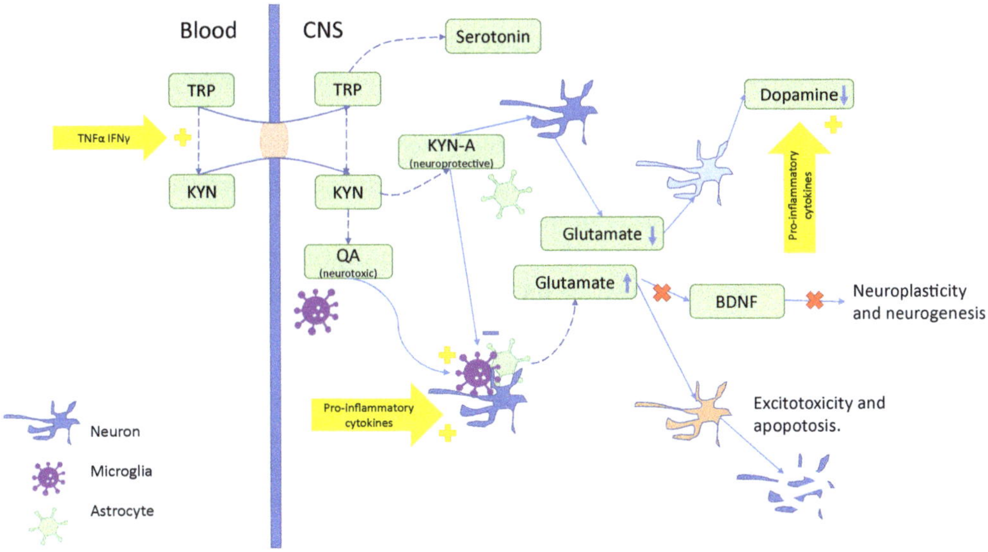

FIGURE 5.2 During inflammation, tryptophan metabolism is shunted down the kynurenine pathway. Both tryptophan and kynurenine can enter the CNS through the large neutral amino acid transporter protein. In the CNS, activated microglia contribute to relatively more neurotoxic kynurenine metabolites such as quinolinic acid being produced. Quinolinic acid stimulates neurons and glia to release glutamate, and increased extracellular glutamate can induce excitotoxicity and apoptoses with reduced production of BDNF and thereby reduced neuroplasticity and neurogenesis as a result. Dopaminergic neurotransmission is reduced among other things by the action of kynurenic acid, a glutamatergic antagonist. Reduced dopamine is thought to lead to anhedonia and lack of motivation, two core features of depression.
BDNF = brain derived neurotropic factor, TRP = tryptophan, KYN = kynurenine, QA = quinolinic acid, KYN-A = kynurenic acid

pathway and microglia for the neurotoxic pathway (Guillemin *et al.*, 2005). In inflammation, microglia can become activated and thus more neurotoxic kynurenine metabolites are produced.

Clinical and preclinical studies support the importance of QA and 3-OHK for neuropathological states (reviewed by Savitz, 2016). Patients suffering from neurodegenerative disease have increased levels of QA in their cerebrospinal fluid (Ting *et al.*, 2009; Vogelgesang *et al.*, 1996) and mice lacking enzymes necessary to produce neurotoxic metabolites were protected against inflammation-induced cognitive deficits (Heisler *et al.*, 2015). The neurotoxic metabolites of kynurenine exert their negative effect on brain functioning by influencing the glutamatergic system and by inducing oxidative stress (Guillemin, 2012; Schwarcz *et al.*, 2012). This increases the efflux of glutamate out of neurons and glia and into the extracellular space. Pro-inflammatory cytokines released into the CNS also contribute to glutamate efflux. An increase in extracellular glutamate can lead to excitotoxicity and cell death, and to a decrease in brain derived neurotrophic factor (BDNF) levels

(Haaron and Miller; 2016). BDNF stimulates neural plasticity and neurogenesis, and inflammation will therefore hamper those processes with dendrite atrophy and decreased neurogenesis as possible results. Ketamine, an anaesthetic drug that works as an antagonist on glutamate receptors, has strong anti-depressive properties (Walker *et al.*, 2013). In human subjects with major depressive disorder, the levels of glutamate (measured by imaging) in the basal ganglia correlated with CRP levels in plasma (as a measure of peripheral inflammation), and with anhedonia and psychomotor slowing (Haaron *et al.*, 2016), indicating that the neurophysiological changes described above are indeed involved in the development of depressive symptoms.

Because tryptophan is the precursor for serotonin, IDO induction could also influence mood and behaviour by decreasing serotonin levels, corresponding to monoamine-centered explanations of depression (discussed by Dantzer, 2016). However, the reported findings are not consistent. Some studies indicate a central role for serotonin, such as the report from Raison *et al.*, indicating that the levels of the serotonin-metabolite 5-HIAA in the cerebrospinal fluid of human subjects treated with IFN-α predicted the degree of depression (Raison *et al.*, 2009). On the other hand, an experiment in mice found that inhibiting IDO activity prevented depressive behaviour but did not influence serotonin metabolism (O'Connor *et al.*, 2009), indicating that at least in that model, kynurenine metabolites were more important. Noradrenaline, the neurotransmitter responsible for modulating alertness, focused attention and mood, is also influenced in experimental inflammation models. The initial response to inflammation is an increase in the activity of noradrenergic neurons in the locus coeruleus and thereby an increase in noradrenaline levels (Linthorst *et al.* 1996; Sekio and Seki 2015). Subsequently, noradrenaline levels drop (in the brain-stem and hypothalamus of rats; Besedovsky *et al.* 1983), and this has also been observed in pigs 72 hours after LPS injection (in the frontal cortex, left hippocampus and hypothalamus; Nordgreen *et al.* 2018).

Returning to the kynurenine metabolites, they do not only influence glutamatergic signalling (through their actions on glutamate receptors and their oxidative effects), but dopaminergic signalling as well. Both pathways of kynurenine metabolism influence dopamine availability negatively. Kynurenic acid (neuroprotective and a glutamate receptor antagonist) inhibits dopamine release through its effect on glutamatergic receptors (Felger 2016). In addition, the oxidative stress that the neurotoxic pathway induces contribute to usurp a cofactor (tetrahydrobiopterin, BH4) that is needed for the enzymes transforming dopamine precursors to dopamine, further decreasing dopamine availability. Behavioural studies support the importance of these biochemical changes. Animals that showed depressive symptoms after IFN-α administration also had reduced levels of dopamine metabolites in their cerebrospinal fluid (Felger *et al.*, 2007). Microdialysis showed that reduced dopamine release in the striatum correlated with reduced efforts to get at a palatable reward (Felger and Lotrich, 2013). Hens (*Gallus gallus domesticus*) administered a dopamine receptor antagonist showed reduced anticipatory behaviour when expecting a highly palatable reward (Moe *et al.* 2011).

In addition to influencing neurotransmission, inflammation can also decrease adult hippocampal neurogenesis (Chesnokova *et al.*, 2016). The mechanism is both direct through cytokines (Goshen *et al.* 2007), and indirect through decreasing the production of BDNF as described earlier in this section. IL-1β signalling mediates the anti-neurogenic effects of stress through the transcription factor NF-κB (Goshen *et al.*, 2007; Koo and Duman 2008; Koo *et al.*, 2010), which is increased both during inflammation and stress (Bierhaus *et al.* 2003; Munhoz *et al.* 2006). Decreased neurogenesis seems to affect both cognition and mood negatively (Chesnokova *et al.*, 2016; Dranovsky 2006). Neurogenesis is necessary for antidepressants to have an effect (Santarelli *et al.*, 2003), the hippocampal volume is decreased in patients with major depressive disorder (Videbech and Ravnkilde 2004) and antidepressant treatment in animals prevent stress-induced decreases in hippocampal volume (Czeh *et al.*, 2001). Thus, changes in neurogenesis during inflammation may contribute to mood changes. The neurogenesis pathway to altered mood in inflammation is particularly interesting as it is affected by chronic inflammation and stress (Poirier *et al.*, 2019). Production animals are often kept under suboptimal conditions that could conceivably lead to a chronic inflammatory state with potential dire consequences for neurophysiology and quality of life. In the last section of this chapter, we will discuss the influence of poor health on the brain in the context of pig physiology and behaviour.

5 The effect of immune activation on pig physiology and behaviour, and the relevance for tail biting.

In this section we review what we know about the effect of immune stimulation on pig neurophysiology and behaviour and discuss the relevance of the theoretical framework from the previous sections for understanding tail biting. The chapter ends with a discussion of the application of this knowledge for those working with pig production, health, and welfare.

Poor health and damaging behaviour are closely linked (reviewed by Boyle *et al.*, 2022). The focus of this chapter has been on discussing how poor health can influence the brain and whether this can help us understand, on a mechanistic level, why pigs perform damaging behaviour. However, it is important to remember that tail biting can also lead to poor health, particularly in the victims (Boyle *et al.*, 2022). While research is still needed on this topic, we know enough to state that improvements in health will almost certainly lower the risk of damaging behaviour, and vice versa (Boyle *et al.*, 2022). Farms with health challenges are at greater risk of having a tail biting outbreak (Moinard *et al.*, 2003; Taylor *et al.*, 2012), while deworming and vaccination seem to reduce tail biting and losses due to damaging behaviours (Almond and Bilkei, 2006; Barnikol, 1978). Positive correlations between leg disorders and respiratory inflammation and tail damage have been shown (Munsterhjelm *et al.*, 2013; Niemi *et al.*, 2011). In addition, several risk factors for tail biting such as an unhygienic environment and psychological stress, can elicit an immune response (reviewed in Nordgreen *et al.*, 2020). While this supports the hypothesis of the immune system having a role in the development of damaging behaviour, the question of what comes first remains; bit tail or inflammation? The fact that several risk factors are described on group or farm level, whereas the effects of inflammation on neurophysiology and behaviour is described at the level of the individual, poses an additional challenge when it comes to relating the two bodies of literature to each other. The use of experimental models of inflammation in pigs can give us some insight into the effects of immune activation on their physiology and behaviour and will be the focus of the next two paragraphs.

In humans and in rodent models, the focus is on understanding the role of inflammation in depression as outlined in the previous sections of this chapter. Core features of depression such as lethargy, lack of motivation and lowered mood have similarities with sickness behaviour but are not easily reconciled with the active biting behaviour of tail biters. However, irritability is listed as a core symptom of depression in adolescents in DSM-V (the diagnostic and statistical manual of mental disorders) and mentioned as a diagnostic feature for other age groups as well (DSM-V). There are discussions as to its importance and prevalence in current scientific literature (discussed in Balbuena *et al.*, 2016; Fava *et al.*, 2010; Pedrelli *et al.*, 2013). This could well be relevant in finisher pigs and was the reason for choosing the title 'sick and grumpy' for a paper in which we studied the effect of LPS injection in pigs housed in triplets. Pigs that received an LPS injection increased the manipulation of the ears of their penmates two days after the injection, ie after any overt signs of sickness had abated (Munsterhjelm *et al.* 2019; Veit *et al.* 2021a), even though no tail biting was observed in these studies. Extreme emotional lability, irritability, short temper, tearfulness and cognitive impairment have been reported in human clinical studies on the effects of treatment with pro-inflammatory cytokines such as IL-2 and interferon α (IFNα) (Capuron *et al.*, 2000; Denicoff *et al.*, 1987; Renault *et al.*, 1987). In psychiatric patients with intermittent explosive disorder, IL-6 and CRP were elevated (Coccaro *et al.* 2014). Patients treated with IFN-α had (in one study) increased hostility/ and increased depression scores (Kraus *et al.* 2003). Thus, cytokines might influence mood in many ways, and even though tail biting is not believed to be an aggressive behaviour, it is possible that the threshold for expressing damaging behaviour could be decreased as a result of immune activation. Pigs that were diagnosed with respiratory disease showed an increased tendency to ear- and tail-bite one to two weeks prior to being diagnosed as sick (Munsterhjelm and Nordgreen *et al.*, 2017), and the levels of some key pro-and anti-inflammatory cytokines correlated with different social and feeding-related behaviours. Pigs provided with environmental enrichment, an intervention known to *reduce* the risk of damaging behaviour, show decreased expression of microglial markers in their frontal cortex one hour after enrichment access (Brown *et al.*, 2018), indicating a healthier, less inflammatory, neurophysiology.

Porcine microglia will produce pro-inflammatory cytokines and reactive oxygen species in vitro in response to high pathogen porcine reproductive and respiratory virus (Chen *et al.*, 2014), and when sows are infected during pregnancy, the fetal microglia of their piglets are altered (Antonson *et al.*, 2019). However, the knowledge of how the response of pig microglia influence neurophysiology and behaviour is sparse, but there are a number of studies on other neurophysiological effects of immune activation that are relevant for discussing the link between the immune system and damaging behaviour. An injection with LPS led to a decrease in noradrenaline levels in the hypothalamus, frontal cortex and hippocampus of gilts at 72 hours after injection (Nordgreen *et al.*, 2018). In addition, the pro-inflammatory cytokine IFN-γ was increased in the frontal cortex. A follow-up study which sampled group-housed pigs at the same time-point did not replicate those findings in their entirety, but did report a decrease in plasma, frontal cortex, and brain stem tryptophan, frontal cortex and brain stem kynurenine, and hippocampal noradrenaline in pigs given LPS, compared to different control groups (Veit *et al.*, 2021). We do not yet know whether neurotransmitters change in the same way with naturally occurring disease or psychological stress, nor do we know how the pigs are affected by the neurophysiological changes we see. However, based on the roles of the neurotransmitters for normal function, we can speculate as to what aspect of behaviour can be influenced. Reduced levels of noradrenaline could impact alertness, mood and cognitive functioning. Reduced noradrenaline levels were believed to be important in depression, and several antidepressants increase noradrenaline availability in the synaptic cleft (eg tricyclic antidepressiva and selective noradrenaline reuptake inhibitors). Mice injected with LPS had a significant drop in frontal cortex noradrenaline concentration 24 hours after injection, and fluoxetine (a selective serotonin reuptake inhibitor) decreased this effect and the depression-like effects as well (Zhu *et al.*, 2015). However, while the effect on noradrenaline level is fast, the time taken for the (human) patients treated with reuptake inhibitors to improve is longer, making it difficult to reconcile reduced noradrenaline levels as directly responsible for depression. Some studies have measured a gradual return of some, but not all functions and then the therapeutic lag is shorter (two weeks with a selective noradrenaline reuptake inhibitor (Katz *et al.*, 2004)), strengthening the evidence for noradrenaline's role in mood regulation. Noradrenaline is also important for our ability to focus on relevant stimuli (Aston-Jones *et al.*, 1999), and for effortful attention (like when you are focusing on reading this chapter) (Milstein *et al.*, 2007). Therefore, noradrenaline reduction in the wake of illness could influence how pigs function in relation to their environment and in their social group. There are some indications that monoaminergic signalling differs between biters, victims, and neutral control pigs. Biters had lower serotonin levels and higher serotonin uptake velocity into the blood platelets (Ursinus *et al.* 2014; discussed in Nordgreen *et al.* 2020), and a tendency towards increased 5-HIAA levels in the prefrontal cortex compared to controls (Valros *et al.* 2015). There was also a correlation between the ratio of tryptophan to other amino acids, and to 5-HIAA and serotonin in the frontal cortex in biters, and this could indicate a stronger dependency of serotonin synthesis on tryptophan levels in biters (*ibid*). As mentioned earlier, tryptophan availability in the diet will most likely influence results, as tryptophan supplementation has decreased damaging behaviour (Martinez-Trejo *et al.*, 2009), and modified the effect of being housed in a suboptimal environment (Van der Meer, 2016). Importantly, the microbiota-gut-brain axis activity will influence tryptophan availability, as recently described by Kobek-Kjeldager and colleagues (Kobek-Kjeldager *et al.*, 2022), but discussing this complex system lies beyond the scope of this chapter.

In conclusion, poor health due to infections or stress influences our brain. This is an important protective response, but if it doesn't resolve it can develop into a pathological state that reduces quality of life and induces mental unhealth in the form of altered motivation, emotions, cognition and behaviour. The neurophysiology of pigs is influenced by inflammation, and the changes are similar to those observed in experimental models in other species. However, we know less about effects of immune activation on neurogenesis and microglial activation in pigs than we do in rodents. An even more important knowledge gap is that of understanding to which degree housing systems and husbandry procedures influence the immune system and the neurophysiology of pigs, both negatively and positively. As mentioned earlier in this chapter, different sources of stress and pathogens will activate the immune system throughout an animal's lifespan, and immune cells are able to remember previous insults. Pigs are subjected to procedures that might well serve as priming events on the maturing immune system. In many countries, pigs are tail docked and/or surgically castrated during their first week(s) of life. These procedures cause pain, stress and tissue damage, and might also open the body up for pathogens. Thus, an inflammatory response can be expected in the wake of tail docking and castration, and changes in gene expression profiles were still present in the dorsal root ganglion of piglets 4 months after tail docking (Sandercock *et al.*, 2019). There is therefore a potential for a priming of the immune system early on in a pig's life, which could increase the strength and severity of the response and consequences to later immune-stimulating events such as disease, psychological stress (mixing, transport), and an unhygienic environment. This is summarized in Figure 5.1. A better understanding of the effects of these early stressors on the immune system is needed.

On the positive side, enrichment and the possibility to exercise (like in some free-range or organic systems) may decrease inflammation (Fablet *et al.*, 2012; Brown *et al.*, 2018), increase adult hippocampal neurogenesis (reviewed by Liu and

Nusslock 2018) increase quality of life through improved mood and reduced anxiety (Lopez-Bueno *et al.*, 2020) and thereby counteract effects of challenges and negative events.

While the importance of healthy livestock is widely acknowledged, the hypotheses presented in this chapter gives an additional dimension to health challenges as a cause of poor animal welfare and provides a framework for discussing and evaluating benefits of alternative housing systems.

6 Take home message

- The immune system can influence brain function through signalling molecules, called cytokines.
- Sickness behaviour shares features with depression, and in both conditions, social motivation is affected.
- Poor health can be a threat to animal welfare not only because of the direct effects on physiological functioning, but also through the effects on the brain leading to changes in social behaviour.
- We need to know more about how different husbandry procedures and production systems influence the immune system and the brain, but it is already known that environmental enrichment and the opportunity to be physically active will have protective and positive effects on neurophysiology and behaviour.

References

American Psychiatric Association: Diagnostic and statistical manual of mental disorders, fifth edition (DSM-V). Arlington, VA, American Psychiatric Association, 2013.

Almond, P.K., and Bilkei, G., 2006. Effects of oral vaccination against *Lawsonia intracellularis* on growing-finishing pig's performance in a pig production unit with endemic porcine proliferative enteropathy (PPE). Deutsche Tierartzliche Wochenschrift. 113, 232–235.

Alzarea, S., Rahman, S., 2019. Alpha-7 nicotinic receptor allosteric modulator PNU120596 prevents lipopolysaccharide-induced anxiety, cognitive deficit and depression-like behaviors in mice. Behavioral Brain Research, 366, 19–28. https://doi.org/10.1016/j.bbr.2019.03.019.

Antonson, A.M., Lawson, M.A., Caputo, M.P., Johnson, R.W. 2019. Maternal viral infection causes global alterations in porcine fetal microglia. PNAS, 116 (40) 20190–20200. https://doi.org/10.1073/pnas.1817014116.

Aston-Jones, G., Rajkowski, J.,Cohen, J., 1999. Role of locus coeruleus in attention and behavioral flexibility, Biol. Psychiatry 46, 1309–1320. https://doi.org/10.1016/s0006-3223(99)00140-7.

Balbuena, L., Bowen, R., Baetz, M., Marwaha, S., 2016. Mood instability and irritability as core symptoms of major depression: an exploration using rasch analysis. 7(174). https://doi.org/10.3389/fpsyt.2016.00174.

Barnikol, M., 1978. Observations of tail biting (cannibalism) amongst fattening hogs and breeding pigs in connection with protein and mineral nourishment, Tierartzliche Umschau, 33, 540–546.

Besedovsky, H., del Rey, A., Sorkin, E., Da Prada, M., Burri, R., Honegger, C., 1983. The immune response evokes changes in brain noradrenergic neurons. Science, 221(4610), 564–566. https://doi.org/10.1126/science.6867729.

Bierhaus, A., Wolf, J., Andrassy, M., Rohleder, N., Humpert, P.M., Petrov, D., Ferstl, R., von Eynatten, M., Wendt, T., Rudofsky, G., Joswig, M., Morcos, M., Schwaninger, M., McEwen, B., Kirschbaum, D., Nawroth, P.P., 2003. A mechanism converting psychosocial stress into mononuclear cell activation. Proceedings of the National Academy of Sciences the United States of America 100, 1920–1925. https://doi.org/10.1073/pnas.0438019100.

Bilbo, S.D., Newsum, N.J., Sprunger, D.B., Watkins, L.R., Rudy, J.W., Maier, S.F., 2007. Differential effects of neonatal handling on early life infection-induced alterations in cognition in adulthood. Brain, Behavior and Immunity 21, 332–342. https://doi.org/10.1016/j.bbi.2006.10.005.

Bilbo, S.D., Schwarz, J.M., 2009. Early-life programming of later-life brain and behavior: a critical role for the immune system. Frontiers in Behavioral Neuroscience 3, 14. https://doi.org/10.3389/neuro.08.014.2009.

Bonaccorso, S., Marino, V., Puzella, A., Pasquini, M., Biondi, M., Artini, M., Almerighi, C., Verkerk, R., Meltzer, H., Maes, M., 2002. Increased depressive ratings in patients with hepatitis C receiving interferon-alpha-based immunotherapy are related to interferon-alpha-induced changes in the serotonergic system. Journal of Clinical Psychopharmacology, 22, 86–90. https:/doi.org/10.1097/00004714-200202000-00014.

Boyle, L.A., Edwards, S.A., Bolhuis, J.E., Pol, F., Zupan Šemrov, M., Schütze, S., Nordgreen, J., Bozakova, N., Sossidou, E.N., Valros, A., 2022. The evidence for a causal link between disease and damaging behavior in pigs. Frontiers in Veterinary Science, 8, 771682. https://doi.org/10.3389/fvets.2021.771682.

Brown, S.M., Bush, S.J., Summers, K.M., Hume, D.A., Lawrence, A.B., 2018. Environmentally enriched pigs have transcriptional profiles consistent with neuroprotective effects and reduced microglial activity. Behavioral Brain Researc, 350, 6–15. https://doi.org/10.1016/j.bbr.2018.05.015.

Capuron, L, Ravaud, A., Dantzer, R. 2000. Early depressive symptoms in cancer patients receiving interleukin 2 and/or interferon alfa-2b therapy. Journal of Clinical Oncology, 18, 2143–2151. https:/doi.org/10.1200/JCO.2000.18.10.2143.

Casaril, A.M., Vichaya, E.G., Rishi, M.R., Ford, B.G., Dantzer, R., 2021. Lipopolysaccharide does not alter behavioural response to successive negative contrast in mice. Psychopharmacology 238, 691–697. https://doi.org/10.1007/s00213-020-05721-7.

Castera, L., Zigante, F., Bastie, A., Buffet, C., Dhumeaux, D., Hardy, P., 2002. Incidence of interferon alfa-induced depression in patients with chronic hepatitis C. Hepatology, 35:978–979 9. https://doi.org/10.1053/jhep.2002.32104.

Cepeda, M.S., Stang, P., Makadia, R., 2016. Depression is associated with high levels of C-reactive protein and low levels of fractional exhaled nitric oxide. Journal of Clinical Psychiatry, 77:12. 1666–1671. https://doi.org/10.4088/JCP.15m10267.

Chen, X-x., Quan, R., Guo, X-k., Gao, L., Shi, J., Feng, W-h., 2014. Up-regulation of pro-inflammatory factors by HP-PRRSV infection in microglia: implications for HP-PRRSV neuropathogenesis. Veterinary Microbiology. 170, 48–57. https://doi.org/10.1016/j.vetmic.2014.01.031.

Chesnokova, V., Pechnickb, R.N., Wawrowskya, K., 2016. Chronic peripheral inflammation, hippocampal neurogenesis, and behavior. Brain Behavior and Immunity, 58, 1–8. https://doi.org/10.1016/j.bbi.2016.01.017.

Czéh, B., Michaelis, T., Watanabe, T., Frahm, J., de Biurrun, G., van Kampen, M., Bartolomucci, A., Fuchs, E., 2001. Stress-induced changes in cerebral metabolites, hippocampal volume, and cell proliferation are prevented by antidepressant treatment with tianeptine. Proceedings of the National Academy of Sciences the United States of America, 98(22), 12796–12801. https://doi.org/10.1073/pnas.211427898.

Coccaro, E.F., Lee, R., Coussons-Read, M., 2014. Elevated plasma inflammatory markers in individuals with intermittent explosive disorder and correlation with aggression in humans. JAMA Psychiatry 71(2), 158–165. https://doi.org/10.1001/jamapsychiatry.2013.3297.

Constant, A., Castera, L., Dantzer, R., Couzigou, P., De Ledinghen, V., Demotes-Mainard, J., Henry, C., 2005. Mood alterations during interferon-alfa therapy in patients with chronic hepatitis C: evidence for an overlap between manic/hypomanic and depressive symptoms. Journal of Clinical Psychiatry, 66, 1050–1057. https://doi.org/10.4088/jcp.v66n0814.

Dantzer, R., O'Connor, J.C., Freund, G.G., Johnason, R.W., Kelley, K.W. 2008. From inflammation to sickness and depression: when the immune system subjugates the brain. Nature Reviews Neuroscience, 9, 46–57. https://doi.org/10.1038/nrn2297.

Dantzer, R., 2016. Role of the kynurenine metabolism pathway in inflammation-induced depression: preclinical approaches. In: Dantzer, R., and Capuron, L. (eds). Inflammation-associated depression: evidence, mechanisms and implications. Current topics in behavioral neurosciences, Springer International Publishing, Switzerland. pp. 117–133. https://doi.org/10.1007/7854_2016_2.

D'eath, R., Arnott, G., Turner, S.P., Jensen, T., Lahrmann, H.P., Busch, M.E., Niemi, J.K., Lawrence, A.B., Sandøe, P. 2014. Injurious tail biting in pigs: how can it be controlled in existing systems without tail docking? Animal. 8:9; 1479–1497. https://doi.org/10.1017/S1751731114001359.

Denicoff, K.D., Rubinow, D.R., Papa, M.Z., Simpson, C., Seipp, C.A., Lotze, M.T., Chang, A.E., Rosenstein, D., Rosenberg, S.A. 1987. The neuropsychiatric effects of treatment with interleukin-2 and lymphokine-activated killer-cells. Annals of Internal Medicine 107; 293–300. https://doi.org/10.7326/0003-4819-107-2-293.

D'Mello, C., Le, T., Swain, M., 2009. Cerebral microglia recruit monocytes into the brain in response to tumor necrosis factor alpha signaling during peripheral organ inflammation. Journal of Neuroscience 29(7), 2089–2102. https://doi.org/10.1523/JNEUROSCI.3567-08.2009.

Dhama, K., Latheef, S.K., Dadar, M., Samad, H.A., Munjal, A., Khandia, R., Karthik, K., Tiwari, R., Yatoo, M.I., Bhatt, P., Chakraborty, S., Singh, K.P., Iqbal, H.M.N., Chaicumpa, W., Joshi, S.K., 2019. Biomarkers in stress related diseases/disorders: diagnostic, therapeutic and prognostic values. Frontiers in Molecular Biosciences 6, 91 https://doi.org/10.3389/fmolb.2019.00091.

Dranovsky, A., Hen, R., 2006. Hippocampal neurogenesis: regulation by stress and antidepressants. Biological Psychiatry, 59, 1136–1143. https://doi.org/10.1016/j.biopsych.2006.03.082.

Eskilsson, A. 2020. Inflammatory signalling across the blood-brain barrier and the generation of fever. Linköping University Medical Dissertation No. 1724; page 34–35.

Fablet, C., Dorenlor, V., Eono, F., Eveno, E., Jolly, J.P., Portier, F., Bidan, F., Madec, F., Rose, N. 2012. Noninfectious factors associated with pneumonia and pleuritis in slaughtered pigs from 143 farrow-to-finish pig farms. Preventive Veterinary Medicine 104, 271–280. https://doi.org/10.1016/j.prevetmed.2011.11.012.

Fava, M., Hwang, I., Rush, A.J., Sampson, N., Walters, E.E., and Kessler, R.C., 2010. The importance of irritability as a symptom of major depressive disorder: results from the National Comorbidity Survey Replication. Molecular Psychiatry 15(8), 856–867. https://doi.org/10.1038/mp.2009.20.

Felger, J.C., Alagbe, O., Hu, F., Mook, D., Freeman, A.A., Sanchez, M.M., Kalin, N.H., Ratti, E., Nemeroff, C.B., Miller, A.H., 2007. Effects of interferon-alpha on rhesus monkeys: a nonhuman primate model of cytokine-induced depression. Biological Psychiatry 62(11): 1324–1333. https://doi.org/10.1016/j.biopsych.2007.05.026.

Felger, J.C., Lotrich, F.E., 2013. Inflammatory cytokines in depression: neurobiological mechanisms and therapeutic implications. Neuroscience, 246, 199–229. https://doi.org/10.1016/j.neuroscience.2013.04.060.

Felger, J.C., Mun, J., Kimmel, H.L., Nye, J.A., Drake, D.F., Hernandez, C.R., Freeman, A.A., Rye, D.B., Goodman, M.M., Howell, L.L., Miller, A.H., 2013. Chronic interferon-alpha decreases dopamine 2 receptor binding and striatal dopamine release in association with anhedonia-like behavior in nonhuman primates. Neuropsychopharmacology 38(11), 2179–2187. https://doi.org/10.1038/npp.2013.115.

Felger, J.C., 2017. The role of dopamine in inflammation-associated depression: mechanisms and therapeutic implications. Current topics in Behavioral Neurosciences 31, 199–219. https://doi.org/ 10.1007/7854_2016_13.

Fitzgerald, K.A., Rowe, D.C., Barnes, B.J., Caffrey, D.R., Visintin, A., Latz, E., Monks, B., Pitha, P.M., Golebock, D.T., 2003., LPS-TLR4 Signaling to IRF-3/7 and NF-κB involves the toll adapters TRAM and TRIF. Journal of Experimental Medicine, 198(7), 1043–1055. https://doi.org/10.1084/jem.20031023.

Fujigaki, H, Saito, K, Fujigaki, S, Takemura, M, Sudo, K, Ishiguro, H, *et al.* 2006. The signal transducer and activator of transcription 1 α and interferon regulatory factor 1 are not essential for the induction of indoleamine 2,3-dioxygenase by lipopolysaccharide: involvement of p38 mitogen-activated protein kinase and nuclear factor-kappaB pathways, and synergistic effect of several proinflammatory cytokines. Journal of Biochemistry. 139: 655–662. https://doi.org/10.1093/jb/mvj072.

Girardet, C., Bonnet, M.S., Jdir, R., Sadoud, M., Thirion, S., Tardivel, C., Roux, J., Lebrun, B., Mounien, L., Trouslard, J., Jean, A., Dalaporta, M., Troadec, J.-D. 2011. Central inflammation and sickness-like behavior induced by the food contaminant deoxynivalenol: a PGE2-independent mechanism. Toxicological Science 124(1): 179–191. https://doi.org/10.1093/toxsci/kfr219.

Goshen, I., Kreisel, T., Ben-Menachem-Zidon, O., Licht, T., Weidenfeld, J., Ben-Hur, T., Yirmiya, R., 2008. Brain interleukin-1 mediates chronic stress-induced depression in mice via adrenocortical activation and hippocampal neurogenesis suppression. Molecular Psychiatry, 13:717–728. https://doi.org/10.1038/sj.mp.4002055.

Gouin, J.-P., Glaser, R., Malarkey, W.B., Beversdorf, D., Kiecolt-Glaser, J., 2012. Chronic stress, daily stressors, and circulating inflammatory markers. Health Psychology, 31(2), 264–268. https://doi.org/10.1037/a0025536.

Guillemin, G.J., Smythe, G., Takikawa, O., Brew, B.J., 2005. Expression of indoleamine 2,3-dioxygenase and production of quinolinic acid by human microglia, astrocytes, and neurons. Glia 49:15–23. https://doi.org/10.1002/glia.20090.

Guillemin, G.J., 2012. Quinolinic acid, the inescapable neurotoxin. Federation of European Biochemical Societies journal 279:1356–1365. https://doi.org/10.1111/j.1742-4658.2012.08485.x.

Haaron, E., Miller, A.H., 2016. Inflammation effects on brain glutamate in depression: mechanistic considerations and treatment implications. In: Dantzer, R., and Capuron, L. (eds). Inflammation-associated depression: evidence, mechanisms and implications. Current topics in behavioral neurosciences, Springer International Publishing, Switzerland. pp. 173–191. https://doi.org/10.1007/7854_2016_2.

Haroon, E., Fleischer, C.C., Felger, J.C., Chen, X., Woolwine, B.J., Patel, T., Hu, X.P., Miller, A.H., 2016. Conceptual convergence: increased inflammation is associated with increased basal ganglia glutamate in patients with major depression. Molecular Psychiatry, 21(10), 1351–1357. https://doi.org/10.1038/mp.2015.206.

Hart, B.L. 1988. Biological basis of the behavior of sick animals. Neuroscience and Biobehavioral Reviews, 12, 123–137. https://doi.org/10.1016/s0149-7634(88)80004-6.

Heisler, J.M., O'Connor, J.C., 2015. Indoleamine 2,3-dioxygenase-dependent neurotoxic kynurenine metabolism mediates inflammation-induced deficit in recognition memory. Brain Behavior and Immunity 50:115–124. https://doi.org/10.1016/j.bbi.2015.06.022.

Katz, M.M., Tekell, J.L., Bowden, C.L., Brannan, S., Houston, J.P., Berman, N., Frazer, A., 2004. Onset and early behavioral effects of pharmacologically different antidepressants and placebo in depression, Neuropsychopharmacology 29, 566–579. https://doi.org/10.1038/sj.npp.1300341.

Kobek-Kjeldager, C., Schönherz, A.A., Canibe, N., Juul Pedersen, L., 2022. Diet and microbiota-gut-brain axis in relation to tail biting in pigs: a review. Applied Animal Behaviour Science, 246, 105514. https://doi.org/10.1016/j.applanim.2021.105514.

Koo, J.W., Duman, R.S., 2008. IL-1beta is an essential mediator of the antineurogenic and anhedonic effects of stress. Proceedings of the National Academy of Sciences of the United States of America, 105, 751–756. https://doi.org/10.1073/pnas.0708092105.

Koo, J.W., Russo, S.J., Ferguson, D., Nestler, N.J., Duman, R.S., 2010. Nuclear factor-κB is a critical mediator of stress-impaired neurogenesis and depressive behavior. Proceedings of the National Academy of Sciences of the United States of America, 107, 2669–2674. https://doi.org/10.1073/pnas.0910658107.

Kraus, M.R., Schafer, A., Faller, H., Csef, H., Scheurlen, M., 2003. Psychiatric symptoms in patients with chronic hepatitis C receiving interferon alfa-2b therapy. Journal of Clinical Psychiatry, 64, 708–714. https://doi/org/10.4088/jcp.v64n0614.

Lasselin, J., Capuron, L., 2014. Chronic low-grade inflammation in metabolic disorders: relevance for behavioral symptoms. Neuroimmunomodulation, 21, 95–101. https://doi.org/10.1159/000356535.

Lawrence, T., 2009. The nuclear factor NF-kappaB pathway in inflammation. Cold Spring Harbor Perspectives in Biology. 1, 1–10. https://doi.org/10.1101/cshperspect.a001651.

Lerner, A.M., Robinson, D.A., Yang, L., Williams, C.F., Newman, L.M., Breen, J.J., Eisinger, R.W., Schneider, J.S., Adimora, A.A., Erbelding, E.J., 2021. Toward understanding COVID-19 recovery: National Institutes of Health workshop on postacute COVID-19. Ann. Intern. Med. M21-1043. https://doi.org/10.7326/M21-1043.

Lestage, J., Verrier, D., Palin, K., and Dantzer, R., 2002. The enzyme indoleamine 2,3-dioxygenase is induced in the mouse brain in response to peripheral administration of lipopolysaccharide and superantigen. Brain Behavior and Immunity 16(5), 596–601. https://doi.org/10.1016/s0889-1591(02)00014-4.

Linthorst, A.C., Flachskamm, C., Holsboer, F., Reul, J.M., 1996. Activation of serotonergic and noradrenergic neurotransmission in the rat hippocampus after peripheral administration of bacterial endotoxin: involvement of the cyclo-oxygenase pathway. Neuroscience, 72:989–997. https://doi.org/10.1016/0306-4522(95)00604-4.

Liu, P.Z., and Nusslock, R. 2018. Exercise-mediated neurogenesis in the hippocampus via BDNF. Frontiers in Neuroscience, 12: 52. https://doi.org/10.3389/fnins.2018.00052.

Maes, M., Scharpé, S., Meltzer, H.Y., Cosyns, P.,1993. Relationships between increased haptoglobin plasma levels and activation of cell-mediated immunity in depression. Biological Psychiatry, 34(10), 690–701. https://doi.org/10.1016/0006-3223(93)90042-C.

Maes, M., Smith, R, Scharpe, S., 1995. The monocyte T-lymphocyte hypothesis of major depression. Psychoneuroendocrinology, 20, 111–116. https://doi.org/10.1016/0306-4530(94)00066-j.

Maes, M., Song, C., Lin, A., De Jongh, R., Van Gastel, A., Kenis, G., Bosmans, E., De Meester, I., Benoy, I., Neels, H., Demedts, P., Janca, A., Scharpé, S., Smith, R.S., 1998. The effects of psychological stress on humans: increased production of pro-inflammatory cytokines and a TH1-like response in stress-induced anxiety. Cytokine 10, 313–318. https://doi.org/10.1006/cyto.1997.0290.

Marazziti, D., Mucci, F., Lombardi, A., Falaschi, V., Dell'Osso, L. 2015. The cytokine profile of OCD: pathophysiological insights. International Journal of Interferon, Cytokine and Mediator Research. https://dx.doi.org/10.2147/IJICMR.S76710.

Martinez-Trejo, G., Zamora, V., Rodarte, L.F., Herrera Haro, J.G., Figueroa, J., Galindo-Maldonado, F., Sanchez-Martinez, O., Bueno, A.L., 2009. Aggressiveness and productive performance of piglets supplemented with tryptophan. Journal of Animal and Veterinary Advances, 8, 608–611.

Mikami, O., Kubo, M., Murata, H., Muneta, Y., Nakajima, Y., Miyazaki, S., Tanimura, N., Katsuda, K. 2011. The effects of acute exposure to deoxynivalenol on some inflammatory parameters in miniature pigs. Journal of Veterinary Medical Science 73:665–71. https://doi.org/10.1292/jvms.10-0461.

Milrad, S.F., Hall, D.L., Jutagir, D.R., Lattie, E.G., Czaja, S.J., Perdomo, D.M., Fletcher, M.A., Klimas, N., Antoni, M.H., 2018. Depression, evening salivary cortisol and inflammation in chronic fatigue syndrome: A psychoneuroendocrinological structural regression model. International Journal of Psychophysiology, 131, 124–130. https://doi.org/10.1016/j.ijpsycho.2017.09.009.

Milstein, J.A., Lehmann, O., Theobald, D.E.H., Dalley, J.W., Robbins, T.W., 2007. Selective depletion of cortical noradrenaline by anti-dopamine beta-hydroxylase-saporin impairs attentional function and enhances the effects of guanfacine in the rat, Psychopharmacologia 190, 51–63. https://doi.org/10.1007/s00213-006-0594-x.

Moe, R.O., Nordgreen, J., Janczak, A.M., Spruijt, B.M., Kostal, L., Skjerve, E., Zanella, A.J., Bakken, M., 2011. Effects of haloperidol, a dopamine D2-like receptor antagonist, on reward-related behaviors in laying hens. Physiology & Behavior, 102, 400–405. https://doi.org/10.1016/j.physbeh.2010.12.008.

Moinard, C., Mendl, M., Nicol, C.J., Green, L.E., 2003. A case control study of on-farm risk factors for tail biting in pigs. Applied Animal Behaviour Science, 81, 333–355. https://doi.org/10.1016/S0168-1591(02)00276-9.

Momtazmanesh, S., Zareh-Shahabadi, A., Rezaei, N. 2019. Cytokine alterations in schizophrenia: an updated review. Frontiers in Psychiatry https://doi.org/10.3389/fpsyt.2019.00892.

Müller, N., Myint, A.M., Schwarz, M.J., 2009. The impact of neuroimmune dysregulation on neuroprotection and neurotoxicity in psychiatric disorders – relation to drug treatment. Dialogues in Clinical Neuroscience 11(3):319–332. https://doi.org/10.31887/DCNS.2009.11.3/nmueller.

Munhoz, C.D., Lepsch, L.B., Kawamoto, E.M., Malta, M.B., de Sá Lima, L., Werneck Avellar, M.C., Sapolsky, R.M., Scavone, C., 2006. Chronic unpredictable stress exacerbates lipopolysaccharide induced activation of nuclear factor-kappaB in the frontal cortex a d hippocampus via glucocorticoid secretion. Journal of Neuroscience, 26, 3813–3820. https://doi.org/10.1523/JNEUROSCI.4398-05.2006.

Munsterhjelm, C., Simola, O., Keeling, L., Valros, A., Heinonen, M. 2013. Health parameters in tail biters and bitten pigs in a case-control study, Animals 7, 814–821. https://doi.org/10.1017/S1751731112002194.

Munsterhjelm, C., Nordgreen, J., Aae, F., Heinonen, M., Olstad, K., Aasmundstad, T., Janczak, A.M., Valros, A., 2017. To be blamed or pitied? The effect of illness on social behavior, cytokine levels and feed intake in undocked boars. Physiology and Behavior 179, 298–307. https://doi.org/10.3389/fvets.2020.617634.

Munsterhjelm, C., Nordgreen, J., Aae, F., Heinonen, M., Valros, A., Janczak, A.M., 2019. Sick and grumpy: changes in social behavior after a controlled immune stimulation in group-housed gilts. Physiology and Behavior. 198, 76–83. https://doi.org/10.1016/j.physbeh.2018.09.018.

Myint, A.M., Kim, Y.K., 2003. Cytokine-serotonin interaction through IDO: a neurodegeneration hypothesis of depression. Medical Hypotheses, 61, 519–525. https://doi.org/10.1016/s0306-9877(03)00207-x.

Myint, A.M., Schwarz, M.J., Steinbusch, H.W.M., Leonard, B.E., 2009. Neuropsychiatric disorders related to interferon and interleukins treatment. Metabolic Brain Disease, 24(1), 55–68. https://doi.org/10.1007/s11011-008-9114-5.

Nalbandian, A., Sehgal, K., Gupta, A., Madhavan, M.V., McGroder, C., Steven, J.S., Cook, J.R., Nordvig, A.S., Shalev, D., Sehrawat, T.S., Ahluwalia, N., Bikdeli, B., Dietz, D., Der-Nigoghossian, C., Liyanage-Don, N., Rosner, G.F., Bernsein, E.J., Mohan, S., Beckley, A.A., Seres, D.S., Choueiri, T.K., Uriel, N., Ausiello, J.C., Accili, D., Freedberg, D.E., Baldwin, M., Schwartz, A., Brodie, D., Garcia, C.K., Elkind, M.S.V., Connors, J.M., Bilezikian, J.P., Landry, D.W., Want, E.Y., 2021. Post-acute COVID-19 syndrome. Nature Medicine 27, 601–615. https://doi.org/10.1038/s41591-021-01283-z.

Neher, J.J. and Cunningham, C., 2019. Priming microglia for innate immune memory in the brain. Trends in Immunology 40, 358–374. https://doi.org/10.1016/j.it.2019.02.001.

Nemeroff, C.B., Goldschmidt-Clermont, P.J., 2012. Heartache and heartbreak-the link between depression and cardiovascular disease. Nature Reviews Cardiology, 9, 526–539. https://doi.org/10.1038/nrcardio.2012.91.

Niemi, J.K., Sinisalo, A., Valros, A., Heinonen, M., 2011. The timing and treatment of tail biting in fattening pigs, Nordic Association of Agricultural Scientists. SLU, Uppsala, Sweden, p. 6.

Nordgreen, J., Munsterhjelm, C., Aae, F., Popova, A., Boysen, P., Ranheim, B., Heinonen, M., Raszplewicz, J., Piepponen, P., Lervik, A., Valros, A., Janczak, A.M. 2018. The effect of lipopolysaccharide (LPS) on inflammatory markers in blood and brain and on behavior in individually-housed pigs. Physiology and Behavior, 195, 98–111. https://doi.org/10.1016/j.physbeh.2018.07.013.

Nordgreen, J., Edwards, S., Boyle, L.A., Bolhuis, E., Veit, C.M., Sayyari, A., Marin, D., Dimitrov, I., Janczak, A.M., Valros, A., 2020. A proposed role for pro-inflammatory cytokines in damaging behaviour in pigs. Frontiers in Veterinary Science, 7. https://doi.org/10.3389/fvets.2020.00646.

O'Connor, J.C., Lawson, M.A., Andre, C., Moreau, M., Lestage, J., Castanon, N., Kelley, K.W., Dantzer, R., 2009. Lipopolysaccharide-induced depressive-like behavior is mediated by indoleamine 2,3-dioxygenase activation in mice. Molecular Psychiatry, 14(5), 511–522. https://doi.org/10.1038/sj.mp.4002148.

Pedrelli, P., Nyer, M., Holt, D., Bakow, B.R., Fava, M., Baer, L., Cassiello, C., Mulligan, M., Cusin, C., Farabaugh, A., 2013. Correlates of irritability in college students with depressive symptoms. Journal of Nervous and Mental Disease 201(11), 953–958. https://doi.org/10.1097/nmd.0000000000000038.

Poirier, C., Bateson, M., Gualtieri, F., Armstrong, E.A., Laws, G.C., Boswell, T., Smulders, T.V., 2019. Validation of hippocampal biomarkers of cumulative affective experience.

Neuroscience and Biobehavioral Reviews, 101, 113–121. https://doi.org/10.1016/j.neubiorev.2019.03.024.

Postal, M., Lapa, A.T., Sinicato, N.A., de Oliveira Peliçari, K., Peres, F.A., Costallat, L.T.L., Fernandes, P.T., Marini, R., Appenzeller, S., 2016. Depressive symptoms are associated with tumor necrosis factor alpha in systemic lupus erythematosus. Journal of Neuroinflammation, 2016, 13, 1–7. https://doi.org/10.1186/s12974-015-0471-9.

Raison, C.L., Borisov, A.S., Majer, M., Drake, D.F., Pagnoni, G., Woolwine, B.J., Vogt, G.J., Massung, B., Miller, A.H., 2009. Activation of central nervous system inflammatory pathways by interferon-alpha: relationship to monoamines and depression. Biological Psychiatry 65(4):296–303. https://doi.org/10.1016/j.biopsych.2008.08.010.

Remus, J.L., Dantzer, R., 2016. Inflammation models of depression in rodents: relevance to psychotropic drug discovery. International Journal of Neuropsychopharmacology, 19, 13. https://doi.org/10.1093/ijnp/pyw028.

Renault, P.F., Hoofnagle, J.H., Park, Y., Mullen, K.D., Peters, M., Jones, B., Rustgi, V., Jones, A., 1987. Psychiatric complications of long-term interferon alfa therapy. Archives of Internal Medicine 147, 1577–80. https://10.1001/archinte.1987.00370090055011.

Reu, P., Khosravi, A., Bernard, S., Mold, J.E., Salehpour, M., Alkass, K., Perl, S., Tisdale, J., Possnert, G., Druid, H., Frisén, J., 2017. The lifespan and turnover of microglia in the human brain. Cell Reports 20, 779–784. https://doi.org/10.1016/j.celrep.2017.07.004.

Rivest, S., Lacroix, S., Vallieres, L., Nadeau, S., Zhang, J., Laflamme, N., 2000. How the blood talks to the brain parenchyma and the paraventricular nucleus of the hypothalamus during systemic inflammatory and infectious stimuli. Proceedings of the Society for Experimental Biology and Medicine, 223(1), 22–38. https://doi.org/10.1046/j.1525-1373.2000.22304.x.

Roque, K., Shin, K.M., Jo, J.H., Lim, G.D., Song, E.S., Shin, S.J., Heo, Y. 2018. Association between endotoxin levels in dust from indoor swine housing environments and the immune responses of pigs. Journal of Veterinary Science, 19:331–8. https://doi.org/10.4142/jvs.2018.19.3.331.

Roth, W., Zedeh, K., Vekariya, R., Ge, Y., Mohamadzadeh, M. 2021. Tryptophan metabolism and gut-brain homeostasis. International Journal of Molecular Sciences, 22(6), 2973. https://doi.org/10.3390/ijms22062973.

Sahlander, K., Larsson, K., Palmberg, L. 2012. Daily exposure to dust alters innate immunity. PLoS ONE. 7:e31646. https://doi.org/10.1371/journal.pone.0031646.

Santarelli, L., Saxe, M., Gross, C., Surget, A., Battaglia, F., Dulawa, S., Weisstaub, N., Lee, J., Duman, R., Arancio, O., Belzung, C., Hen, R., 2003. Requirement of hippocampal neurogenesis for the behavioral effects of antidepressants. Science, 301(5634), 805–9. https://doi/org/10.1126/science.1083328.

Savitz, J., 2016. Role of kynurenine metabolism pathway activation in major depressive disorders. In: Dantzer, R., and Capuron, L. (eds). Inflammation-associated depression: evidence, mechanisms and implications. Current topics in behavioral neurosciences, Springer International Publishing, Switzerland. pp. 251–261. https://doi.org/10.1007/7854_2016_2.

Schildkraut, J.J., 1965. The catecholamine hypothesis of affective disorders-a review of supporting evidence. American Journal of Psychiatry, 122, 509–522. https://doi.org/10.1176/ajp.122.5.509.

Schwarcz, R., Bruno, J.P., Muchowski, P.J., Wu, H.Q., 2012. Kynurenines in the mammalian brain: when physiology meets pathology. Nature Reviews Neuroscience, 13:465–477. https://doi.org/10.1038/nrn325746.

Sekio, M., Seki, K. 2014. Lipopolysaccharide-induced depressive-like behavior is associated with α1-adrenoceptor dependent downregulation of the membrane GluR1 subunit in the mouse medial prefrontal cortex and ventral tegmental area. International Journal of Neuropsychopharmacology, 1–12. https://doi.org/10.1093/ijnp/pyu005.

Smith, R.S. 1991. The macrophage theory of depression. Medical Hypotheses, 35, 298–306. https://doi.org/10.1016/0306-9877(91)90272-z.

Takikawa, O., 2005. Biochemical and medical aspects of the indoleamine 2,3-dioxygenase-initiated L-tryptophan metabolism. Biochemical and Biophysical Research Communications 338, 12–19. https://doi.org/10.1016/j.bbrc.2005.09.032.

Taylor, N.R., Parker, R.M.A., Mendl, M., Edwards, S.A., Main, D.C.J., 2012. Prevalence of risk factors for tail biting on commercial farms and intervention strategies, Veterinary Journal, 194, 77–83. https://doi.org/10.1016/j.tvjl.2012.03.004.

Ting, K.K., Brew, B.J., Guillemin, G.J., 2009. Effect of quinolinic acid on human astrocytes morphology and functions: implications in Alzheimer's disease. Journal of Neuroinflammation 6:36. https://doi.org/10.1186/1742-2094-6-36.

Ursinus, W.W., Van Reenen, C.G., Reimert, I., Bolhuis, J.E., 2014. Tail biting in pigs: blood serotonin and fearfulness as pieces of the puzzle? Plos One 9(9). https://doi.org/10.1371/journal.pone.0107040.

Valros, A., Palander, P., Heinonen, M., Munsterhjelm, C., Brunberg, E., Keeling, L., Piepponen, P., 2015. Evidence for a link between tail biting and central monoamine metabolism in pigs (*Sus scrofa domestica*), Physiology & Behavior, 143, 151–157. https://doi.org/10.1016/j.physbeh.2015.02.049.

Van der Meer, Y., Lammers, A., Jansman, A.J., Rijnen, M.M., Hendricks, W.H., Gerrits, W.J. 2016. Performance of pigs kept under different sanitary conditions affected by protein intake and amino acid supplementaion. Journal of Animal Science 94, 4704–4719. https://doi.org/10.2527/jas.2016-0787.

Veit, C.M., Janczak, A.M., Ranheim, B., Vas, J., Valros, A., Sandercock, D.A., Piepponen, P., Dulgheriu, D., Nordgreen, J., 2021a. The effect of LPS and ketoprofen on cytokines, brain nomoamines, and social behavior in group-housed pigs. Frontiers in Veterinary Science, 7(617634). https://doi.org/10.3389/fvets.2020.617634.

Videbech, P., Ravnkilde, B., 2004. Hippocampal volume and depression: a meta-analysis of MRI studies. American Journal of Psychiatry, 161, 1957–1966. https://doi.org/10.1176/appi.ajp.161.11.1957.

Vogelgesang, S.A., Heyes, M.P., West, S.G., Salazar, A.M., Sfikakis, P.P., Lipnick, R.N., Klipple, G.L., Tsokos, G.C., 1996. Quinolinic acid in patients with systemic lupus erythematosus and neuropsychiatric manifestations. Journal of Rheumatology 23:850–855.

Walker, A.K., Budac, D.P., Bisulco, S., Lee, A.W., Smith, R.A., Beenders, B., Kelley, K.W., Dantzer R., 2013. NMDA receptor blockade by ketamine abrogates lipopolysaccharide-induced depressive-like behavior in C57BL/6J mice. Neuropsychopharmacology 38(9), 1609–1616. https://doi.org/10.1038/npp.2013.71.

Weiss, G., Murr, C., Zoller, H., Haun, M., Widner, B., Ludescher, C., Fuchs, D., 1999. Modulation of neopterin formation and tryptophan degradation by Th1- and Th2-derived cytokines in human monocytic cells. Clinical and Experimental Immunology 116(3), 435–440. https://doi.org/10.1046/j.1365-2249.1999.00910.x.

Wyns, H., Croubels, S., Vandekerchove, M., Demeyere, K., De Backer, P., Goddeeris, B.M., Mayer, E., 2015. Multiplex analysis of pro-inflammatory cytokines in serum of Actinobacillus pleuropneumoniae-infected pigs. Research in Veterinary Science 102:45–8. https://doi.org/10.1016/j.rvsc.2015.07.006.

Zdilar, D., Franco-Bronson, K., Buchler, N., Locala, J.A., Younossi, Z.M., 2000. Hepatitis C, interferon alpha, and depression. Hepatology, 31, 1207–1211. https://doi.org/10.1053/jhep.2000.7880.

Zhu, L.P., Wei, T.T., Gao, J., Chang, X.Y., He, H., Miao, M.X., Yan, T., 2015. Salidroside attenuates lipopolysaccharide (LPS) induced serum cytokines and depressive-like behavior in mice. Neuroscience Letters 606, 1–6. https://doi.org/10.1016/j.neulet.2015.08.025.

CHAPTER 6

The social environment

Inonge Reimert[1,] and Winanda W. Ursinus[2]*

[1]Adaptation Physiology Group, Department of Animal Sciences, Wageningen University & Research, P.O. Box 338, 6700 AH, Wageningen, The Netherlands; [2]Office for Risk Assessment & Research (BuRO), Netherlands Food and Consumer Product Safety Authority (NVWA), P.O. Box 43006, 3540 AA, Utrecht, The Netherlands; *Correspondence: inonge.reimert@wur.nl

Abstract

The social environment of domesticated pigs is clearly different from the natural social environment of their wild counterparts. This difference could frustrate pigs and, thereby, lead to damaging behaviours such as tail biting. However, much about this presumed relationship between tail biting in pigs and the social environment is unknown. In this chapter we present available knowledge about tail biting behaviour in its social context and identify knowledge gaps to be explored. We discuss practical procedures that change the social environment such as cross-fostering and weaning and we debate whether several individual characteristics such as sex and body size may influence tail biting behaviour by how they are distributed in the group as a whole. Based on the studies described, the social environment seems to affect the degree of tail biting and being bitten. Though, results of studies do generally not point into the same direction. For example, cross-fostering is suggested as a risk factor, but not consistently so. Similarly, more tail biting is suggested to occur in male only compared to female only groups, but also vice versa. This stresses that the social environment with respect to tail biting remains to be further elucidated. Hence, we propose that more knowledge is needed on the effect of a more natural distribution of different types of pigs in a group and more should be explored about the collective personality of a group of pigs with respect to undesired consequences such as tail biting behaviour and tail damage.

Keywords

collective personality – cross-fostering – group composition – group modification – litter size

1 Introduction

Pigs are highly social animals (Stolba and Wood-Gush, 1989) and therefore tail biting in pigs may as much be a social as it is an individual matter.

Wild or feral pigs live in groups of several sows with their latest offspring (Gabor *et al.*, 1999). Large groups can split in smaller subgroups and exchange of pigs between these subgroups is possible. Just before farrowing, a sow leaves her group to build a nest in which up to about seven piglets are born (D'Eath and Turner, 2009). The piglets stay close to the nest for the first two weeks, after which they reunite with the group (Jensen, 1986). Group life is based on a strong hierarchical structure (Puppe *et al.*, 2008) and piglets learn from both the sow and other group members how to socially engage with other pigs (D'Eath and Turner, 2009), both in an affiliative and competitive way (Miguel-Pacheco and Seddon, 2021; Telkänranta and Edwards, 2018). In addition, offspring receive care not only from their dam, but also from other group members (Fraser *et al.*, 1995). Weaning is a gradual process and takes places between 14 and 17 weeks of age (Jensen and Recén, 1989). Boars leave the group from about 16 months of age and only join a group to mate.

In contrast, a domesticated sow does 'leave' the group before farrowing, but she generally gives birth to 15–16 piglets (Van der Peet *et al.*, 2018) and they are born in a pen (4–5 m^2) that is not made of natural materials (Van Nieuwamerongen *et al.*, 2015). Moreover the sow and piglets do not join other sows and piglets after two weeks, but stay in the pen. Weaning is an abrupt process and it already takes place when piglets are about 3–4 weeks of age (Oostindjer *et al.*, 2014). The sow will be moved back to the service barn, while the piglets are usually moved to a nursery barn where they are housed with familiar and unfamiliar piglets of similar size and sometimes also of similar sex. A second move may happen at around 25 kg at which pigs are again mixed, and sometimes pigs are moved and mixed a third time (Garcia and McGlone, 2021). Group size is dependent on farm management and may go from groups of 10 up to 2000 animals (Turner and Edwards, 2004).

As the social environment of pigs is so different in practice compared to nature, this 'unnatural' social environment may be difficult to cope with at some period in life and hence could result in tail biting behaviour (Li *et al.*, 2018; Telkänranta and Edwards, 2018). However, much about this presumed relationship needs yet to be explored. In this chapter, therefore, we will present available knowledge about tail biting behaviour in its social context, identify knowledge gaps and end with practical tips for creating a good social environment.

2 The early social environment and modifications of the (early) social environment

2.1 *The early social environment*

The first social environment of pigs consists of pre- and postnatal litter mates and their dam. Prenatal litters may differ from postnatal litters in terms of size and genetics due to mortality, and mixing of unfamiliar piglets from different litters (see 2.2 'Cross-fostering'). Litter dynamics including tail biting behaviour may stem from both the prenatal and postnatal (social) environment.

Prenatal stress can impact social behaviour later in life, e.g. socially stressed sows may produce more anxious gilts by altered brain tissue, which eventually results in abnormal maternal care for the next generation (Rutherford *et al.*, 2014). Also litter size may be a prenatal stressor for piglets. The larger the litter the less space is available for each piglet and their placenta as there is limited uterine capacity (König *et al.*, 2021), meaning that feed competition may already start in the womb. Litter size of modern pig breeds increased over the recent years which partly explains overall lower birth weight and variable piglet sizes (see also '3.4 Body size'). In Ursinus *et al.* (2014a), we hypothesized that increasing production of pigs by increasing litter sizes and growth may interfere with the metabolic demand of pigs possibly affecting foraging behaviour and thereby also tail biting behaviour. In that study both genotypic and phenotypic litter size were indeed associated with increased damaging biting behaviours of rearing gilts (6–11 weeks of age). Thus, the prenatal environment may affect social behaviour later in life (Prunier *et al.*, 2020), but there is no evidence found in literature of social mediation in the prenatal litter by the piglets themselves. In human twin foetuses, however, movement directed to the co-twin has been found to occur as early as the 14th week of gestation (Castiello *et al.*, 2010). This could imply that intra-uterine competition, and possible other interactions, may affect offspring neural development, which could potentially predispose pigs to be more sensitive in later life to develop aberrant behaviour such as tail biting.

Postnatal social interactions occur in the early days of piglets between litter mates, and between piglets and their sow in order to regulate social development, thermoregulation and nutrition among others (Telkänranta and Edwards, 2018). Not surprisingly, social bonding is closest between litter mates and their dam (Newberry and Wood-Gush, 1986; Petersen *et al.*, 1989). The ability to interact socially may, however, be affected by the farrowing environment (Miguel-Pacheco and Seddon, 2021) such as lack of space, overcrowding and presence of obstacles like the frame of the sow's crate when conventionally housed. A large litter size affects nursing behaviours (Ocepek *et al.*, 2017) and may induce competition between piglets. This leads to social stress that may play a role in the development of tail biting behaviour later in life (Prunier *et al.*, 2020). Tail biting behaviour starts as early as the farrowing phase (Hakansson and Bolhuis, 2021; Ursinus *et al.*, 2014b), it was profoundly present in conventionally housed piglets of three weeks of age (Ursinus *et al.*, 2014b) and may temporarily persist post-weaning. Weaned piglets who performed tail biting had a higher chance to stem from a litter with relatively high levels of tail biting behaviours. However, this was not the case with tail damage, the receipt of tail biting (Ursinus *et al.*, 2014b). Also according to Canario and Flatres-Grall (2018), the litter is not a huge explaining factor with respect to tail damage variation. This stresses the importance of acknowledging the difference between giving and receiving tail biting behaviour at least with respect to litter effects. By mixing piglets of one litter into multiple groups after weaning, tail biting may be spread out through an entire pig facility. Note, other factors like the absence or presence of environmental enrichment may confound litter effects (Ursinus *et al.*, 2014a; 2014b).

All in all, it is not fully clear why pigs originating from larger litters may be prone to display more tail biting. It may be related to initial crowding and competition over resources in the litter, but may also be caused by a higher intrinsic motivation to explore and forage due to a high genetic production level which requests more energy input.

2.2 *Cross-fostering*

Cross-fostering is a strategy performed on commercial or research farms to improve the chance of survival for piglets born in a litter in which the number of live born piglets exceeds the number of functional teats or born in a heterogenous (e.g. in size, vigour and sex) litter (Baxter *et al.*, 2013). With cross-fostering, the group composition of a litter is (temporarily) changed as (some) piglets are removed from their litter and added to another litter. If cross-fostering is performed after colostrum uptake from the biological sow and within 48h after birth, (long-term) consequences for welfare and productivity of the whole litter seem to be minimal (Baxter *et al.*, 2013; Telkänranta and Edwards, 2018). It can, however, also be performed later in life (e.g. between one and two weeks of age) and then consequences may be more severe (reviewed in Baxter *et al.*, 2013; Telkänranta and Edwards, 2018). For here it is interesting whether changing the litter composition due to this strategy has effects on tail biting and/or tail damage. This may be expected as cross-fostering can lead to (acute) separation stress in the piglet(s) that is(are) removed as well as in (acute) social stress in all piglets as they likely need to fight to establish a new teat order (Telkänranta and Edwards, 2018), and stress has been proposed to underly tail biting behaviour (e.g. Taylor *et al.*, 2010; Valros, 2018).

To our knowledge, only few studies investigated whether cross-fostering has an effect on tail biting or tail damage, but the ones found sum up to being inconclusive on this subject. In the epidemiological study of Moinard *et al.* (2003), the incidence of tail biting after weaning was higher on farms where cross-fostering was practised compared to farms where it was hardly or not practised. However, it is not clear whether the higher incidence of tail biting was found in the cross-fostered or non-cross-fostered piglets, nor whether it was a direct result of cross-fostering or whether it was due to another factor such as a large litter size (Prunier *et al.*, 2020). In another epidemiological study by Calderón Díaz *et al.* (2017) cross-fostering – whether done once or more often before 24h after birth or later on in lactation – had no effect on tail lesions of docked pigs when scored at slaughter. Using the same data, but defining the timing of cross-fostering differently, Calderón Díaz *et al.* (2018) also did not find any effect of early (within a week after birth) and late cross-fostering (performed when piglets were between 2–3 weeks of age) on the presence of tail lesions between weaning and slaughter. The cross-fostered pigs were, however, found to be more at risk of dying or being euthanised and having severe tail lesions was one the reasons for euthanasia (Calderon Diaz *et al.*, 2018), which may explain why no effect of cross-fostering was found later in life. This suggests that cross-fostering may promote severe tail biting (Prunier *et al.*, 2020). In line with this, Hakansson and Houe (2020) showed that being cross-fostered before weaning is a risk factor for having tail damage between four and six weeks of age, but this risk disappeared for tail damage present later in life. Thus, more research is needed to get a better insight whether cross-fostering leads to more tail biting and hence damage.

2.3 *Pre-weaning socialisation or group housing*

Another procedure that can change the social environment is pre-weaning socialisation. This is a procedure that can be applied to reduce part of the weaning stress caused by mixing pigs at weaning (D'Eath, 2005; Prunier *et al.*, 2020). Hereto, usually litters from adjacent pens are mixed from about 1–2 weeks of age up to weaning by removing the partition between pens (D'Eath, 2005; Prunier *et al.*, 2020). Pre-weaning socialisation also occurs in group housing of sows and piglets during lactation as in group housing sows and piglets socialise naturally (reviewed in Van Nieuwamerongen *et al.*, 2014).

For pre-weaning socialisation, we could only find two studies that focused on the tail. The study of Klein *et al.* suggested that pre-weaning socialisation resulted in more pigs with intact tails at day 100 of finishing and in longer tails (Klein *et al.*, 2016 as cited in Prunier *et al.*, 2020). On the other hand, Van Kerschaver (2021) found no effect of pre-weaning socialisation on tail damage scored at day 42 after weaning. Similarly, both positive and negative as well as no effects of group housing during lactation have been found for tail biting and damage. For example, Van Nieuwamerongen *et al.* (2015) found that group housed piglets showed less manipulative behaviour, including tail biting, at day 15 and 26 of age and in the first five weeks after weaning compared to piglets raised in conventional farrowing pens. Tail damage the day before weaning and 33 days after weaning was, however, not different between both housing conditions. In contrast, more piglets raised in a group housing system were found to have lost a part of their tail due to tail biting after weaning compared to piglets raised in conventional farrowing pens (Lange *et al.*,

2021). But, also a lack of difference between group housed piglets and conventional reared piglets on tail damage or loss has been found (Gentz *et al.*, 2020; Lange *et al.*, 2021; Naya *et al.*, 2019).

At present, it is therefore not clear whether such a change of the social environment really affects tail biting and/or damage. It could be that early socialisation is more beneficial to reduce aggressive behaviour rather than oral manipulative behaviour after weaning (Gentz *et al.*, 2020; Ko *et al.*, 2020; Prunier *et al.*, 2020). Or it could be that effects are caused by other differences between housing conditions such as more space and/or enrichment in group housing as compared to conventional housing during lactation (Naya *et al.*, 2019; Van Nieuwamerongen *et al.*, 2015).

2.4 *Weaning*

Weaning is a procedure that affects the social environment as well as during this procedure, one change is that the sow and piglets are separated from each other. In practice, it is performed abruptly at maximum 3–4 weeks of age in conventional housing. As this is an age at which pigs are still developing behaviourally and physiologically (D'Eath and Turner, 2009; Miguel-Pacheco and Seddon, 2021; Newberry and Wood-Gush, 1988), being separated (suddenly) from their mother at such a young age could lead to tail biting behaviour due to the caused stress going along with it (Baxter *et al.*, 2013; Valros *et al.*, 2015; Weary *et al.*, 2008). Furthermore, it might also be true that piglets learn proper pig behaviour from their mother and without a good example they may develop aberrant behaviours such as tail biting (D'Eath and Turner, 2009; Telkänranta and Edwards, 2018). Indeed, in chickens, studies have shown that the absence of the mother hen resulted in more feather pecking behaviour (Riber *et al.*, 2007; Rodenburg, 2010; Rodenburg *et al.*, 2009). However, weaning as early as seven days of age or weaning later than 3–4 weeks of age does not seem to clearly increase or respectively decrease tail biting behaviour and/or tail damage after weaning (ESFA, 2007; Moinard *et al.*, 2003; Naya *et al.*, 2019; reviewed in Prunier *et al.*, 2020; Van Nieuwamerongen *et al.*, 2017).

Artificial rearing can be considered as weaning at an even younger age (Telkänranta and Edwards, 2018). This is a procedure carried out to increase the survival of surplus piglets in large litters and it may become more common in the future (Baxter *et al.*, 2013; Schmitt *et al.*, 2019). In arteficial rearing, piglets are raised without the sow from as early as three days old (Baxter *et al.*, 2013) or even earlier when the sow dies and no foster sow is available. In general, artificial rearing is found to be disadvantageous for piglet growth, behaviour and welfare (Baxter *et al.*, 2013; Schmitt *et al.*, 2019) and therefore it may support tail biting problems early or later on in life. To our knowledge, three studies have investigated tail biting by and tail damage of artificially reared piglets. The frequency and percentage of time of manipulation of pen mates, which included tail biting, was found to be higher between seven and 24 days of age in artificially reared (between three and six days of age) compared to sow reared piglets (Rzezniczek *et al.*, 2015). It should be said, though, that the artificial pen differed much from the sow pen which makes it difficult to conclude whether this difference was solely due to the absence of the sow or other factors (e.g. space allowance and quality and amount of rooting material) (Rzezniczek *et al.*, 2015). In a more recent study, piglets reared artificially from seven days of age also performed more tail manipulation and had more tail lesions at 12 and 19 days of age compared to piglets reared in conventional farrowing pens (Schmitt *et al.*, 2019). The cause of this difference is, however, not so clear. It could have been stress, the absence of maternal learning and care, or other differences between both rearing conditions (Schmitt *et al.*, 2019). Interestingly, Gentz (2020) studied long-term effects of artificial rearing on tail lesions. They found that fewer artificially reared pigs from two days of age had tail lesions in the period between four and 22 weeks of age and fewer artificial reared pigs had lost a part of their tail at the end of finishing compared to piglets reared in conventional farrowing pens. This result is in contrast with the other two studies and also not in line with the expectation that artificial rearing leads to more tail damage. It could be that the difference is caused by studying tail damage at different ages or due to other differences between both housing conditions in the latter study (e.g. at four weeks of age artificial reared pigs were not mixed, while the conventional reared pigs were mixed with other pigs in new pens). More research is therefore needed to be able to conclude whether and to what extent this change in the social environment caused by artificial rearing affects tail biting and tail damage in pigs of different ages.

Besides the sow and piglets being separated at weaning, piglets are also often mixed with other unfamiliar piglets at weaning, resulting in a different social environment as well (Weary *et al.*, 2008). This aspect of weaning likely induces stress in the pigs too, resulting in tail biting behaviour (Dybkjær, 1992; Sinisalo *et al.*, 2012; Valros *et al.*, 2015). It may be hypothesized that minimizing weaning stress by not mixing pigs results in less tail biting. Buijs and Muns (2019) concluded, however, that there is not much evidence in favour of this hypothesis. For instance, Veit *et al.* (2017) found no difference in tail damage nor in tail loss in the weeks following weaning between pigs that were and were not mixed at weaning. In addition, also in a recent study not mixing pigs at weaning did not result in less oral manipulation, including tail biting, in pigs of four to seven weeks of age (Camerlink *et al.*, 2021). On the other hand, both Gent *et al.* (2020) and Lange *et al.* (2021) found less tail damage during the rearing and fattening stage in pigs when they were not mixed at weaning. However, in both latter studies the non-mixed pigs also had more space, especially during rearing, than the mixed piglets. Therefore their results on tail lesions and losses may also be (partly) due to this difference (Gent *et al.*, 2020; Lange *et al.*, 2021). Overall, tail biting and damage seem not clearly reduced by keeping litter mates together upon weaning and more research is needed to find out which other social stressors or changes influence tail biting behaviour after weaning (Veit *et al.*, 2017).

3 The effect of different group compositions on tail biting

A group of pigs can consist of only females, for instance, while another group contains both females and males. A group of pigs can also consist out of pigs with a uniform body size (e.g. only runt pigs together in a pen) or out of pigs with various body sizes. Moreover, it could be that in one group there are more pigs with personality type A than B and in another group it is the other way around. Would such a difference have an influence on tail biting or would it not matter? We will explore this for several characteristics in this part of the chapter.

3.1 *Sex*

There are indications that there is a difference between individual male and female pigs in performing tail biting or becoming a victim. For instance, Wallgren and Lindahl (1996) and others (Hakansson and Houe, 2020; Valros, 2018) found that barrows were more often a victim of tail biting than gilts and the same has been found for boars (Calderón Díaz *et al.*, 2017; Harley *et al.*, 2014). In addition, female pigs have been (anecdotally) reported to perform more tail biting behaviour compared to male pigs (Li *et al.*, 2017; Valros, 2018). However there are also studies that did not find a difference between male and female pigs in tail biting or tail damage (e.g. Naya *et al.*, 2019; Sinisalo *et al.*, 2012; Ursinus *et al.*, 2014b).

When groups of different sex compositions were studied, Penny and colleagues found that boar only groups performed more tail biting and were hence also more bitten compared to gilt only groups (Penny *et al.*, 1981 as cited in Schrøder-Petersen and Simonsen, 2001). In another study, the probability of having tail damage at the slaughter line was found to be higher for pigs coming from single-sex groups compared to mixed-sex groups and this result was – not surprisingly – most clear in non-docked pigs (Hunter *et al.*, 2001). The study did not mention whether the male pigs were castrated or not nor whether both male types were included. Yet another result was found by Schrøder-Petersen *et al.* (2004). In this study, tail manipulation was less frequently scored in male only groups compared to female only and mixed-sex groups. Tail manipulation was similar in the mixed-sex and female only groups. The male pigs in this study were castrated. Kritas and Morrison (2004) found that the proportion of barrows to gilts in a pen did not influence the incidence or the severity of tail damage, but they did find a positive correlation between the incidence of barrows with tail damage and the percentage of gilts in the pen. The authors suggested that the latter may mean that gilts are more inclined to bite the tail of barrows than other gilts in mixed pens. However, in an earlier study, sex ratio did not affect the number of biters or bitten pigs in mixed-sex (barrows and gilts) pens (Blackshaw, 1981). Zonderland *et al.* (2010) studied the effect of sex on tail damage in both mixed- and single-sex pens as well. Male pigs were not castrated. At the beginning of the experiment, they found that tail damage became more severe quicker in female only pens compared to mixed or male only pens. However this difference had disappeared at the end of the experiment. In contrast, in yet another study tail damage was found to be more common in groups of entire males than in mixed-sex (boars and gilts and barrows and gilts) groups, but only during the beginning of the finishing period (Holinger *et al.*, 2015). Single female and single barrow groups were not included in this study. In an epidemiological study conducted on commercial farms, pigs from mixed-sex and single-sex groups did not differ in their levels of tail damage (Scollo *et al.*, 2016). In this study, males were castrates, but no further distinction was given whether the single-sex groups were only male, only female or both. Interestingly, however, within a subset of these studied farms, more tail damage occurred in single-sex pens than mixed-sex pens (Scollo *et al.*, 2017). Finally, Haigh *et al.* (2019) and Vermeer *et al.* (2017) did not find a difference in tail damage between boar only and gilt only groups.

Based on these studies, we cannot unequivocally conclude whether group composition with respect to sex is an underlying causal factor for tail biting and damage. By studying groups of pigs of various group sizes and different sex compositions (single-sex, and mixed-sex groups of various sex distributions) including castrated and non-castrated males, a final conclusion may be made.

3.2 *Body size*

Body size has been associated with dominance, mainly (though not exclusively so) the heavier the pig, the higher in rank (Mesarec *et al.*, 2021), and between more symmetrical/uniform pigs (similar in body weight) hierarchy may be less stable (Prevolnik Povše *et al.*, 2021). By means of a social network analysis it was shown that in pigs from 29 pens, 21.1% of pigs represented the dominant, 57.5% the subordinate, and 21.4% represented the isolated pigs (Agha *et al.*, 2020). Instability of social hierarchy may go hand in hand with behaviours such as tail biting (Ewbank, 1976).

Due to increased litter sizes, body size of piglets at birth may be more variable as birth weight variability is higher (König *et al.*, 2021). Previously it has been suggested that litter size may play a role in the development of tail biting and, therefore, it might be expected that litter birth weight is too, but this seems not to be the case except for the genetic potential for a higher litter birth weight and ear biting behaviour (Ursinus *et al.*, 2014a). Also birth weight (reviewed in Prunier *et al.*, 2020) and weaning weight (Van de Weerd *et al.*, 2005) of individual piglets seems not related to tail biting behaviours, although Munsterhjelm *et al.* (2016) did find a relationship between early life tail manipulation and a slightly lower birth weight, but not weaning weight. The lack of a consistent relationship between (litter) birth weight and tail biting may be explained by the findings that in later life both the smaller, lighter pigs or runts (e.g. Breuer *et al.*, 2005; Van de Weerd *et al.*, 2005; Zupan *et al.*, 2012) and the heavier, faster growing pigs (Ursinus *et al.*, 2014a) are prone to develop tail biting behaviour. Also in victim pigs, consistency lacks. They may start with a lower growth potential, but they may also be larger just before a tail biting outbreak (reviewed in Valros, 2018) and weight gain may slow down as a consequence of being bitten, eventually leading to a lack of difference in body size between biters and victims (Buijs and Muns, 2019). Interestingly, at least for a poorer pen growth rate an association was found with an increased risk for tail biting (Grümpel *et al.*, 2018) and also lower average daily gain at farm level was correlated with a higher percentage of tail lesions (Pandolfi *et al.*, 2018). Furthermore, Taylor *et al.* (2012) observed that an uneven distribution of body sizes in weaner pigs resulted in a higher risk for tail biting in weaners.

We hypothesize therefore that mainly groups with uniform body sizes either relatively light or heavy, and groups with a higher variability in body sizes may suffer from tail biting, instead of uniform groups with medium-weight pigs. Furthermore, these differences are likely to exist mainly in the early growing period and not in the first weeks of life and not anymore in older (finishing) pigs.

3.3 *Genetics*

Pigs can be housed with different indirect genetic effects (IGE) compositions in pens (please see chapter 4 for an explanation of indirect or social genetic effects). Camerlink *et al.* (2015) found that biting behaviour in general (biting in e.g. ears, tails and enrichment material) was lower between eight and 23 weeks of age in pens with pigs with only high or positive IGE for the growth of their pen mates compared to pens with pigs with only low or negative IGE for the growth of their pen mates. Tail biting behaviour itself was not different between both types of pig groups. One explanation could have been that for ethical reasons control measures (e.g. treatment of tails and provision of enrichment material) were taken to keep tail biting to a minimum which redirected biting on tails to biting on the enrichment provided, resulting in a low occurrence of tail biting in all groups and therefore a statistical difference could not be found (Camerlink *et al.*, 2015). Interestingly, pigs with a positive IGE for the growth of their pen mates were found to have less tail damage than pigs with a negative IGE for growth. This may suggest that the positive IGE pigs did tail bite less after all or bit more gently (Camerlink *et al.*, 2015). However, studies with different IGE compositions (e.g. positive IGE only, negative IGE only and mixed positive and negative IGE for growth pens) investigating tail biting have not been performed as far as we know. Therefore it remains to be investigated how the social environment with respect to IGE for growth or for other traits (such as tail damage: Canario and Flatres-Grall, 2018) exactly affects tail biting.

3.4 *Personality*

The personality of individual pigs has been studied in relation to tail biting, but literature is not consistent here as well. Pigs that perform (tail) biting behaviour are suggested to be either more fearful (Ursinus *et al.*, 2014c), less fearful (Zupan *et al.*, 2012), more proactive (Koolhaas *et al.*, 2007; Korte *et al.*, 2009), more reactive

(Bolhuis *et al.*, 2006; Luo *et al.*, 2020; Melotti *et al.*, 2011), or not (consistently) differing in their reaction to novelty (reviewed in Buijs and Muns, 2019; Chou *et al.*, 2017) or coping strategy (Ursinus, 2014).

Victim pigs have been suggested to be more reactive (Korte *et al.*, 2009), although this was found to be not consistently so throughout life (only above 16 weeks of age) (Ursinus, 2014). Type of housing (barren or enriched) was seen to affect the outcome, with seemingly the more challenging (barren) environment providing insight in differences in personality types (e.g. Luo *et al.*, 2019 with respect to coping strategy), possibly as the level of tail manipulation is too low in less challenging situations to be able to notify any differences in coping strategy (Prunier *et al.*, 2020).

Although personality traits seem at least to a certain degree consistent throughout life (correlation of 0.19–0.45 and repeatability of 0.25–0.39, Zebunke *et al.*, 2015) and are modestly heritable (estimates of 0.37–0.56, Iversen *et al.*, 2017; Velie *et al.*, 2009), nothing is 100% certain. Indeed, in Ursinus *et al.* (2014b) we show that tail biting behaviour in individual pigs is not consistent throughout life (there may be more consistency in being a victim (e.g. Ursinus *et al.*, 2014b; Hakansson and Houe, 2020), but it remains to be seen whether being a victim is a type of personality or whether the consistency has more to do with being attractive to other pigs), meaning that associations found with personality traits are probably bound to differ throughout life as well. Making it even more complicated is the existence of behavioural flexibility, as animals may shift to a more proactive coping strategy in challenging situations (Camerlink *et al.*, 2014). In addition, Zebunke *et al.* (2017) suggested that backtest behaviour in piglets is not necessarily a fixed strategy, and also Bengston and Jandt (2014) suggested that throughout development (especially adolescence) expression of behaviours may shift in animals, thereby also stressing that behaviour in juveniles should not be compared with behaviour in adults with respect to personality. If individual pig(let)s may change or vary in their behavioural strategies (e.g. learning by doing) it may be highly relevant to not only focus on personality of individual pigs, but also on personality variation in the group as a whole.

To the best of our knowledge, in pigs, studies lacked to consider the (natural) variation in personalities within a group in relation to damaging biting behaviour. Bengston and Jandt (2014) suggested the existence of a collective personality in a social group, which means that this is not merely the sum or average of individual personalities within that group. The formation of a group may reduce individual differences within the group e.g. by social facilitation (reviewed in Planas-Sitja *et al.*, 2018). Planas-Sitja *et al.* (2018) confirmed this in cockroaches by showing that phenotypes with low behavioural plasticity entrained the ones with a higher behavioural plasticity and thereby changed group performance. A collective personality may determine the success of a group and it may evolve over time (Bengston and Jandt, 2014). One can imagine that in pigs, selection pressure by humans and man-made groups of pigs affect collective personality and thereby affect the behaviour displayed by the group as a whole. Therefore, a better insight is warranted into the exact relationship between tail biting and individual personalities (Prunier *et al.*, 2020), but maybe even more interesting between damaging behaviours and a collective personality.

3.5 *Health status*

Activation of the immune system has been proposed to affect group dynamics with tail biting as consequence (reviewed in Nordgreen *et al.*, 2020). Sick animals may out of frustration or discomfort disturb other individuals by biting on their tails leading to altered resting and social behaviour of the group (Li *et al.*, 2017; Munsterhjelm *et al.*, 2017; Munsterhjelm *et al.*, 2019). In addition, tail biting by sick animals may also be expressed as a form redirected foraging behaviour as sick animals try to compensate for depletion of nutrients due to immune activation (Nordgreen *et al.*, 2020). Sick animals may also be the ones being bitten as they may spend more time resting and are therefore prone to become a victim (Munsterhjelm *et al.*, 2017; Munsterhjelm *et al.*, 2019; Nieme *et al.*, 2011). Thus, general activity will be altered when sick animals are present with generally more resting behaviour and less feeding behaviour, albeit dependent on the ratio of sick and healthy pigs in a group; the larger the group the less the relative effect one individual may have on the group as whole. This hypothesis may withstand if tail biting is as simple as that. However, the complexity with spreading of damaging behaviour through a group either by 'seeing is doing' (contagion/attraction to wounds) or by frustrating others and thereby creating an amenable group cannot by excluded. So, health issues in a group can affect group dynamics leading to more tail biters and/or victims, but the outcome will likely depend on the combination of other (social) factors present within the housing system (Munsterhjelm *et al.*, 2017; Nordgreen *et al.*, 2020) and/or on the type of health issues present and the phase of illness the pigs are in (Buijs and Muns, 2019). To conclude, the focus should not only be on sick individuals but also on the effect health issues have on the total group (Buijs and Muns, 2019; Veit *et al.*, 2021).

3.6 *Group activity, dynamics, and synchronisation*

Pigs are known to preferably synchronise their daily activities such as exploring, feeding and resting (Docking *et al.*, 2008; Merlot *et al.*, 2004; Schouten, 1986; Zwicker *et al.*, 2015). In a study of the behaviour of free-ranging Iberian pigs it was found that more than half of the daytime occupation consisted of resting behaviour (lying). More than a quarter of the daytime consisted of exploring and foraging, although this depends on the type of feed available as feeding on natural resources is more time consuming compared to feeding on commercial feed. The remainder of daytime consisted of behaviours such as walking and running, wallowing (only during warm periods), gazing, drinking, and social behaviours (slightly more negative than positive social behaviours) (Martínez-Macipe *et al.*, 2020). Pigs perform social nosing behaviours towards each other (mainly nose-to-nose contact rather than nosing ear, groin, tail or ano-genital regions) which is thought to function as affiliative behaviour, for recognition or to satisfy an intrinsic need to nose. But social nosing is largely unrelated to harmful behaviours (Camerlink and Turner, 2013). It is suggested that a shift in time budget will occur when less feed is available, where resting time will be replaced by time to explore and forage (Martínez-Macipe *et al.*, 2020). Here tail biting may be added to the story.

Tail biting behaviour is considered to be related to the motivation to explore and forage (EFSA, 2007; Wood-Gush and Vestergaard, 1989; Zonderland, 2010). A heightened motivation to explore and forage which cannot be fulfilled by the right means (i.e. sufficient exploration materials and feed) may lead to damaging biting behaviours including tail biting (Ursinus, 2014). Although tail biting may seem an individual matter as underlying states and traits of the individual are involved, it is as good a group matter as well. Also, a disbalance in synchronisation could lead to more aggression and/or more tail biting (Zwicker *et al.*, 2015). Where individual behaviours other than a low/tucked tail posture may not always be sufficient/consistent predictors of upcoming tail biting behaviour of individual pigs (reviewed in Edwards and Valros, 2021), a changed (type of) activity pattern (such as in explorative and feeding behaviours; reviewed in Larsen *et al.*, 2016; Valros, 2018) in a group of pigs may be a more sufficient forebode of a tail biting outbreak (Larsen *et al.*, 2019; Statham *et al.*, 2009; Ursinus *et al.*, 2014b; reviewed in Valros, 2018; Zonderland *et al.*, 2011). Though, the direction of changes (increased or decreased activity) differed in studies (reviewed in Buijs and Muns, 2019) and when using easy to measure parameters such as changes in body postures (Wedin *et al.*, 2018) or changes in optical flow as indicator of group movement (Larsen *et al.*, 2020), no such evidence for a predictive value of group activity was found. Therefore, also changes in total group activity may not be a reliable predictor for tail biting in pigs (reviewed in Buijs and Muns, 2019). Synchronisation of behaviours was not found to be different between tail biting and non-biting pens (Wedin *et al.*, 2018). Though, not being able to synchronise behaviours due to the housing or management system is a risk factor for tail biting as for instance the inability to feed together can induce tail biting behaviour (reviewed in Valros, 2018). Furthermore, the development of tail biting behaviour is largely explained by a lack of sufficient enrichment materials, but it remains a multifactorial welfare problem (e.g. Godyń *et al.*, 2019; Ursinus, 2014), i.e. several risk factors may lead to tail biting behaviour. It may, therefore, not be surprizing that social dynamics around tail biting is not as straightforward. Although experiments are well-designed to test certain factors, it is likely that the combination of risk factors at farms differs as it is not always known whether other factors may also have played a role in the development of tail biting. If so, behavioural patterns are likely to be different and results in individual and group activity in relation to tail biting to differ as well.

If pigs need to fight over resources frequently (not enough feeders or drinkers or feeder or drinker is not working properly for instance (Edwards and Valros, 2021)), this could lead to more sudden-forceful biters in a pen or just more tail biting as an outlet of frustration and thereby more and quicker damage on tails. For example, Smulders *et al.* (2008) found that pigs on farms where pens for weaners have a moderate number of pigs per feeding place show less tail biting at a later age, as compared to farms where weaned pigs need to compete more over feed. Possibly, social network analysis may help to learn more about the dynamics of tail biting behaviour (Canario *et al.*, 2020). Additionally if the analysis includes multi-partite networks it may also help connecting risk factors such as refused access to feed or enrichment materials, to tail biting in groups of pigs (Büttner *et al.*, 2015).

In sum, tail biting may follow changes in group activity, but the direction and type of changes are likely dependent on the (combination of) underlying risk factors. Social network analysis is suggested to be a useful tool to understand social dynamics, but at this point social dynamics among pigs and specifically with respect to (tail) biting behaviour of pigs is far from understood yet. Nonetheless, with the combination of camera vision and other sensor technology, social network analysis and algorithms, changes in group activity and dynamics can be tracked and potential relationships with tail biting behaviour better understood.

3.7 *Group size*

Apart from that groups of pigs can differ in the ratio of e.g. males and females, groups can also differ in size. And it could be that being housed in a group size that deviates from the natural group size results in social tension or stress and hence in tail biting (Rodenburg and Koene, 2007). Another explanation for a relation between group size and tail biting could be that a tail biter could do damage to more tails with more animals in the group and the chances of other pigs to start biting as well become larger in bigger groups (Buijs and Muns, 2019). As pen size usually does not change for a group of pigs in a certain stage of the production cycle, studies that investigated the effect of stocking density on tail biting have been taken into account here as well.

Unfortunately, some studies report a positive relationship between group size and tail biting, while others report a negative or no relationship (reviewed by: Buijs and Muns, 2019; Edwards and Valros, 2021; Valros, 2018). For example, group size was reviewed to have no effect on tail damage due to tail biting by D'Eath *et al.* (2014), while Palander (2016, as cited in Valros, 2018) found that the risk for tail biting increased in pigs when housed in groups of more than 10 animals. Perhaps the results are inconsistent, because there is also no fixed natural group size and thus pigs can generally cope quite well with various group sizes (at least likely up to the maximum natural group size which seems to be about 30 animals, i.e. sows with offspring (Gabor *et al.*, 2009; Podgórksi *et al.*, 2018)). It could, however, also be that relationships were caused by confounding factors such environmental enrichment (Edwards and Valros, 2021; Valros, 2018). So at present, the social environment with respect to group size does not clearly positively or negatively affect tail biting behaviour and tail damage.

4 More social aspects to consider

Aside from the previously mentioned aspects – such as litter size, body size, and sex – that constitute part of social dynamics and may relate to tail biting behaviour, aspects such as social contagion, tail communication and frequent regrouping may influence group dynamics and thereby tail biting as well, but these topics are still in their infancy in pig research.

It is known that tail biting behaviour including tail damage can spread very slowly (kind of simmering during the pre-damage stage of two-stage type biters, Taylor *et al.*, 2010), but also very quickly through a group of pigs (the epidemic-type; Valros, 2018). Explanations to date are that pigs are attracted to blood (Fraser, 1987), a wound at the tail (Moinard *et al.*, 2003), scent and behaviour of victim pigs, and increased activity in both biters and victims (Buijs and Muns, 2019) which are all very plausible. Chou *et al.* (2019) indeed stressed that once bleeding tails were present in finishing pigs, a tail biting outbreak was very difficult to control and stop. Also Naya *et al.* (2019) suggested that contagion in tail biting behaviour is likely to occur. However, what remains to be elucidated is whether pigs can otherwise socially contaminate each other with respect to tail biting; is it possible that the state or emotion a pig experiences is contagious for other pigs and thereby make way for other pigs to start tail biting? Reimert *et al.* (2013) suggested that pigs are sensitive to emotional contagion. In a study of Larsen *et al.* (2020), they seemed to observe social influence but then the opposite as we hypothesized and the question remains if it was emotionally regulated or by other factors; they found that pigs in a pen with a tail biting outbreak changed their behaviour before the outbreak, and that the neighbouring pen without a tail biting outbreak did the same, but other pens without tail biting did not.

The tail is used for intraspecific communication in several species. Pigs keep their tail in various combinations of postures and movements in different (emotional) situations (reviewed in Camerlink and Ursinus, 2020) and a low or tucked tail posture may indicate the presence of tail biting in a group of pigs (e.g. Ursinus *et al.*, 2014b, Zonderland *et al.*, 2009). It is, however, unknown whether the pig tail serves a similar communicative role as in other species, but it definitely deserves attention to its social functioning with regard to tail biting and tail docking. Docking and thereby a loss in certain postures may lead to hindered communication (reviewed in Buijs and Muns, 2019), and in turn to more tail biting. Or, that due to wounds on the tail, some tail postures are not possible anymore and thus hampers communication.

Management procedures such as repeated mixing and removing some animals out for slaughter may disrupt the social structure of the group and thereby may change dynamics in the group possibly affecting the occurrence of tail biting behaviour as well. Therefore, such practical implications that affect social dynamics should be considered more too.

5 Conclusion

It is clear that the social environment is associated with tail biting behaviour. However, when combining research findings cause and effect varies; i.e. the social environment can facilitate tail biting behaviour, but counter wise tail biting pigs have an effect on the social environment too. The discussed aspects of the social environment, whether they concern modifications to or the composition of the social group, point in different directions. More tail biting, for instance, is suggested to occur in male only compared to female only groups, but also the other way around. This seems to indicate that situations differ, and each farm is unique, leading to a different outcome as well. However, this does not detract from the fact that the social environment as a whole with respect to tail biting remains to be further elucidated and that the focus should not only be on individual animals. Considering more natural distributions of different types of pigs in a group may be beneficial for pig welfare. This implies that formation of groups of pigs should not merely be based on increasing or decreasing variation of pig characteristics such as uniformity, sex or personality.

6 Short summary of the main hands-on/practical application of the chapter subject

- The social environment can affect tail biting behaviour
- Research on the social environment is contradictory with respect to tail biting
- Large litters may lead to tail biting
- Modifications to the original group structure may trigger tail biting
- Uniformity in group composition may be key, but be aware that some forms of uniformity may be advantageous, while other forms may be detrimental for tail biting
- A natural distribution of pig types might be beneficial for group dynamics and thereby tail biting behaviour

References

Agha, S., Fàbrega, E., Quintanilla, R. and Sánchez, J.P., 2020. Social network analysis of agonistic behaviour and its association with economically important traits in pigs. Animals 10: 2123. https://doi.org/10.3390/ani10112123.

Baxter, E.M., Rutherford, K.M.D., D'eath, R.B., Arnott, G., Turner, S.P., Sandøe, P., Moustsen, V.A., Thorup, F., Edwards, S.A. and Lawrence, A.B., 2013. The welfare implications of large litter size in the domestic pig II: management factors. Animal Welfare 22: 219–238. https://doi.org/10.7120/09627286.22.2.219.

Bengston, S.E. and Jandt, J.M., 2014. The development of collective personality: the ontogenetic drivers of behavioral variation across groups. Frontiers in Ecology and Evolution 2: 81. https://doi.org/10.3389/fevo.2014.00081.

Blackshaw, J.K., 1981. Some behavioural deviations in weaned domestic pigs: persistent inguinal nose thrusting, and tail and ear biting. Animal Science 33: 325–332. https://doi.org/10.1017/S000335610003172X.

Bolhuis, J.E., Schouten, W.G., Schrama, J.W. and Wiegant, V.M., 2006. Effects of rearing and housing environment on behaviour and performance of pigs with different coping characteristics. Applied Animal Behaviour Science 101: 68–85. https://doi.org/10.1016/j.applanim.2006.01.001.

Breuer, K., Sutcliffe, M.E.M., Mercer, J.T., Rance, K.A., O'Connell, N.E., Sneddon, I.A. and Edwards, S.A., 2005. Heritability of clinical tail-biting and its relation to performance traits. Livestock Production Science 93: 87–94. https://doi.org/10.1016/j.livprodsci.2004.11.009.

Buijs, S. and Muns, R., 2019. A review of new knowledge on tail biting and tail docking. Update on the 2015 report 'Practical solutions to reduce tail biting in NI pig herds'. Agri-food & Biosciences Institute, Belfast, Northern Ireland, 63 pp.

Büttner, K., Scheffler, K., Czycholl, I. and Krieter, J., 2015. Network characteristics and development of social structure of agonistic behaviour in pigs across three repeated rehousing and mixing events. Applied Animal Behaviour Science 168: 24–30. https://doi.org/10.1016/j.applanim.2015.04.017.

Calderón Díaz, J.A., Boyle, L.A., Diana, A., Leonard, F.C., Moriarty, J.P., McElroy, M.C., McGettrick, S., Kelliher, D. and Manzanilla, E.G., 2017. Early life indicators predict mortality, illness, reduced welfare and carcass characteristics in finisher pigs. Preventive Veterinary Medicine 146: 94–102. https://doi.org/10.1016/j.prevetmed.2017.07.018.

Calderón Díaz, J.A., Garcia Manzanilla, E., Diana, A. and Boyle, L.A., 2018. Cross-fostering implications for pig mortality, welfare and performance. Frontiers in Veterinary Science 5: 123. https://doi.org/10.3389/fvets.2018.00123.

Camerlink, I., Proßegger, C., Kubala, D., Galunder, K. and Rault, J.L., 2021. Keeping littermates together instead of social mixing benefits pig social behaviour and growth post-weaning. Applied Animal Behaviour Science 235: 105230. https://doi.org/10.1016/j.applanim.2021.105230.

Camerlink, I. and Turner, S.P., 2013. The pig's nose and its role in dominance relationships and harmful behaviour. Applied Animal Behaviour Science 145: 84–91. https://doi.org/10.1016/j.applanim.2013.02.008.

Camerlink, I. and Ursinus, W.W., 2020. Tail postures and tail motion in pigs: a review. Applied Animal Behaviour Science 230: 105079. https://doi.org/10.1016/j.applanim.2020.105079.

Camerlink, I., Ursinus, W.W., Bijma, P., Kemp, B. and Bolhuis, J.E., 2015. Indirect genetic effects for growth rate in domestic pigs alter aggressive and manipulative biting behaviour. Behavior Genetics 45: 117–126. https://doi.org/10.1007/s10519-014-9671-9.

Camerlink, I., Ursinus, W.W. and Bolhuis, J.E., 2014. Struggling to survive: early life challenges in relation to the backtest in pigs. Journal of Animal Science 92: 3088–3095. https://doi.org/10.2527/jas.2013-7537.

Canario, L., Bijma, P., David, I., Camerlink, I., Martin, A., Rauw, W.M., Flatres-Grall, L., Van der Zande, L., Turner, S.P., Larzul, C. and Rydhmer, L., 2020. Prospects for the analysis and reduction of damaging behaviour in group-housed livestock, with application to pig breeding. Frontiers in Genetics 11: 611073. https://doi.org/10.3389/fgene.2020.611073.

Canario, L. and Flatres-Grall, L., 2018. Genetics of tail-biting receipt in gilts from the Tai Zumu line. In: 11th World Congress on Genetics Applied to Livestock Production. February 11–16, 2018. Auckland, New Zealand, p. 1130.

Castiello, U., Becchio, C., Zoia, S., Nelini, C., Sartori, L., Blason, L., D'Ottavio, G., Bulgheroni, M. and Gallese, V., 2010. Wired to be social: the ontogeny of human interaction. PloS ONE 5: e13199. https://doi.org/10.1371/journal.pone.0013199.

Chou, J.Y., D'Eath, R.B., Sandercock, D.A. and O'Driscoll, K., 2017. Investigation of response to the novel object test in pigs in relation to tail biting phenotypes. In: Bak Jensen, M., Herskin, M.S. and Malmkvist, J. (eds.) Proceedings of the 51st Congress of the International Society for Applied Ethology. August 7–10, 2017. Aarhus, Denmark, p. 141.

Chou, J.Y., O'Driscoll, K., D'Eath, R.B., Sandercock, D.A. and Camerlink, I., 2019. Multi-step tail biting outbreak intervention protocols for pigs housed on slatted floors. Animals 9: 582. https://doi.org/10.3390/ani9080582.

D'Eath, R.B., 2005. Socialising piglets before weaning improves social hierarchy formation when pigs are mixed post-weaning. Applied Animal Behaviour Science 93: 199–211. https://doi.org/10.1016/j.applanim.2004.11.019.

D'Eath, R.B., Arnott, G., Turner, S.P., Jensen, T., Lahrmann, H.P., Busch, M.E., Niemi, J.K., Lawrence, A.B. and Sandøe, P., 2014. Injurious tail biting in pigs: how can it be controlled in existing systems without tail docking? Animal 8: 1479–1497. https://doi.org/10.1017/S1751731114001359.

D'Eath, R.B. and Turner, S.P., 2009. The natural behaviour of the pig. In: Marchant-Forde, J.N. (ed.) The welfare of pigs. Springer Science + Business Media B.V., Dordrecht, the Netherlands, pp. 13–45. https://doi.org/10.1007/978-1-4020-8909-1_2.

Docking, C.M., Van de Weerd, H.A., Day, J.E.L. and Edwards, S.A., 2008. The influence of age on the use of potential enrichment objects and synchronisation of behaviour of pigs. Applied Animal Behaviour Science 110: 244–257. https://doi.org/10.1016/j.applanim.2007.05.004.

Dybkjaer, L., 1992. The identification of behavioural indicators of 'stress' in early weaned piglets. Applied Animal Behaviour Science 35: 135–147. https://doi.org/10.1016/0168-1591(92)90004-U.

Edwards, S. and Valros, A., 2021. Understanding and preventing tail biting in pigs. In: Edwards, S. (ed) Understanding the behaviour and improving the welfare of pigs. Burleigh Dodds Science Publishing Limited, Cambridge, UK, pp. 361–400. https://dx.doi.org/10.19103/AS.2020.0081.10.

European Food Safety Authority (EFSA), 2007. Scientific opinion of the Panel on Animal Health and Welfare on a request from commission on the risks associated with tail biting in pigs and possible means to reduce the need for tail docking considering the different housing and husbandry systems. The EFSA Journal 611: 1–13. https://doi.org/10.2903/j.efsa.2007.611.

Ewbank, R., 1976. Social hierarchy in suckling and fattening pigs: a review. Livestock Production Science 3: 363–372. https://doi.org/10.1016/0301-6226(76)90070-1.

Fraser, D., 1987. Attraction to blood as a factor in tail-biting by pigs. Applied Animal Behaviour Science 17: 61–68. https://doi.org/10.1016/0168-1591(87)90008-6.

Fraser, D., Kramer, D.L., Pajor, E.A. and Weary, D.M., 1995. Conflict and cooperation: sociobiological principles and the behaviour of pigs. Applied Animal Behaviour Science 44: 139–157. https://doi.org/10.1016/0168-1591(95)00610-5.

Gabor, T.M., Hellgren, E.C., Van Den Bussche, R.A. and Silvy, N.J., 1999. Demography, sociospatial behaviour and genetics of feral pigs (*Sus scrofa*) in a semi-arid environment. Journal of Zoology 247: 311–322. https://doi.org/10.1111/j.1469-7998.1999.tb00994.x.

Garcia, A. and McGlone, J.J., 2021. Optimizing pig welfare in the growing and finishing stage. In: Edwards, S. (ed) Understanding the behaviour and improving the welfare of pigs. Burleigh Dodds Science Publishing Limited, Cambridge, UK, pp. 209–246. http://dx.doi.org/10.19103/AS.2020.0081.06.

Gentz, M., 2020. Pig tail biting in different farrowing and rearing systems with a focus on tail lesions, tail losses and activity monitoring. PhD Thesis, University of Göttingen, Göttingen, Germany, 112 pp.

Gentz, M., Lange, A., Zeidler, S., Lambertz, C., Gauly, M., Burfeind, O. and Traulsen, I., 2020. Tail lesions and losses of docked and undocked pigs in different farrowing and rearing systems. Agriculture 10: 130. https://doi.org/10.3390/agriculture10040130.

Godyń, D., Nowicki, J. and Herbut, P., 2019. Effects of environmental enrichment on pig welfare – a review. Animals 9: 383. https://doi.org/10.3390/ani9060383.

Grümpel, A., Krieter, J., Veit, C. and Dippel, S., 2018. Factors influencing the risk for tail lesions in weaner pigs (*Sus scrofa*). Livestock Science 216: 219–226. https://doi.org/10.1016/j.livsci.2018.09.001.

Haigh, A., Yun-Chou, J. and O'Driscoll, K., 2019. An investigation into the effectiveness of compressed straw blocks in reducing abnormal behaviour in growing pigs. Animal 13: 2576–2585. https://doi.org/10.1017/S1751731119000715.

Hakansson, F. and Bolhuis, J.E., 2021. Tail-biting behaviour pre-weaning: association between other pig-directed and general behaviour in piglets. Applied Animal Behaviour Science 241: 105385. https://doi.org/10.1016/j.applanim.2021.105385.

Hakansson, F. and Houe, H., 2020. Risk factors associated with tail damage in conventional non-docked pigs throughout the lactation and rearing period. Preventive Veterinary Medicine 184: 105160. https://doi.org/10.1016/j.prevetmed.2020.105160.

Harley, S., Boyle, L.A., O'Connell, N.E., More, S.J., Teixeira, D.L. and Hanlon, A., 2014. Docking the value of pigmeat? Prevalence and financial implications of welfare lesions in Irish slaughter pigs. Animal Welfare 23: 275–285. https://doi.org/10.7120/09627286.23.3.275.

Holinger, M., Früh, B. and Hillmann, E., 2015. Group composition for fattening entire male pigs under enriched housing conditions – influences on behaviour, injuries and boar taint compounds. Applied Animal Behaviour Science 165: 47–56. https://doi.org/10.1016/j.applanim.2015.01.016.

Hunter, E.J., Jones, T.A., Guise, H.J., Penny, R.H.C. and Hoste, S., 2001. The relationship between tail biting in pigs, docking procedure and other management practices. The Veterinary Journal 161: 72–79. https://doi.org/10.1053/tvjl.2000.0520.

Iversen, M.W., Bolhuis, J.E., Camerlink, I., Ursinus, W.W., Reimert, I. and Duijvesteijn, N., 2017. Heritability of the backtest response in piglets and its genetic correlations with production traits. Animal 11: 556–563. https://doi.org/10.1017/S1751731116001853.

Jensen, P., 1986. Observations on the maternal behaviour of free-ranging domestic pigs. Applied Animal Behaviour Science 16: 131–142. https://doi.org/10.1016/0168-1591(86)90105-X.

Jensen, P. and Recén, B., 1989. When to wean – observations from free-ranging domestic pigs. Applied Animal Behaviour Science 23: 49–60. https://doi.org/10.1016/0168-1591(89)90006-3.

Ko, H.L., Chong, Q., Escribano, D., Camerlink, I., Manteca, X. and Llonch, P., 2020. Pre-weaning socialization and environmental enrichment affect life-long response to regrouping in commercially-reared pigs. Applied Animal Behaviour Science 229 105044. https://doi.org/10.1016/j.applanim.2020.105044.

König, N.L., Wähner, M., Seeger, J., Sigmarsson, H.L. and Kauffold, J., 2021. An investigation into uterine capacity based on litter and placental characteristics in two sow lines with different prolificacy (Danish Landrace × Danish Yorkshire versus German Saddleback). Reproduction in Domestic Animals 56: 34–45. https://doi.org/10.1111/rda.13847.

Koolhaas, J.M., De Boer, S.F., Buwalda, B. and Van Reenen, K., 2007. Individual variation in coping with stress: a multidimensional approach of ultimate and proximate mechanisms. Brain, Behavior and Evolution 70: 218–226. https://doi.org/10.1159/000105485.

Korte, S.M., Prins, J., Vinkers, C.H. and Olivier, B., 2009. On the origin of allostasis and stress-induced pathology in farm animals: Celebrating Darwin's legacy. The Veterinary Journal 182: 378–383. https://doi.org/10.1016/j.tvjl.2009.08.023.

Kritas, S.K. and Morrison, R.B., 2004. An observational study on tail biting in commercial grower-finisher barns. Journal of Swine Health and Production 12: 17–22.

Lange, A., Hahne, M., Lambertz, C., Gauly, M., Wendt, M., Janssen, H. and Traulsen, I., 2021. Effects of different housing systems during suckling and rearing period on skin and tail lesions, tail losses and performance of growing and finishing pigs. Animals 11: 2184. https://doi.org/10.3390/ani11082184.

Larsen, M.L.V., Andersen, H.M.L. and Pedersen, L.J., 2016. Can tail damage outbreaks in the pig be predicted by behavioural change? The Veterinary Journal 209: 50–56. https://doi.org/10.1016/j.tvjl.2015.12.001.

Larsen, M.L.V., Andersen, H.L. and Pedersen, L.J., 2019. Changes in activity and object manipulation before tail damage in finisher pigs as an early detector of tail biting. Animal 13: 1037–1044. https://doi.org/10.1017/S1751731118002689.

Larsen, M.L.V., Pedersen, L.J., Edwards, S., Albanie, S. and Dawkins, M.S., 2020. Movement change detected by optical flow precedes, but does not predict, tail-biting in pigs. Livestock Science 240: 104136. https://doi.org/10.1016/j.livsci.2020.104136.

Li, Y., Zhang, H., Johnston, L.J. and Martin, W., 2018. Understanding tail-biting in pigs through social network analysis. Animals 8: 13. https://doi.org/10.3390/ani8010013.

Li, Y.Z., Zhang, H.F., Johnston, L.J., Martin, W., Peterson, J.D. and Coetzee, J.F., 2017. Effects of tail docking and tail biting on performance and welfare of growing–finishing pigs in a confinement housing system. Journal of Animal Science 95: 4835–4845. https://doi.org/10.2527/jas2017.1571.

Luo, L., Reimert, I., De Haas, E.N., Kemp, B. and Bolhuis, J.E., 2019. Effects of early and later life environmental enrichment and personality on attention bias in pigs (*Sus scrofa domesticus*). Animal Cognition 22: 959–972. https://doi.org/10.1007/s10071-019-01287-w.

Luo, L., Reimert, I., Middelkoop, A., Kemp, B. and Bolhuis, J.E., 2020. Effects of early and current environmental enrichment on behavior and growth in pigs. Frontiers in Veterinary Science 7: 268. https://doi.org/10.3389/fvets.2020.00268.

Martínez-Macipe, M., Mainau, E., Manteca, X. and Dalmau, A., 2020. Environmental and management factors affecting the time budgets of free-ranging Iberian pigs reared in Spain. Animals 10: 798. https://doi.org/10.3390/ani10050798.

Melotti, L., Oostindjer, M., Bolhuis, J.E., Held, S. and Mendl, M., 2011. Coping personality type and environmental enrichment affect aggression at weaning in pigs. Applied Animal Behaviour Science 133: 144–153. https://doi.org/10.1016/j.applanim.2011.05.018.

Merlot, E., Meunier-Salaün, M.C. and Prunier, A., 2004. Behavioural, endocrine and immune consequences of mixing in weaned piglets. Applied Animal Behaviour Science 85: 247–257. https://doi.org/10.1016/j.applanim.2003.11.002.

Mesarec, N., Skok, J., Škorjanc, D. and Povše, M.P., 2021. Group dynamics in a spontaneously established group of newly weaned piglets. Applied Animal Behaviour Science 238: 105317. https://doi.org/10.1016/j.applanim.2021.105317.

Miguel-Pacheco, G. and Seddon, Y.M., 2021. Developmental influences on pig behaviour. In: Edwards, S. (ed) Understanding the behaviour and improving the welfare of pigs. Burleigh Dodds Science Publishing Limited, Cambridge, UK, pp. 27–83. http://dx.doi.org/10.19103/AS.2020.0081.02.

Moinard, C., Mendl, M., Nicol, C.J. and Green, L.E., 2003. A case control study of on-farm risk factors for tail biting in pigs. Applied Animal Behaviour Science 81: 333–355. https://doi:10.1016/S0168-1591(02)00276-9.

Munsterhjelm, C., Heinonen, M. and Valros, A., 2016. Can tail-in-mouth behaviour in weaned piglets be predicted by behaviour and performance? Applied Animal Behaviour Science 184: 16–24. https://doi.org/10.1016/j.applanim.2016.08.013.

Munsterhjelm, C., Nordgreen, J., Aae, F., Heinonen, M., Olstad, K., Aasmundstad, T., Janczak, A.M. and Valros, A., 2017. To be blamed or pitied? The effect of illness on social behavior, cytokine levels and feed intake in undocked boars. Physiology & Behavior 179: 298–307. http://dx.doi.org/10.1016/j.physbeh.2017.06.024.

Munsterhjelm, C., Nordgreen, J., Aae, F., Heinonen, M., Valros, A. and Janczak, A.M., 2019. Sick and grumpy: changes in social behaviour after a controlled immune stimulation in group-housed gilts. Physiology & Behavior 198: 76–83. https://doi.org/10.1016/j.physbeh.2018.09.018.

Naya, A., Traulsen, I., Gertz, M., Hasler, M., Burfeind, O., große Beilage, E. and Krieter, J., 2019. Is tail biting in growing pigs reduced by a prolonged suckling period? Applied Animal Behaviour Science 211: 41–46. https://doi.org/10.1016/j.applanim.2018.10.020.

Newberry, R.C. and Wood-Gush, D.G.M., 1986. Social relationships of piglets in a semi-natural environment. Animal Behaviour 34: 1311–1318. https://doi.org/10.1016/S0003-3472(86)80202-0.

Newberry, R.C. and Wood-Gush, D.G.M., 1988. Development of some behaviour patterns in piglets under semi-natural conditions. Animal Science 46: 103–109. https://doi.org/10.1017/S0003356100003160.

Niemi, J.K., Sinisalo, A., Valros, A. and Heinonen, M., 2011. The timing and treatment of tail biting in fattening pigs. In: Hultgren, J., Persson, P., Nadeau, E. and Fogelberg, F. (eds.) Proceedings of the 24th Congress of the Nordic Association of Agricultural Scientists. June 14–16, 2011. Uppsala, Sweden, p. 50.

Nordgreen, J., Edwards, S.A., Boyle, L.A., Bolhuis, J.E., Veit, C., Sayyari, A., Marin, D.E., Dimitrov, I., Janczak, A.M. and Valros, A., 2020. A proposed role for pro-inflammatory cytokines in damaging behavior in pigs. Frontiers in Veterinary Science 7: 646. https://doi.org/10.3389/fvets.2020.00646.

Ocepek, M., Newberry, R.C. and Andersen, I.L., 2017. Trade-offs between litter size and offspring fitness in domestic pigs subjected to different genetic selection pressures. Applied Animal Behaviour Science 193: 7–14. https://doi.org/10.1016/j.applanim.2017.03.008.

Oostindjer, M., Kemp, B., Van den Brand, H. and Bolhuis, J.E., 2014. Facilitating 'learning from mom how to eat like a pig' to improve welfare of piglets around weaning. Applied Animal Behaviour Science 160: 19–30. https://doi.org/10.1016/j.applanim.2014.09.006.

Pandolfi, F., Edwards, S.A., Maes, D. and Kyriazakis, I., 2018. Connecting different data sources to assess the interconnections between biosecurity, health, welfare, and performance in commercial pig farms in Great Britain. Frontiers in Veterinary Science 5: 41. https://doi.org/10.3389/fvets.2018.00041.

Petersen, H.V., Vestergaard, K. and Jensen, P., 1989. Integration of piglets into social groups of free-ranging domestic pigs. Applied Animal Behaviour Science 23: 223–236. https://doi.org/10.1016/0168-1591(89)90113-5.

Planas-Sitjà, I., Nicolis, S.C., Sempo, G. and Deneubourg, J.L., 2018. The interplay between personalities and social interactions affects the cohesion of the group and the speed of aggregation. PloS ONE 13: e0201053. https://doi.org/10.1371/journal.pone.0201053.

Podgórski, T., Apollonio, M. and Keuling, O., 2018. Contact rates in wild boar populations: Implications for disease transmission. The Journal of Wildlife Management 82: 1210–1218. https://doi.org/10.1002/jwmg.21480.

Prevolnik Povše, M., Mesarec, N., Skok, J. and Škorjanc, D., 2021. Agonistic interactions between littermates reappear after mixing multiple litters at weaning in pigs. Agriculture 11: 844. https://doi.org/10.3390/agriculture11090844.

Prunier, A., Averos, X., Dimitrov, I., Edwards, S.A., Hillmann, E., Holinger, M., Ilieski, V., Leming, R., Tallet, C., Turner, S.P., Zupan, M. and Camerlink, I., 2020. Review: early life predisposing factors for biting in pigs. Animal 14: 570–587. https://doi.org/10.1017/S1751731119001940.

Puppe, B., Langbein, J., Bauer, J. and Hoy, S., 2008. A comparative view on social hierarchy formation at different stages of pig production using sociometric measures. Livestock Science 113: 155–162. https://doi.org/10.1016/j.livsci.2007.03.004.

Reimert, I., Bolhuis, J.E., Kemp, B. and Rodenburg, T.B., 2013. Indicators of positive and negative emotions and emotional contagion in pigs. Physiology & Behavior 109: 42–50. https://doi.org/10.1016/j.physbeh.2012.11.002.

Riber, A.B., Wichman, A., Braastad, B.O. and Forkman, B., 2007. Effects of broody hens on perch use, ground pecking, feather pecking and cannibalism in domestic fowl (*Gallus gallus domesticus*). Applied Animal Behaviour Science 106: 39–51. https://doi.org/10.1016/j.applanim.2006.07.012.

Rodenburg, T.B., 2010. Maternal care reduces cannibalistic toe pecking in domestic laying hens. In: Lidfors, L., Blokhuis, H. and Keeling, L. (eds.) Proceedings of the 44th Congress of the International Society for Applied Ethology. August 4–7, 2010. Uppsala, Sweden, p. 8.

Rodenburg, T.B. and Koene, P., 2007. The impact of group size on damaging behaviours, aggression, fear and stress in farm animals. Applied Animal Behaviour Science 103: 205–214. https://doi:10.1016/j.applanim.2006.05.024.

Rodenburg, T.B., Uitdehaag, K.A., Ellen, E.D. and Komen, J., 2009. The effects of selection on low mortality and brooding by a mother hen on open-field response, feather pecking and cannibalism in laying hens. Animal Welfare 18: 427–432.

Rutherford, K.M.D., Piastowska-Ciesielska, A., Donald, R.D., Robson, S.K., Ison, S.H., Jarvis, S., Brunton, P.J., Russel, J.A. and Lawrence, A.B., 2014. Prenatal stress produces anxiety prone female offspring and impaired maternal behaviour in the domestic pig. Physiology & Behavior 129: 255–264. https://doi.org/10.1016/j.physbeh.2014.02.052.

Rzezniczek, M., Gygax, L., Wechsler, B. and Weber, R., 2015. Comparison of the behaviour of piglets raised in an artificial rearing system or reared by the sow. Applied Animal Behaviour Science 165: 57–65. https://doi.org/10.1016/j.applanim.2015.01.009.

Schmitt, O., O'Driscoll, K., Boyle, L.A. and Baxter, E.M., 2019. Artificial rearing affects piglets pre-weaning behaviour, welfare and growth performance. Applied Animal Behaviour Science 210: 16–25. https://doi.org/10.1016/j.applanim.2018.10.018.

Schouten, W.G.P., 1986. Rearing conditions and behaviour in pigs. PhD Thesis, Landbouwhogeschool Wageningen, Wageningen, the Netherlands, 151 pp.

Schrøder-Petersen, D.L., Heiskanen, T. and Ersbøll, A.K., 2004. Tail-in-mouth behaviour in slaughter pigs, in relation to internal factors such as: age, size, gender, and motivational background. Acta Agriculturae Scandinavica, Section A-Animal Science 54: 159–166. https://doi.org/10.1080/09064700410003835.

Schrøder-Petersen, D.L. and Simonsen, H.B., 2001. Tail biting in pigs. The Veterinary Journal 162: 196–210. https://doi.org/10.1053/tvjl.2001.0605.

Scollo, A., Contiero, B. and Gottardo, F., 2016. Frequency of tail lesions and risk factors for tail biting in heavy pig production from weaning to 170 kg live weight. The Veterinary Journal 207: 92–98. https://doi.org/10.1016/j.tvjl.2015.10.056.

Scollo, A., Gottardo, F., Contiero, B. and Edwards, S.A., 2017. A cross-sectional study for predicting tail biting risk in pig farms using classification and regression tree analysis. Preventive Veterinary Medicine 146: 114–120. https://doi.org/10.1016/j.prevetmed.2017.08.001.

Sinisalo, A., Niemi, J.K., Heinonen, M. and Valros, A., 2012. Tail biting and production performance in fattening pigs. Livestock Science 143: 220–225. https://doi:10.1016/j.livsci.2011.09.019.

Smulders, D., Hautekiet, V., Verbeke, G. and Geers, R., 2008. Tail and ear biting lesions in pigs: an epidemiological study. Animal Welfare 17: 61–69.

Statham, P., Green, L., Bichard, M. and Mendl, M., 2009. Predicting tail-biting from behaviour of pigs prior to outbreaks. Applied Animal Behaviour Science 121: 157–164. https://doi.org/10.1016/j.applanim.2009.09.011.

Stolba, A. and Wood-Gush, D.G.M., 1989. The behaviour of pigs in a semi-natural environment. Animal Science 48: 419–425.

Taylor, N.R., Main, D.C., Mendl, M. and Edwards, S.A., 2010. Tail-biting: a new perspective. The Veterinary Journal 186: 137–147. https://doi.org/10.1016/j.tvjl.2009.08.028.

Taylor, N.R., Parker, R.M., Mendl, M., Edwards, S.A. and Main, D.C., 2012. Prevalence of risk factors for tail biting on commercial farms and intervention strategies. The Veterinary Journal 194: 77–83. https://doi.org/10.1016/j.tvjl.2012.03.004.

Telkänranta, H. and Edwards, S.A., 2018. Lifetime consequences of the early physical and social environment of piglets. In: Špinka, M. (ed.) Advances in pig welfare. Woodhead Publishing, Duxford, UK, pp. 101–136. https://doi.org/10.1016/B978-0-08-101012-9.00013-7.

Turner, S.P. and Edwards, S.A., 2004. Housing immature domestic pigs in large social groups: implications for social organisation in a hierarchical society. Applied Animal Behaviour Science 87: 239–253. https://doi.org/10.1016/j.applanim.2004.01.010.

Ursinus, W.W., 2014. A tale too long for a tail too short? Identification of characteristics in pigs related to tail biting and other oral manipulations directed at conspecifics. PhD Thesis, Wageningen University & Research, Wageningen, the Netherlands, 248 pp.

Ursinus, W.W., Van Reenen, C.G., Kemp, B. and Bolhuis, J.E., 2014b. Tail biting behaviour and tail damage in pigs and the relationship with general behaviour: predicting the inevitable? Applied Animal Behaviour Science 156: 22–36. https://doi.org/10.1016/j.applanim.2014.04.001.

Ursinus, W.W., Van Reenen, C.G., Reimert, I. and Bolhuis, J.E., 2014c. Tail biting in pigs: blood serotonin and fearfulness as pieces of the puzzle? PloS ONE 9: e107040. https://doi.org/10.1371/journal.pone.0107040.

Ursinus, W.W., Wijnen, H.J., Bartels, A.C., Duijvesteijn, N., Van Reenen, C.G. and Bolhuis, J.E., 2014a. Damaging biting behaviors in intensively kept rearing gilts: The effect of jute sacks and relations with production characteristics. Journal of Animal Science 92: 5193–5202. https://doi.org/10.2527/jas2014-7918.

Valros, A., 2018. Tail biting. In: Špinka, M. (ed.) Advances in pig welfare. Woodhead Publishing, Duxford, UK, pp. 137–166. https://doi.org/10.1016/B978-0-08-101012-9.00004-6.

Valros, A., Palander, P., Heinonen, M., Munsterhjelm, C., Brunberg, E., Keeling, L. and Piepponen, P., 2015. Evidence for a link between tail biting and central monoamine metabolism in pigs (*Sus scrofa domestica*). Physiology & Behavior 143: 151–157. https://dx.doi.org/10.1016/j.physbeh.2015.02.049.

Van de Weerd, H.A., Docking, C.M., Day, J.E.L. and Edwards, S.A., 2005. The development of harmful social behaviour in pigs with intact tails and different enrichment backgrounds in two housing systems. Animal Science 80: 289–298. https://doi.org/10.1079/ASC40450289.

Van der Peet, G., Leenstra, F., Vermeij, I., Bondt, N., Puister, L. and van Os, J., 2018. Feiten en cijfers over de Nederlandse veehouderijsectoren 2018 (No. 1134). Wageningen Livestock Research, Wageningen, the Netherlands, 109 pp. https://doi.org/10.18174/464128.

Van Kerschaver, C., 2021. Co-mingling of non-littermate suckling piglets to easy the weaning transition. PhD Thesis, Ghent University, Ghent, Belgium, 155 pp.

Van Nieuwamerongen, S.E., Bolhuis, J.E., Van der Peet-Schwering, C.M.C. and Soede, N.M., 2014. A review of sow and piglet behaviour and performance in group housing systems for lactating sows. Animal 8: 448–460. https://doi:10.1017/S1751731113002280.

Van Nieuwamerongen, S.E., Soede, N.M., Van der Peet-Schwering, C.M.C., Kemp, B. and Bolhuis, J.E., 2015. Development of piglets raised in a new multi-litter housing system vs. conventional single-litter housing until 9 weeks of age. Journal of Animal Science 93: 5442–5454. https://doi.org/10.2527/jas2015-9460.

Van Nieuwamerongen, S.E., Soede, N.M., Van der Peet-Schwering, C.M.C., Kemp, B. and Bolhuis, J.E., 2017. Gradual weaning during an extended lactation period improves performance and behavior of pigs raised in a multi-suckling system. Applied Animal Behaviour Science 194: 24–35. http://dx.doi.org/10.1016/j.applanim.2017.05.005.

Veit, C., Büttner, K., Traulsen, I., Gertz, M., Hasler, M., Burfeind, O., große Beilage, E. and Krieter, J., 2017. The effect of mixing piglets after weaning on the occurrence of tail-biting during rearing. Livestock Science 201: 70–73. http://dx.doi.org/10.1016/j.livsci.2017.05.003.

Veit, C., Foister, S., Valros, A., Munsterhjelm, C., Sandercock, D.A., Janczak, A.M., Ranheim, B. and Nordgreen, J., 2021. The use of social network analysis to describe the effect of immune activation on group dynamics in pigs. Animal 15: 100332. https://doi.org/10.1016/j.animal.2021.100332.

Velie, B.D., Maltecca, C. and Cassady, J.P., 2009. Genetic relationships among pig behavior, growth, backfat, and loin muscle area. Journal of Animal Science 87:2767–2773. https://doi.org/10.2527/jas.2008-1328.

Vermeer, H.M., Dirx-Kuijken, N.C. and Bracke, M., 2017. Exploration feeding and higher space allocation improve welfare of growing-finishing pigs. Animals 7: 36. https://doi.org/10.3390/ani7050036.

Wallgren, P. and Lindahl, E., 1996. The influence of tail biting on performance of fattening pigs. Acta Veterinaria Scandinavica 37: 453–460. https://doi.org/10.1186/BF03548085.

Weary, D.M., Jasper, J. and Hötzel, M., 2008. Understanding weaning distress. Applied Animal Behaviour Science 110: 24–41. https://doi.org/10.1016/j.applanim.2007.03.025.

Wedin, M., Baxter, E.M., Jack, M., Futro, A. and D'Eath, R.B., 2018. Early indicators of tail biting outbreaks in pigs. Applied Animal Behaviour Science 208: 7–13. https://doi.org/10.1016/j.applanim.2018.08.008.

Wood-Gush, D.G. and Vestergaard, K., 1989. Exploratory behavior and the welfare of intensively kept animals. Journal of agricultural ethics 2: 161–169.

Zebunke, M., Nürnberg, G., Melzer, N. and Puppe, B., 2017. The backtest in pigs revisited – Inter-situational behaviour and animal classification. Applied Animal Behaviour Science 194: 7–13. https://doi.org/10.1016/j.applanim.2017.05.011.

Zebunke, M., Repsilber, D., Nürnberg, G., Wittenburg, D. and Puppe, B., 2015. The backtest in pigs revisited – an analysis of intra-situational behaviour. Applied Animal Behaviour Science 169: 17–25. https://doi.org/10.1016/j.applanim.2015.05.002.

Zonderland, J.J., 2010. Talking tails: quantifying the development of tail biting in pigs. PhD Thesis, Wageningen University & Research, Wageningen, the Netherlands, 176 pp.

Zonderland, J.J., Bracke, M.B., den Hartog, L.A., Kemp, B. and Spoolder, H.A., 2010. Gender effects on tail damage development in single-or mixed-sex groups of weaned piglets. Livestock Science 129: 151–158. https://doi.org/10.1016/j.livsci.2010.01.018.

Zonderland, J.J., Schepers, F., Bracke, M.B.M., Den Hartog, L.A., Kemp, B. and Spoolder, H.A.M., 2011. Characteristics of biter and victim piglets apparent before a tail-biting outbreak. Animal 5: 767–775. https://doi.org/10.1017/S1751731110002326.

Zonderland, J.J., Van Riel, J.W., Bracke, M.B., Kemp, B., den Hartog, L.A. and Spoolder, H.A., 2009. Tail posture predicts tail damage among weaned piglets. Applied animal Behaviour Science 121: 165–170. https://doi.org/10.1016/j.applanim.2009.09.002.

Zupan, M., Janczak, A.M., Framstad, T. and Zanella, A.J., 2012. The effect of biting tails and having tails bitten in pigs. Physiology & Behavior 106: 638–644. https://doi.org/10.1016/j.physbeh.2012.04.025.

Zwicker, B., Weber, R., Wechsler, B. and Gygax, L., 2015. Degree of synchrony based on individual observations underlines the importance of concurrent access to enrichment materials in finishing pigs. Applied Animal Behaviour Science 172: 26–32. https://doi.org/10.1016/j.applanim.2015.08.037.

PART 2

Risk factors associated with the physical environment

CHAPTER 7

The physical environment

Keelin O'Driscoll[1,*], *Melissa Cupido*[1] *and Jen-Yun Chou*[1,2,3]

[1]Pig Development Department, Teagasc, Animal and Grassland Research and Innovation Centre, Moorepark, Fermoy, Co. Cork, P61 C996, Ireland; [2]Swine Teaching and Research Center, Department of Clinical Studies – New Bolton Center, University of Pennsylvania, 382 West Street Road, Kennett Square, PA, 19348, USA; [3]Prairie Swine Centre, Ethology, #1 Floral Road, S7K 3J8 Saskatoon, Canada; *Correspondence: keelin.odriscoll@teagasc.ie

Abstract

Pigs that are reared for human consumption are generally managed within a controlled and or confined environment. In the wild, the area that pigs range can be up to thousands of hectares, and generally varies widely in topography and vegetation. Thus commercial pigs are generally restricted in some way, whether by quantity or quality of space, or both. Although the space that pigs are confined to can range from being extensive and relatively similar to the natural environment of the pig (e.g. agroforestry systems), most pigs are reared in intensive indoor systems, with little access to natural materials underfoot or in their surroundings, and a barren environment in general (other than close confinement with other pigs). Although tail biting is generally associated with the latter type of system, domestic pigs in all settings have been known to tail bite. As covered in previous chapters, biting occurs for a myriad of reasons, but the root cause is thought to be that the pig is experiencing some level of stress, which manifests as expression of damaging behaviour. As such, any aspect of the physical environment that increases stress levels due to impaired welfare has potential to be a risk factor for tail biting. The reason that intensive systems are more likely to result in tail biting is that their very nature means that more of these contributing risk factors are present. This chapter describes a range of commercial pig production systems that are currently in use internationally, and the risks associated with each for tail biting.

Keywords

pig production systems – extensive pig farming – pen design – floor – atmosphere

1 Introduction

The environment that an animal lives in has a significant effect on its welfare, the most obvious being via a direct effect on the animals' physical or functional state. If this effect is negative, the animal is also likely to experience negative subjective experiences (or affects) that have been described as 'survival-critical' (Mellor, 2016). These negative affects stimulate specific behaviours to improve the animals' chances of maintaining good functioning, and ultimately surviving (e.g., thirst stimulates the animal to drink; Mellor, 2017). In addition, an animals' environment can impact its welfare if the animal is capable of forming a cognitive assessment of its situation. Animals take in information about their environment through their sensory systems, and how they consequently interpret this information can produce 'situation-related' affects, which can involve conscious behavioural choices (Mellor, 2016, 2017). Both survival-critical and situation-related affects contribute to the overall welfare state of the animals.

These concepts were initially introduced in the Five Domains model of animal welfare, first published in 1994 (Mellor and Reid, 1994). Since then, the model has been refined and updated several times as the science of animal welfare has advanced (e.g. Mellor *et al.*, 2020), but in general, the 5 original domains have remained fairly consistent in their meanings over time. The first four domains are considered physical or functional ('Nutrition', 'Environment', 'Health 'and 'Behaviour'), with the fifth domain of 'Mental State' representing the animals' overall welfare status, which comprises the affective outcome of both negative and positive experiences captured by the first four domains (Mellor *et al.*, 2020).

2 The five domains model and the risk of biting behaviour in relation to the environment

Although not in all cases, in general the environment in modern commercial pig production is barren, invariant, and restricted, with little opportunity for pigs to engage in highly motivated species-specific behaviours such as exploration and foraging (Kells *et al.*, 2021). Domain 2, the 'Environment' incorporates both the solid, physical aspects of the environment, and non-tactile aspects such as air quality, noise and temperature, and should evidently be considered in the context of tail biting behaviour. In commercial systems, all of these aspects are generally significantly different from the environment that the wild ancestor of the pig was adapted to living in (wooded, scrub areas). Secondly, we must consider Domain 4, 'Behaviour', as this relates to how the environment affects an animals ability to achieve its goals (i.e. to exercise agency; Mellor, 2017). Agency is defined as 'inner motivated behavioural engagement with the environment' (from Spinka, 2019), and considered a central adaptive characteristic of animals' life. Levels of agency can range from passive/reactive agency, through action-driven agency (the animal pursues desirable outcomes), competence-building agency (the animal engages with the environment to gain skills and information for future use) and finally aspirational agency (the animal achieves long-term goals through planning and reflection) (Spinka, 2019). Regardless of the level involved, an environment which does not provide for the animals' behavioural needs, and allow them to exercise agency, poses risks for their welfare. Finally, Domain 5, Mental State, is relevant, as this integration of the animals' experiences provides an estimation of overall welfare status.

In this chapter, we will use the Five Domains model to consider the evidence about how aspects of the environment in commercial production systems impact pig welfare, with a specific focus on the risk of tail biting occurring. We have taken this approach because even though several aspects of the environment are considered to be risks for tail biting by both producers and researchers (See Chapter 10), in many cases there is a dearth of literature investigating the specific relationship. One reason for this is that the 'triggers' for tail biting outbreaks tend to be farm specific, and outbreaks tend to be temporally sporadic, so it can be difficult to experimentally induce outbreaks by manipulating individual aspects of the environment. A second, related reason, is that it is thought that tail biting outbreaks occur due to an accumulation of stressors (rather than any one specific stressor) overwhelming the animals coping ability (See Chapter 15). We can thus consider what we know about how the environment impacts stress and welfare in general, and potentially have the knock-on effect of triggering biting.

3 Farming systems

Pig production is carried out using a huge variety of methods worldwide. For instance, in the EU farm size ranges from back-yard farms with just one or two pigs (3% of the EU pig herd), to farms with thousands of pigs; in 2020, 97% of pigs in Denmark were kept on farms with 1,000 or more animals (Augère-Granier, 2020). Likewise, in the US 91% of pigs are produced by large operations (5,000 or more pigs on-site), whereas only 3% of agricultural land is used in organic farming (Park *et al.*, 2017). Globally, and even in developing countries, production has shifted significantly from smaller farms towards medium to large scale operations in the past 30 years (Woonwong *et al.*, 2020). For example, in the EU there were 3.8 million pig holdings in 2005, and under 2.2 million in 2013, without an equivalent drop in pig numbers (Augère-Granier, 2020). Similarly, in China during the decade between 2006 and 2016, the proportion of pigs raised by backyard producers reduced from two thirds to one third (Qiao *et al.*, 2016), and this has reduced further since the significant African Swine Fever (ASF) outbreaks that occurred during 2018–2019 (Woonwong *et al.*, 2020). This transition has meant that an increasing number of pigs worldwide are now managed in intensive indoor systems. Although this type of system is economically efficient, unfortunately, with regard to pig welfare, the associated barren environments, high stocking densities, and lack of enrichment contribute to unavoidable negative physical conditions (Domain 2) and impeding of agency (Domain 4; Mellor *et al.*, 2020; Sonoda *et al.*, 2013). No system is perfect, and outdoor or extensive systems can also pose risks for welfare (Ludwiczak *et al.*, 2021), but in general the intensive indoor systems that are increasingly predominant in pig production are considered to be the most challenging when it comes to rearing pigs without docking their tails.

Nevertheless, for animal welfare (to provide animals with more agency) and environmental (to reduce emissions such as ammonia) reasons, there is an emerging reversal with regard to pig farm intensification. In more developed countries, policy has begun to shift towards encouraging systems that allow animals greater agency and improved environmental conditions (e.g., Proposition 12 in California mandates all non-lactating sows have at least space to turn around, from January 2022). The European Commission, as part of its farm to fork strategy, has set a target of 'at least 25% of the EU's agricultural land under organic farming and a significant increase in organic aquaculture by 2030, while remaining commercially viable'. This trend means that pig farming systems are still evolving, and more extensive farming systems (such as organic) are being adapted so that they can be run at a larger scale. Thus there is scope for our increased understanding of animal welfare, and of the risk factors for tail biting to be taken into account as important considerations for the design of new systems. This chapter will describe predominant pig farming systems, ranging from highly extensive to intensive, in relation to the risks associated with tail biting.

3.1 *Outdoor pig farming*

There is significant diversity when it comes to classifying 'outdoor' pig farms relative to the quantity or quality of outdoor access that the pigs have available to them. Indeed in a 2021 EFSA report on ASF in Europe (EFSA, 2021), outdoor systems were classified into 20 different types. There were three main categories; whether the pigs had access to forests/woodlands, pasture/fields, or concrete fenced yards. Within these it was considered whether the outdoor areas were fenced or unfenced (for concrete yards, whether the buildings were open or closed), and whether pigs were outside for their entire lives, only part of their lives and whether or not there was shelter provided for them. Even besides the level and type of outdoor access provided, farms can also be categorised in other ways; for instance whether they have organic status or not, or by the breed of pig that is produced.

Thus when considering factors that affect tail biting occurring in 'outdoor' systems, we need to be cognisant that there is no single type of outdoor system. Although it is often assumed that outdoor production systems have lower or even no tail biting incidents than indoor (Sonoda *et al.*, 2013), this is not necessarily always the case; for instance, Walker and Bilkei (2006) found that the prevalence of tail biting on five outdoor pig farms was between 14% and 20%, and Alban *et al.* (2015) found a greater percentage of tail lesions in meat inspection of pigs from organic/free range pigs than conventional. However, in the study by Alban *et al.* (2015), the pigs in the organic/free range systems weren't docked, whereas the conventional were. Kongsted and Sorensen (2017) reported a similar result, more lesions in outdoor pigs, but these were undocked, whereas the indoor ones were. Thus it is not a 'like for like' comparison; even though that work showed that biting exists in outdoor systems, it was yet possible to produce pigs without routine docking. Pandolfi *et al.* (2017) probed a database of over 1,900 farms in the UK and found the prevalence of severe tail biting damage in pigs in outdoor pens to be half that of pigs indoors (0.07 *v's* 0.17, respectively). This is likely related to factors such as a greater space allowance and improved ability to satisfy exploratory needs and express natural behaviours such as rooting, over conventional systems (Edwards, 2005; Papatsiros, 2011).

Many of the risks for pig welfare in outdoor systems relate to risks towards animal health (Domain 3). For instance, in contrast to a reduced risk for tail biting, Pandolfini *et al.* (2017) found a greater risk of lameness in outdoor systems. Outdoor production in general poses more challenges for biosecurity than indoor (e.g. Juardo *et al.*, 2018), and pigs are generally more exposed to parasites such as worms; the eggs can remain viable in the soil and continuously re-infect the pigs (Edwards, 2005). Likewise outdoor environments also often accommodate a variety of bacteria and viruses which can be difficult to control, and not only pose a risk to the health of the animals reared in outdoor systems, but also consumers (Kijlstra and Eijck, 2006). As well as an increased risk for tail bites, Alban *et al.* (2015) found that outdoor pigs had increased risk of healed fractures, skin disorders (eczema/insect bites/skin corrosion), infectious arthritis, milk spots (evidence of worms), osteomyelitis (bone infection related to tail damage), and abscesses in the mid and hind part of the body. These disorders are all related to exposure to outdoor conditions or are associated with tail biting. Similarly, Kongsted and Sorensen (2017) found that regardless of whether classified as organic or conventional, free range pigs had a greater odds of having white liver-spots, tail lesion, arthritis, skin lesions, bone fracture, septicaemia and abscesses than indoor conventional pigs, again suggesting more challenges in controlling disease in outdoor pigs.

Thus although outdoor systems are perceived to be better for pig welfare, and often are with regard to Domain 2, for the other domains they may not always have benefits over conventional indoor systems. The primary negative factor associated with Domain 2 could be an increased risk of thermal stress (both hot and cold). For the others, outdoor systems could involve potentially increased social competition (Domain 4); although also a risk in indoor systems, if there is limited human assistance or management, low ranking animals may be disadvantaged in outdoor systems when it comes to attaining adequate access to resources such as shelter and food (Domain 1; Edwards, 2005). There is also a risk of inadequate provision of food and water during winter due to ice formation and poor pasture growth etc. Stress caused by these risks in any system have been speculated to contribute to tail biting risk (Papatsiros, 2011), and thus a well-designed outdoor system should aim to minimise the challenges associated with all domains as much as possible. Pandolfi *et al.* (2017) concluded that the prevalence of negative welfare outcomes such as tail biting are not specifically connected to any particular system, but rather inappropriate features of the environment. We will now outline some of the more common types of outdoor pig production, considering their benefits and risks for pig welfare.

3.1.1 Agroforestry

An agroforestry system is one in which the production of livestock (or crops) occurs simultaneously on the same land as the production of trees or other woody vegetation. Livestock production combined with trees and forage is referred to as silvo-pasture or livestock agroforestry and is one of the most prominent forms of agroforestry (Mosquera-Losada *et al.*, 2009). The integration of trees and pigs in the same unit of land can improve environmental and economic sustainability, and although there is not much research in the area, it has the potential to improve pig welfare. Trees in agroforestry systems provide shelter from wind, heat and sunburn, and remove water from the soil (Bealey *et al.*, 2016; Burbi and Olave, 2018; Manevski *et al.*, 2019), which reduces poaching risk (the degradation of soil structure and vegetation cover caused by the trampling of livestock), resulting in soil compaction and loss of grassland productivity and its knock-on effects on the pigs and environment. One of the largest examples of this system is found in Dehesa in Spain, where Iberian pigs are reared amongst evergreen oak trees (Wealleans, 2013). These pigs spend more than 50% of their time in foraging activities such as grazing (Rodríguez-Estévez *et al.*, 2009a). Besides the shelter and enriched environment that the trees provide, in this system the acorns provided by the oak trees form part of the diet of the pigs. During the 'montanera' season (acorn mast feeding season; November to February) pigs are fed exclusively on grazed grass and acorns which fall from trees, which is sufficient to feed them. Pigs in these systems selectively choose acorns depending on their size and nutrient composition, at various parts of the season, suggesting that the pigs are capable of forming associations between variables when choosing to eat or reject the acorns from specific trees (Rodríguez-Estévez *et al.*, 2009a). Thus this type of system also provides pigs with the opportunity to exercise agency when it comes to their diet.

With regard to tail biting (and pig welfare in general), there is very little published in the peer-reviewed literature in relation to agroforestry. However, due to the system's close relation to the feral pig and wild boar's natural habitat preferences, many factors might aid in reducing stress and thus the risk of damaging behaviours. A forestry system has immediate benefits over simply providing access to outdoor paddocks or concrete yards, in that trees and shrub cover provide a natural and stimuli-rich environment for pigs and provide shade and shelter from hot and cold weather respectively (Jakobsen and Kongsted, 2015). The availability of trees, even in paddocks (as opposed to 'wild' or semi-managed forested areas), creates a more enriched environment than seen in most conventional systems (Brownlow, 1994; Brownlow *et al.*, 2005), providing pigs with greater opportunities to express natural behaviours. To our knowledge there is only one study, from Germany, where pigs in an agroforestry system, with ad-libitum food provided by humans, were studied. That body of work used the Welfare Quality® protocol for pigs to investigate animal health, as well as carrying out detailed animal behaviour observations. Out of 289 pigs, only 2 had slight injuries to the tip of the tail, and 259 had no scratches or injuries to any part of the body (Stäbler *et al.*, 2020). The pigs spent about 13% of the day eating green fodder and concentrate that was provided to them, but yet another 24% rooting, indicating that rooting appears to be a highly motivated behaviour even in the absence of hunger (Stäbler *et al.*, 2022). Without any other scientific literature it is not possible to draw final conclusions on the risk of tail biting in agro-forestry systems; however, they can be considered to be the commercial system most like the environment that wild boars evolved in, and as such hold promise as a system that has potential to meet many of the pigs' behavioural and physical needs.

3.1.2 Organic pig farming

Organic animal farming systems aim to develop sustainable and environmentally friendly farming systems, while allowing farm animals a better quality of life than in conventional systems. According to the International Federation of Organic Agricultural Movements (IFOAM), 'organic livestock husbandry is based on the harmonious relationship between land, plants and livestock, respect for the physiological and behavioural needs of livestock and the feeding of good-quality organically grown feedstuffs' (IFOAM, 2019; from Lund, 2006). Thus the concept of 'naturalness' is fundamentally important to both producers and consumers in the organic sector (Lund, 2006); this includes the animal being able to perform natural behaviour, consuming feed adapted to its physiology, and living its life in an environment similar to that which its non-domesticated ancestor was adapted to (Lund, 2006). Pigs in these systems are fed organic feed and roughage, and usually mutilations such as teeth clipping and tail docking are prohibited (Hämeenoja, 2002; Ichim, 2012; Kijlstra and Eijck, 2006). Thus the majority of organic pigs are reared with intact tails. A description of the most prevalent systems for organic pig farming in Europe can be found in Leeb *et al.* (2019).

In general organic systems are designed with greater consideration of the environment than conventional (Kijlstra and Eijck, 2006), with outdoor access for pigs as a requirement. Bracke *et al.* (2013) found that there was no reported difference in tail biting prevalence reported between conventional and organic farmers in the Netherlands, but the crucial difference was that tails on organic farms were not docked, whereas they were in conventional. Thus, considering not docking pigs' tails is the most significant risk factor for biting, it appears that most other risk factors, at least in in Dutch organic pig farms, are likely lower than in conventional. An important aspect of organic farming is the requirement to provide straw, and in fact in Switzerland, it is required to provide roughage (e.g. hay, grass etc.) as well, with the aim of improving behavioural profile and health further. Holinger *et al.* (2018) found that provision of roughage in the form of grass silage, in addition to straw bedding, occupied pigs for longer and reduced severe damage to the intestinal mucosa including gastric ulcers. Indeed lower incidences of tail biting is expected from pigs reared in organic systems (Hämeenoja, 2002), and in fact according to a survey of the 8 main organic pig producing systems in Europe, tail biting was not considered a significant health issue in any of them (Fruh *et al.*, 2014).

Nevertheless, some tail-biting has been recorded under organic conditions (Bracke *et al.*, 2013; Heldmer and Lundeheim, 2010; Sonoda *et al.*, 2013). The causes in general relate to Domains 1 (e.g.body condition loss; Ichim, 2012), 2 (e.g. climatic challenges; Ichim, 2012) and 3 (e.g. parasitic infections; Ichim, 2012; Papatsiros, 2011). As tail biting is an indication of poor welfare (Hämeenoja, 2002) and because most organic systems prohibit the practice of tail docking, welfare problems may be more easily noticed in these systems. Bracke *et al.* (2013) studied the attitudes of Dutch pig farmers toward tail biting and docking and found that organic farmers place more importance on the animal's ability to express natural behaviour. They also found that the majority of conventional farmers do not acknowledge that important welfare problems exist in pig farming, in contrast to their organic counterparts. These conventional farmers considered tail docking to be a more efficient and necessary procedure than environmental enrichment. Thus the reduced level of tail biting observed in organic systems may also be due to the human attitude aspect, as the farmers involved are likely to place more attention to the pigs needs than those working in conventional systems (Hämeenoja, 2002; see Chapter 12).

3.1.3 Outdoor yards and runs

Even for pigs that are housed indoors, access to the outdoors can be provided via yards or runs located adjacent to the pig building. Indeed outdoor access is sometimes mandatory for welfare friendly certification schemes (Heinola *et al.*, 2021), and a yard or run can allow this while minimising risk of poaching of the soil, and where land and labour resources are tight. Indeed concrete yards and runs are an integral component of organic systems across the EU (Früh *et al.*, 2014), not least because organic pigs are legally required to have access to an open air area (EU, 2018). Although outdoor runs provide access to fresh air, in organic production they are not considered best practice as they have the disadvantage of not fully permitting natural behaviour (relative to access to a paddock with soil, or pasture; Domain 4). Nevertheless, they are often used as alternatives to paddocks because organic farming also places importance on protection of the environment, and thus nitrate leaching, and poaching of the land should be minimised. Managing pigs in indoor systems with an attached yard also has benefits for pig performance, compared with fully outdoor access (Leeb *et al.*, 2019)

These outdoor areas are usually designed so that they are permanently accessible from the indoor house, have a concrete floor, and are at least partially roofed (Wimmler *et al.*, 2022a). There is limited peer-reviewed information about not only how provision or design of an outdoor run affects tail biting behaviour, but how it can affect welfare in general. However, Leeb *et al.*, (2019) carried out an epidemiological study comparing the welfare of pigs in organic farms where they were either permanently outdoors, kept indoors with access to a partly roofed outdoor run, or where the systems were combined depending on the production stage. They found that when grower pigs had permanent access to the outdoor run, there was a lower prevalence of respiratory problems, diarrhoea and lameness (Domain 3). Likewise, Ocepek *et al.* (2020) found that positioning drinkers in an outside pen area reduced manipulation of pen-mates. A recent review by Wimmler *et al.* (2022b) concluded that outdoor runs have potential to provide separate functional areas for resting, exploration and elimination at least, and could also facilitate access to a rooting area equipped with material such as straw, or compost. A rooting area with appropriate material has a good chance of reducing the risk of tail biting behaviour (Studnitz *et al.*, 2007; Chapter 8 of this book). Recently, practical guidelines on how to best design runs for organic pig systems were developed and published, from a large EU project, POWER (Dilling, 2022). The project gathered pig producers views on challenges and innovations and researched a range of options. The compiled examples of best practice form an invaluable resource for anyone considering designing a pig system that permits pigs a good level of agency when it comes to their living arrangements.

3.2 *Indoor systems*

It is estimated that over 75% of EU pigs are currently managed in large commercial holdings (Augère-Granier, 2020). Like outdoor systems, the specific design of these systems varies widely; however, in general the modern day pig pen in a typical commercial farm has evolved into a relatively featureless rectangular shaped enclosure, with a concrete floor (either solid or slatted). Where the floor is at least partially solid, some form of drying material or bedding is sometimes added. As well as feeders and drinkers, pens are often equipped with some form of enrichment (normally 'point source' items such as chains, plastic/rubber hanging toys, or racks of organic material; Chapter 8).

Welfare challenges in indoor commercial systems are different to those with access to the outdoors, with more of an impact on Domain 2 (Environment) and 4 (Behaviour). Because tails have been more or less routinely docked in intensive indoor systems for several decades, pigs have evolved to allow a significant accumulation of stressors before the eruption of a tail biting outbreak; nevertheless, tail biting outbreaks occur regularly in indoor systems even in docked pigs (Haigh and O'Driscoll, 2019), and in fact when tails are docked very short, biting behaviour is sometimes re-directed towards ears (Goossens *et al.*, 2008). Nowadays, it is extremely difficult to rear pigs in typical indoor systems without tail docking. Thus it is necessary to re-think the design and management of indoor systems if we are to reduce the risk of tail biting to the point where routine docking is no longer necessary (see Chapter 10). The following sections are primarily factors associated with indoor systems that could affect the risk of biting behaviour, although they are also often relevant to systems where pigs are allowed some access to the outdoors.

3.2.1 Pen size and shape

Although pen size can vary from being only large enough for one litter (approx. 12 pigs) up to pens that can house over 100 pigs, there is little controlled research on the impact of pen size on pig welfare or tail biting risk, and much of it is confounded by stocking density. In a large case control study carried out on commercial units, Moinard *et al.* (2003) found that group size had no effect on tail biting (although as the number of pens, and number of pens per stockperson increased, the risk of biting also did). Contrarily, Misra *et al.* (2021) found that pigs in groups of 48 performed less damaging and aggressive behaviours than those in groups of 12. The authors theorised that when stocking density is kept constant, pigs in larger groups have more shared space per pig, which could provide a more complex and engaging environment for them. In general, stocking density rather than group size *per se* is considered to increase the risk of damaging and biting behaviour (Grümpel *et al.* 2018; Moinard *et al.*, 2003, Scollo *et al.*, 2016), although this is often confounded by access to resources within the pen. Wiegand *et al.* (1994) reported more aggressive behaviour as space allowance decreased, and as it increased, pigs forming more mid-size and fewer large social groups. This could indicate again that more available space per pig allows pigs to self-organise and select their group partners more easily, which could lower stress levels. Indeed in countries where pigs are routinely reared without docking, stocking densities are generally lower than the EU minimum (e.g. Sweden, Wallgren *et al.*, 2019; Finland, Valros, 2021).

When it comes to pen shape, there is very little in the literature. One significant factor that constrains the shape and size of the pen, as well as number of pigs it can accommodate, is the feed system. When we consider how the feed system can impact the risk of tail biting, it is useful to consider how pigs are normally fed in a country where docking does not occur, such as Finland. There, the most common feed system is a long trough where all pigs can line up to feed simultaneously at scheduled times (Valros, 2021). In fact Finnish pig producers ranked 'enough feeding space' (Domain 1) as the most important preventative measure for tail biting (Valros *et al.*, 2016). As it is recommended that each pig have 32cm of space per head (Valros, 2021), this means that in a pen of 20 pigs, at least one pen wall must be 6.4m in length to accommodate the trough (0.35m × 20). For a pen of 20 pigs, if we apply the minimum EU space allowance of 0.65m^2/pig, a pen size of 13m^2 is needed. Thus to include a single long trough where all pigs can feed at once, at the EU space allowance, we would end up with a pen with a length of 6.4m, and a width of just over 2m; an extremely long, narrow pen, with virtually no space away from the feeding area. Thus adoption of recommendations from other countries that do not dock must be considered holistically, and in consideration of how they interact with other aspects of the system, so that the pen functions as a whole. In the example here, at the current space allowance, it would make more sense to have two troughs in the pen rather than one, so that they do not end up so long and narrow. However, it is also likely that we need to rethink completely about the current minimum space allowance in the legislation, as the standards may only be suitable for docked but not undocked pigs.

3.2.2 Functional areas

This leads to the concept of functional areas; that it is not just the quantity, but also the quality of space that is important. The ancestor of the commercial pig, the wild boar, utilised distinct areas for various behaviours such as resting, exploration, elimination, thermoregulation and social behaviour (Stolba and Wood-Gush, 1989). Moreover, recent ethological studies on pigs in both semi-natural and confined conditions have also demonstrated that pigs synchronise behaviours such as feeding, resting, and interaction with enrichment (Rodríguez-Estévez *et al.*, 2009b; Zwicker *et al.*, 2012). Thus to adequately accommodate the animals' natural behaviours and inclinations, pens should be designed so that there are clearly delineated areas that can be used for different functions, and so that these areas are large enough for several pigs to use at once (Domain 4). At the least, all pigs should be able to lie down in an area away from the feeder and dunging area at the same time. Jensen *et al.* (2012) hypothesised that one reason why high stocking densities could be a risk factor for tail biting, is because functional separation of areas is more difficult to achieve.

3.2.3 Priority resource location

To achieve this, it is essential that the location of any infrastructure within the pen is carefully considered so that pigs can use the available space appropriately. One nice example of how inappropriate resource location can inadvertently affect pig behaviour is demonstrated by Misra *et al.* (2021). In that study, three different pen sizes (small = 12 pigs; medium = 24 pigs and large = 48 pigs) were created by simply removing partitions between multiple small pens, which had single space feeders in one corner of the pen; thus the medium and large pens had their feeders located at different corners, not next to one another in a single feeding area as in the small pens. Likewise, in the larger pens the enrichment items (a rack of grass, hanging rubber toy and plank of wood) were distributed at random throughout the pen, rather than in a single area. Although all pens had the same space allowance and ratio of pigs to feeding spaces, the daily feed intake reduced as pen size increased. The authors found that this was because of the huge variation in feeder use within the large pens, with some feeders dispensing more than twice that of others; 75% of the feeders in the small and medium pens were in the middle two quartiles when it came to the amount of feed dispensed during the trial, whereas 70% of the feeders in the large pens were either in the top or bottom quartiles. In the pens with more than one feeder, the coefficient of variation between feeders within large pens was over twice that of the medium. Anecdotally, the authors observed that because resources were scattered around the pen, pigs in the larger pens often created dunging and lying areas near some of the feeders. This inhibited feeding behaviour in some, whereas the remaining ones were over-subscribed. It is likely that if the feeders had been moved to be adjacent to each other the pattern of use for each may have been more uniform, and the pigs may have been more efficient in forming lying and elimination areas.

3.2.4 Hides and barriers

One way to obviously separate areas of the pen is to introduce internal walls or 'hides'. These have several effects; although these have been researched more as a way to reduce aggression (e.g. McGlone and Curtis (1985) found that small hides reduced aggression after regrouping), they have also been shown in some instances to reduce damaging behaviour and can help to demarcate a lying area. Pigs prefer to lie close to solid pen partitions, which provide protection from draughts (Jackson *et al.*, 2020). Bulens *et al.* (2017) added hide walls in the shape of a 'T' in finisher pens, with the leg of the T attached at its base to either the side or back wall. In pens with a hide wall pigs had fewer tail lesions, and performed less pen manipulation and belly nosing, which the authors hypothesised was indicative of lower stress-levels.

3.2.5 Elimination areas

The elimination behaviour of pigs in semi-natural environments and outdoor housing systems is organised so that it occurs in defined areas, away from feeding and lying areas (Salomon *et al.*, 2007; Stolba and Wood Gush, 1989). In confinement systems this also generally holds true; it normally occurs in corners, with pigs in circular pens even using the edges of the feeder (Gou *et al.*, 2015; Wiegand *et al.* 1994). The propensity for site selection appears to be innate (Domain 4), and starts early in life, as piglets start to use distinct areas of the pen for elimination within a few days of age (Andersen *et al.*, 2020). Pigs turn their hind quarters towards a wall before they eliminate, which is hypothesised to occur so that they can avoid being disturbed during elimination (Guo *et al.*, 2015; Weschler *et al.*, 1998); this could be why corners are preferred, as the pig is protected from two sides.

Generally speaking, the places that are selected for elimination tend to be those which are not preferred for lying or feeding (Nannoni *et al.*, 2020). It is not yet known whether pigs choose a site for elimination, and avoid resting there, or rather choose the most comfortable place for resting, and consequently avoid eliminating there. Whichever is the case, in general, favoured elimination areas include well-lit, draughty, wet, and safe places, such as an outside yard when pigs have access to one (Guo *et al.*, 2015). Poor site selection, or widespread elimination can result in soiling of the pen, and as a consequence the pigs. The main causes of pen soiling due to poor site selection are thought to be 'inadequate thermoregulation (overheated/draughty lying area), faulty pen design (disturbance during elimination), and flooring issues (dirty/slippery floors) (Nannoni *et al.*, 2020).

Inappropriate elimination behaviour has potential to cause two primary problems that could impact upon pig welfare, and consequently tail biting behaviour. First, the obvious consequence of widespread elimination is the issue of hygiene (Domain 3), and the fact that pigs may no longer have a dry clean lying area if too much of the pen is soiled (Domain 2). However in addition to this, pig production is responsible for up to 25 % of livestock ammonia emission in Europe with half estimated to originate from decomposition of manure from pigs within housing systems, and the rest from storage and spreading of manure (Philippe *et al.*, 2011). As well as the impact on the environment, elevated levels of ammonia can have negative effects on animal health (Domain 3) and is associated with an increased risk of biting (See later section in this chapter). Thus, minimising the area soiled with faeces is important for both animal welfare and environmental reasons, and pig housing systems should be designed so that the majority of eliminations will be located in a small area of the pen.

The natural tendencies of pigs to eliminate away from their lying area, as well as their intelligence, means that there is potential for pen design to promote at least minimal area for elimination, and potentially separate defecation and urination areas (Tillmans *et al.*, 2022); it is the mixing of urine and faeces that triggers ammonia production. As well as adapting the system to the pigs' natural behaviour, facility design can also have a strong role to play to this end; for instance, Ye *et al.* (2007) investigated how slat profile and width, and floor slope gradient and surface roughness, could effectively separate out urine. They found that a slope gradient of 20° with a fine cement surface performed best. Flooring and pen designs that encourage optimal elimination practices and allow pigs to form well laid out functional areas is an area that warrants further research.

3.2.6 Floors and bedding

Floor type, whether fully concrete, partly slatted at different slat to solid area ratios, or fully slatted, is generally considered the major external risk factor for tail biting (Henry *et al.*, 2021; Schrøder-Petersen and Simonsen, 2001; Valros, 2018). This is in part due to the fact that bedding cannot be provided on fully slatted floors. In the EU in 2005 it was estimated that fully-slatted floors were in 67% of weaner and 44% of finisher housing (EFSA, 2005). In the past 17 years, the number may have decreased since Sweden, Germany, Denmark, the Netherlands, Finland and Switzerland, have either abolished the use of fully-slatted floors or entered a transition in phasing them out (European Commission DG SANTE, 2016; Mul *et al.*, 2010). However, the fully-slatted system is still the predominant system used on a global scale.

Fully slatted systems have the benefits of easy manure removal and maintaining hygiene in the pen, as pigs' urine and faeces directly drop down through the slats without accumulating on the surface and are stored underground. However, the biggest drawback of these systems for pigs, is that it hinders the provision of any loose rooting materials on the floor (Domain 4) as they can block the slats and create hygienic issues. Slatted slurry systems often do not have mechanisms to deal with loose materials that fall through the slats and accumulate in the slurry tank, which can cause congestion. This limitation contributes to the main reason why slatted flooring is considered the greatest risk factor (D'Eath *et al.*, 2014). More detailed discussions on the effect of loose rooting materials on tail biting can be found in Chapter 8 of this book.

Another welfare concern issue with fully slatted systems is that manure stored beneath the pen can become the source of higher ammonia levels in the housing than a straw bedded system, if not removed regularly (Philippe *et al.*, 2011). Although there is some conflicting evidence which suggests bedding can accumulate a higher level of ammonia (EFSA, 2005; Pereira *et al.*, 2011), this could be somewhat due to different types of bedding material being associated with different emission levels (Zhang *et al.*, 2016). Regardless, slatted flooring is associated with a higher occurrence of respiratory diseases (Domain 3; EFSA, 2005; Scott *et al.*, 2006), and ammonia irritates the respiratory tract at concentrations exceeding 15 ppm (Banhazi *et al.*, 2008). Respiratory disorders are themselves considered risk factors associated with tail biting (Kritas and Morrison, 2007; van Staaveren *et al.*, 2016). Peak ammonia levels occur during manure removal in fully slatted systems (EFSA, 2005), and therefore well-designed drainage, good ventilation and climate control are essential in these systems (EFSA, 2005; Pereira *et al.*, 2011; Van Ransbeeck *et al.*, 2013). More details of the relationship between air quality and tail biting will be discussed later in this chapter, but it is clear that different elements of any housing environment are interlinked and can converge to increase the risk of tail biting.

Most experimental studies comparing tail biting in pigs reared on flooring ranging from fully solid to fully slatted are confounded with the provision of straw bedding (as reviewed by Schrøder-Petersen and Simonsen, 2001; Spoolder *et al.*, 2011). In fact, we are only aware of one that did not have this confounding factor; McKinnon *et al.* (1989) compared pigs housed on fully and partly slatted floors and found more pigs massaging other pigs and chewing pen fittings in fully slatted pens, but with no difference in tail/ear/feet biting. As such, much of what we know comes from evidence from on-farm observational studies. In England, tail biting was found to be higher when farms had either partly or fully slatted floors both in the farrowing and in the weaning-finishing units, with a 3.2 fold increase of tail biting risk for the latter (Moinard *et al.*, 2003). In Belgium, a higher proportion of slats in the farrowing unit was associated with a higher percentage of tail and ear lesions (Smulders *et al.*, 2008), and in Finland, the risk of tail biting also increased with an increasing percentage of slatted area in the piglet, weaner and finishing units (Kallio *et al.*, 2018). It must be noted though that many observational studies also pointed out the correlation between floor type and the provision of bedding or loose rooting materials on the floor (Kallio *et al.*, 2018; Moinard *et al.*, 2003; Valros, 2018). Thus solid flooring *per se* may not be entirely preventative. Detailed discussion of how bedding and its relationship with the risk of tail biting can be found in Chapter 8.

In summary, fully slatted systems have some unamendable limitations when it comes to provision of floored rooting materials, creation of thermal comfort zones, and separation of functional areas. These are all important parts of the multifactorial blueprint that could help reduce tail biting as mentioned above (D'Eath *et al.*, 2014). Although no study yet has identified the most suitable slat to solid area ratio, the Finnish experience in managing tail biting in undocked pigs may give us some hints. In Finland, where tail docking is completely banned, more than 80% of farms have partly slatted systems. Finnish farmers still manage to keep 70–95% of tails intact when scored on farm, and only 2.5% of tails had major lesions when scored in the abattoirs (Valros, 2021). Their industry guidelines recommended a minimum of 50% solid flooring in grower and finisher pens (Valros, 2021). This is not the only factor to consider, of course, as this book is illustrating, but it still gives a good example on how appropriate flooring systems may bring benefits to reducing the risk of tail biting without the need to tail dock.

4 The non-physical environment

Many reviews on tail biting have all identified that air flow and quality (ventilation, Schrøder-Petersen and Simonsen, 2001; Taylor *et al.*, 2010; Valros, 2018), temperature (D'Eath *et al.*, 2014; Edwards, 2011; Schrøder-Petersen and Simonsen, 2001), and humidity (Taylor *et al.*, 2010) are the common external risk factors for tail biting (Taylor *et al.*, 2012). However, few experimental studies to date have provided conclusive evidence on how these climatic elements can trigger tail biting. It is difficult to establish a direct causal link between the climatic factors and tail biting, but Edwards (2011) suggested that changes in climatic parameters can create competition for more comfortable resting spaces and indirectly increase the likelihood of sudden-forceful tail biting events. Changes in environmental conditions can also induce acute heat or cold stress and lead to another type of 'epidemic' tail biting (chapter 3; Valros, 2018).

4.1 *Air flow and quality*

Across pig farmers surveyed in various European countries, air flow and quality is often considered one of the top risk factors for tail biting (no. 1 'stable climate' in the Netherlands, Bracke *et al.*, 2013; no. 5 'air movement' and 6 'air quality' in Ireland, Haigh and O'Driscoll, 2019; no. 3 'air movement' and 7 'air quality' in Finland, Valros *et al.*, 2016; no. 5 'Draught' and 7 'air quality' in the UK, Valros and Barber, 2019). On the contrary, Swedish farmers did not rank this factor as high (Wallgren *et al.*, 2016) and neither did UK farmers who signed up for a high welfare assurance scheme (Paul *et al.*, 2007). This suggests that climatic factors might be considered 'high risk' amongst producers who are reluctant to recognise the importance of other risk factors, requiring more cost and labour to fix (e.g. appropriate provision of environmental enrichment, or floor type).

Draught (sudden and high-velocity air flow) increased pigs' general activity level and also stimulated excessive rooting behaviour, pen-mate directed behaviours and aggression (Scheepens *et al.*, 1991). Some studies also found a higher air velocity in the pen increased the amount of tail lesions observed (Sallvik *et al.*, 1984; for docked pigs: Holling *et al.*, 2017). In indoor pig housing, air flow is usually controlled by ventilation systems, whether natural, or artificially controlled natural (ACNV), or mechanical forced ventilation. Surprisingly, on-farm observational studies did not show significant effects of the ventilation system on the risk of tail biting (Pandolfi *et al.*, 2017; Scollo *et al.*, 2016; Smulders *et al.*, 2008). However, Hunter *et al.* (2001) found natural ventilation reduced the probability of tail biting compared with ACNV, artificial and mixed systems, with a stronger effect when pigs were not docked.

In terms of specific components of air quality, currently there is most knowledge on how pigs react to ammonia (NH_3). Pigs avoid ammonia-filled chambers during preference tests (Smith *et al.*, 1996), and are also averse to CO_2 (Raj and Gregory, 1996). Van Putten (1969) managed to induce a tail biting outbreak in finishing pigs (16–17 weeks of age) with both reduced ventilation at 0.30 percent of CO_2, and 0.17 percent of ammonia, at an ambient temperature of 23°C. Interestingly, under the same conditions pigs provided with straw had a lower level of tail biting. However, in another study, 0.13 percent of CO_2 and as high as 80+ p.p.m. of ammonia concentration (mild irritation to human eyes) with an ambient temperature of 21°C over two days did not induce tail biting in growing pigs (Ewbank, 1973).

In on-farm observational studies, higher ammonia levels (> 10 ppm) and poor perceived air quality in the weaner house was found to be associated with higher risk of tail biting in heavy pig production in Italy (Scollo *et al.*, 2016). Using the same data with a more sophisticated modelling, ammonia level was still identified as a major risk factor, with a level higher than 28 ppm increasing tail biting frequency to 3.8% (Scollo *et al.*, 2017). That study also found that on farms with a lower ammonia level, the risk of tail biting was further lowered if farms had fully concrete rather than slatted floors (Scollo *et al.*, 2017). The effects of other particulate matters in the air like dust or hydrogen sulphide that may affect pig health and therefore tail biting are currently unknown.

4.2 *Temperature*

Maintaining a suitable temperature in pig housing is essential for their health, welfare and performance. When the ambient temperature is outside of pigs' thermal comfort zone, their behaviour starts to change (Bus *et al.*, 2021), and extreme temperature fluctuations can also cause cold/heat stress of pigs (Domain 3; Valros, 2018), which can all increase the risk of tail biting.

Early work conducted on partly slatted commercial systems suggested the optimal temperature with the lowest frequency of tail biting for growing pigs (30–40 kg) was 20–22°C. When the temperature increased to 22–24°C, the chances of tail biting also increased (Geers *et al.*, 1989). They also found more tail biting when the temperature was lower than 16°C for pigs at 40–50 kg (Geers *et al.*, 1989). Temperature close to the floor level in the weaner housing (8–20 kg of weight) was shown to correlate positively with the presence of pigs' tail lesions (Smulders *et al.*, 2008). Other work has also pointed to a higher temperature increasing the risk of tail biting, although with some inconsistent results (as reviewed by Schrøder-Petersen and Simonsen, 2001). Indeed, Temple *et al.* (2012) found that pigs (29–180 days of age) housed in partly or fully slatted conventional systems had higher risk of tail biting when the temperature was 16–20.9°C compared to a slightly higher temperature at 25–29°C. A negative relationship between temperature and tail biting was also observed in another study with pigs on fully slatted system (Holling *et al.*, 2017). This could be interpreted that pigs on slatted floors find it difficult to thermoregulate when the temperature is low, as cold air can flow from the slurry pits below. Interestingly, more tail biting was found in straw-bedded systems when the temperature was the highest, which again demonstrates an interaction between the housing system and temperature (Temple *et al.*, 2012).

During the suckling stage, one study found that a higher temperature in the creep area was associated with lower tail lesions and a lower percentage of piglets showing tucked/hanging tails (Vitali *et al.*, 2020). In contrast, a higher temperature in the farrowing pen was associated with more tail lesions and a lower percentage of piglets showing curly tails, which could be linked to the lower body condition score and higher incidences of diarrhoea recorded (Vitali *et al.*, 2020). It shows the difficulty of maintaining an optimal temperature in the farrowing unit as the sows' thermal comfort zone is different from piglets' and this mismatch can lead to compromises in their health (Domain 3) and consequently piglets' tail biting behaviours.

The effect of temperature can also be linked to seasonal changes. In a study involving simulated summer conditions in France (25–30°C), providing a water spraying system to cool down the ambient temperature actually increased the overall number of body lesions recorded on pigs, but reduced the number of severe body lesions and tail biting events (Courboulay *et al.*, 2008). A more recent study using surveillance data collected at national level in abattoirs in England and Wales found a pattern of increasing risk of tail biting when climatic temperature decreased, but without statistical significance (Lee *et al.*, 2020). In summary, both too high or too low a temperature can influence the risk of tail biting. What is key is to maintain a temperature suitable for the pigs' age, and if possible, all the housing systems to permit pigs themselves exercise agency to achieve the optimal thermal comfort (Domain 4).

4.3 *Humidity*

Although humidity has been identified as a risk factor for tail biting (Taylor *et al.*, 2010), no experimental studies on how changes in humidity impacts the occurrence of tail biting have been carried out. Some indirect studies showed that humidity can affect pig behaviour; higher levels of precipitation (and therefore humidity) increased outdoor pigs' rooting behaviours (Andresen and Redbo, 1999), and it also decreased the lying duration of pigs kept in pens with outdoor runs (Olsen *et al.*, 2001). Less lying behaviour may indirectly contribute to more tail biting. However, indoor housed pigs showed longer lying behaviour with higher humidity (Huynh *et al.*, 2005). Humidity and temperature also have an interactive effect on pig behaviours; at a higher humidity level, pigs tended to lie more on slatted floor at a lower temperature threshold (Huynh *et al.*, 2005). These changes in general behaviour could also affected the occurrence of tail biting.

4.4 *Noise and light*

Pigs show similar responses to long-term noise exposure to humans. When pigs of 25–30 kg were exposed to 108 dB of noise (equal to live rock music) continuously for 72h, their plasma catecholamine concentrations increased, with a prolonged increased plasma cortisol even after the noise ceased, which signifies inhibition (Kemper *et al.*, 1976). However, only one experimental study to date has studied the potential effect of noise on tail biting. Surveying 755 pig compartments from 214 finishing farms (with mostly docked pigs) in Germany, Wegner *et al.* (2019) found the average on-farm noise level was 70.2 ± 5.2 dB (52.5–92.4 dB). Compartments with older finishing pigs (17–22 weeks of age) were noisier than those with younger ones (11–16 weeks of age). Wet feed systems were noisier than dry feed systems, and a slightly higher noise level was found in partly slatted than fully slatted systems. However, they did not find a difference in tail lesion between different noise levels recorded (Wegner *et al.*, 2019). On the other hand, it has been suggested that some level of noise made by farm staff can act as a buffer for tail biting as anecdotally tail biting happened more often during the weekends when staff was not present (Schrøder-Petersen and Simonsen, 2001). However although the effect of noise on tail biting is still unclear, pig producers considered it the least important measure in dealing with tail biting on commercial farms (Haigh and O'Driscoll, 2019; Valros *et al.*, 2016; Valros and Barber, 2019).

Another rarely researched topic is how light provision may affect tail biting. Some anecdotal evidence suggests that dimming light can stop pigs from tail biting (Bracke *et al.*, 2013; Schrøder-Petersen and Simonsen, 2001); similar reports can be found in relation to feather pecking in laying hens (Rodenburg and Koene, 2004). This can be generally attributed to a known preference for pigs to rest in a darker space (Taylor *et al.*, 2006), and therefore the assumption that darkness induces inactivity, and hence less tail biting. However, appropriate lighting is important for pigs' day-time behaviour as they prefer to have access to light during the day (Taylor *et al.*, 2006). Future research into how different aspects of lighting can affect pig welfare could also shed more light on how it may affect tail biting (Scaillierez *et al.*, 2021).

5 Designing the pig farm of the future

Historically, research into infrastructure in pig housing tended to be reductive, investigating the effect of single factors on a range of multidisciplinary measures, whether behavioural, health, or performance based. However, as demonstrated above, many of the design features of pig houses interact with each other, and measurements should not always be interpreted in isolation. Thus as well as ongoing targeted research, we hope to see an increasing proportion of inter- and transdisciplinary research, into system-based solutions as to how to manage pigs routinely in a way that means there is no need to tail dock. One recent example is work by Brandt *et al.* (2020), who carried out a system study that compared two pen designs, one 'standard' and the other a design that incorporated several aspects of design that are known to reduce the risk of biting (larger proportion of solid floor area, reduced group size, increased space allowance and straw on the floor), and found significantly less biting in the enhanced pens. We would like to see more work like this, and moreover, that as well as considering just the issue of pig welfare, other societal and environmental considerations are taken into account. Pig production could place significant strain upon the environment when it comes to emissions, and the public are becoming increasingly aware of, and concerned about, farm animal husbandry conditions and welfare standards. Thus development of new systems needs to take these issues into account, as well as economics. An interesting long term area for exploration is whether the opportunity for pigs to have some access to the outdoors, and once outdoors provided with the opportunity to root etc. can be facilitated at large commercial scale level. However, in the immediate term at least, research into and adoption of facility design that accommodates the pigs' natural behaviours, and allow animals to thrive, not just survive, should be promoted (Mellor, 2016).

6 Summary

- There are knowledge gaps in terms of alternative, non-indoor farming systems and the risks of tail biting
- Systems with some access to the outdoors, and ideally a rooting area with appropriate material provided (e.g. compost, soil etc.), are more likely to satisfy pigs' behavioural needs than current conventional indoor systems
- In indoor housing systems, slatted flooring is the main risk factor for tail biting, although this is usually confounded with the availability of bedding and rooting materials
- There are interactive effects between the physical and non-physical environment on the risk of tail biting

References

Alban, L., Petersen, J.V. and Busch, M.E., 2015. A comparison between lesions found during meat inspection of finishing pigs raised under organic/free-range conditions and conventional, indoor conditions. Porcine Health Management 1:4. https://doi.org/10.1186/2055-5660-1-4.

Andersen, H.M.L., Kongsted, A.G. and Jakobsen, M., 2020. Pig elimination behavior – A review. Applied Animal Behaviour Science, 222, p. 104888. https://doi.org/10.1016/j.applanim.2019.104888.

Andresen, N. and Redbo, I., 1999. Foraging behaviour of growing pigs on grassland in relation to stocking rate and feed crude protein level. Applied Animal Behaviour Science, 62(2–3):183–197. https://doi.org/10.1016/S0168-1591(98)00221-4.

Augère-Granier, M.-L.; European Parliament: European Parliamentary Research Service, 2020. European Parliament Briefing 'The EU pig sector' PE 652.044. https://www.europarl.europa.eu/RegData/etudes/BRIE/2020/652044/EPRS_BRI(2020)652044_EN.pdf Accessed 09 February 2022.

Banhazi, T.M., Seedorf, J., Rutley, D.L. and Pitchford, W.S., 2008. Identification of risk factors for sub-optimal housing conditions in Australian piggeries: Part 2. Airborne pollutants. J. Agric. Saf. Health, 14:21–39. https://doi.org/10.13031/2013.24122.

Bealey, W.J., Dore, A.J., Dragosits, U., Reis, S., Reay, D.S. and Sutton, M.A., 2016. The potential for tree planting strategies to reduce local and regional ecosystem impacts of agricultural ammonia emissions. Journal of environmental management, 165:106–116. https://doi.org/10.1016/j.jenvman.2015.09.012.

Bracke, M.B.M., de Lauwere, C.C., Wind, S.M.M. and Zonderland, J.J., 2013. Attitudes of Dutch pig farmers towards tail biting and tail docking. Journal of Agricultural and Environmental Ethics 26(4): 847–868. https://doi.org/10.1007/s10806-012-9410-2.

Brandt., P., Hakansson, F., Jensen,T., Nielsen, M.B.F, Lahrmann, H.P., Hansen, C.F. and Forkman, B., 2020. Effect of pen design on tail biting and tail-directed behaviour of finishing pigs with intact tails. Animal 14(5):1034–1042. https://doi.org/10.1017/S1751731119002805.

Brownlow, M., Dorward, P. and Carruthers, S., 2005. Integrating natural woodland with pig production in the United Kingdom: an investigation of potential performance and interactions. Agroforestry Systems 64: 251–263. https://doi.org/10.1007/s10457-004-0250-6.

Brownlow, M.J.C., 1994. Towards a framework of understanding for the integration of forestry with domestic pig (*Sus scrofa domestica*) and European wild boar (*Sus scrofa scrofa*) husbandry in the United Kingdom. Forestry: An International Journal of Forest Research 67(3): 189–218. https://doi.org/10.1093/forestry/67.3.189.

Buijs, S., Muns, R., 2019. A review of the effects of non-straw enrichment on tail biting in pigs. Animals 9, 824. https://doi.org/10.3390/ani9100824.

Bulens, A., Van Beirendonck, S., Van Thielen, J., Buys, N. and Driessen, B., 2017. Hiding walls for fattening pigs: Do they affect behavior and performance? Applied Animal Behaviour Science, 195:32–37. https://doi.org/10.1016/j.applanim.2017.06.009.

Burbi, S. and Olave, R.J., 2018. Supporting farmers in the transition to agroecology to promote carbon sequestration from silvopastoral systems. International Journal of Agricultural Extension, 6(3):17–27.

Bus, J.D., Boumans, I.J., Webb, L.E. and Bokkers, E.A., 2021. The potential of feeding patterns to assess generic welfare in growing-finishing pigs. Applied Animal Behaviour Science, 241, p. 105383. https://doi.org/10.1016/j.applanim.2021.105383.

Courboulay, V., Souffran, H., Guingand, N., Massabie, P. and Dubois, A., 2008. Intérêt de la brumisation en engraissement pour améliorer le confort des porcs en croissance (Fogging system: a way to improve the welfare of growing-finishing pigs during summer periods). Journées Recherche Porcine, 40:243–246.

D'Eath, R.B., Arnott, G., Turner, S.P., Jensen, T., Lahrmann, H.P., Busch, M.E., Niemi, J.K., Lawrence, A.B., Sandøe, P., 2014. Injurious tail biting in pigs: how can it be controlled in existing systems without tail docking? Animal 8, 1479–1497. https://doi.org/10.1017/s1751731114001359.

Dilling, C., 2022. POWER: Power to strengthen welfare and resilience in organic pig production. Available at https://projects.au.dk/coreorganiccofund/core-organic-cofund-projects/power.

Edwards, S., 2005. Product quality attributes associated with outdoor pig production. Livestock Science 94:5–14 https://doi.org/10.1016/j.livprodsci.2004.11.028.

Edwards, S., 2011. What do we know about tail biting today? Pig Journal 66, 81–86.

EFSA (European Food Safety Authority), 2005. The welfare of weaners and rearing pigs – effects of different space allowances and floor types. (Scientific Report; EFSA-Q-2004-077) Eur. Food Safety Authority (EFSA)–AHAW Panel (Animal Health and Welfare), (Annex to EFSA-Journal No. 268, 1–19).

EFSA (European Food Safety Authority), 2021. Scientific opinion on African swine fever and outdoor farming of pigs. EFSA Journal 19(6):6639 doi: 10.2903/j.efsa.2021.6639.

EU, 2018. European Parliament and Council Regulation (EU) 2018/848 of the European Parliament and of the Council of 30 May 2018 on organic production and labelling of organic products and repealing Council Regulation (EC) No 834/2007 Official Journal of the European Union, L150, :1–92.

European Commission DG SANTE, 2016. Final report of a study visit carried out in Switzerland from 04 April 2016 to 08 April 2016 in order to share best practice on rearing pigs with intact tails (Health and food audits and analysis No. DG(SANTE) 2016-8774-MR). Directorate-General For Health And Food Safety, Brussels, Belgium.

Ewbank, R., 1973. Abnormal behaviour and pig nutrition. An unsuccessful attempt to induce tail biting by feeding a high energy, low fibre vegetable protein ration. British veterinary journal, 129(4):366–369. https://doi.org/10.1016/S0007-1935(17)36440-0.

Früh, B., Bochicchio, D., Edwards, S., Hegelund, L., Leeb, C., Sundrum, A., Werne, S., Wiberg, S., Prunier, A., 2014. Description of organic pig production in Europe. Organic Agriculture 4, 83–92. https://doi.org/10.1007/s13165-013-0056-9.

Geers, R., Dellaert, B., Goedseels, V., Hoogerbrugge, A., Vranken, E., Maes, F. and Berckmans, D., 1989. An assessment of optimal air temperatures in pig houses by the quantification of behavioural and health-related problems. Animal Science, 48(3):571–578. https://doi.org/10.1017/S0003356100004098.

Goossens, X., Sobry, L., Ödberg, F., Tuyttens, F., Maes, D., De Smet, S., Nevens, F., Opsomer, G., Lommelen, F. and Geers R., 2008. A population-based on-farm evaluation protocol for comparing the welfare of pigs between farms. Animal Welfare17:35–41. https://doi.org/10.1017/S0962728600031961.

Grümpel, A., Krieter, J., Veit, C. and Dippel, S., 2018. Factors influencing the risk for tail lesions in weaner pigs (Sus scrofa). Livestock Science 216:219–226. https://doi.org/10.1016/j.livsci.2018.09.001.

Guo, Y., Lian, X. and Yan. P., 2015. Diurnal rhythms, locations and behavioural sequences associated with eliminative behaviours in fattening pigs. Applied Animal Behaviour Science, 168:18–23. https://doi.org/10.1016/j.applanim.2015.01.011.

Haigh, A. and O'Driscoll, K., 2019. Irish pig farmer's perceptions and experiences of tail biting. Porcine Health Management 5:30 https://doi.org/10.1186/s40813-019-0135-8.

Hämeenoja, P., 2002. Animal health and welfare – pig production. Proceedings of the 14th Internordic Symposium of the Nordic Committee for Veterinary Scientific Cooperation (NKVet). Acta Veterinaria Scandinavica 43(1): 1–4.

Heinola, K., Kauppinen, T., Niemi, J.K., Wallenius, E. and Raussi, S., 2021. Comparison of 12 different animal welfare labeling schemes in the pig sector. Animals 11:2430 https://doi.org/10.3390/ani11082430.

Heldmer, E. and Lundeheim, N., 2010. Gross lesions at slaughter among organic pigs in Sweden. 21st IPVS Congress International Pig Veterinary Society, Vancouver, Canada.

Henry, M., Jansen, H., Amezcua, M.D.R., O'Sullivan, T.L., Niel, L., Shoveller, A.K. and Friendship, R.M., 2021. Tail-biting in pigs: a scoping review. Animals, 11(7):2002. https://doi.org/10.3390/ani11072002.

Hunter, E.J., Jones, T.A., Guise, H.J., Penny, R.H.C. and Hoste, S., 2001. The relationship between tail biting in pigs, docking procedure and other management practices. The Veterinary Journal, 161(1):72–79. https://doi.org/10.1053/tvjl.2000.0520.

Ichim, O., 2012. An overview of organic pig farming in Romania. Porcine Research 2(2): 50–65.

Holinger, M., Früh, B., Stoll, P., Kreuzer, M. and Hillmann, E., 2018. Grass silage for growing-finishing pigs in addition to straw bedding: Effects on behaviour and gastric health. Livestock Science, 218:50–57. https://doi.org/10.1016/j.livsci.2018.10.012.

Holling, C., Vidondo, B. and Nathues, C., 2017. Provision of straw by a foraging tower – effect on tail biting in weaners and fattening pigs. Porcine health management, 3(1):1–14. https://doi.org/10.1186/s40813-017-0052-7.

Huynh, T.T.T., Aarnink, A.J.A., Gerrits, W.J.J., Heetkamp, M.J.H., Canh, T.T., Spoolder, H.A.M., Kemp, B. and Verstegen, M.W.A., 2005. Thermal behaviour of growing pigs in response to high temperature and humidity. Applied animal behaviour science, 91(1–2):1–16. https://doi.org/10.1016/j.applanim.2004.10.020.

IFOAM, 2019. IFOAM Norms version 2014. IFOAM basic standards for organic production and processing. International Federation of Organic Movements, Tholey-Theley, Germany (http://www. ifoam.org/standard/norms/ibs.pdf). Accessed 8 March 2022.

Jakobsen, M. and Kongsted, A.C., 2015. Free-range pigs integrated with agroforestry. International conference on pig welfare: improving pig welfare – what are the ways forward? Copenhagen, Denmark, 29–30 April 2015 p. 109.

Jackson, P., Nasirahmadi, A., Guy, H.J., Bull, S., Avery, J.P., Edwards, A.S., Sturm, B., 2020. Using CFD modelling to relate pig lying locations to environmental variability in finishing pens. Sustainability 12, 1928. https://doi.org/10.3390/su12051928.

Jensen, T., Kold Nielsen, C., Vinther, J. and D'Eath, R.B., 2012. The effect of space allowance for finishing pigs on productivity and pen hygiene. Livestock Science 149, 33–40. https://doi.org/10.1016/j.livsci.2012.06.018.

Jurado, C., Martínez-Avilés, M., De La Torre, A., Štukelj, M., de Carvalho Ferreira, H.C., Cerioli, M., Sánchez-Vizcaíno, J.M., Bellini, S., 2018. Relevant measures to prevent the spread of African swine fever in the European Union domestic pig sector. Frontiers in Veterinary Science 5:77 https://doi.org/10.3389/fvets.2018.00077.

Kallio, P.A., Janczak, A.M., Valros, A.E., Edwards, S.A. and Heinonen, M., 2018. Case control study on environmental, nutritional and management-based risk factors for tail-biting in long-tailed pigs. Animal Welfare 27(1):21–34 https://doi.org/10.7120/09627286.27.1.021.

Kells, N.J., 2021. Review: The Five Domains model and promoting positive welfare in pigs. Animal 16:100378 https://doi.org/10.1016/j.animal.2021.100378.

Kemper, A., Wildenhahn, V. and Lyhs, L., 1976. Effect of long-lasting noise on the plasma concentration of catecholamines, glucocorticosteroids and PBI in pigs. Archiv fur Experimentelle Veterinarmedizin, 30(4):619–625.

Kijlstra, A. and Eijck, I., 2006. Animal health in organic livestock production systems: a review. NJAS-Wageningen Journal of Life Sciences 54(1): 77–94. https://doi.org/10.1016/S1573-5214(06)80005-9.

Kongsted, H. and Sørensen, J.T., 2017. Lesions found at routine meat inspection on finishing pigs are associated with production system. The Veterinary Journal 223:21–26. https://doi.org/10.1016/j.tvjl.2017.04.016.

Kritas, S.K., Morrison, R.B., 2007. Relationships between tail biting in pigs and disease lesions and condemnations at slaughter. Veterinary Record 160, 149–152. https://doi.org/10.1136/vr.160.5.149.

Lee, H., Perkins, C., Gray, H., Hajat, S., Friel, M., Smith, R.P., Williamson, S., Edwards, P. and Collins, L.M., 2020. Influence of temperature on prevalence of health and welfare conditions in pigs: time-series analysis of pig abattoir inspection data in England and Wales. Epidemiology & Infection, 148:e20. https://doi.org/10.1017/S0950268819002085.

Leeb, C., Rudolph, G., Bochicchio, D., Edwards, S., Früh, B., Holinger, M., Holmes, D., Illmann, G., Knop, D., Prunier, A., Rousing, T., Winckler, C., Dippel, S., 2019. Effects of three husbandry systems on health, welfare and productivity of organic pigs. Animal 13, 2025–2033. https://doi.org/10.1017/S1751731119000041.

Ludwiczak, A., Skrzypczak, E., Składanowska-Baryza, J., Stanisz, M., Sl'ósarz, P. and Racewicz, P., 2021. How housing conditions determine the welfare of pigs. Animals 11, 3484. https://doi.org/10.3390/ani11123484.

Lund, V., 2006. Natural living – a precondition for animal welfare in organic farming. Livestock Science 100:71–83. https://doi.org/10.1016/j.livprodsci.2005.08.005.

Manevski, K., Jakobsen, M., Kongsted, A.G., Georgiadis, P., Labouriau, R., Hermansen, J.E. and Jørgensen, U., 2019. Effect of poplar trees on nitrogen and water balance in outdoor pig production – a case study in Denmark. Science of the total environment, 646:1448–1458. https://doi.org/10.1016/j.scitotenv.2018.07.376.

McGlone, J.J. and Curtis, S.E., 1985. Behavior and performance of weanling pigs in pens equipped with hide areas. Journal of Animal Science, 60(1):20–24. https://doi.org/10.2527/jas1985.60120x.

McKinnon, A.J., Edwards, S.A., Stephens, D.B. and Walters, D.E., 1989. Behaviour of groups of weaner pigs in three different housing systems. British Veterinary Journal, 145(4):367–372. https://doi.org/10.1016/0007-1935(89)90035-3.

Mellor, D.J., 2016. Updating animal welfare thinking: moving beyond the "Five Freedoms" to "A Life worth Living". Animals 6:21. https://doi.org/10.3390/ani6030021.

Mellor, D.J., 2017. Operational details of the five domains model and its key applications to the assessment and management of animal welfare. Animals 7:60. https://doi.org/10.3390/ani7080060.

Mellor, D.J., Beausoleil, N.J., Littlewood, K.E., McLean, A.N., McGreevy, P.D., Jones, B. and Wilkins, C., 2020. The 2020 Five Domains model: including human–animal interactions in assessments of animal welfare. 10:1870 https://doi.org10.3390/ani10101870.

Mellor, D.J. and Reid, C.S.W., 1994. Concepts of animal well-being and predicting the impact of procedures on experimental animals. In Improving the well-being of animals in the research environment; Australian and New Zealand Council for the Care of Animals in Research and Teaching (ANZCCART): Glen Osmond, SA, Australia:3–18.

Misra, S., Bokkers, E.A., Upton, J., Quinn, A.J. and O'Driscoll, K., 2021. Effect of environmental enrichment and group size on the water use and waste in grower-finisher pigs. Scientific Reports, 11(1):1–10. https://doi.org/10.1038/s41598-021-95880-0.

Moinard, C., Mendl, M., Nicol, C.J. and Green, L.E., 2003. A case control study of on-farm risk factors for tail biting in pigs. Applied Animal Behaviour Science 81(4): 333–355. https://doi.org/10.1016/S0168-1591(02)00276-9.

Mosquera-Losada, M.R., McAdam, J.H., Romero-Franco, R., Santiago-Freijanes, J.J. and Rigueiro-Rodróguez, A., 2009. Definitions and components of agroforestry practices in Europe. Agroforestry in Europe 3–19. https://doi.org/10.1007/978-1-4020-8272-6_1.

Mul, M., Vermeij, I., Hindle, V., Spoolder, H., 2010. EU-Welfare legislation on pigs (No. 273). Wageningen UR Livestock Research, Lelystad, the Netherlands.

Nannoni, E., Aarnink, A.J., Vermeer, H.M., Reimert, I., Fels, M. and Bracke, M., 2020. Soiling of pig pens: A review of eliminative behaviour. Animals, 10(11):2025. https://doi.org/10.3390/ani10112025.

Ocepek, M., Goold, C.M., Busančić, M., Aarnink, A.J.A., 2020. Maize silage as enrichment material improves the welfare of growing-finishing pigs in environmentally-friendly pens. Applied Animal Behaviour Science 230:105043. https://doi.org/10.1016/j.applanim.2020.105043.

Olsen, A.W., Dybkjær, L. and Simonsen, H.B., 2001. Behaviour of growing pigs kept in pens with outdoor runs: II. Temperature regulatory behaviour, comfort behaviour and dunging preferences. Livestock Production Science, 69(3):265–278. https://doi.org/10.1016/S0301-6226(01)00173-7.

Park, H.-S., Min, B. and Oh, S.-H., 2017. Research trends in outdoor pig production – A review. Asian-Australasian Journal of Animal Sciences 30:1207–1214 https://doi.org/10.5713/ajas.17.0330.

Pandolfi, F., Kyriazakis, I., Stoddart, K., Wainwright, N. and Edwards, S.A., 2017. The "Real Welfare" scheme: identification of risk and protective factors for welfare outcomes in commercial pig farms in the UK. Preventive Veterinary Medicine 146: 34–43. https://doi.org/10.1016/j.prevetmed.2017.07.008.

Papatsiros, V.G., 2011. Impact of animal health management on organic pig farming in Greece. Biotechnology in Animal Husbandry 27(1):115–125. https://doi.org/10.2298/BAH1101115P.

Paul, E.S., Moinard, C., Green, L.E. and Mendl, M., 2007. Farmers' attitudes to methods for controlling tail biting in pigs. Veterinary record, 160(23):803. https://doi.org/10.1136/vr.160.23.803.

Pereira, J., Fangueiro, D., Misselbrook, T.H., Chadwick, D.R., Coutinho, J., Trindade, H., 2011. Ammonia and greenhouse gas emissions from slatted and solid floors in dairy cattle houses: A scale model study. Biosystems Engineering 109, 148–157. https://doi.org/10.1016/j.biosystemseng.2011.02.011.

Qiao, F., Huang, J., Wang, D., Liu, H. and Lohmar, B., 2016. China's hog production: from backyard to large-scale. China Economic Review, 38:199–208. https://doi.org/10.1016/j.chieco.2016.02.003.

Raj, A.B.M. and Gregory, N.G., 1996. Welfare implications of the gas stunning of pigs 2. Stress of induction of anaesthesia. Animal Welfare, 5(1):71–78. https://doi.org/10.1017/S0962728600018352.

Philippe, F.-X., Cabaraux, J.-F., Nicks, B., 2011. Ammonia emissions from pig houses: influencing factors and mitigation techniques. Agriculture, Ecosystems and the Environment 141:245–260. https://doi.org/10.1016/j.agee.2011.03.012.

Rodenburg, T.B. and Koene, P., 2004. Feather pecking and feather loss. In Welfare of the laying hen (Vol. 27). Perry, G.C. ed., CABI. p. 227.

Rodríguez-Estévez, V., García, A., Peña, F. and Gómez, A.G., 2009a. Foraging of Iberian fattening pigs grazing natural pasture in the dehesa. Livestoci Sci. 120:135–143. https://doi.org/10.1016/j.livsci.2008.05.006.

Rodríguez-Estévez, V., García, A., Peña, F. and Gómez, A.G., 2009b. Characteristics of the acorns selected by free range Iberian pigs during the *montanera* season. Livestock Science 122:169–176. https://doi.org/10.1016/j.livsci.2008.08.010.

Sällvik, K. and Walberg, K., 1984. The effects of air velocity and temperature on the behaviour and growth of pigs. Journal of Agricultural Engineering Research 30:305–312. https://doi.org/10.1016/S0021-8634(84)80031-1.

Salomon, E., Åkerhielm, H., Lindahl, C. and Lindgren, K., 2007. Outdoor pig fattening at two Swedish organic farms – spatial and temporal load of nutrients and potential environmental impact. Agriculture, ecosystems & environment, 121(4):407–418. https://doi.org/10.1016/j.agee.2006.11.017.

Scaillierez, A.J., Boumans, I.J.M.M., van Nieuwamerongen, S.E., van der Tol, P.P.J. and Bokkers, E., 2021, April. Enlighted pigs: throwing a light on pig welfare. In WIAS Annual

Conference 2021: Resilience:46. Wageningen University & Research. https://edepot.wur.nl/549419.

Scheepens, C.J.M., Hessing, M.J.C., Laarakker, E., Schouten, W.G.P. and Tielen, M.J.M., 1991. Influences of intermittent daily draught on the behaviour of weaned pigs. Applied Animal Behaviour Science, 31(1–2):69–82. https://doi.org/10.1016/0168-1591(91)90154-P.

Schrøder-Petersen, D.L. and Simonsen, H.B., 2001. Tail biting in pigs. The Veterinary Journal, 162(3):196–210.

Scott, K., Chennells, D.J., Campbell, F.M., Hunt, B., Armstrong, D., Taylor, L., Gill, B.P., Edwards, S.A., 2006. The welfare of finishing pigs in two contrasting housing systems: Fully-slatted versus straw-bedded accommodation. Livestock Science 103:104–115. https://doi.org/10.1016/j.livsci.2006.01.008.

Scollo, A., Contiero, B., and Gottardo, F., 2016. Frequency of tail lesions and risk factors for tail biting in heavy pig production from weaning to 170 kg live weight. The Veterinary Journal 207:92–98. https://doi.org/10.1016/j.tvjl.2015.10.056.

Scollo, A., Gottardo, F., Contiero, B. and Edwards, S.A., 2017. A cross-sectional study for predicting tail biting risk in pig farms using classification and regression tree analysis. Preventive veterinary medicine, 146:114–120. https://doi.org/10.1016/j.prevetmed.2017.08.001.

Smith, J.H., Wathes, C.M. and Baldwin, B.A., 1996. The preference of pigs for fresh air over ammoniated air. Applied Animal Behaviour Science, 49(4):417–424. https://doi.org/10.1016/0168-1591(96)01048-9.

Smulders, D., Hautekiet, V., Verbeke, G. and Geers, R., 2008. Tail and ear biting lesions in pigs: an epidemiological study. Animal Welfare, 17(1):61–69. https://doi.org/10.1017/S0962728600031997.

Sonoda, L.T., Fels, M., Oczak, M., Vranken, E., Ismayilova, G., Guarino, M., Viazzi, S., Bahr, C., Berckmans, D. and Hartung, J., 2013. Tail biting in pigs – causes and management intervention strategies to reduce the behavioural disorder. A review. Berl Munch Tierarztl Wochenschr 126(3–4):104–112. https://doi.org/10.2376/0005-9366-126-104.

Spinka, M., 2019. Animal agency, animal awareness and animal welfare. Animal Welfare 28:11–20. https://doi.org/10.7120/09627286.28.1.011.

Stäbler, R., Patzkewitsch, D., Reese, S., Erhard, M. and Hartmannsgruber, S., 2020. Alternative husbandry system for fattening pigs in the forest – assessment of animal health and carcass quality. Berl Munch Tierarztl Wochenschr 133. https://doi.org/10.2376/1439-0299-2020-11.

Stäbler, R., Patzkewitsch, D., Reese, S., Erhard, M. and Hartmannsgruber, S., 2022. Behavior of domestic pigs under near-natural forest conditions with ad libitum supplementary feeding. Journal of Veterinary Behaviour 48:20–35 https://doi.org/10.1016/j.jveb.2021.10.011.

Stolba, A. and Wood-Gush, D.G.M., 1989. The behaviour of pigs in a semi-natural environment. Animal Science 48:419–425. https://doi.org/10.1017/S0003356100040411.

Studnitz, M., Jensen, M.B. and Pedersen, L., 2007. Why do pigs root and in what will they root? A review on the exploratory behaviour of pigs in relation to environmental enrichment. Applied Animal Behaviour Science 107:183–197. https://doi.org/10.1016/j.applanim.2006.11.013.

Spoolder, H., Bracke, M., Mueller-Graf, C., Edwards, S., 2011. Technical report submitted to EFSA preparatory work for the future development of animal based measures for assessing the welfare of pigs. Report 2: preparatory work for the future development of animal based measures for assessing the welfare of weaned, growing and fattening pigs including aspects related to space allowance, floor types, tail biting and need for tail docking (No. EN-181). EFSA, Parma.

Taylor, N.R., Main, D.C., Mendl, M. and Edwards, S.A., 2010. Tail-biting: a new perspective. The Veterinary Journal, 186(2):137–147. https://doi.org/10.1016/j.tvjl.2009.08.028.

Taylor, N.R., Parker, R.M., Mendl, M., Edwards, S.A. and Main, D.C., 2012. Prevalence of risk factors for tail biting on commercial farms and intervention strategies. The Veterinary Journal, 194(1):77–83. https://doi.org/10.1016/j.tvjl.2012.03.004.

Taylor, N., Prescott, N., Perry, G., Potter, M., Le Sueur, C. and Wathes, C., 2006. Preference of growing pigs for illuminance. Applied Animal Behaviour Science, 96(1–2):19–31. https://doi.org/10.1016/j.applanim.2005.04.016.

Temple, D., Courboulay, V., Velarde, A., Dalmau, A. and Manteca, X., 2012. The welfare of growing pigs in five different production systems in France and Spain: assessment of health. Animal Welfare 21(2):257–271. https://doi.org/10.1017/S1751731111001868.

Tillmanns, M., Scheepens, K., Stolte, M., Herbrandt, S., Kemper, N., and Fels, M., 2022. Implementation of a pig toilet in a nursery pen with a straw-littered lying area. Animals. 12:113. https://doi.org/ 10.3390/ani12010113.

Valros, A., 2018. Tail biting, in: Spinka, M. (Ed.), Advances in pig welfare, the advances in farm animal welfare series. Elsevier, Duxford, UK, :137–166. https://doi.org/10.1016/B978-0-08-101012-9.00004-6.

Valros, A., 2021. The tale of the Finnish pig tail – how to manage non-docked pigs? Animal, p. 100353. https://doi.org/10.1016/j.animal.2021.100353.

Valros, A. and Barber, C., 2019. Producer perceptions of the prevention of tail biting on UK farms: association to bedding use and tail removal proportion. Animals, 9(9):628. https://doi.org/10.3390/ani9090628.

Valros, A., Munsterhjelm, C., Hänninen, L., Kauppinen, T. and Heinonen, M., 2016. Managing undocked pigs – on-farm prevention of tail biting and attitudes towards tail biting and docking. Porcine health management, 2(1):1–11. https://doi.org/10.1186/s40813-016-0020-7.

van Putten, G., 1969. An investigation into tail-biting among fattening pigs. British Veterinary Journal, 125(10): 511–517. https://doi.org/10.1016/S0007-1935(17)48710-0.

Van Ransbeeck, N., Van Langenhove, H., Demeyer, P., 2013. Indoor concentrations and emissions factors of particulate matter, ammonia and greenhouse gases for pig fattening facilities. Biosystems Engineering 116, 518–528. https://doi.org/10.1016/j.biosystemseng.2013.10.010.

van Staaveren, N., Vale, A.P., Manzanilla, E.G., Teixeira, D.L., Leonard, F.C., Hanlon, A., Boyle, L.A., 2016. Relationship between tail lesions and lung health in slaughter pigs. Preventive Veterinary Medicine 127, 21–26. https://doi.org/10.1016/j.prevetmed.2016.03.004.

Vitali, M., Santacroce, E., Correa, F., Salvarani, C., Maramotti, F.P., Padalino, B. and Trevisi, P., 2020. On-farm welfare assessment protocol for suckling piglets: a pilot study. Animals, 10(6):1016. https://doi.org/10.3390/ani10061016.

Walker, P.K. and Bilkei, G., 2006. Tail-biting in outdoor pig production. The Veterinary Journal. 171(2): 367–369. https://doi.org/10.1016/j.tvjl.2004.10.011.

Wallgren, T., Lundeheim, N., Wallenbeck, A., Westin, R. and Gunnarsson, S., 2019. Rearing pigs with intact tails – experiences and practical solutions in Sweden. Animals 9:812 https://doi.org/10.3390/ani9100812.

Wallgren, T., Westin, R. and Gunnarsson, S., 2016. A survey of straw use and tail biting in Swedish pig farms rearing undocked pigs. Acta Veterinaria Scandinavica, 58(1):1–11. https://doi.org/10.1186/s13028-016-0266-8.

Wealleans, A.L., 2013. Such as pigs eat: the rise and fall of the pannage pig in the UK. Journal of the Science of Food and Agriculture 93(9): 2076–2083. https://doi.org/10.1002/jsfa.6145.

Weschler, B.; Bachmann, I., 1998. A sequential analysis of eliminative behaviour in domestic pigs. Applied Animal Behaviour Science 56:29–36. https://doi.org/10.1016/S0168-1591(97)00075-0.

Wegner, B., Spiekermeier, I., Nienhoff, H., Große-Kleimann, J., Rohn, K., Meyer, H., Plate, H., Gerhardy, H., Kreienbrock, L., Grosse Beilage, E. and Kemper, N., 2019. Status quo analysis of noise levels in pig fattening units in Germany. Livestock Science, 230:103847. https://doi.org/10.1016/j.livsci.2019.103847.

Wiegand, R.M., Gonyou, H. and Curtis, S.E., 1994. Pen shape and size: effects on pig behavior and performance. Applied Animal Behaviour Science 39:49–61. https://doi.org/10.1016/0168-1591(94)90015-9.

Wimmler, C., Knoll, M., Leeb C., and Vermeer, H., 2022a. Welfare and environmental impact of organic pig production: A collection of Factsheets. Research Institute of Organic Agriculture FiBL, Switzerland. https://www.fibl.org/fileadmin/documents/shop/1300-hb-power-en.pdf Accessed 9 March 2022.

Wimmler, C., Vermeer, H.M., Leeb, C., Salomon, E. and Andersen, H.M.-L., 2022b. Review: concrete outdoor runs for organic growing-finishing pigs – a legislative, ethological and environmental perspective. Animal 16:100435 https://doi.org/10.1016/j.animal.2021.100435.

Woonwang, Y., Do Tein, D. and Thanawongnuwech, R., 2020. The future of the pig industry after the introduction of African swine fever into Asia. Animal Frontiers 10(4):30–37 https://doi.org/10.1093/af/vfaa037.

Ye, Z., Li, B., Cheng, B., Chen, G., Zhang, G., Shi, Z, Wei, X. and Xi, L., 2007. A concrete slatted floor system for separation of faeces and urine in pig houses. Biosystems Engineering 98:206–214 https://doi.org/10.1016/j.biosystemseng.2007.07.007.

Zhang, L., Sheng, J., Zhang, Y., Chen, L., Sun, G. and Zheng, J., 2016. Ammonia and greenhouse gas emissions from different types of deep litter used for pig rearing. Livestock Science 188:166–173. https://doi.org/10.1016/j.livsci.2016.04.012.

Zwicker, B., Weber, R., Wechsler, B. and Gyax, L., 2012. Degree of synchrony based on individual observations underlines the importance of concurrent access to enrichment materials in finishing pigs. Applied Animal Behaviour Science 172:26–32 https://doi.org/10.1016/j.applanim.2015.08.037.

CHAPTER 8

Environmental enrichment

Stephanie Buijs[1,*], *Jen-Yun Chou*[2,3] *and Heleen van de Weerd*[4]

[1]Agri-Food and Biosciences Institute, Sustainable Agri-Food Sciences Division, Livestock Production Sciences Branch, Large Park, Hillsborough BT26 6DR, UK; [2]Pig Development Department, Teagasc, Animal & Grassland Research and Innovation Centre, Moorepark, Fermoy, Co. Cork, P61 C996, Ireland; [3]Institute of Animal Welfare Science, University of Veterinary Medicine, Veterinärplatz 1, Vienna, 1210, Austria; [4]Cerebrus associates, The White House, 2 Meadrow, Godalming, Surrey GU7 3HN, UK; *Correspondence: stephanie.buijs@afbini.gov.uk

Abstract

Environmental enrichment is commonly used to reduce tail biting. The most used types of enrichment for pigs are organic substrates (e.g., straw or wood shavings) and 'point source' objects (e.g., hanging wooden blocks or plastic toys). These can stimulate activity and give pigs some opportunity to engage in rewarded action, whilst stimulating their different senses to varying degrees. Amongst the organic substrates, straw has been studied most. It is most effective in large quantities, although smaller quantities can still reduce tail biting. Several less-studied substrates seem at least as effective as straw, offering opportunities where straw is unavailable. Many different 'point source' objects have been tested, of which fresh wood and hessian sacks were found to reduce tail biting to some extent. Other objects were generally ineffective unless changed regularly. As questions around the causation of tail biting are not fully solved, designing objects that addresses these causes remains challenging. Although most research has focused on preventive enrichment, enrichment can also be introduced when signs of tail biting have just started, to minimise the impact of biting. Pre-weaning or even pre-natal enrichment may also be beneficial, but knowledge on this is currently scarce. Similarly, little is known about the biosecurity risks, although unsafe levels of bacteria and toxins have occasionally been reported for some substrates. Poor compatibility with slatted floors, costs and biosecurity concerns currently limit the wider application of substrates in commercial practice, whereas cheaper but less effective point-source enrichment is more common. In conclusion, appropriate enrichment provision can reduce tail biting substantially. Nonetheless, it does not eliminate the behaviour to such an extent that no other measures to reduce tail biting are necessary.

 | DOI:10.3920/9789004694989_010

Keywords

pig – tail biting – lesion – substrate – point source

1 Introduction

Environmental enrichment is any modification of the environment that improves animal welfare, for instance by improving an animal's biological functioning, or its capacity to cope with stressors or to perform internally motivated behaviours (Newberry, 1995; Baumans, 2000). In intensive production systems, pigs are often confined within simple and invariant housing, compromising their physiological and behavioural needs (van de Weerd and Day, 2009). Enrichment can improve such housing systems, but not all changes to the farming environment are necessarily enriching (as changes may not actually improve welfare even if provided with that intention). Furthermore, it is important to note that enrichment is not a uniform category. Many different types of enrichment exist, and their value to the pig differs greatly. Effective enrichment can improve pigs' emotional state (Douglas *et al.* 2012) and there is general agreement that enrichment has some capacity to reduce tail biting. However, opinions on the importance of enrichment relative to other tail biting reduction strategies differ (Valros *et al.*, 2016; Benard *et al.*, 2014; Van de Weerd and Day, 2009).

Enrichment can be divided into five categories (Bloomsmith *et al.*, 1991), as described below. Pig enrichment often combines characteristics of multiple categories, as this is likely to make it more appealing (EU, 2016).

1.1 *Categories of enrichment*

Occupational and physical enrichment are the most commonly studied types of enrichment in the context of tail biting. Whilst occupational enrichment encourages activity and/or stimulates animals to apply cognitive abilities to gain control over their environment, physical enrichment increases environmental complexity by adding objects, substrates or permanent structures (Bloomsmith *et al.*, 1991). In the pig context, these categories tend to overlap as objects and substrates are provided both to stimulate activity and to provide a sense of control over the environment (Van de Weerd *et al.*, 2006). Examples of occupational/physical enrichment discussed in this chapter are bedding materials (which can stimulate the pig to forage, root, burrow and build nests), feed/substrate dispensers that provide access to their contents when manipulated correctly, and chewable point-source items. Another common form of physical enrichment is to increase the (subdivision of the) space that is available to the pigs.

Sensory enrichment stimulates the visual, auditory, olfactory, tactile or taste sense (Bloomsmith *et al.*, 1991). The types of enrichment mentioned in the previous categories will stimulate the pig's visual and tactile sense at least to some extent, as they are designed to be seen, manipulated and chewed. Furthermore, fresh organic materials (e.g. grass, freshly cut wood) or scented toys stimulate the pig's olfactory sense, and suspending enrichment from chains that jangle upon manipulation results in some auditory stimulation. Although this means that sensory enrichment has been studied implicitly, it is usually confounded with other aspects. It remains unknown how much stimulation of specific senses contributes to the effectiveness of enrichment, and what the most effective stimulus is for each sense.

Nutritional enrichment consists of feeding varied or novel feeds or varying how feed is delivered (Bloomsmith *et al.*, 1991). In the pig context this often means offering roughage in addition to the normal diet. Feed scattering and nutrient blocks have been investigated as well. Providing novel feed types is a largely unexplored area, likely because pigs are prone to feed neophobia (Middelkoop *et al.*, 2020).

Social enrichment is the provision of contact with other animals (Bloomsmith *et al.*, 1991). As growing pigs are kept in groups, this provides opportunities for both positive and negative interactions. When and how pigs are grouped with conspecifics may affect tail biting, as discussed in the chapter "Social environment".

1.2 *Mechanisms of action*

Although enrichment can reduce tail biting, the underlying mechanism is often unclear (Buijs and Muns, 2019). Some have hypothesised that enrichment reduces tail biting by alleviating boredom (Benard *et al.*, 2004). Others have theorised that it allows the satisfaction of an innate foraging motivation, which would otherwise be redirected to conspecifics (Pedersen *et al.*, 2005). Yet another theory is that interacting with enrichment provides an outlet for stress due to any other factor, thereby reducing tail biting as a result of such stress (Benard *et al.*, 2004; Pearce *et al.*, 1993). A lack of specific nutrients (Fraser, 1987) or of nest-like stimuli during early life (Telkanranta *et al.*, 2014b) have also been suggested as a possible causes for tail biting.

Understanding the underlying mechanism is important, as it determines which enrichment is useful. For instance, if tail biting is caused by a lack of specific nutrients, only enrichment that provides such nutrients would be expected to be beneficial. Conversely, if tail biting is due to stress in general, any change in housing or husbandry practises that reduces stress would be expected to reduce tail biting. Furthermore, the mechanism determines how the effectiveness of enrichment can be evaluated. Evaluating enrichment just by studying its direct effects on tail biting can be challenging, as tail biting can vary greatly between pens and batches even under seemingly identical conditions. This means that experiments that only measure tail biting (damage) require a very large sample size which is not always feasible or ethically acceptable. Therefore, the time spent interacting with the enrichment is often taken as a proxy of its effectiveness. This is appropriate if enrichment reduces tail biting by alleviating boredom. However, more time spent in interaction does not necessarily indicate a greater satisfaction of specific motivations, as pigs may continue to interact precisely because the interaction is not fully satisfactory. It has even been shown that some biting toys may increase the motivation for biting without allowing satisfactory interactions, thus resulting in increased tail biting (Van de Weerd *et al.*, 2006; Lahrmann *et al.*, 2019, Chou *et al.*, 2020b).

1.3 *In this chapter*

We describe existing knowledge on how enrichment affects tail biting behaviour and tail damage if provided preventively (i.e., before any signs of the behaviour) and correctively (in response to early signs of tail biting, Zonderland *et al.*, 2008). Generally, enrichment is supplied during life stages where tail biting is more likely to occur (i.e., the weaner or grower phase), but the few studies on pre-natal and pre-weaning enrichment are described in a separate section. Finally, we discuss aspects that currently limit the implementation of enrichment.

2 Preventive enrichment – organic substrates provided as bedding or in dispensers

Preventive enrichment is supplied to pigs ahead of any indications of tail biting, with the aim to minimise the development of this behaviour. As such, it is usually provided to (or withheld from) all weaned pigs on a farm, regardless of their behaviour. Two commonly studied types of preventive enrichment are organic substrates (discussed here) and point-source objects (discussed in the next section).

Several organic substrates (consisting of particulate materials) have been studied in the context of reducing tail biting: e.g., hay, silage, grass, beets, peat, and sawdust. However, the substrate that has attracted by far the most attention is straw. Straw featured in the earliest experimental studies on tail biting, where it was found to result in a marked reduction of tail biting (Van Putten, 1969; Fraser *et al.*, 1991). This has been confirmed in many later studies, which, for instance, showed that providing straw bedding reduced tail damage substantially compared to providing no enrichment or hanging toys, hoses, or chains (as reviewed in D'Eath *et al.*, 2014). However, straw bedding is not compatible with all current housing systems and the quantity and the way in which it is provided affect its effectiveness.

2.1 *Difficulties with providing straw as bedding*

Many current husbandry systems are poorly suited to the provision of straw bedding, because they have fully or partially slatted floors. More in-depth discussion on flooring can be found in the chapter "Infrastructure" in this book. Briefly, the gaps in the floor facilitate urine and manure removal, but straw can fall through the gaps as well. This not only means that the straw is no longer accessible to the pigs, but also that these openings can become blocked leading to dirty, unhygienic pens.

Furthermore, straw blockages can impede drainage of the slurry tank underneath the floor of the pig house. This risk of blockage may be overestimated, as 81% of finishing pig farms and 56% of nursery farms surveyed in Sweden indicated that they never experienced manure handling problems when providing limited quantities of straw on partially slatted floors (Wallgren *et al.*, 2016). Where periodic blockages do occur, these can be dealt with by using slurry agitators (European Commission, 2016). Nonetheless, current manure handling systems may differ between countries, as do floor types. Fully slatted floors are commonly applied in those countries that allow their use, and may be at greater risk of blockage. Both systems that minimise the spillage of straw into the slurry tank and slurry flushing systems that are able to deal with substantial straw spillage have been developed, but are not commonly applied (yet). Therefore, there has been considerable interest in the effectiveness of providing small quantities of straw, sometimes in a rack or dispenser, with a catching tray, or pressed into a block to prevent it from falling down the slats too quickly. This can also reduce the costs associated with straw provision, which can also hold implementation back.

2.2 *Providing straw in smaller quantities, chopped, or in dispensers*

Small quantities of straw provided loose on the floor have been found to reduce tail biting, with even quantities as small as 20 g/pig/day being effective (D'Eath *et al.*, 2014). However, several studies show that increasing this straw ration decreases tail biting further (Wallgren *et al.*, 2019; Munsterhjelm *et al.*, 2015; Pedersen *et al.*, 2014). Where very small quantities are used, their lower effectiveness may be simply because pigs consume the straw rapidly and are subsequently left without enrichment. Furthermore, pigs may not be able to manipulate smaller portions of straw in an equally satisfactory way as larger quantities. Whilst an amount of 80 g/pig/day appears to be sufficient to ensure that some straw is left when the next day's ration is provided, the beneficial effects of straw continue to rise until approximately 400 g/pig/day (Pedersen *et al.*, 2014). Full straw bedding can be used in several ways (e.g., foraging/exploring/rooting in and underneath it, smelling and chewing it, carrying it around and constructing nests with it). It can also be manipulated from different postures as it has a greater spatial presence (e.g. pigs can manipulate it, while lying on it, Van de Weerd *et al.*, 2006). In contrast, the possible interactions with small quantities are much more limited. When the quantity of straw limits its suitability for foraging, exploring and rooting, this is likely to impact on the straw's capacity to reduce tail biting, as one theory suggests that tail biting is a re-direction of such behaviours to pen mates in the absence of satisfactory possibilities to direct them to the environment (Pedersen *et al.*, 2005).

Chopping the straw into smaller pieces before provision is another way to reduce the risk of blocking the floor and slurry system. However, chopping straw affects the way the pig can interact with it, compared to long straw. Chopping limits the ability of a pig to pick up bits of straw and interactions are limited to licking (Day *et al.*,

2008). This potentially explains why some studies found that chopped straw was less effective at reducing tail biting. However, others found no such effect (Lahrman *et al.*, 2015) and further studies on the effect of chopping might be required.

Yet another way to provide straw is to put it into racks or dispensers. In the last case the straw is usually chopped or pressed into compact straw blocks. This means that the pig has to work to get the straw out before it can interact with it. Thus, it is less likely that straw falls down between the slats before the pig actually uses it, especially if there is a tray at the bottom that catches any spilled straw. Working to obtain a preferred resource (here: straw) can be intrinsically rewarding (De Jonge *et al.*, 2008). However, depending on how easy it is for the pig to obtain larger quantities from the rack/dispenser, it also means that they are likely to interact with very small portions of straw obtained. As discussed previously, this affects the way they can interact with the straw. In addition, providing straw in a dispenser or rack means that competition may occur as these are often not large enough for several animals to interact with simultaneously or the rack is provided in the wrong location (Van de Weerd and Ison, 2019). The issue of competition is discussed in more detail in the next section on point-source enrichment. It seems that when straw is provided in a pressed block or in a dispenser that requires extensive manipulation, this reduces its effectiveness. Although one study found that pressed straw blocks resulted in less tail biting than chains (Bulens *et al.*, 2016b), no such effect was found in several other studies where pressed straw blocks (Bulens *et al.*, 2016a; Haigh *et al.*, 2019) or 'foraging towers' (a plastic tubular container filled with chopped short straw fixed on the floor, Holling *et al.*, 2017) were compared to plastic or metal objects. It is possible that in addition to competition, there may be an element of frustration and therefore increased stress in some cases, if getting straw out of the devices is too difficult.

2.3 *Other organic substrates and their effectiveness*

Although straw has certainly been studied most amongst the organic substrates, there are many alternatives to straw. In choice tests, pigs seem to prefer certain alternative substrates over straw (peat, compost, sand, sawdust, wood shavings, branches, bark, beets and silage, Beattie *et al.*, 1998; Studnitz *et al.*, 2007). Similarly, pigs spent more time exploring a dispenser filled with chopped hay than when it was filled with chopped straw (Kauselmann *et al.*, 2021). This may indicate that these other organic substrates may also be more effective in reducing tail biting if they can be supplied in sufficient quantities. Alfalfa hay, maize silage, compost and beets have been shown to reduce tail biting, and maize silage was shown to outperform chopped straw in its effectiveness (Buijs and Muns, 2019). Such substrates may be especially advantageous to enrich pig pens in regions or periods when straw is expensive, hard to come by, or poses a biosecurity risk. It would also be worthwhile to determine the effectiveness of materials that are more easily available outside of Europe (e.g., rice husks).

2.4 *Implementation in commercial practice*

Although the use of organic substrates is common on commercial farms in some European countries (Finland 72% of farms, United Kingdom 68%) it is far less common in others (Germany 34%, Italy 0%) where point-source enrichment is much more common (Buijs and Muns, 2019). In China, the provision of substrates is very rare and in the USA it is mostly confined to niche market production (Van de Weerd and Ison, 2019). In some cases it is even prohibited due to biosecurity concerns (personal communications). Apart from non-compatibility with slatted floor systems, the implementation of substrates is reduced due to the labour involved in their regular replacement and/or replenishment. In addition, some preferred substrates can form a biosecurity risk. How these factors limit substrate implementation is discussed later in the section on "Limitations".

3 Preventive enrichment – point source items

As discussed in the previous section, substrates provided on the floor have been identified to be the most suitable and effective enrichment for pigs (EFSA, 2007; Spoolder *et al.*, 2011; Tuyttens, 2005). However, providing floored substrates or beddings can be problematic in the slatted flooring system, which is still commonly used around the world. Therefore, alternative types of enrichment that are compatible with the slatted systems have been studied.

Van de Weerd *et al.* (2006) first termed "point-source" enrichment items, which are objects that are "often restricted to a single location in a pen and they are limited in size, in that they generally do not allow all animals in a group simultaneous access." Van de Weerd *et al.* (2003) systematically investigated materials/items that may be suitable to be used as enrichment for pigs. They included an extensive list, out of which many are point-source items, for example, a ball, hessian sack, hose pipe, rope and rubber toys. These items can be provided hanging on the side/in the middle of the pen, loose/fixed on the floor of the pen or fixed on the pen walls, which may partially satisfy pigs' chewing and rooting needs (Fraser *et al.*, 1991). Hanging ropes, for example, have been shown to be preferred over substrates provided in a box (i.e., wood shavings or sawdust; Guy *et al.*, 2013). Most commercial farms use these types of point-source enrichment due to the convenience of their maintenance (Buijs and Muns, 2019).

In contrast to substrates that can be spread over a large area in the enclosure, a major limitation for most point-source items is their accessibility. As Van de Weerd *et al.* (2006) pointed out, because they are limited in size, usually only a single pig can interact with it at one time, which may increase competition. Therefore, it is important to consider the quantity and location of point-source items (Van de Weerd and Ison, 2019) to reduce competition. Some moveable items can be located in the middle of the pen and create interactions between pigs which can also improve accessibility.

Determining how many point-source enrichments need to be provided to maximise their effect on tail biting is challenging. Although it may be expected that providing enough objects for all pigs in the pen to engage with simultaneously would be optimal, this may not always be feasible. Furthermore, pigs do not necessarily prefer to engage with the enrichment simultaneously (rather than taking turns) if provided with the opportunity (Larsen *et al.*, 2019). Nonetheless, interaction with point source enrichment has been reported to increase when more objects were provided even when simultaneous use was unaffected (Larsen *et al.*, 2019, comparing 4 against 2 wooden beams per 16 pigs). This is in contrast with an earlier study that reported toy interaction did not differ between groups provided with 1 or 4 toys (suspended crosses of plastic pipes, Scott *et al.*, 2007), possibly because these toys simply were not very attractive to the pigs. Importantly, the latter study did not find an effect of the number of toys on how much the pigs manipulated each other. This is more relevant in the context of tail biting, because as previously stated greater interaction times do not necessarily mean a greater reduction in tail biting. Scientific information on how many point-source objects need to be provided to reduce tail biting is lacking and is likely to depend on the type and size of objects, including whether the object can be used by more than one pig. Rather than setting an applicable number for all different types of enrichment, it has been recommended to assess if sufficient enrichment is provided by dividing the number of pigs manipulating the enrichment by the number of pigs manipulating the enrichment, pen fittings, and other pigs. If this ratio is lower than 0.18, more (effective) enrichment has to be provided (EU, 2016). However, a ratio of 0.18 still means for every pig manipulating the enrichment there are approximately 4 pigs manipulating the pen or each other, and some producer organisations have advocated a ratio of at least 0.5 (AHDB, 2017).

In addition to competition, the shape, presentation and location of items may also affect their accessibility. For instance, knotting ropes to prolong their durability reduces how much pigs interact with these because they become more difficult to chew on (Apple and Craig, 1992). Wood posts provided in a vertical position were found to be less easy for pigs to chew on than horizontal or diagonal ones because of their neck and snout features (Chou *et al.*, 2020b; Telkänranta *et al.*, 2014a).

Maintaining hygiene of the objects is also important as items covered in faeces may lose pigs' interest quickly (Blackshaw *et al.*, 1997; Bracke, 2007; Trickett *et al.*, 2009). However, a recent study showed that cleaned, identical objects did not attract pigs' attention more compared to uncleaned ones, although the dirty items in this study were not completely soiled with faeces (Beaudoin *et al.*, 2019). Point-source items should not be placed in areas where using these would obstruct other behaviour (e.g., the sleeping area) or where it is likely to get soiled (dunging area, Van de Weerd and Ison, 2019).

3.1 *Wood as organic enrichment*

Wood is one of the most researched point-source materials to-date in relation to tail biting. It is versatile because it can be freshly collected from the tree or processed in bulk, it comes in different shapes (e.g., long post, short plank, round, square, etc.), it is an organic material, and there are many different species of wood. More importantly, it can be cheaply and easily acquired in most regions. However, the European statutory body classified wooden materials as "suboptimal," and recommended their use as essential items in conjunction with other enrichment items to provide different types of stimulation (European Commission, 2016b).

Research showed that, depending on the different methods to provide wood as an enrichment item for pigs, the level of interaction fluctuates (Elkmann and Hoy, 2009; Trickett *et al.*, 2009; Zonderland *et al.*, 2003). Wood may not always generate more interactions from pigs, even when compared to a dog toy in the shape of a rubber bar, which is not considered an appropriate enrichment item (Canning *et al.*, 2013). This comparison was made when both items were hung horizontally on a lever device. More recent studies started to emphasise different species of wood with unique characteristics such as hardness, moisture and odour, which can affect its effectiveness as enrichment as pigs may be able to chew or smell them differently (Chou *et al.*, 2018). Softer wood species increased pigs' interaction but did not necessarily reduce tail biting if only one wood post was present in a pen (Chou *et al.*, 2020b; Nannoni *et al.*, 2016). Commercially sourced wood has nearly always been kiln-dried and stored for several years before deployment. This may alter its smell, taste, and physical properties. Freshly felled, undried wood was found to have a beneficial effect on tail biting: fewer tail and ear lesions were observed in pens with hanging fresh birch wood than in pens with hanging rubber pipes or a branched chain (all types of pens also had a chain, straw rack and wood shavings) (Telkänranta *et al.*, 2014a). As this is the only study assessing the effect of fresh wood after weaning, this promising result requires further evaluation in other contexts (e.g., when provided as the only enrichment).

3.2 *Other single items and combining items*

Although point-source items are preferred by pig producers (Nalon and De Briyne, 2019), a recent review showed only a few types that were able to reduce tail biting in pigs, and none when pigs' tails were not docked (Buijs and Muns, 2019). In docked pigs, a wall-mounted hessian sack reduced severe tail damage in gilts (Ursinus *et al.*, 2014). By alternating one hanging item every week (in total seven items tested including a rope, ribbon, hose, ball, etc.), van de Perre *et al.* (2011) also reduced tail damage in docked pigs, compared to pigs that only received a hanging metal chain. This could be the result of maintaining the novelty effect (see below section on novelty for further discussion). However, no studies to date managed to decrease tail biting behaviour or tail injuries by only providing a single point-source item to a group of undocked pigs. Both the quantity and the quality of enrichment are important to satisfy pigs' different exploratory needs, improve access and reduce competition. Combining different kinds of enrichment in such a way that the combination allows the pig to manipulate, investigate, eat/smell and chew is recommended by the EU (EU, 2016c). However, satisfying this set of criteria without the use of substrates is very difficult. There are currently no studies that showed a reduction in tail biting as a result of applying a combination of objects that satisfied these four criteria without including substrates.

3.3 *Using dividers to create diversity in the pen*

Another type of enrichment is to increase the complexity of the enclosure by dividing it into separate areas. For example, a second level (platform/mezzanine) or additional wall can be added. This also alters the layout of the pen, providing clear functional areas and fragmenting space allowance. More discussions on the pen design and stocking density can be found in the chapters "Infrastructure" and "Social environment" in this book. Bulens *et al.* (2017a) added a T-shape hiding wall fixed to one side of the pen and found that undocked finishing pigs showed fewer mild and severe tail wounds (although they had more ear wounds). The same authors also provided a hanging toy, a compressed straw dispenser and the same hiding wall to pens of undocked finishing pigs and again found a lower percentage of severe tail lesion scores compared to the control pigs with only a hanging toy. However, the pigs in the more "enriched" pens had a higher frequency of tail biting (Bulens *et al.*, 2018). These authors suggested that the higher incidence of tail biting could be due to increased competition for lying space, which was fragmented by the wall. Another study increased complexity of the enclosure by introducing platforms with slopes and also effectively reduced tail lesions in docked pigs (Bulens *et al.*, 2017b).

4 Corrective enrichment

Despite years of study, it appears that it is almost impossible to prevent tail biting fully, due to its multifactorial and sporadic nature. Albeit at lower levels, tail lesions are also recorded in some but not all herds with ample straw provision (Statham *et al.*, 2011), and in outdoor and organic production (Alban *et al.*, 2015; Kongsted and Sørensen, 2017; Rudolph *et al.*, 2017).

Enrichment has been considered an important remedial method when some level of tail biting is already observed on commercial farms (Bracke *et al.* 2013; Valros *et al.* 2016), although it may not be the sole solution as other factors triggering tail biting need to be considered as well (see other chapters on different risk factors). When enrichment is used as a remedial measure, items that are not normally provided to the pigs are considered more useful due to the novelty effect (see below section on novelty). Furthermore, such corrective enrichment may allow the use of some enrichment types that would be impractical if given routinely (e.g., straw in some regions).

Zonderland *et al.* (2008) found that providing long straw on the solid floor (20 g/pig/day) once a tail biting outbreak occurred lowered the percentage of pigs with fresh blood on the tails in the following days, although tail biting still persisted. Lahrmann *et al.* (2019) compared using a rope, a plastic hanging Bite-Rite toy (with four extending plastic pipes for pigs to chew on) or a small amount of chopped straw (around 7 g/pig/day) provided on the floor to mitigate tail biting outbreaks. They found that all interventions reduced tail-directed behaviours compared with control groups without corrective enrichment. Furthermore, straw was more effective than the other interventions in curbing an escalation of tail biting, in line with other studies. Nevertheless, the authors also reported that even with straw, there was still a probability of 25% of tail biting escalating. Thus, additional steps would be needed to consistently halt tail biting outbreaks. Chou *et al.* (2019b) used a multi-step intervention protocol, where three measures (removing biters, removing victims and providing three hemp ropes) were randomly ordered and combined as sets of intervention measures and deployed according to the progression of outbreaks. They found that adding three ropes (1m hemp ropes hanging from three sides of the pen) was the fastest method to overcome bad tail biting outbreaks.

When applying corrective enrichment, it is crucial to keep in mind that the timing of the corrective measure will impact its outcome. It is important to closely monitor the onset of tail biting by regularly checking the pigs' tail postures manually (Zonderland *et al.*, 2009; Lahrmann *et al.*, 2018; Larsen *et al.*, 2018; Wedin *et al.*, 2018) or by automated monitoring (D'Eath *et al.*, 2018). It is also worth noting that if a tail biting outbreak was triggered by unsatisfactory foraging options, the corrective enrichment used to halt the outbreak would need to be more biologically stimulating for the pigs for it to be effective. For example, the enrichment used should be more chewable, destructible or ingestible, which are all characteristics that were found to be preferred by pigs (van de Weerd and Day, 2009).

5 Pre-weaning and pre-natal enrichment

Both tail biting and tail damage have been observed pre-weaning (Ursinus *et al.*, 2014a; Schmitt *et al.*, 2019). However, at this age the behaviour is relatively rare. Consequently, where research has been devoted to pre-weaning enrichment, this was more commonly done aiming to prevent tail biting later in life rather than in the pre-weaning period itself. This is because both the early life environment and the prenatal environment can impact on animal behaviour throughout life (as reviewed in Prunier *et al.*, 2020).

Prior to birth, the piglet's brain development is highly dependent on endocrine signals from its mother. Stressors that impact on the sow may therefore have long-term consequences on the cognitive and behavioural development of her piglets. For instance, stressed sows were shown to give birth to piglets with increased responsiveness to pain, which may affect their propensity to become a victim of tail biting (Rutherford *et al.*, 2009). However, such research has mainly focused on the effect of averse social environments rather than assessing any beneficial effects of an enriched environment. Thus, more research on this topic is necessary. The same goes for prenatal stimulation of the piglet's senses. Re-exposing weaned piglets to a flavour they had previously experienced pre-natally was shown to reduce oral manipulation of pen mates post-weaning (Oostindjer *et al.*, 2010). Thus enriching the post-weaning environment with such flavours may be expected to reduce tail biting as well, although this has not been studied directly yet.

Once the piglet is born its brain development continues and can be affected by stressors that work directly on the pig, for instance a lack of sensory stimuli. However, it is presently unclear if early life enrichment can reduce tail biting later in life substantially. Three studies (on wood shavings + straw, ropes + newspapers, and peanuts + celery + cereal honey loops, respectively) found that enrichment prior to weaning reduced tail damage post-weaning (Munsterhjelm *et al.*, 2009; Telkanranta *et al.*, 2014; Middelkoop *et al.*, 2019). In contrast, several other studies on tail biting post-weaning found no beneficial effect of pre-weaning enrichment, or an effect that was much less pronounced than that of the post-weaning environment itself (as reviewed in Prunier *et al.*, 2019). Nonetheless, only a very limited number of different types and quantities of enrichment has been evaluated and further research on pre-weaning enrichment is required. The most studied type of enrichment during the pre-weaning period is actually social enrichment, which is dealt with in the chapter "Social environment" in this book.

6 Limitations and challenges

6.1 *Biosecurity*

If enrichment is to improve an animal's welfare, it is of course highly important that it is safe for the pig to use. Safety concerns around point source enrichments mainly focus on the object itself being unsafe (EU, 2016), for instance by causing injuries when swallowed (e.g. pieces of synthetic rope) or bitten (metal strips in tyres). In contrast, for substrates the main concerns are around contamination of the substrate with harmful pathogens. Although contamination is of importance in any region, it is an especially big concern in China (Van de Weerd and Ison, 2019).

Knowledge on the presence of harmful contamination of substrates for pigs is limited. Making direct comparisons between different types of substrates to identify the safest ones is difficult, because contamination depends on numerous physical, chemical, and biological factors during harvest and storage. As these factors differ between batches and farms, so does the risk of contamination (Tenbrink *et al.*, 2020).

Nonetheless, there are some studies that show that unsafe levels of contamination of substrates do occur. High levels of mycobacteria (which can cause tuberculous lesions leading to carcass condemnation at slaughter) have been detected in peat in multiple studies (Wagner *et al.*, 2018; Johansen *et al.*, 2014; Trckova *et al.*, 2006). Although peat can be treated to remove the mycobacteria, such processes are considered unsuitable for the bulk amounts required under field conditions (Matlova *et al.*, 2012). In addition, maize pellets and maize silage were found to contain high levels of two types of mycotoxins that pigs are particularly sensitive to. These fungal metabolites can impair growth, reproduction and immunity (Tenbrink *et al.*, 2020). Although on average mycotoxin concentrations in straw would be too low to cause animal health problems, these concentrations were found to differ markedly between farms. In combination with mycotoxin contamination in the feed this may cause safe levels to be exceeded (Nordkvist and Häggblom, 2014). Improperly stored maize silage has also been associated with morbidity and mortality due to listeriosis in fattening pigs (Stein *et al.*, 2018).

Assessing the exact risks of contaminated substrates is hampered by a lack of knowledge on how much substrate is actually eaten by the pig. Furthermore, this risk also depends on the animal's overall immune status and additional contamination in the feed. However, it is clear that there is a potential risk. It has therefore been recommended to source substrates based on the same hygiene and quality criteria as applied to feed (Tenbrink *et al.*, 2020).

6.2 *Revisiting the concept of novelty*

One of the difficulties in providing indoor-housed pigs with constant access to environmental enrichment is how to maintain their interest in the materials offered. Pigs' novelty seeking behaviour has been well documented and shown to correlate with play behaviour and positive emotions (Wood-Gush and Vestergaard, 1991; Hemsworth *et al.*, 1996). Furthermore, expression of curiosity to novel things was also used as an indicator of good welfare in pigs housed in an enriched environment compared to those in an impoverished environment (Wemelsfelder *et al.*, 2000). Van de Weerd *et al.* (2003) found pigs to be more attracted to items with a "renewed" characteristic. They suggested "novelty" to be understood as "*a degree of uncertainty on how an object's properties are to be controlled, so enrichment may be best provided by objects that suggest controllability, but which actually resist complete control, thus maintaining responsiveness.*" Studies on environmental enrichment often reported this novelty effect, when pigs' interest in the objects was the highest when they were newly introduced and waned over time, especially with point-source enrichment items (Trickett *et al.*, 2009).

In contrast, this novelty effect (i.e. decreasing interest over time) was seldom reported when substrates were provided. Pigs still engage with straw when it is replenished with straw of the same type (rather than fully replaced with a different type of straw), although we can assume that such pigs have already become familiar with straw's properties. Whereas they would more easily lose interest in destructible point-source items such as rubber toys, even though these also change in shape while being chewed away (thus creating some novelty). On the other hand, studies have shown that when compost and corn pellets were given as rooting materials, the level of engagement from pigs was the highest when fresh materials were dispensed (Knoll *et al.*, 2021). This suggests that pigs also value "freshness" in daily enrichment provisions, and they may be able to pick up fresh olfactory or tactile cues from newly replenished substrates. Topping up substrates on a regular basis or rotating point-source items to keep a certain level of "freshness" for the pigs and therefore maintain pigs' interest, may also improve their effectiveness in preventing damaging behaviours.

6.3 *Economics*

According to multiple surveys, pig producers consider enrichment one of the highest ranked solutions when dealing with tail biting outbreaks (Hunter *et al.*, 2001; Haigh and O'Driscoll, 2019). However, challenges still arise when encouraging producers to use adequate daily environmental enrichment to prevent tail biting, and one of their main concerns is the additional cost associated with it (Van de Weerd and Ison, 2019). D'Eath *et al.* (2015) estimated costs for four different scenarios in relation to tail docking, space allowance and enrichment provision. Switching from providing point-source enrichment items to providing a small amount of chopped straw (100g/day/pig) required the least extra cost of materials and additional labour, and would see an increase of production cost at €1.5/slaughtered pig. Switching to an even higher welfare scenario of 200g/day/pig of straw provision saw a €4/slaughtered pig increase in total as it also required a higher labour input to clean up the leftover, soiled straw daily (D'Eath *et al.*, 2015). Another study modelling the costs of different ways to reduce tail biting estimated the costs of straw provision between €1.67–4.17/pig and the provision of point-source enrichment items alone (e.g., a fresh piece of wood, a processed wood post, a jute sack to growing-finishing pigs or 10 sisal ropes to piglets) between €0.03–1.78/pig (Niemi *et al.*, 2021). If using multiple point-source items plus a dispenser with grass, the enrichment material cost (without the labour cost), could add up to €1.62 to €2.00/slaughtered pig (Chou *et al.*, 2020c).

The discrepancy between the potential cost of using straw and simple point-source items seems large, considering the small margin of profit in pig production (D'Eath *et al.*, 2015). Nevertheless, it is worth noting the different effectiveness in reducing the risk of tail biting between different levels of enrichment provision as mentioned earlier, especially in undocked pigs. Moreover, in countries where tail docking is banned, the cost of non-compliance should also be considered. For example, the EU is strengthening its enforcement of the ban on tail docking, and it will be vital to implement effective solutions to prevent tail biting in undocked pigs. There could be direct financial consequences for producers to ignore the tail docking ban, e.g., fines, losing subsidies or negative reputation among consumers.

Indirectly, ineffective enrichment provision can increase the risk of tail biting and lead to reduced growth and carcass condemnation, which also have negative financial implications (Harley *et al.*, 2014; Valros, 2018). Studies have shown that straw provision can be cost-effective when the risk of tail biting is high (Niemi *et al.*, 2021). More efforts have been made to develop lower-cost enrichment items that can be used on commercial farms and that effectively prevent tail biting (Telkänranta *et al.*, 2014c).

The direct and indirect beneficial effects of enrichment on growth (and thus economics) also need to be stressed. It has been shown that better enrichment provision can enhance pigs' performance and health (Van de Weerd and Day, 2009; Tuyttens, 2005; van Dixhoorn *et al.*, 2016; Chou *et al.*, 2020c). Furthermore, better enrichment reduces the need for tail docking, which has been shown to exert acute negative effects on piglets' physiology and behaviour as their response to pain after the procedure (Nannoni *et al.* 2014). Zhou *et al.* (2013) also reported a reduction in pigs' weight gain up to 70 days of age due to teeth reduction and tail docking. The long-term effect of tail docking on pigs' performance is less understood, but preliminary data showed that raising pigs with intact tails in an enriched environment can have a positive effect on their growth and therefore a better profit margin over the whole production period, compared to docked pigs in an impoverished environment (Chou, 2020). This calls for more investigations into the long-term economical benefits of raising undocked pigs in an enriched environment.

7 Conclusions

Both substrates and point source objects can enrich the pig's environment, thereby contributing to reducing levels of tail biting although not eliminating the behaviour fully. In general, substrates are more effective in reducing tail biting, but for both types of enrichment it is important that they are supplied in the correct way (i.e. accessible to the pigs without creating competition, in a way that avoids soiling). Enrichment can have many goals, of which reducing tail biting is only one. Thus, lack of an effect on tail biting does not mean that an enrichment does not improve overall pig welfare. Conversely, high interaction times with an enrichment may result from a continued unsatisfied motivation, and should therefore not be taken as evidence that the enrichment reduces tail biting.

Substrates are more effective in greater quantities, but have some effects when supplied sparsely (e.g., straw is effective at 20 g/pig/day, but it's effectiveness increases with quantity, up to 400 g/pig/day). Small quantities provide opportunities in situations where large quantities are incompatible with the manure handling system, or result in unacceptable costs or labour. However, providing minimal quantities of substrate as pressed blocks or in dispensers that require extensive manipulation was only rarely found to reduce tail biting.

Of the wide variety of point source objects that has been tested as preventive enrichment, a significant reduction in tail biting has only been reported for hessian sacks, fresh wood, and objects that were regularly changed. When supplied in response to an outbreak, ropes and scented plastic biting toys reduced tail biting, although to a lesser extent than straw did. Incomplete understanding of the causes of tail biting, and of how enrichment can interface with these, currently hampers the design of more effective objects. Although a set of recommendations has been made (enrichment is to be manipulable, edible, chewable and investigable), it is very hard to satisfy these criteria using objects only (i.e., without including substrates).

In addition to the type of enrichment, the number of enrichments is relevant where point-source objects or substrate dispensers are used. Providing a greater number reduces the risk of competition and allows the pigs to explore more items. However, knowledge on how many items need to be provided per group of pigs to achieve a meaningful reduction in tail biting is currently lacking. Instead it is recommended to assess the ratio between pigs manipulating enrichment and those manipulating enrichment, the pen, or each other. There is no uniformly accepted cut-off value for this ratio (below which more, or more effective, enrichment needs to be supplied). This is an important knowledge gap.

Knowledge on the capacity of pre-weaning and pre-natal enrichment to reduce tail biting is currently very limited, but suggests that both may have beneficial effects. Thus, this area requires further research.

The costs and labour associated with providing enrichment can be a barrier to implementation, especially for substrates. Nonetheless, reducing tail biting has a positive effect on farm economies which can partly or even fully offsets these costs, depending on the risk of tail biting on a particular farm. Furthermore, enrichment strategies can aid the transition to successfully rearing undocked pigs. Not docking can improve growth and thus profitability.

Finally, it is important to realise that enrichment is only one component in a strategy to reduce tail biting, with several other elements being described throughout this book.

8 Summary

- Organic substrates (e.g. straw, hay, silage) are usually more effective in reducing tail biting than point-source objects.
- Quantities of straw (or other similar substrates) as small as 20 g per day per pig can reduce tail biting, but larger quantities (up to 400 g/pig/day) are more effective. Pressed straw blocks and dispensers that require extensive manipulation to obtain the substrate are usually ineffective in reducing tail biting.
- Fresh wood and hessian sacks can reduce tail biting, but other tested point-source objects appear to be ineffective in reducing tail biting unless exchanged

regularly. Nonetheless, such objects may still have a beneficial effect on other aspects of pig welfare.
- Straw, ropes and scented plastic biting toys can be used to minimise the effects of an ongoing tail biting outbreak.
- Providing the right number and placement of enrichment is important to avoid competition, as this reduces enrichment's effectiveness. The ratio between pigs manipulating enrichment and those manipulating enrichment, the pen, or each other has been used to evaluate if enrichment is sufficient.
- Select enrichment based on the same hygiene and quality criteria as feed to avoid biosecurity risks.
- Soiled enrichment may reduce a pigs' interest in enrichment, likely decreasing its effectiveness in reducing tail biting.
- Enrichment should not be placed in the dunging or sleeping area, to avoid soiling and disruption of sleep.
- Depending on the risk of tail biting on a farm, the added costs of providing enrichment may be offset by improved growth.

References

AHDB, 2017. A Practical Guide to Environmental Enrichment for Pigs – A handbook for pig farmers

Alban, L., Petersen, J.V., Busch, M.E. 2015. A comparison between lesions found during meat inspection of finishing pigs raised under organic/free-range conditions and conventional, indoor conditions. Porcine Health Management 1, 4. https://doi.org/10.1186/2055-5660-1-4.

Apple, J.K., Craig, J.V. 1992. The influence of pen size on toy preference of growing pigs. Applied Animal Behaviour Science 35, 149–155. https://doi.org/10.1016/0168-1591(92)90005-V.

Baumans, V. 2000. Environmental enrichment: A right of rodents! In: Balls, M., Van Zeller, A.-M., Halder, M., eds. Progress in the reduction, refinement and replacement of animal experimentation. Amsterdam: Elsevier BV. pp. 1251–1255.

Beattie, V., Walker, N., Sneddon, I. 1998. Preference testing of substrates by growing pigs. Animal Welfare 7, 27–34. https://doi.org/10.1017/S0962728600020236.

Beaudoin, J.M., Bergeron, R., Devillers, N., Laforest, J. 2019. Growing pigs' interest in enrichment objects with different characteristics and cleanliness. Animals 9, 85. https://doi.org/10.3390/ani9030085.

Benard, M., Schuitmaker, T.J., de Cock Buning, T. 2014. Scientists and Dutch pig farmers in dialogue about tail biting: unravelling the mechanism of multi-stakeholder learning. Journal of Agricultural and Environmental Ethics 27, 431–452. https://doi.org/10.1007/s10806-013-9471-x.

Blackshaw, J.K., Thomas, F.J., Lee, J.-A. 1997. The effect of a fixed or free toy on the growth rate and aggressive behaviour of weaned pigs and the influence of hierarchy on initial investigation of the toys. Applied Animal Behaviour Science 53, 203–212. https://doi.org/10.1016/S0168-1591(96)01087-8.

Bloomsmith, M.A., Brent, L.Y., Schapiro, S.J. 1991. Guidelines for developing and managing an environmental enrichment program for nonhuman primates. Laboratory Animal Science 41, 372–377.

Bracke, M.B.M. 2007. Multifactorial testing of enrichment criteria: pigs 'demand' hygiene and destructibility more than sound. Applied Animal Behaviour Science 107, 218–232. https://doi.org/10.1016/j.applanim.2006.10.001.

Bracke, M.B.M. 2017. Chains as proper enrichment for intensively-farmed pigs? In: Spinka, M. (ed) Advances in pig welfare. Woodhead publishing, Duxford, United Kingdom, pp. 167–192.

Buijs, S., Muns, R. 2019. A review of the effects of non-straw enrichment on tail biting in pigs. Animals 9, 824. https://doi.org/10.3390/ani9100824.

Bulens, A., Biesemans, C., Van Beirendonck, S., Van Thielen, J., Driessen, B. 2016a. The effect of a straw dispenser on behavior and lesions in weanling pigs. Journal of Veterinary Behavior 12, 49–53. https://doi.org/10.1016/j.jveb.2016.02.001.

Bulens, A., Van Beirendonck, S., Van Thielen, J., Buys, N., Driessen, B. 2016b. Long-term effects of straw blocks in pens with finishing pigs and the interaction with boar type. Applied Animal Behaviour Science 176, 6–11. https://doi.org/10.1016/j.applanim.2016.01.008.

Bulens, A., Van Beirendonck, S., Van Thielen, J., Buys, N. and Driessen, B. 2017a. Hiding walls for fattening pigs: do they affect behavior and performance? Applied Animal Behaviour Science, 195, 32–37. https://doi.org/10.1016/j.applanim.2017.06.009.

Bulens, A., Van Beirendonck, S., Van Thielen, J., Buys, N. and Driessen, B. 2017b. A two-level pen for fattening pigs: effects on behavior, performance, and postslaughter measurements. Journal of Animal Science, 95(2), 616–625. https://doi.org/10.2527/jas.2016.0831.

Bulens, A., Van Beirendonck, S., Van Thielen, J., Buys, N., Driessen, B. 2018. Rearing finishing pigs with intact tails: do they benefit from an enriched environment? Journal of Veterinary Behavior 24, 1–8. https://doi.org/10.1016/j.jveb.2017.12.003.

Bracke, M.B.M., De Lauwere, C.C., Wind, S.M. and Zonerland, J.J. 2013. Attitudes of Dutch pig farmers towards tail biting and tail docking. Journal of Agricultural and Environmental Ethics, 26(4), 847–868. https://doi.org/10.1007/s10806-012-9410-2.

Canning, P., Robbins, K., Kains, F., Blackwell, T. 2013. Sensory devices for nursery and grow-finisher pigs. Presented at the Centralia Swine Research Update, Kirkton, Ontario, 43–44.

Chou, J.Y. 2020. Strategies to manage tail biting in pigs housed in fully-slatted systems. University of Edinburgh PhD thesis.

Chou, J.Y., D'Eath, R.B., Sandercock, D.A. and O'Driscoll, K. 2020b. Enrichment use in finishing pigs and its relationship with damaging behaviours: Comparing three wood species and a rubber floor toy. Applied Animal Behaviour Science, 224, 104944. https://doi.org/10.1016/j.applanim.2020.104944.

Chou, J.Y., D'Eath, R.B., Sandercock, D.A., Waran, N., Haigh, A. and O'Driscoll, K. 2018. Use of different wood types as environmental enrichment to manage tail biting in docked pigs in a commercial fully-slatted system. Livestock Science, 213, 19–27. https://doi.org/10.1016/j.livsci.2018.04.004.

Chou, J.Y., O'Driscoll, K., D'Eath, R.B., Sandercock, D.A. and Camerlink, I. 2019b. Multi-step tail biting outbreak intervention protocols for pigs housed on slatted floors. Animals, 9(8), 582. https://doi.org/10.3390/ani9080582.

Chou, J.Y., Sandercock, D.A., D'Eath, R.B. and O'Driscoll, K. 2020c. A high enrichment replenishment rate reduces damaging behaviors and increases growth rate in undocked pigs kept in fully slatted pens. Frontiers in veterinary science, 7. https://doi.org/10.3389/fvets.2020.584706.

D'Eath, R.B.; Arnott, G.; Turner, S.P.; Jensen, T.; Lahrmann, H.P.; Busch, M.E.; Niemi, J.K.; Lawrence, A.B.; Sandoe, P. 2014. Injurious tail biting in pigs: How can it be controlled in existing systems without tail docking? Animal 8, 1479–1497. https://doi.org/10.1017/S1751731114001359.

D'Eath, R.B., Jack, M., Futro, A., Talbot, D., Zhu, Q., Barclay, D., Baxter, E.M. 2018. Automatic early warning of tail biting in pigs: 3D cameras can detect lowered tail posture before an outbreak. PLoS ONE 13, e0194524. https://doi.org/10.1371/journal.pone.0194524.

D'eath, R.B., Niemi, J.K., Ahmadi, B.V., Rutherford, K.M.D., Ison, S.H., Turner, S.P., Anker, H.T., Jensen, T., Busch, M.E., Jensen, K.K. and Lawrence, A.B. 2016. Why are most EU pigs tail docked? Economic and ethical analysis of four pig housing and management scenarios in the light of EU legislation and animal welfare outcomes. Animal, 10(4), 687–699. https://doi.org/10.1017/S1751731115002098.

Day, J.E.L, Van de Weerd, H.A., Edwards, S.A. 2008. The effect of varying lengths of straw bedding on the behaviour of growing pigs, Applied Animal Behaviour Science 109, 249–260. https://doi.org/10.1016/j.applanim.2007.02.006.

De Jonge, F.H., Tilly, S., Baars, A.M., Spruijt, B.M. 2008. On the rewarding nature of appetitive feeding behaviour in pigs (Sus scrofa): do domesticated pigs contrafreeload? Applied Animal Behaviour Science 114, 359–372. https://doi.org/10.1016/j.applanim.2008.03.006.

Douglas, C., Bateson, M., Walsh, C., Bedue, A., Edwards, S.A. 2012. Environmental enrichment induces optimistic cognitive biases in pigs. Applied Animal Behaviour Science 139, 65–73. https://doi.org/10.1016/j.applanim.2012.02.018.

EFSA, 2005. The welfare of weaners and rearing pigs effects of space allowances and floor types, The EFSA Journal.

EFSA, 2007. Scientific opinion of the Panel on Animal Health and Welfare on a request from commission on the risks associated with tail biting in pigs and possible means to reduce the need for tail docking considering the different housing and husbandry systems (No. The EFSA Journal 611), The EFSA Journal.

Elkmann, A., Hoy, S. 2009. Frequency of occupation with different simultaneously offered devices by fattening pigs kept in pens with or without straw. Livestock Science 124, 330–334. https://doi.org/10.1016/j.livsci.2008.12.008.

European Commission, 2016. Overview report. Study visits on rearing pigs with intact tails. Directorate-General for Health and Food Safety. DG(SANTE)/2016-8987-MR. http://ec.europa.eu/food/audits-analysis/overview_reports/act_getPDF.cfm?PDF_ID=790.

European Commission, 2016b. Commission staff working document on best practices with a view to the prevention of routine tail-docking and the provision of enrichment materials to pigs accompanying the document commission recommendation on the application of council directive 2008/120/EC laying down minimum standards for the protection of pigs as regards measures to reduce the need for tail-docking. https://ec.europa.eu/food/system/files/2016-12/aw_practice_farm_pigs_stfwrkdoc_en.pdf&sa=D&source=docs&ust=1642948063277089&usg=AOvVaw1Y6129igOvnGK61v8StVKw.

European Commission, 2016c. COMMISSION RECOMMENDATION (EU) 2016/336 of 8 March 2016 on the application of Council Directive 2008/120/EC laying down minimum standards for the protection of pigs as regards measures to reduce the need for tail-docking. Official Journal of the European Union. L 62/20.

Fraser, D. 1987. Mineral-deficient diets and the pig's attraction to blood: implications for tail-biting. Canadian Journal of Animal Science. 67: 909–918. https://doi.org/10.4141/cjas87-096.

Fraser, D., Phillips, P.A., Thompson, B.K., Tennessen, T. 1991. Effect of straw on the behaviour of growing pigs. Applied Animal Behaviour Science 30, 307–318. https://doi.org/10.1016/0168-1591(91)90135-K.

Guy, J.H., Meads, Z.A., Shiel, R.S., Edwards, S.A. 2013. The effect of combining different environmental enrichment materials on enrichment use by growing pigs. Applied Animal Behaviour Science 144, 102–107. https://doi.org/10.1016/j.applanim.2013.01.006.

Haigh, A. and O'Driscoll, K. 2019. Irish pig farmer's perceptions and experiences of tail and ear biting. Porcine health management, 5(1), 1–10. https://doi.org/10.1186/s40813-019-0135-8.

Haigh, A., Chou, J.-Y., & O'Driscoll, K. 2019. An investigation into the effectiveness of compressed straw blocks in reducing abnormal behaviour in growing pigs. Animal, 13(11), 2576–2585. https://doi.org/10.1017/S1751731119000715.

Harley, S., Boyle, L.A., O'Connell, N.E., More, S.J., Teixeira, D.L., Hanlon, A. 2014. Docking the value of pigmeat? Prevalence and financial implications of welfare lesions in Irish slaughter pigs. Animal Welfare 23, 275–285. https://doi.org/10.7120/09627286.23.3.275.

Hemsworth, P.H., Price, E.O. and Borgwardt, R. 1996. Behavioural responses of domestic pigs and cattle to humans and novel stimuli. Applied animal behaviour science, 50(1), 43–56. https://doi.org/10.1016/0168-1591(96)01067-2.

Holling, C., grosse Beilage, E., Vidondo, B., Nathues, C. 2017. Provision of straw by a foraging tower – effect on tail biting in weaners and fattening pigs. Porcine Health Management 3:4. https://doi.org/10.1186/s40813-017-0052-7.

Hunter, E.J., Jones, T.A., Guise, H.J., Penny, R.H.C., Hoste, S. 2001. The relationship between tail biting in pigs, docking procedure and other management practices. The Veterinary Journal 161, 72–79. https://doi.org/10.1053/tvjl.2000.0520.

Johansen, T.B., Agdestein, A., Lium, B., Jørgensen, A., Djønne, B. 2014. Mycobacterium avium subsp. Hominis suis infection in swine associated with peat used for bedding. BioMed Research International 2014, 189649 https://doi.org/10.1155/2014/189649.

Kongsted, H., Sørensen, J.T. 2017. Lesions found at routine meat inspection on finishing pigs are associated with production system. The Veterinary Journal 223, 21–26. https://doi.org/10.1016/j.tvjl.2017.04.016.

Larsen, Mona Lilian Vestbjerg, Andersen, H.M.-L., Pedersen, L.J. 2018. Tail posture as a detector of tail damage and an early detector of tail biting in finishing pigs. Applied Animal Behaviour Science 209, 30–35. https://doi.org/10.1016/j.applanim.2018.08.016.

Lahrmann, H.P., Oxholm, L.C., Steinmetz, H., Nielsen, M.B.F., D'Eath, R.B. 2015. The effect of long or chopped straw on pig behaviour. Animal 9, 862–870. https://doi.org/10.1017/S1751731114003024.

Lahrmann, H.P., Faustrup, J.F., Hansen, C.F., D'Eath, R.B., Nielsen, J.P., Forkman, B. 2019. The effect of straw, rope, and bite-rite treatment in weaner pens with a tail biting outbreak. Animals 9, 365. https://doi.org/10.3390/ani9060365.

Lahrmann, H.P., Hansen, C.F., D'Eath, R., Busch, M.E., Forkman, B. 2018. Tail posture predicts tail biting outbreaks at pen level in weaner pigs. Applied Animal Behaviour Science 200, 29–35. https://doi.org/10.1016/j.applanim.2017.12.006.

Matlova, L., Kaevska, M., Moravkova, M., Beran, V., Shitaye, J.E., Pavlik, I. 2012. Mycobacteria in peat used as a supplement for pigs: failure of different decontamination methods to eliminate the risk. Veterinarni Medicina, 57, 212–217. https://doi.org/10.17221/5924-VETMED.

Middelkoop, A., van Marwijk, M.A., Kemp, B. and Bolhuis, J.E. 2019. Pigs like it varied; feeding behavior and pre- and post-weaning performance of piglets exposed to dietary diversity and feed hidden in substrate during lactation. Frontiers in Veterinary Science 6, 408. https://doi.org/ 10.3389/fvets.2019.00408.

Middelkoop, A., Kemp, B., Bolhuis, J.E. 2020. Early feeding experiences of piglets and their impact on novel environment behaviour and food neophobia. Applied Animal Behaviour Science 232, 105142. https://doi.org/10.1016/j.applanim.2020.105142.

Munsterhjelm, C., Heinonen, M., Valros, A. 2015. Application of the Welfare Quality (R) animal welfare assessment system in Finnish pig production, part II: associations between animal-based and environmental measures of welfare. Animal Welfare 24, 161–172. https://doi.org/10.7120/09627286.24.2.161.

Munsterhjelm, C., Peltoniemi, O.A.T., Heinonen, M., Hälli, O., Karhapää, M., Valros, A. 2009. Experience of moderate bedding affects behaviour of growing pigs. Applied Animal Behaviour Science 118, 42–53. https://doi.org/10.1016/j.applanim.2009.01.007.

Nalon, E., De Briyne, N. 2019. Efforts to ban the routine tail docking of pigs and to give pigs enrichment materials via EU law: Where do we stand a quarter of a century on? Animals 9, 132. https://doi.org/10.3390/ani9040132.

Nannoni, E., Sardi, L., Vitali, M., Trevisi, E., Ferrari, A., Barone, F., Bacci, M.L., Barbieri, S., Martelli, G. 2016. Effects of different enrichment devices on some welfare indicators of post-weaned undocked piglets. Applied Animal Behaviour Science 184, 25–34. https://doi.org/10.1016/j.applanim.2016.08.004.

Nannoni, E., Valsami, T., Sardi, L. and Martelli, G. 2014. Tail docking in pigs: a review on its short-and long-term consequences and effectiveness in preventing tail biting. Italian Journal of Animal Science, 13(1), 3095. https://doi.org/10.4081/ijas.2014.3095.

Newberry, R.C. 1995. Environmental enrichment – increasing the biological relevance of captive environments. Applied Animal Behaviour Science 44, 229–243. https://doi.org/10.1016/0168-1591(95)00616-Z.

Niemi, J.K., Edwards, S.A., Papanastasiou, D.K., Piette, D., Stygar, A.H., Wallenbeck, A. and Valros, A. 2021. Cost-effectiveness analysis of seven measures to reduce tail biting lesions in fattening pigs. Frontiers in Veterinary Science, 1020. https://doi.org/10.3389/fvets.2021.682330.

Nordkvist, E., Häggblom, P. 2014. Fusarium mycotoxin contamination of cereals and bedding straw at Swedish pig farms. Animal Feed Science and Technology, 198. 231–237. https://doi.org/10.1016/j.anifeedsci.2014.10.002.

Oostindjer, M., Bolhuis, J.E., van den Brand, H., Roura, E., Kemp, B. 2010. Prenatal flavor exposure affects growth, health and behavior of newly weaned piglets. Physiology and Behavior 99, 579–586 https://doi.org/10.1016/j.physbeh.2010.01.031.

Paoli, M., Lahrmann, H., Jensen, T., D'Eath, R. 2016. Behavioural differences between weaner pigs with intact and docked tails. Animal Welfare 25, 287–296. https://doi.org/10.7120/09627286.25.2.287.

Pearce, G.P., Paterson, A.M. 1993. The effect of space restriction and provision of toys during rearing on the behavior, productivity and physiology of male pigs. Applied Animal Behaviour Science 36, 11–28. https://doi.org/10.1016/0168-1591(93)90095-7.

Pedersen, L.J., Herskin, M.S., Forkman, B., Halekoh, U., Kristensen, K.M., Jensen, M.B. 2014. How much is enough? The amount of straw necessary to satisfy pigs' need to perform exploratory behaviour. Applied Animal Behaviour Science 160, 46–55. https://doi.org/10.1016/j.applanim.2014.08.008.

Pedersen, L.J., Holm, L., Jensen, M.B., Jorgensen, E. 2005. The strength of pigs' preferences for different rooting materials measured using concurrent schedules of reinforcement. Applied Animal Behaviour Science 94, 31–48. https://doi.org/10.1016/j.applanim.2004.11.023.

Prunier, A., Averos, X., Dimitrov, I., Edwards, S.A., Hillmann, E., Holinger, M., Ilieski, V., Leming, R., Tallet, C., Turner, S.P., Zupan, M., Camerlink, I. 2020. Early life predisposing factors for biting in pigs Animal 14: 570–587 https://doi.org/10.1017/S1751731119001940.

Rudolph, G., Dippel, S., Bochicchio, D., Edwards, S., Früh, B., Holinger, M., Holmes, D., Illmann, G., Knop, D., Prunier, A., Rousing, T., Winckler, C., Leeb, C. 2017. Does animal health and welfare of organic pigs differ between husbandry systems?, in: The 7th International Conference on the Assessment of Animal Welfare at Farm and Group Level. Presented at the the 7th International Conference on the Assessment of Animal Welfare at Farm and Group level, Wageningen Academic Publishers, Ede, the Netherlands.

Rutherford, K.M.D., Robson, S.K., Donald, R.D., Jarvis, S., Sandercock, D.A., Scott, E.M., Nolan, A.M., Lawrence, A.B. 2009. Pre-natal stress amplifies the immediate behavioural responses to acute pain in piglets. Biology Letters 5, 452–454. https://doi.org/10.1098/rsbl.2009.0175.

Schmitt, O., O'Driscoll, K., Boyle, L.A., Baxter, E.M. 2019. Artificial rearing affects piglets pre-weaning behaviour, welfare and growth performance. Applied Animal Behaviour Science, 210, 16–25. https://doi.org/10.1016/j.applanim.2018.10.018.

Spoolder, H., Bracke, M., Mueller-Graf, C., Edwards, S. 2011. Technical report submitted to EFSA preparatory work for the future development of animal based measures for assessing the welfare of pigs. Report 2: preparatory work for the future development of animal based measures for assessing the welfare of weaned, growing and fattening pigs including aspects related to space allowance, floor types, tail biting and need for tail docking (No. EN-181). EFSA, Parma.

Statham, P., Green, L., Mendl, M. 2011. A longitudinal study of the effects of providing straw at different stages of life on tail-biting and other behaviour in commercially housed pigs. Applied Animal Behaviour Science 134, 100–108. https://doi.org/10.1016/j.applanim.2011.08.009.

Stein, H., Stessl, B., Brunthaler, R., Loncaric, I., Weissenböck, H., Ruczizka, U., Ladinig, A., Schwarz, L. 2018. Listeriosis in fattening pigs caused by poor quality silage – a case report. BMC Veterinary Research 14, 362 https://doi.org/10.1186/s12917-018-1687-6.

Studnitz, M. Jensen, M.B., Pedersen, L.J. 2007. Why do pigs root and in what will they root? A review on the exploratory behaviour of pigs in relation to environmental enrichment. Applied Animal Behaviour Science 107, 183–197 https://doi.org/10.1016/j.applanim.2006.11.013.

Telkänranta, H., Bracke, M.B.M., Valros, A. 2014a. Fresh wood reduces tail and ear biting and increases exploratory behaviour in finishing pigs. Applied Animal Behaviour Science 161, 51–59 https://doi.org/10.1016/j.applanim.2014.09.007.

Telkänranta, H., Swan, K., Hirvonen, H., Valros, A. 2014b. Chewable materials before weaning reduce tail biting in growing pigs. Applied Animal Behaviour Science 157, 14–22. https://doi.org/10.1016/j.applanim.2014.01.004.

Telkanranta, H., Bracke, M.B.M. and Valros, A. 2014c. Developing chewable and rootable low-cost objects for commercial pig farms. In Proceedings of the 25th Nordic Regional Symposium of the International Society for Applied Ethology (10-10).

Tenbrink, K.M., Schulz, J. and Kemper, N. 2020. Mycotoxin contamination of selected organic enrichment materials used in pig farming. Agriculture 10, 565 https://doi.org/10.3390/agriculture10110565.

Trckova, M., Zraly, Z., Matlova, I., Beran, V., Moravkova, M., Svobodova, J., Pavlik, I. 2006. Effects of peat feeding on the performance and health status of fattening pigs and environmentally derived mycobacteria. Veterinarni Medicina 51, 533–543. https://doi.org/10.17221/5587-VETMED.

Trickett, S.L., Guy, J.H., Edwards, S.A. 2009. The role of novelty in environmental enrichment for the weaned pig. Applied Animal Behaviour Science 116, 45–51 https://doi.org/10.1016/j.applanim.2008.07.007.

Tuyttens, F.A.M. 2005. The importance of straw for pig and cattle welfare: a review. Applied Animal Behaviour Science 92, 261–282 https://doi.org/10.1016/j.applanim.2005.05.007.

Ursinus, W.W., Van Reenen, C.G., Kemp, B., Bolhuis, J.E. 2014a. Tail biting behaviour and tail damage in pigs and the relationship with general behaviour: Predicting the inevitable? Applied Animal Behaviour Science 156, 22–36. https://doi.org/10.1016/j.applanim.2014.04.001.

Ursinus, W.W., Wijnen, H.J., Bartels, A.C., Dijvesteijn, N., van Reenen, C.G., Bolhuis, J.E. 2014b. Damaging biting behaviors in intensively kept rearing gilts: the effect of jute sacks and relations with production characteristics. Journal of Animal Science 92, 5193–5202. https://doi.org/10.2527/jas.2014-7918.

Valros, A., Munsterhjelm, C., Hänninen, L., Kauppinen, T., Heinonen, M. 2016. Managing undocked pigs – on-farm prevention of tail biting and attitudes towards tail biting and docking. Porcine Health Management 2, 2. https://doi.org/10.1186/s40813-016-0020-7.

Van de Perre, V., Driessen, B., Van Thielen, J., Verbeke, G. and Geers, R. 2011. Comparison of pig behaviour when given a sequence of enrichment objects or a chain continuously. Animal Welfare, 20(4), 641–649. https://doi.org/10.1017/S0962728600003286.

Van de Weerd, H.A., Ison, S. 2019. Providing effective environmental enrichment to pigs: how far have we come? Animals 9, 254. https://doi.org/10.3390/ani9050254.

Van de Weerd, H.A., Day, J.E.L. 2009. A review of environmental enrichment for pigs housed in intensive housing systems. Applied Animal Behaviour Science 116, 1–20. https://doi.org/10.1016/j.applanim.2008.08.001.

Van de Weerd, H.A., Docking, C.M., Day, J.E.L, Breuer, K., Edwards, S.A. 2006. Effects of species-relevant environmental enrichment on the behaviour and productivity of finishing pigs. Applied Animal Behaviour Science 99, 230–247. https://doi.org/10.1016/j.applanim.2005.10.014.

Van de Weerd, H.A., Docking, C.M., Day, J.E.L., Avery, P.J., Edwards, S.A. 2003. A systematic approach towards developing environmental enrichment for pigs. Applied Animal Behaviour Science 84, 101–118. https://doi.org/10.1016/s0168-1591(03)00150-3.

Van Dixhoorn, I.D.E., Reimert, I., Middelkoop, J., Bolhuis, J.E., Wisselink, H.J., Groot Koerkamp, P.W.G., Kemp, B., Stockhofe-Zurwieden, N. 2016. enriched housing reduces disease susceptibility to co-infection with porcine reproductive and respiratory virus (PRRSV) and Actinobacillus pleuropneumoniae (A. pleuropneumoniae) in Young Pigs. PLoS ONE 11, e0161832. https://doi.org/10.1371/journal.pone.0161832.

Van Putten, G. 1969. An investigation into tail-biting among fattening pigs. British Veterinary Journal 125, 511–517. https://doi.org/10.1016/S0007-1935(17)48710-0.

Wagner, K.M., Schulz, J. and Kemper, N. 2018. Examination of the hygienic status of selected organic enrichment materials used in pig farming with special emphasis on pathogenic bacteria. Porcine Health Management 4, 24, https://doi.org/10.1186/s40813-018-0100-y.

Wallgren, T., Larsen, A., Lundeheim, N., Westin, R., Gunnarsson, S. 2019. Implication and impact of straw provision on behaviour, lesions and pen hygiene on commercial farms rearing undocked pigs. Applied Animal Behaviour Science 210, 26–37. https://doi.org/10.1016/j.applanim.2018.10.013.

Wallgren, T., Westin, R., Gunnarsson, S. 2016. A survey of straw use and tail biting in Swedish pig farms rearing undocked pigs. Acta Veterinaria Scandinavia 58, 11. https://doi.org/10.1186/s13028-016-0266-8.

Wedin, M., Baxter, E.M., Jack, M., Futro, A., D'Eath, R.B. 2018. Early indicators of tail biting outbreaks in pigs. Applied Animal Behaviour Science 208, 7–13. https://doi.org/10.1016/j.applanim.2018.08.008.

Wemelsfelder, F., Haskell, M., Mendl, M.T., Calvert, S., and Lawrence, A.B. 2000. Diversity of behaviour during novel object tests is reduced in pigs housed in substrate-impoverished conditions. Animal Behaviour, 60(3), 385–394. https://doi.org/10.1006/anbe.2000.1466.

Wood-Gush, D.G.M. and Vestergaard, K. 1991. The seeking of novelty and its relation to play. Animal Behaviour, 42(4), 599–606. https://doi.org/10.1016/S0003-3472(05)80243-X.

Zhou, B., Yang, X.J., Zhao, R.Q., Huang, R.H., Wang, Y.H., Wang, S.T., Yin, C.P., Shen, Q., Wang, L.Y. and Schinckel, A.P. 2013. Effects of tail docking and teeth clipping on the physiological responses, wounds, behavior, growth, and backfat depth of pigs. Journal of Animal Science, 91(10), 4908–4916. https://doi.org/10.2527/jas.2012-5996.

Zonderland, J.J., van Riel, J.W., Bracke, M.B., Kemp, B., den Hartog, L.A. and Spoolder, H.A. 2009. Tail posture predicts tail damage among weaned piglets. Applied animal behaviour science, 121(3–4), 165–170. https://doi.org/10.1016/j.applanim.2009.09.002.

Zonderland, J., Vermeer, H., Vereijken, P., Spoolder, H. 2003. Measuring a pig's preference for suspended toys by using an automated recording technique. Agricultural Engineering International: the CIGR Journal of Scientific Research and Development.

Zonderland, J.J., Wolthuis-Fillerup, M., van Reenen, C.G., Bracke, M.B.M., Kemp, B., Hartog, L.A., Spoolder, H.A.M. 2008. Prevention and treatment of tail biting in weaned piglets. Applied Animal Behaviour Science 110, 269–281. https://doi.org/10.1016/j.applanim.2007.04.005.

CHAPTER 9

Feeds and feeding

Sandra Edwards

School of Natural and Environmental Sciences, Newcastle University, Newcastle upon Tyne NE1 7RU, UK
sandra.edwards@ncl.ac.uk

Abstract

Farmers in many countries rank deficiencies in feed composition and method of provision as among the most important risk factors for outbreaks of tail biting. Insufficient feeding space for the group when pigs are either meal-fed in long troughs or fed ad libitum from hoppers can result in sudden-forceful tail biting due to aggressive competition and frustration from inability to access feed. This is exacerbated by dividing the daily ration into many small meals, or by breakdowns or outages precluding an expected feed delivery. Deficits in the supply of energy, protein, minerals or micronutrients will induce a state of general or specific metabolic hunger which increases foraging motivation. In the absence of appropriate substrate in the environment, foraging behaviour can be redirected to penmates to cause two-stage tail biting. Hunger may be mitigated by dietary inclusion of some forms of fibre, but excessive dietary fibre may impair intake and digestion of other nutrients. Deficiencies or imbalances of some key amino acids can impair the synthesis of influential brain neurotransmitters regulating feeding, exploration and mood. This might exacerbate conditions predisposing biting and may contribute to development of pathological obsessive tail biting. Amino acid requirements are increased by high genetic lean growth potential or the presence of dietary antinutrients, and deficiencies or imbalances may also be induced by health challenges and gut inflammatory conditions. Other hypothesised mechanisms for dietary effects on tail biting risk include stress-induced changes in sodium metabolism, and behaviour and mood modification induced by microbiome dysbiosis via the microbiome-gut-brain axis. Further research is needed to elucidate these mechanisms and to indicate potential dietary interventions which can prevent and terminate damaging tail biting.

Keywords

diet composition – feed provision method – feed form – feed ingredients

 | DOI:10.3920/9789004694989_011

1 Introduction

Feeds and feeding infrastructure have long been recognised as one of the important sources of risk for a tail biting outbreak. Indeed, one of the first scientific experiments on tail biting sought to validate the dietary basis of the problem reported anecdotally by farmers (Ewbank, 1973). In more recent surveys of farmers' opinions on tail biting, 'feed and water', including aspects of both feed composition and feeding method, was rated as the most important category of risk factors by Finnish farmers (Valros *et al.*, 2016) and Swedish farmers (Wallgren *et al.*, 2016). Water provision and feed quality were also ranked highly as risk factors in similar surveys of British (Valros and Barber, 2019) and Irish farmers (Haigh and O'Driscoll, 2019), while 'nutrient poor foods' was by far the most commonly mentioned cause of tail biting in a survey of Brazilian farmers (Albernaz-Gonçalves *et al.*, 2021). The quality of food and system of feeding were also highlighted as important factors by Swiss (Sell *et al.*, 2021) and Dutch farmers (Bracke *et al.*, 2013), though ranked lower than other risk factors in these counties. Diet and feeding system have also featured highly in scientific assessments of tail biting risk (Bracke *et al.*, 2004).

The important association of feeding with tail biting outbreaks is not surprising when the hypothesised motivations for a pig to initiate tail biting are considered. Taylor *et al.* (2010) suggested, based on the phenotypic characteristics of the behaviour, that three different forms of tail biting existed with possibly different motivational bases. 'Two-stage biting', in which animals progress from gentle oral manipulation of tails (tail-in-mouth behaviour) to eventual injurious biting, has been proposed to arise from redirected foraging behaviour, which is plainly linked to the animal's perception of need for the nutrients necessary to survive and thrive. 'Sudden-forceful biting', which is seen as sudden injurious biting with no prior manipulation, was proposed to arise from frustration associated with the thwarting of access to a desired resource. Since feed is a very highly valued resource for pigs, this represents a focus of such risk. Finally, 'obsessive biting', in which pigs repeatedly seek out and bite the tails of others for extended periods of time to the exclusion of all other behaviours, was suggested to result from a pathological change in behaviour, possibly associated with neurochemical changes in the brain. Since amino acid nutrition makes an important contribution to neurotransmitter biology, feeding is likely once again to play a role. The mechanistic pathways linking feed-related factors with tail biting are outlined in Figure 9.1 and discussed in detail in the following sections. First the more general issues relating to the total supply of feed and method of provision are discussed. Then, the role of specific nutrients in the diet and their hypothesised effects on mechanistic pathways for tail biting are reviewed, and finally the risks associated with specific feed ingredients are considered.

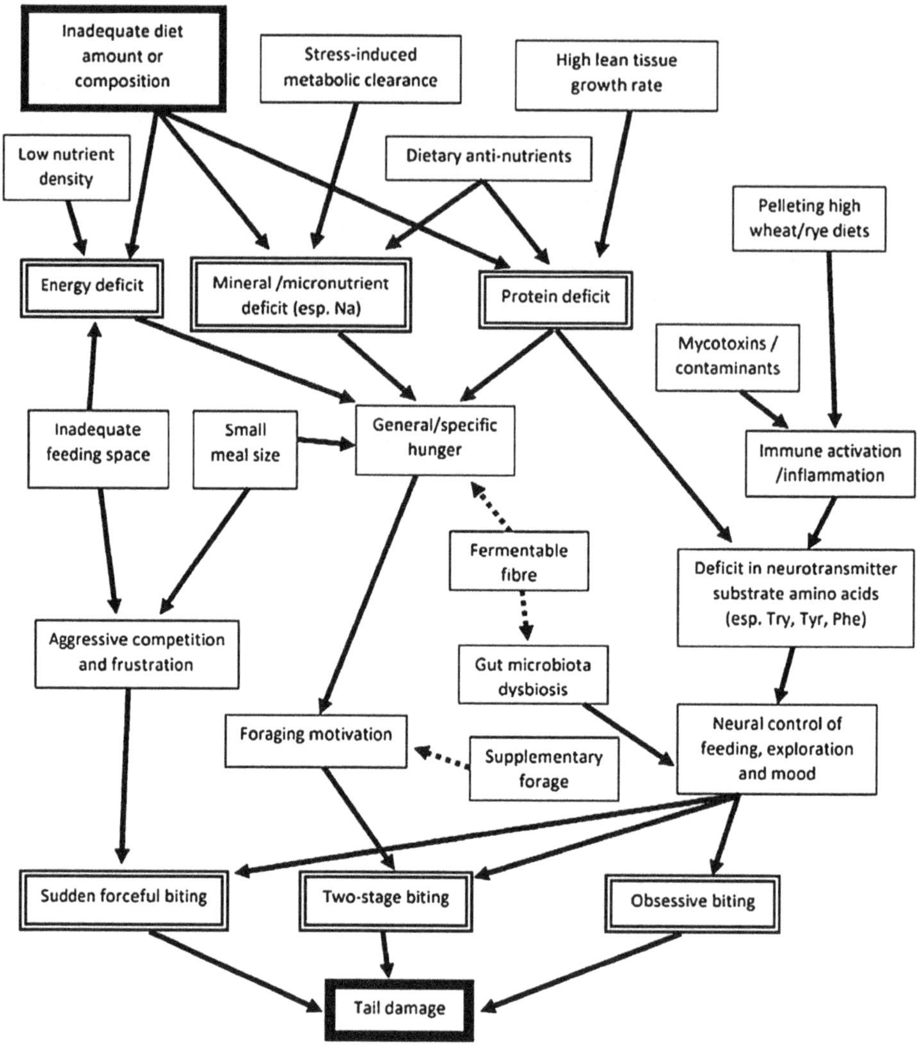

FIGURE 9.1 The proven and hypothesised pathways involved in feeding-related causes of tail biting

2 The amount and pattern of feed supply

Given the aforementioned considerations, it might be expected that tail biting risk would be related to the amount of feed supplied. Reducing the feed supply results in an increase in appetitive foraging behaviour (Day *et al.*, 1995) which, in barren environments, is directed towards pen-mates (Robert *et al.*, 1991) and may result in two-stage biting. An increased level of hunger may also give more aggressive interactions within the group (Vargas Vargas *et al.*, 1987), more competition when feed does become available, and thus increase the risk of sudden-forceful biting.

Although early anecdotal reports suggested that pigs on a restricted feeding regime were at higher risk of tail biting than those fed ad libitum (Jericho and Church, 1972), this has generally not been substantiated in subsequent epidemiological studies (Hunter *et al.*, 2001; Sell *et al.*, 2021). However, since such studies often confound feed restriction with feed form and method of provision, it can be difficult to interpret the true causal factor. Furthermore, the degree of restriction applied to commercial pigs of modern genotype in the growing and finishing stages is generally not severe, since fast growth is the primary production target. As a result, the most common commercial practice is to feed growing pigs ad libitum from hoppers, although feeding in discrete meals is also widely practiced, especially in liquid feeding systems.

Systems in which the access to feed was time-restricted were associated with more severe tail wounds than ad libitum feeding systems in the study of Temple *et al.* (2012) and they also identified systems with more frequent feeding (2–4×/day) as giving more wounds than those with less frequent feeding (1–2×/day). Kallio *et al.* (2018) also reported feeding in more than 2 meals/day as a risk factor for tail biting in an epidemiological study of Finnish farms. Provision of feed in small meals leaves animals unsatisfied after eating; indeed feeding motivation after ingesting a small amount may be higher than if not fed at all (de Leeuw *et al.*, 2008; Terlouw *et al.*, 1993). Furthermore, the increased frequency of competition for feed at the onset of each new meal might induce sudden-forceful tail biting. Hessel *et al.* (2006) demonstrated that a greater feeding frequency (9 vs 3×/day) increased disturbance in the pen and resulted in a more competitive feeding environment, giving an increase in caudal lesions on the pigs.

The risk of sudden-forceful biting due to frustrated feeding motivation is also increased in situations where feed is expected but fails to appear. Variations and delays in feeding time have been identified as an important risk factor for tail biting (Scollo *et al.*, 2017) and breakdown of automatic feeding systems has often been anecdotally implicated in tail biting outbreaks (Paul *et al.*, 2007). Taylor *et al.* (2010) suggested that this factor might underlie the epidemiological finding that hand-feeding is associated with lower tail biting prevalence than feeding by automated systems (Chambers *et al.*, 1995; Moinard *et al.*, 2003). Hand-fed pigs may become more tolerant of daily variation and delays in food provision than pigs fed on automated systems, where feed delivery can normally be predicted very precisely.

3 Feed form and method of feed and water provision

Whilst the absolute amount of feed provided has seldom emerged as a risk factor in published reports, the form in which the feed is given is frequently discussed. Several epidemiological studies have suggested a higher risk of tail biting when pigs are given liquid feed rather than dry feed (Chambers *et al.*, 1995; Kallio *et al.*, 2018; Moinard *et al.*, 2003; Pandolfi *et al.*, 2017; Temple *et al.*, 2012). However, such a comparison often contains many confounding factors, since liquid feeds are more frequently given in discrete meals, in long troughs designed for all pigs to feed simultaneously. They often contain by-products with variability and imbalances in nutrient composition, and may have a high degree of dilution with water to aid pumping which can impose physical limits on nutrient intake. In contrast to these epidemiological indications, Smulders *et al.* (2008) found less tail biting when the finishing feed was fed wet, but in this case wet feed was achieved by the presence of a water nipple within the trough, rather than being true liquid feed. They suggested that a reduction in dust and respiratory irritation might explain this outcome, but this was only partially supported by the experiment of Scott *et al.* (2007). In this, more pigs were removed to hospital pens due to respiratory conditions when fed a dry rather than a liquid diet, but with no differences in the frequency of veterinary treatments for respiratory conditions or in the post-slaughter assessment of lung damage. When each form of feed was offered ad libitum, Scott *et al.* (2007) further reported that pigs receiving liquid feed spent more time resting, and performed fewer investigatory behaviours (nosing, chewing, rooting or biting any available substrate) and pen-mate directed behaviours, suggesting that greater gut fill might have reduced foraging motivation.

The result of any comparison between liquid and dry feed may also depend on the nature of the dry feed, since Hunter *et al.* (2001) reported less tail biting with liquid feed than with pelleted dry feed, though not when compared to dry feed given in meal form. Pelleted feed also increased tail biting damage in weaned piglets when compared to a crumb diet (Watanabe *et al.*, 2015). Eating speed will be faster with a pelleted diet and it has also been suggested that pelleting of a diet may induce tail biting through gastrointestinal discomfort (Taylor *et al.*, 2010). As a result of the associated heat treatment and modification of particle size, pelleting is a known risk factor for gastric ulceration (Nielsen and Ingvartsen, 2000) and colitis (Thomson, 2009) which can give rise to inflammatory responses (see later).

The form of the feed is frequently confounded with the feed delivery system, since liquid feed is commonly provided in discrete meals in long troughs, whilst pelleted dry feed is most commonly provided ad libitum from hoppers (Hunter *et al.*, 2001). In univariate analyses, ignoring these confounds, they found an increased risk of tail biting when undocked pigs were fed on the floor or from a single space hopper, compared to a long trough or hopper with multiple feeding spaces. In contrast, Moinard *et al.* (2003) found a higher risk with trough feeding compared to hopper feeding. Temple *et al.* (2012) also showed that liquid feeding in meals in a long trough gave greater risk of tail biting than ad libitum delivery through a wet-feed hopper, with a dry feed hopper being intermediate. Whilst it is difficult to disentangle the effects of feed form, feeding frequency and delivery method in such studies, it is likely that the degree of competition for feed in individual farm circumstances may be the most critical factor. This relates to whether the feeding space provided is appropriate for the animals and feeding method in each case.

3.1 *Feeding space*

Providing adequate feeding space was the highest ranked preventive measure against tail biting by Finnish farmers, where most finishing farms use long troughs and (liquid) meal feeding (Valros *et al.*, 2016). It is widely recommended to allow 1.1× pig shoulder widths to allow all animals to feed simultaneously. Doubling this feeding space did not reduce the level of feed-related aggression, but conflicts were almost abolished when head, or especially head + shoulder, partitions were installed in the trough (Baxter, 1986). In a Swedish study involving restricted feeding in troughs, tail biting prevalence was greater when trough space was less than 30 cm per pig (Holmgren and Lundeheim, 2004). Since shoulder width can be estimated by the allometric equation: width (m) = $0.064 \times \text{liveweight}^{0.33}$ (kg) (Petherick, 1983), 30 cm would equate to the approximate shoulder width of a 105 kg pig. It is, however, possible that a genetic change in conformation of meat-type pigs in recent decades has increased their width relative to liveweight (Doeschl *et al.*, 2004), and feeding space requirements may need to be re-evaluated.

Where feed is offered ad libitum, the number of pigs per feeding place is an important factor which might influence tail biting risk, although there is no consensus on what this value should be. This is not surprising, since the rate of nutrient ingestion, and hence the feeding time necessary for a pig to achieve an intake that satisfies its appetite, will vary with both the nutrient density and the form of the feed (Manteca and Edwards, 2009). Ingestion rate is slowest for dry meal, higher for pelleted feeds and fastest for wet feed. In a multivariate model of factors associated with the incidence of tail and ear biting lesions on Belgian farms, Smulders *et al.* (2008) found that the number of feeding places per animal in the nursery was the most significant determinant. The relationship appeared progressive over the range from 0.1–0.8 (from 10.0 down to 1.2 pigs per space). Feeding space may be particularly critical in the nursery phase, since the synchronisation of suckling means that feeding behaviour is highly socially facilitated prior to weaning and this may be carried over into the post-weaning behaviour. In a nursery study comparing 3.75, 5.00, 6.25, or 7.50 pigs per feeder hole, achieved by changing group size in a pen with 4 feeding places, Laskowski *et al.* (2018) found the lowest occurrence of tail lesions in the treatment with 3.75 pigs per feeder hole. The benefit of a further reduction in feed competition is still uncertain. A German study comparing weaned piglets given ad libitum access to dry pellets in a long trough (animal-to-feeding-place ratio of 1:1) or short trough (ratio of 4:1), found no significant overall effect on tail-biting during the 6 weeks after weaning, although in two of the 4 experimental batches significantly fewer tail-losses were present at the end of the rearing period in the long trough treatment (Honeck, 2019).

Moinard *et al.* (2003) reported that one of the most important risk factors for tail biting in a multivariate analysis of wean-to-finish farms was having >5 pigs per feed space, although it is unclear in exactly which growth phase this was critical. In the grow/finish stage, a survey with variable group sizes (Hunter *et al.*, 2001) showed that the probability of tail biting was greater with a single space feeder than a double or multi-space feeder. Similarly, in a controlled experiment, the number of tail-bitten pigs was significantly greater in pens of 8 pigs with only one feeding place at an ad libitum hopper, than in pens with 4–5 feeding places (Hansen *et al.*, 1982). They noted that the tail-biters were seen to bite directly at the tails of their victims while the latter were feeding (sudden-forceful biting), with no association between tail biting and general nosing of penmates (two-stage biting). In a later experiment, Botermans and Svendsen (2000) observed no difference in tail biting prevalence between pens of 16 pigs fed from one or 4 hoppers, despite greater competition at the feeder and increased body lesions with only a single hopper, but they did see a tendency for increased tail biting when pigs were dry fed in a long trough. Feed was initially offered ad libitum, but later restricted and given once/day in the hoppers or twice/day in the trough, thus confounding feeding system with meal frequency. Similarly, Georgsson and Svendsen (2002) found no difference in tail biting prevalence in pens of 16 (undocked) pigs with 1 or 2 self-feeders, which were feed stations with head/neck protective sides, when feed was initially ad libitum but later restricted. The extent to which the feeding stall influenced this outcome is unknown, but Morrow and Walker (1994) reported that fitting a full-length feeding stall to a single space feeder for a group of 20 pigs reduced the number of feeding visits, forced withdrawals, frequency of queuing and the prevalence of tail biting.

It should be noted in passing that the competition for feed (milk) experienced by the piglet in early life might have a predisposing role in later tail biting. Hakansson and Houe (2020) highlighted that tail lesions already present at weaning were the major risk factor for later lesions and were associated with weight variability in the litter. Since the heavier pigs at weaning were more likely to have damage, it might be suggested that pigs with poorer access to milk developed damaging behaviour. This is in accord with the recent review of Prunier *et al.* (2019) which concluded that there was clear, though not unambiguous, evidence of a predisposing effect of undernutrition during the lactation period on subsequent chewing and biting behaviours towards pen-mates. However, they could find no evidence to support any earlier effect of prenatal nutrition and birthweight on subsequent predisposition to tail biting.

3.2 *Provision of water*

Consideration of the method of feed provision cannot be dissociated from the subject of water. Water may be supplied in association with feed, either in liquid feed systems or from nipples in wet-dry feeders, or may be supplied separately through troughs, bowls or nipple drinkers. A lack of water resulting from drinker malfunction, low flow rate, competition for access or poor water quality can cause frustration (sudden-forceful biting), as well as reduced feed intake and thus increased foraging motivation (two-stage biting). Whilst there have been no experimental studies of the relationship between inadequate water supply and tail biting, this has been identified as the most important risk factor by British farmers (Valros and Barber, 2019). Moreover, 'problems with the drinking system' were rated as more important than 'problems with the feeding system' in an earlier British survey (Paul *et al.*, 2007). Water availability was also rated amongst the important risk factors by farmers in other countries (Haigh and O'Driscoll, 2019; Sell *et al.*, 2021; Valros *et al.*, 2016). However, when included as a possible risk factor in epidemiological studies, neither type of drinker nor number of pigs/drinker proved to have a significant effect (Moinard *et al.*, 2003). Low water flow rate (<1 l/min for growing/finishing pigs) was identified as a common problem on farms, and absence of good access to drinking water (expressed as a low ratio of pigs to drinkers, clean drinkers and good water pressure) contributed significantly to tail biting risk score (Taylor *et al.*, 2012).

4 Energy supply and dietary fibre

For pigs on restricted feeding, the amount of energy they ingest is determined by the amount of ration they receive. When feed is offered ad libitum, the energy ingested depends on their voluntary feed intake. Pigs adjust their voluntary intake in response to changes in the nutrient density of the diet, such that they ingest the same daily energy intake. However, increasing intake to compensate for a lower density diet involves an increased daily feeding time until limits are eventually reached due to digesta transit time and physical gut capacity (Forbes, 1995). Controlled scientific study of the effect of diet density on tail biting risk is lacking, although this aspect is inherent in many of the studies of the role of dietary fibre (see later). However, one large commercial study involving grow/finish pigs from 30 to 130 kg has been reported, in which the removal rate for vice (i.e. tail bite removals) increased progressively from 1.3 to 5.2% as the net energy of the diet reduced from 2.52 to 2.11 Mcal/kg (PIC, 2021). The reduced live-weight gain of pigs on the lower density diets suggests that, despite partial intake compensation, pigs failed to ingest the same daily nutrients because of physical or behavioural limitations. Tail biting may therefore have resulted from increased foraging motivation, but it is also possible that the extra feeding time associated with their increased intake created greater competition at the feeder.

The nature of the energy substrate provided in the diet may be just as important as the absolute energy density. More concentrated diets supply energy in the form of readily digestible carbohydrate, such as cereal starch, and added fats of animal or vegetable origin. Less dense diets use bulky raw materials with a high content of dietary fibre. For example, in the study reported by PIC (2021), reducing energy density was associated with an increase in neutral detergent fibre from 9.5 to 24.2%. Interestingly, the removal rate for vice was also slightly higher (2.6–2.9%) for more nutrient dense diets up to 2.73 Mcal/kg, raising the question of whether a low fibre diet might also be a risk factor. Differences in dietary energy substrate generate very different metabolic profiles for the animal. Since the pig produces no enzymes for fibre digestion, this substrate must be broken down by bacteria in the gut which produce volatile fatty acids that can subsequently be absorbed and utilised by the pig. High starch diets give a rapid postprandial increase in glucose and insulin which fall away again after approximately 2h. In contrast, high fibre diets give a lower but sustained supply of nutrients over many hours (Rérat, 1996). This characteristic, together with the increased mastication, prolonged eating time and greater gut fill associated with high fibre diets, has been shown to influence the level and time course of feeding motivation. Growing pigs fed ad libitum on diets with high fibre content eat fewer meals per day and have longer feeding bouts associated with a slower ingestion rate, resulting in a lower daily nutrient intake despite greater daily feeding time (Kallabis and Kaufmann, 2012). Furthermore, animals on restricted feed show reduced physical activity when given high fibre diets (de Leeuw *et al.*, 2008). It has been suggested that, for these reasons, pigs given high fibre diets might be less predisposed to tail biting. Early anecdotal reports linked diets low in fibre with tail biting (Gadd, 1967), but this could not be proven experimentally (Ewbank, 1973). More recently, a retrospective field study of the risk factors for high tail biting prevalence in slaughter pigs on German farms noted that the crude fibre content of diet samples taken from high-risk farms was lower than in samples from low-risk farms (Schumacher, 2019). Pigs from high-risk farms also had a higher prevalence of pathological alterations in the gastric mucosa, including gastric ulceration – a problem which is known be alleviated by increased dietary fibre and might underlie the association (Taylor *et al.*, 2010). Experimental studies have failed to demonstrate any effect on tail biting prevalence when dietary fibre was increased by i) addition of soya hulls, sugarbeet pulp or oat fibre to raise crude fibre from 40 to 50 or 60 g/kg for weaned piglets between 4 and 10 weeks of age (Honeck *et al.*, 2020), ii) addition of unspecified fibre sources to raise soluble dietary fibre (NDF minus ADF content) from 8.1% to 8.6% (stage I diet) or 9.7% (stage II diet) between 4 and 8 weeks of age (Naya *et al.*, 2019). In both studies, large between-batch variability in tail biting was reported to make made demonstration of diet effects difficult. However Chou *et al.* (2020) found that contrary to their original hypothesis, addition of soya hulls to raise crude fibre from 37 to 53 g/kg (weaner diet) and 59 to 116 g/kg (finisher diet) between 8 weeks of age and slaughter resulted in significantly more pigs removed from their pen due to a tail biting outbreak, along with more observations of tail biting behaviour, and a tendency for worse tail damage scores (Chou *et al.*, 2020).

The conflicting indications regarding the influence of dietary fibre on tail biting may be because, in experiments to date, fibre has usually been described in terms of the simpler chemical determinations of crude fibre, neutral detergent fibre and acid detergent fibre. These measures do not give a true description of the nature of dietary fibre and its likely physical and physiological properties within the gut (de Leeuw *et al.*, 2008). Some fibre sources, such as straw, have a high content of relatively inert structural compounds such as lignin. These will increase mastication time and physical bulk in the gut, but make little contribution to microbial fermentation. Other fibres, such as sugarbeet pulp and soyabean hulls, increase gutfill because of their water holding capacity, and are also highly fermentable. Soluble fibres, as also found in oats and some legumes, can increase digesta viscosity and reduce gastric emptying rate. Any highly fermentable fibre will increase the microbial yield of short chain fatty acids, which can be absorbed from the gut over an extended period of time and influence postprandial satiety. Thus, the exact chemical nature of a fibre source can have a major influence on its functional properties in modifying feeding motivation and consequent tail biting risk. Moreover, it can influence the bioavailability of other nutrients in the diet including minerals, amino acids and micronutrients.

One of the principal reasons for a suggested role of dietary fibre in tail biting has been the repeated observation that pigs housed with access to fibrous materials such straw, hay or silage as supplementary forages have reduced tail biting risk (EFSA, 2007). In experimental studies, the presence of straw, either as bedding or recreational material (Larsen *et al.*, 2018; Zonderland *et al.*, 2008), or of hay or silage as a recreational material or nutritional supplement (Holinger *et al.*, 2018; Olsen, 2001; Veit *et al.*, 2016) has been shown to reduce the prevalence or severity of tail damage. Whilst a contribution of fibre ingestion to ulcer reduction, gut health, toxin binding, and microbiome modification have all been suggested as being implicated in the outcome, there is no possibility to dissociate nutritional effects from occupational effects as a foraging substrate in such studies.

4.1 *The role of the gut microbiome*

In addition to its direct influences on feeding motivation and gut integrity, dietary fibre may also have other indirect effects through modification of the gut microbiome (Brunberg *et al.*, 2016; Kobek-Kjeldager *et al.*, 2022), since it provides an important substrate to promote the proliferation of beneficial bacteria. Bi-directional signalling via the microbiota-gut-brain axis involves neural, endocrine and microbial-derived metabolites. Studies in a range of species have demonstrated a role for the gut microbiota in influencing brain functions, including emotional processing, mood and behaviour, and two recent studies have highlighted an association between the gut microbiota profile and tail biting. Rabhi *et al.* (2020) reported that beta-diversity of faecal microbiota differed between pigs in pens where tail biting was ongoing, in comparison with a negative control group from unaffected pens, where *Lactobacillus sp* were significantly more abundant. However, no significant difference was found between tail-biting or tail-bitten individuals within the outbreak pens. Verbeek *et al.* (2021) also demonstrated a difference in the faecal microbiota composition between tail-biting pigs and control pigs that were located in separate unaffected pens, with greater microbial diversity and a higher relative abundance of Firmicutes in biters; victims within the same pen as biters did not differ significantly from either group. They further detected differences in plasma and faecal short chain fatty acid profiles between biter and victim pigs in the same pen, indicating the existence of physiological differences. In both cases, however, the cause-effect relationship between tail biting and the gut microbiota cannot be determined, since stress responses can both modify the microbiota composition and be modified by it (Kraimi *et al.,* 2019). A further observation relevant to consideration of the role of the microbiota is that in-feed antibiotics, which reduce microbial diversity, tended to increase the risk of tail biting lesions in weaned piglets (Diana *et al.*, 2017). Such an association between in-feed antibiotics and tail biting has not always been found (Moinard *et al.*, 2003), and the role of antibiotics on microbiota change and gut health is complex (Lange *et al.*, 2016). Thus, whilst nutritional influences on the microbiota-gut-brain axis offer an attractive mechanism to mediate effects of the diet and feeding system on tail biting behaviour, and the possibility to design interventions based on functional fibres in the diet which can promote a beneficial microbiota, much remains to be elucidated.

5 Protein supply and specific amino acids

A deficiency in the quantity or quality of dietary protein was one of the factors implicated in many of the early anecdotal reports on tail biting (Gadd, 1967; Jericho and Church, 1972), and is still reported by farmers in recent surveys (Albernaz-Gonçalves *et al.*, 2021). Pressures to minimise the risk of diarrhoea in weaned piglets, particularly after removal of antimicrobials, and to reduce environmental pollution from nitrogen excretion in finishing pigs have forced a reduction in the total protein level in diets, which can result in an increase in penmate-directed biting (Van der Meer *et al.*, 2017). In that study the addition of synthetic amino acids to mitigate any deficiency in essential amino acids in the low protein diet gave some benefits, but this strategy has not always been effective in preventing increased tail biting (MLC, 2005b; Table 1). However, in both the MLC and another study (Presto Åkerfeldt *et al.*, 2019), lowering dietary protein did not increase tail biting when pigs were housed with the presence of straw which reduced overall risk. A better matching of dietary protein supply to the needs of the animal at different growth stages by adopting a phase feeding regime has been shown to reduce the prevalence of tail biting in both epidemiological (Holmgren and Lundeheim, 2004) and controlled experimental studies (MLC, 2004b; Table 1).

Many other indirect pieces of evidence point to an important role for dietary protein in tail biting causation. Pigs observed to tail bite have been reported as having low serum protein concentration (Barnikol, 1978; Holmgren *et al.*, 2004), although the more detailed study of Palander *et al.* (2013) failed to find any difference in

TABLE 9.1 The effect of different dietary treatments on tail biting in finishing pigs in a series of large UK experiments

Experimental comparison[a]	% pigs died or removed due to tail damage	Vet treatments for tail damage (pig days)[b]	Oro-nasal behaviour directed towards other pigs (% time)
MLC 2004a:			
Pelleted dry feed from a 5-space hopper	1.8%	0	9.4
Liquid feed (22% DM) from a sensor-controlled 4-space trough	1.0% (ns)	23	7.4 ($P < 0.01$)
MLC 2004b:			
Single liquid diet 0.7g lysine/MJ DE (18.5% CP in meal equiv.)	4.5%	188	7.4
Phased liquid diets 0.91–0.60 g lysine/MJ DE (21.4–17.5% CP)	0.6% ($P < 0.001$)	118	7.2 (ns)
MLC 2005a:			
Fresh liquid feed (DE 13.5 MJ, CP 17.95% in meal equiv.)	2.1%	34	6.3
Fermented liquid feed	1.4% (ns)	8	6.9 (ns)
MLC 2005b:			
Control diet 8.5g SID lysine/kg (20% CP in meal equiv.)	0%	20	6.0
Low protein diet 8.5g SID lysine/kg (15% CP in meal equiv.)	2.0% ($P < 0.01$)	69	5.4 (ns)

a Each experiment involved 16 pens (512 pigs) per treatment, divided equally between fully-slatted and straw-based housing. Pigs were offered their experimental diets ad libitum throughout and monitored from approximately 35–100 kg liveweight.

b Veterinary treatment involved a topical spray to deter further biting or antibiotic injection if infection present plasma amino acid concentrations in tail-biters, with the exception of elevated histidine. Furthermore, pigs with a higher genetic potential for lean tissue deposition rate (Breuer *et al.*, 2005; Ursinus *et al.*, 2014a) or showing a phenotypically higher growth rate (Ursinus *et al.*, 2014a; Valros *et al.*, 2021), and hence greater protein requirement, also have a greater predisposition to tail biting. Dietary protein deficiency, resulting in failure to meet amino acid requirements, might affect both the general foraging behaviour of pigs and their attraction to blood. Pigs can detect deficiencies in specific nutrients, even when energy supply is not limiting. In choice feeding experiments, they show the ability to make an appropriate selection from different diets to achieve the optimum protein intake for their gender and genetic growth potential (Kyriazakis *et al.*, 1993) and to compensate for any previous deficiencies (Kyriazakis and Emmans, 1991). Reduction in protein content in a diet fed ad libitum accordingly stimulated increased foraging behaviour despite no lack of energy (Jensen *et al.*, 1993).

Fraser *et al.* (1991) used an experimental model to examine in more detail the effect of nutrition on aspects of tail biting behaviour. Presenting pigs simultaneously with a pair of model tails soaked in either blood or water enabled them to examine both responses in general chewing behaviour, and in the specific attraction to blood in response to dietary change. When fed a cereal-based diet from which the protein source (soyabean meal) had been removed, preference score (chewing of blood model less that of water model) was increased within one week and continued to increase over the following 3 weeks. Partial restoration of the dietary protein by supplementation of the basal cereal component with synthetic amino acids, either lysine alone (the first limiting amino acid) or a package of 7 essential amino acids (lysine, leucine, isoleucine, methionine, threonine, tryptophan and valine), gave a preference score which was intermediate between the positive and negative control, probably because other essential amino acids subsequently became limiting. Both between and within diets, there was a strong negative correlation between live-weight gain and preference score, indicating the importance of protein intake adequacy at the individual level when diets were offered ad libitum. In an attempt to tease apart the importance of energy and protein intake, a later study employing a similar model tail test (McIntyre and Edwards, 2002a) fed controlled daily amounts of energy and protein designed to bring about a similar growth rate reduction. In contrast to the earlier study, no dietary effect was detected until the final week of a 5-week feeding period, when both low energy and low protein regimes showed a higher blood preference than control pigs. They suggested that the greater age of the pigs and relatively lower degree of growth retardation in their study might explain the different results.

5.1 *The neural role of specific amino acids*

It has been hypothesised that the reason for high sensitivity to dietary protein deficiency lies in the link between dietary protein and brain chemistry. Amino acids provide the substrate for synthesis of neurotransmitters involved in the CNS pathways regulating feeding, exploration and emotional state (Wurtman *et al.*, 1980). This may be particularly pertinent to the case of obsessive tail biting, where a pathological neural state as demonstrated in other species has been suggested (Nordgreen *et al.*, 2020). Phenylalanine and tyrosine serve as precursors for brain catecholamines, such as dopamine and norepinephrine, while tryptophan is a precursor of the neurotransmitter serotonin. Changes in dietary supply of these precursors can directly affect neurotransmitter levels and there can also be indirect effects of other dietary constituents, such as other large neutral amino acids which compete with tryptophan for binding sites at the blood-brain barrier (Wurtman and Fernstrom, 1975). McIntyre and Edwards (2002a) revealed a significant correlation between plasma tryptophan concentration and blood preference scores in a model tail test when pigs were fed a low protein diet. They subsequently showed that diets deficient in tryptophan increased activity and foraging behaviours in a home pen situation and attraction to blood in a model tail test (McIntyre and Edwards, 2002b). The link between tryptophan and tail biting was reinforced by Ursinus *et al.* (2014b), who demonstrated that pigs performing tail biting behaviour had lower whole blood and blood platelet serotonin levels. Furthermore, Valros *et al.* (2015), when comparing the brain chemistry of pigs that were identified as a tail biter, a victim, or a control pig from an unaffected pen, found a positive correlation between the level of plasma tryptophan and brain serotonin only in tail biters. Since they also found an increased level of 5-HIAA, the main metabolite of serotonin, in the prefrontal cortex of these animals, they suggested that tail biting pigs might have an enhanced serotonin metabolism which caused tryptophan to become a limiting factor influencing their behaviour. Dietary preference studies suggest that piglets are able to detect tryptophan deficiency (Ettle and Roth, 2004), and tryptophan supplementation has been experimentally related to decreased ear biting and tail biting (Martinez-Trejo *et al.*, 2009). However, the application of tryptophan supplementation as a strategy to reduce tail biting must take account of the levels of other amino acids which compete with it for binding sites at the blood-brain barrier.

Deficiencies in these key amino acids may also arise, without dietary change, if they are repartitioned to deal with other important functions in the body. Immune activation occurring in situations of antigenic challenge and inflammatory states results in increased synthesis of acute phase proteins, which have a higher content of tryptophan, tyrosine and phenylalanine than muscle protein. A change in health status can therefore modify metabolic requirements and induce deficiencies, stimulating exploratory oral behaviours in pigs (Pastorelli *et al.*, 2005) and leading to increased injurious biting (van der Meer *et al.*, 2017). A detailed discussion of the links between immune activation, amino acid metabolism and neurotransmitter systems is given in the recent review of Nordgreen *et al.* (2020).

6 Mineral supply

One of the earliest nutritional factors to be linked anecdotally to tail biting was mineral supply (Gadd, 1967). Old textbooks of pig keeping frequently recommended the addition of salt as a remedy for the problem (Barron, 1964) and this is still a remedy still regularly applied today (Sell *et al.*, 2021; Valros *et al.*, 2016). Sprinkling of salt on the floor reduced the incidence of biting behaviour in one small experimental study (Tsourgiannis *et al.*, 2002), while Watanabe *et al.* (2015) reported that young growing pigs in a barren environment had reduced tail damage after being offered 1.8% sodium chloride solution in addition to fresh water. The role of mineral nutrition in tail biting was first investigated experimentally by Fraser (1987). Using the model system with artificial tails, he showed that removal of the mineral supplement from the diet over a period of four weeks gave a significant increase in chewing of the blood-soaked tail, while chewing of the plain model remained unchanged. This attraction to the bloodied tail reduced again after reinstatement of the minerals in the diet. In a follow-up experiment, he further showed that removal of the iodised salt alone from the diet gave an equivalent effect, whereas omission of all other components of the mineral supplement resulted in a much smaller and non-significant response.

When considering this result, Fraser (1987) suggested that a sodium appetite was responsible for a strong attraction to blood. Given the known effect of ACTH in increasing sodium appetite in other species, he suggested that a physiological change in sodium metabolism might represent a final common pathway for the mechanism by which many different stressors appear to trigger tail biting. When testing this theory experimentally, administration of long-acting ACTH failed to increase ingestion of a salt solution (Jankevicius and Widowski, 2003) or to increase chewing of either salt or blood-soaked model tails (Jankevicius and Widowski, 2004). However, they acknowledged that exogenous ACTH might represent a somewhat limited model for stress. Beattie *et al.* (2001) reported that feeding pigs a salt-deficient diet did not increase their attraction to salt-soaked tail models, but did increase activity and foraging behaviours. It may therefore be that a generalised response to any specific nutrient deficiency increases foraging behaviour, which may be redirected to tails in certain situations.

Another mineral sometimes anecdotally linked to alleviation of tail biting is magnesium (García-Mochales *et al.*, 2018). Pigs in tail biting pens showed a lower serum magnesium and higher red blood cell magnesium concentration in comparison with control non-biting pens, indicating a shift of magnesium from extracellular to cellular fluid (Holmgren *et al.*, 2004). In humans, magnesium can play a role in reducing stress, anxiety and depression by neuroendocrine modification. A systematic review of publications on pigs found inconsistent results, but concluded that supplementary magnesium can have beneficial effects on reducing measures of stress and aggression (Bushby *et al.*, 2021). However, when examined experimentally, magnesium-rich supplements have had no effect on the occurrence or severity of tail-biting outbreaks or on tail lesion scores (Krider *et al.*, 1975; O'Driscoll *et al.*, 2013).

Deficiencies in other dietary minerals, such as iron, copper, calcium or phosphorous, as well as calcium/phosphorus imbalance, have also been suggested anecdotally as possible causes of tail biting (Ewbank, 1973), but no epidemiological or experimental evidence regarding these exists. However, in line with the known effects of specific nutrient deficiencies on foraging behaviour, such causes are plausible.

7 Raw materials, anti-nutritive factors and contaminants

Epidemiological studies have sometimes indicated that specific raw materials may be implicated in tail biting risk. Regarding the cereal base, wheat (Kallio *et al.*, 2018) and triticale (Holmgren and Lundeheim, 2004) have been suggested to increase risk. This might be linked to the demonstrated effect of diets with high inclusion of wheat, especially of varieties carrying the rye gene, to increase risk of gut inflammatory conditions including gastric ulcers (Smith and Edwards, 1996) and colitis (Thomson, 2009). However, the finding of Holmgren and Lundeheim (2004) of increased tail biting risk with diets where barley constituted >50% of the grain component then seems surprising, and they could offer no explanation for this.

Cereals may also carry a risk of ergot or mycotoxin contamination. Mycotoxins are secondary metabolites produced by different types of fungi and can have pro-inflammatory effects. Mycotoxin contamination has been anecdotally linked with tail biting (e.g. Hubbard Feeds, 2021), although clear evidence is still lacking. Deoxynivalenol (DON) is one of the most prevalent Fusarium mycotoxins in grain and has been shown to influence brain neurochemistry, causing changes in both serotoninergic and catecholaminergic activities (see Nordgreen *et al.*, 2020, for a more detailed discussion). Whilst the evidence in relation to this risk factor is still anecdotal, cleaning of cereals before use in feed or the addition of mycotoxin binders could be potential mitigation strategies.

Regarding the protein component of the diet, both liquid whey (Holmgren and Lundeheim, 2004) and whey powder (Kallio *et al.*, 2018) have been implicated in risk. This may reflect their variable mineral composition. Distillers' waste, which has poor protein quality due to its cereal source and risk of heat damage, has also been implicated (Holmgren and Lundeheim 2004). Vegetable protein sources have been reported to be more risky than animal proteins (Gadd, 1967), which might have been associated with their lower amino acid digestibility and ideal protein composition at this early time when diet formulation was less sophisticated. Furthermore, there is also a possibility that other anti-nutritive factors associated with vegetable proteins, such as their content of tannins, metal chelators, protease inhibitors, indigestible oligosaccharides and antigenic proteins (Samtiya *et al.*, 2020), might adversely affect nutrient bioavailability, giving rise to deficiencies, or compromise gut health, and result in microbial dysbiosis or inflammatory responses (Nordgreen *et al.*, 2020). There has been no experimental investigation of the effects of such compounds on tail biting risk, but the mechanisms by which they might induce tail biting are present.

Other contaminants or accidental errors in diet formulation have also been suggested anecdotally to induce a tail biting outbreak. For example, Kavanagh (1992) associated a tail biting outbreak with accidental dietary inclusion of the antimicrobial additive Carbadox at twice the recommended rate (100 mg/kg), noting that serum sodium levels were reduced in the affected pigs.

Another practical consideration which may relate to feed ingredients is the observation that tail biting is often associated with the time of feed transition in a multiphase feed program. Consecutive feeds are often not well aligned, with a sudden change in nutrient specification and the use of different raw materials. This means that pigs have to adjust feeding and digestion (especially to new fibre sources), and a reduction in feed (and nutrient) intake for several days is common. The associated risk of tail biting can be reduced by a more gradual phasing in of a new diet.

Finally, a range of propriety feed additives to reduce tail biting risk or mitigate outbreak severity are now present in the market, many of which target the hypothesised mechanisms identified in this review. However, there is little independent evidence to demonstrate their efficacy. For example, evaluation of a product with wheat bran, linseed cake and additional salts (NaCl, $NaSO_2$, $NaCO_3$), failed to demonstrate effective control of acute tail biting outbreaks (Palander *et al.* 2012). However, further investigation of the potential for such approaches is certainly merited.

8 Conclusions

Both the composition of the diet and the method of feeding play an important role in the occurrence of tail biting. Farmers have the biggest role to play in optimising feeding management and reducing the risk of sudden-forceful biting. This risk is increased by any circumstances which induce competition for feed or frustration at failing to achieve a desired intake. Feed nutrition companies have an important role to play in minimising risk of two-stage biting by producing feeds which fit the requirements of the pigs, and making sure that there are no detrimental factors in the feed. Any nutrient deficiencies or imbalances could increase foraging activity which can be redirected to chewing of penmates, especially in the absence of appropriate environmental substrate. Hypothesised mechanisms for dietary effects on tail biting risk include (i) deficiencies or imbalances of some key amino acids which can impair the synthesis of influential brain neurotransmitters regulating feeding, exploration and mood, (ii) stress-induced changes in sodium metabolism, and (iii) behaviour and mood modification induced by microbiome dysbiosis via the microbiome-gut-brain axis. Further research is needed to elucidate these mechanisms and to indicate potential dietary interventions which can prevent and terminate damaging tail biting in pigs of all ages.

9 Key points for practice

- Ensure that pigs have adequate feeding space to avoid competition for access, taking account of feed form, diet density and delivery system. Observe animal behaviour rather than relying on standard recommendations.
- Avoid feeding in frequent small meals.
- Carry our regular maintenance of automatic feed delivery systems and ensure timely delivery of new supplies to avoid feed outages.
- Ensure that drinkers are clean, correctly positioned and deliver an adequate flow rate.
- Obtain professional nutritional advice to correctly formulate diets for the growth stage, genetic potential and health status of the animals, with particular attention to amino acid composition.
- Ensure that raw materials are analysed for nutrient composition, stored correctly and free from contamination.
- In case of a tail biting outbreak, increase sodium chloride in the diet (to 0.5%, but not more than 1%), sprinkle salt on the pen floor or provide a mineral block, in all cases ensuring plentiful availability of fresh water

References

Albernaz-Gonçalves, R., Olmos, G. and Hötzel, M.J., 2021. My pigs are ok, why change? – animal welfare accounts of pig farmers. Animal 15: 100154. https://doi.org/10.1016/j.animal.2020.100154.

Barnikol, M., 1978. Observations of tail biting (cannibalism) amongst fattening hogs and breeding pigs in connection with protein and mineral nourishment. Tieraerztliche Umschau 33: 540.

Barron, N., 1964. The pig farmer's veterinary book, 4th edition. Farming Press (Books) Ltd., Ipswich, p. 140.

Baxter, M.R., 1986. The design of the feeding environment for the pig. PhD thesis, University of Aberdeen.

Beattie, V.E., Breuer, K., Dunne, L.M., Slade, E.C., O'Connell, N.E., Mercer, J.T., Rance, K.A., Sneddon, I.A., Sutcliffe, M.E.M. and Edwards, S.A., 2001. The effect of salt deficiency on the behaviour of finishing pigs in a tail chew test. Proceedings of the British Society of Animal Science, PC2.

Botermans, J.A.M., and Svendsen, J., 2000. Effect of feeding environment on performance, injuries and behaviour in growing-finishing pigs: group-based studies. Acta Agriculturae Scandinavica – Section A 50:237–249. https://doi.org/10.1080/090647000750069430.

Bracke, M.B., Hulsegge, B., Keeling, L., and Blokhuis, H.J., 2004. Decision support system with semantic model to assess the risk of tail biting in pigs: 1. Modelling. Applied Animal Behaviour Science 87: 31–44. https://doi.org/10.1016/j.applanim.2003.12.005.

Bracke, M.B.M., De Lauwere, C.C., Wind, S.M.M. and Zonderland, J.J., 2013. Attitudes of Dutch pig farmers towards tail biting and tail docking. Journal of Agricultural and Environmental Ethics 26: 847–868. https://doi.org/10.1007/s10806-012-9410-2.

Breuer, K., Sutcliffe, M., Mercer, J., Rance, K., O'Connell, N., Sneddon, I. and Edwards, S., 2005. Heritability of clinical tail-biting and its relation to performance traits. Livestock Production Science 93: 87–94. https://doi.org/10.1016/j.livprodsci.2004.11.009.

Brunberg, E.I., Rodenburg, T.B., Rydhmer, L., Kjaer, J.B., Jensen, P. and Keeling, L.J., 2016. Omnivores going astray: a review and new synthesis of abnormal behavior in pigs and laying hens. Frontiers in Veterinary Science 3: 57. https://doi.org/10.3389/fvets.2016.00057.

Bushby, E.V., Dye, L. and Collins, L.M., 2021. Is magnesium supplementation an effective nutritional method to reduce stress in domestic pigs? a systematic review. Frontiers in Veterinary Science 7: 1100. https://doi.org/10.3389/fvets.2020.596205.

Chambers, C., Powell, L., Wilson, E. and Green, L.E., 1995. A postal survey of tail biting in pigs in south-west England. Veterinary Record 136: 147–148. https://doi.org/10.1136/vr.136.6.147.

Chou, J.Y., O'Driscoll, K., Sandercock, D.A. and D'Eath, R.B., 2020. Can increased dietary fibre level and a single enrichment device reduce the risk of tail biting in undocked growing-finishing pigs in fully slatted systems? PLoS ONE 15: e0241619. https://doi.org/10.1371/journal.pone.0241619.

Day, J.E.L., Kyriazakis, I. and Lawrence, A.B., 1995. The effect of food deprivation on the expression of foraging and exploratory behaviour in the growing pig. Applied Animal Behaviour Science 42: 193–206. https://doi.org/10.1016/0168-1591(95)93889-9.

de Leeuw, J.A., Bolhuis, J.E., Bosch, G. and Gerrits, W.J., 2008. Effects of dietary fibre on behaviour and satiety in pigs. Proceedings of the Nutrition Society 67: 334–342.

Diana, A., Manzanilla, E.G., Calderon Diaz, J.A., Leonard, F.C. and Boyle, L.A., 2017. Do weaner pigs need in-feed antibiotics to ensure good health and welfare? PLoS ONE 12: e0185622. https://doi.org/10.1371/journal.pone.0185622.

Doeschl, A., Green, D., Whittemore, C., Schofield, C., Fisher, A. and Knap, P., 2004. The relationship between the body shape of living pigs and their carcass morphology and composition. Animal Science 79: 73–83. https://doi.org/10.1017/S1357729800054540.

EFSA, 2007. Scientific report on the risks associated with tail biting in pigs and possible means to reduce the need for tail docking considering the different housing and husbandry systems. EFSA Journal 611: 1–13.

Ettle, T. and Roth, F.X., 2004. Specific dietary selection for tryptophan by the piglet. Journal of Animal Science 82: 1115–1121. https://doi.org/10.2527/2004.8241115x.

Ewbank, R., 1973. Abnormal behaviour and pig nutrition. An unsuccessful attempt to induce tail biting by feeding a high energy, low fibre vegetable protein ration. British Veterinary Journal 129: 366–369. https://doi.org/10.1016/S0007-1935(17)36440-0.

Forbes, J.M., 1995. Voluntary food intake and diet selection in farm animals. CAB International, Wallingford, 532 pp.

Fraser, D., 1987. Mineral-deficient diets and the pig's attraction to blood, implications for tail-biting. Canadian Journal of Animal Science 67: 904–918.

Fraser, D., Bernon, D.E. and Ball, R.O., 1991. Enhanced attraction to blood by pigs with inadequate dietary protein supplementation. Canadian Journal of Animal Science 71: 603–619. https://doi.org/10.4141/cjas91-076.

Gadd, J., 1967. Tail-biting. Pig Farming 15: 57–58.

García-Mochales, C.A., Böhne, I., Pitkin, A., Heras, A.V. and Wang, B., 2018. Tail biting outbreaks: strategies to minimize the problem and on-going preventative measures. Available at: https://www.pig333.com/articles/tail-biting-outbreaks-strategies-to-minimize-the-problem_14056/.

Georgsson, L. and Svendsen, J., 2002. Degree of competition at feeding differentially affects behavior and performance of group-housed growing-finishing pigs of different relative weights. Journal of Animal Science 80: 376–383. https://doi.org/10.2527/2002.802376x.

Haigh, A. and O'Driscoll, K., 2019. Irish pig farmer's perceptions and experiences of tailbiting. Porcine Health Management 5: 30. https://doi.org/10.1186/s40813-019-0135-8.

Hakansson, F. and Houe, H., 2020. Risk factors associated with tail damage in conventional non-docked pigs throughout the lactation and rearing period. Preventive Veterinary Medicine 184: 105160. https://doi.org/10.1016/j.prevetmed.2020.105160.

Hansen, L.L., Hagelso, A.M. and Madsen, A., 1982. Behavioural results and performance of bacon pigs fed 'ad lib' from one or several feeders. Applied Animal Ethology 8: 307–333. https://doi.org/10.1016/0304-3762(82)90065-7.

Hessel, E.F., Wulbers-Mindermann, M., Berg, C., Van den Weghe, H.F.A. and Algers, B., 2006. Influence of increased feeding frequency on behavior and integument lesions in growing-finishing restricted-fed pigs. Journal of Animal Science 84: 1526–1534. https://doi.org/10.2527/2006.8461526x.

Holinger, M., Früh, B., Stoll, P., Kreuzer, M. and Hillmann, E., 2018. Grass silage for growing-finishing pigs in addition to straw bedding: effects on behaviour and gastric health. Livestock Science 218: 50–57. https://doi.org/10.1016/j.livsci.2018.10.012.

Holmgren, N. and Lundeheim, N., 2004. Risk factors for tail biting. Proceedings of the 18th IPVS Congress, Hamburg, p. 786.

Holmgren. N., Keeling, L. and Lundeheim, N., 2004. Blood parameters in relation to tail biting in pigs. Proceedings of the 18th IPVS Congress, Hamburg, p. 786.

Honeck, A., 2019. Influence of crude fibre in piglets' rations and the animal-to-feeding-place ratio on tail-biting in weaning pigs. Doctoral dissertation, Stiftung Tierärztliche Hochschule Hannover.

Honeck, A., Ahlhorn, J., Burfeind, O., Gertz, M., Grosse Beilage, E., Hasler, M., Tölle, K., Visscher, C. and Krieter, J., 2020. Influence on tail-biting in weaning pigs of crude fibre content and different crude fibre components in pigs' rations. Journal of Agricultural Science 158: 233–240. https://doi.org/10.1017/S0021859620000404.

Hubbard Feeds. 2021. Tail biting. Available at: https://www.hubbardfeeds.com/sites/default/files/Finisher%20Management%20Guide%20Tail%20Biting%20Final.pdf.

Hunter, E.J., Jones, T.A., Guise, H.J., Penny, R.H.C. and Hoste, S., 2001. The relationship between tail biting in pigs, docking procedure and other management practices. Veterinary Journal 161: 72–79. https://doi.org/10.1053/tvjl.2000.0520.

Jankevicius, M.L. and Widowski, T.M., 2003. Exogenous adrenocorticotrophic hormone does not elicit a salt appetite in growing pigs. Physiology and Behavior 78: 277–284. https://doi.org/10.1016/S0031-9384(02)00970-8.

Jankevicius, M.L. and Widowski, T.M., 2004. The effect of ACTH on pigs' attraction to salt or blood-flavoured tail-models. Applied Animal Behaviour Science 87: 55–58. https://doi.org/10.1016/j.applanim.2003.12.012.

Jensen, M.B., Kyriazakis, I. and Lawrence, A.B., 1993. The activity and straw directed behaviour of pigs offered foods with different crude protein content. Applied Animal Behaviour Science 37: 211–221. https://doi.org/10.1016/0168-1591(93)90112-3.

Jericho, K.W.F. and Church, T.L., 1972. Cannibalism in pigs. Canadian Veterinary Journal 13: 156–159.

Kallabis, K.E. and Kaufmann, O., 2012. Effect of a high-fibre diet on the feeding behaviour of fattening pigs. Archives of Animal Breeding 55: 272–284. https://doi.org/10.5194/aab-55-272-2012.

Kallio, P.A., Janczak, A., Valros, A.E., Edwards, S.A. and Heinonen, M., 2018. Case control study on environmental, nutritional and management-based risk factors for tail-biting in long-tailed pigs. Animal Welfare 27: 21–34. https://doi.org/10.7120/09627286.27.1.021.

Kavanagh, N.T., 1992. Carbadox toxicity associated with an outbreak of tail, ear and flank biting and urine sucking. Pig Veterinary Journal 28: 125–128.

Kobek-Kjeldager, C., Schönherz, A.A., Canibe, N. and Pedersen, L.J., 2022. Diet and microbiota-gut-brain axis in relation to tail biting in pigs: a review. Applied Animal Behaviour Science 246: 105514. https://doi.org/10.1016/j.applanim.2021.105514.

Kraimi, N., Dawkins, M., Gebhardt-Henrich, S.G., Velge, P., Rychlik, I., Volf, J., Creach, P., Smith, A., Colles, F. and Leterrier, C., 2019. Influence of the microbiota-gut-brain axis on behavior and welfare in farm animals: a review. Physiology and Behaviour 210: 112658. https://doi.org/10.1016/j.physbeh.2019.112658.

Krider, J.L., Albright, J.L., Plumlee, M.P., Conrad, J.H., Sinclair, C.L., Underwood, L., Jones, R.G. and Harrington, R.B., 1975. Magnesium supplementation, space and docking effects on swine performance and behavior. Journal of Animal Science 40: 1027–1033. https://doi.org/10.2527/jas1975.4061027x.

Kyriazakis, I. and Emmans, G., 1991. Diet selection in pigs: dietary choices made by growing pigs following a period of underfeeding with protein. Animal Science 52: 337–346. https://doi.org/10.1017/S0003356100012861.

Kyriazakis, I., Leus, K., Emmans, G., Haley, C. and Oldham, J., 1993. The effect of breed (Large White × Landrace v. purebred Meishan) on the diets selected by pigs given a choice between two foods that differ in their crude protein contents. Animal Science 56: 121–128. https://doi.org/10.1017/S0003356100006231.

Lange, K., Buerger, M., Stallmach, A. and Bruns, T., 2016. Effects of antibiotics on gut microbiota. Digestive Diseases 34: 260–268. https://doi.org/10.1159/000443360.

Larsen, M.L.V., Andersen, H.M.L. and Pedersen, L.J., 2018. Which is the most preventive measure against tail damage in finisher pigs: tail docking, straw provision or lowered stocking density? Animal 12: 1260–1267. https://doi.org/10.1017/S175173111700249X.

Laskoski, F., Faccin, J.E., Vier, C.M., Gonçalves, M.A., Orlando, U.A., Kummer, R., Mellagi, A.P., Bernardi, M.L., Wentz, I. and Bortolozzo, F.P., 2019. Effects of pigs per feeder hole and group size on feed intake onset, growth performance, and ear and tail lesions in nursery pigs with consistent space allowance. Journal of Swine Health and Production 27: 12–18.

Manteca, X. and Edwards, S., 2009. Feeding behaviour and social influences on feed intake. In: Torrallardona, D. and Roura, E. (eds.) Voluntary feed intake in pigs. Wageningen Academic Publishers, Wageningen. pp. 293–306.

Martínez-Trejo, G., Ortega-Cerrilla, M.E., Rodarte-Covarrubias, L.F., Herrera-Haro, J.G., Figueroa-Velasco, J.L., Galindo-Maldonado, F., Sánchez-Martínez, O. and Lara-Bueno, A., 2009. Aggressiveness and productive performance of piglets supplemented with tryptophan. Journal of Animal and Veterinary Advances 8: 608–611. Available at: https://medwelljournals.com/abstract/?doi=javaa.2009.608.611.

McIntyre, J. and Edwards, S.A., 2002a. An investigation into the effect of different protein and energy intakes on model tail chewing behaviour of growing pigs. Applied Animal Behaviour Science 77: 93–104. https://doi.org/10.1016/S0168-1591(02)00044-8.

McIntyre, J. and Edwards, S.A., 2002b. An investigation into the effect of tryptophan on tail chewing behaviour of growing pigs. Proceedings of the British Society of Animal Science, York, p. 34.

MLC 2004a. Finishing pigs: systems research. Production trial 1. Dry versus liquid feeding in two contrasting finishing systems (fully slatted versus straw based housing). Meat and Livestock Commission. Milton Keynes.

MLC 2004b. Finishing pigs: systems research. Production trial 2. Evaluation of phase feeding in two contrasting finishing systems (fully slatted versus straw based housing). Meat and Livestock Commission. Milton Keynes.

MLC 2005a. Finishing pigs: systems research. Production trial 3. Controlled fermentation of cereals in liquid diets fed to pigs in two contrasting finishing systems (fully slatted versus straw based housing). Meat and Livestock Commission. Milton Keynes.

MLC 2005b. Finishing pigs: systems research. Production trial 4. Reducing the protein content of liquid diets fed to pigs in two contrasting finishing systems (fully slatted versus straw based housing). Meat and Livestock Commission. Milton Keynes.

Moinard, C., Mendl, M., Nicol, C.J. and Green, L.E., 2003. A case control study of on-farm risk factors for tail biting in pigs. Applied Animal Behaviour Science 81: 333– 355. https://doi.org/10.1016/S0168-1591(02)00276-9.

Morrow, A.T.S. and Walker, N., 1994. A note on changes to feeding behaviour of growing pigs by fitting stalls to single-space feeders. Animal Science 59: 151–153. https://doi.org/10.1017/S0003356100007613.

Naya, A., Gertz, M., Hasler, M., große Beilage, E. and Krieter, J., 2019. Does a higher content of fibre in the piglet diet have an influence on tail biting in growing pigs? Livestock Science 223: 133–137. https://doi.org/10.1016/j.livsci.2019.03.010.

Nielsen, E.K. and Ingvartsen, K.L., 2000. Effects of cereal disintegration method, feeding method and straw as bedding on stomach characteristics including ulcers and performance in growing pigs. Acta Agriculturae Scandinavica – Section A 50: 30–38. https://doi.org/10.1080/090647000423906.

Nordgreen, J., Edwards. S.A., Boyle, L.A., Bolhuis, J.E., Veit, C., Sayyari, A., Marin, D.E., Dimitrov, I., Janczak, A.M. and Valros, A., 2020. A proposed role for pro-inflammatory cytokines in damaging behavior in pigs. Frontiers in Veterinary Science 7: 646. https://doi.org/10.3389/fvets.2020.00646.

O'Driscoll, K., O'Gorman, D.M., Taylor, S., Boyle, L.A., 2013. The influence of a magnesium-rich marine extract on behaviour, salivary cortisol levels and skin lesions in growing pigs. Animal 7: 1017–1027. https://doi.org/10.1017/S1751731112002431.

Olsen, A.W., 2001. Behaviour of growing pigs kept in pens with outdoor runs. I. Effect of access to roughage and shelter on oral activities. Livestock Production Science 69: 255–264. https://doi.org/10.1016/S0301-6226(01)00172-5.

Palander, P., Heinonen, M., Simpura, I., Edwards, S. and Valros, A., 2013. Jejunal morphology and blood metabolites in tail biting, victim and control pigs. Animal 7: 1523–1531. https://doi.org/10.1017/S1751731113000669.

Palander, P.A., Valros, A., Heinonen, M. and Edwards, S.A., 2012. Supplementary feed offered to a tail biting pen changes feeding behaviour, feed intake, growth and tail health of fattening pigs. Proceedings 46th Congress of the International Society for Applied Ethology, Vienna, Wageningen Academic Publisher, 156.

Pandolfi, F., Stoddart, K., Wainwright, N., Kyriazakis, I. and Edwards, S.A., 2017. The 'Real Welfare' scheme: benchmarking welfare outcomes for commercially farmed pigs. Animal 11: 1816–1824. https://doi.org/10.1017/S1751731117000246.

Pastorelli, H., Le Floc'h, N., Merlot, E., Meunier-Salaün, M.C., van Milgen, J. and Montagne, L., 2012. Sanitary housing conditions modify the performance and behavioural response of weaned pigs to feed- and housing-related stressors. Animal 6: 1811–1820. https://doi.org/10.1017/S1751731112001231.

Paul, E., Moinard, C., Green, L.E. and Mendl, M., 2007. Farmers' attitudes to methods for controlling tail biting in pigs. Veterinary Record 160: 803–805. https://doi.org/10.1136/vr.160.23.803.

Petherick, J.C., 1983. A biological basis for the design of space in livestock housing. In: Baxter, S.H., Baxter, M.R. and MacCormack, J.A.C. (eds.) Farm animal housing and welfare. Martinus Neuhoff Publishers, Lancaster, pp. 103–120.

PIC 2021. PIC® nutrition and feeding guidelines. PIC Hendersonville, Tennessee. Available at: https://www.pic.com/resources/nutrition.

Presto Åkerfeldt, M., Lindberg, J.E., Göransson, L. and Andersson, K., 2019. Effects of reducing dietary content of crude protein and indispensable amino acids on performance and carcass traits of single-phase- and 2-phase-fed growing-finishing pigs. Livestock Science 224: 96–101. https://doi.org/10.1016/j.livsci.2019.04.014.

Prunier, A., Averos, X., Dimitrov, I., Edwards, S.A., Hillmann, E., Holinger, M., Ilieski, V., Leming, R., Tallet, C., Turner, S.P., Zupan, M. and Camerlink, I., 2020. Review on early life predisposing factors for biting in pigs. Animal 14: 570–587. https://doi.org/10.1017/S1751731119001940.

Rabhi, N., Thibodeau, A., Côté, J-C., Devillers, N., Laplante, B., Fravalo, P., Larivière-Gauthier, G., Thériault, W.P., Faucitano, L., Beauchamp, G. and Quessy, S., 2020. Association between tail-biting and intestinal microbiota composition in pigs. Frontiers in Veterinary Science 7: 563762. https://doi.org/10.3389/fvets.2020.563762.

Rérat, A., 1996. Influence of the nature of carbohydrate intake on the absorption chronology of reducing sugars and volatile fatty acids in the pig. Reproduction Nutrition Development 36: 3–19. https://doi.org/10.1051/rnd:19960101.

Robert, S., Matte, J.J. and Girard, C.L., 1991. Effect of feeding regimen on behavior of growing-finishing pigs supplemented or not supplemented with folic acid. Journal of Animal Science 69: 4428–4436. https://doi.org/10.2527/1991.69114428x.

Samtiya, M., Aluko, R.E. and Dhewa, T., 2020. Plant food anti-nutritional factors and their reduction strategies: an overview. Food Production, Processing and Nutrition 2: 6. https://doi.org/10.1186/s43014-020-0020-5.

Schumacher, S., 2019. Retrospektive Feldstudie bei Schlachtschweinen zu den Risikofaktoren der Caudophagie. Doctoral dissertation, Freien Universität Berlin.

Scollo, A., Gottardo, F., Contiero, B. and Edwards, S.A., 2017. A cross-sectional study for predicting tail biting risk in pig farms using classification and regression tree analysis. Preventive Veterinary Medicine 146: 114–120. https://doi.org/10.1016/j.prevetmed.2017.08.001.

Scott, K., Chennells, D.J., Armstong, D., Taylor, L., Gill, B.P. and Edwards, S.A., 2007. The welfare of finishing pigs under different housing and feeding systems: liquid versus dry feeding in fully-slatted and straw-based housing. Animal Welfare 16: 53–62. https://doi.org/10.1017/S0962728600030931.

Sell, A., Vidondo, B., Nathues, H., Burla, J-B. and Wechsler, B., 2021. Lésions de la queue chez les porcs d'engraissement: Évolution en cours d'engraissement et facteurs de risques. Transfer Agroscope N° 359. Agroscope, Tänikon. Available at: https://doi.org/10.34776/at359f.

Smith, W.J. and Edwards, S.A., 1996. Ulceration of the pars oesophagea – the role of a factor in wheat. Pig Journal 36: 194–200.

Smulders, D., Hautekiet, V., Verbeke, G. and Geers, R., 2008. Tail and ear biting lesions in pigs: an epidemiological study. Animal Welfare 17: 61–69. https://doi.org/10.1017/S0962728600031997.

Taylor, N.R., Main, D.C.J., Mendl, M. and Edwards, S.A., 2010. Tail-biting: a new perspective. Veterinary Journal 186: 137–147. https://doi.org/10.1016/j.tvjl.2009.08.028.

Taylor, N.R., Parker, R.M.A., Mendl, M., Edwards, S.A. and Main, D.C.J., 2012. Prevalence of risk factors for tail biting on commercial farms and intervention strategies. Veterinary Journal 194: 77–83. https://doi.org/10.1016/j.tvjl.2012.03.004.

Temple, D., Courboulay, V., Velarde, A., Dalmau, A. and Manteca, X., 2012. The welfare of growing pigs in five different production systems in France and Spain: assessment of health. Animal Welfare 21: 257–271. https://doi.org/10.7120/09627286.21.2.257.

Terlouw, E.M.C., Wiersma, A., Lawrence, A.B. and Macleod, H.A., 1993. Ingestion of food facilitates the performance of stereotypies in sows. Animal Behaviour 46: 939–950. https://doi.org/10.1006/anbe.1993.1275.

Thomson, J., 2009. Feed-associated colitis of growing pigs and its interaction with enteric infections. Acta Scientiae Veterinariae 37 (Suppl. 1): s1–s9.

Tsourgiannis, C.A., Robertson, J.F. and Fowler, V.R., 2002. Influence of salt application on biting by growing–finishing pigs. Proceedings of the British Society of Animal Science, York, p. 223.

Ursinus, W.W., Wijnen, H.J., Bartels, A.C., Dijvesteijn, N., van Reenen, C.G. and Bolhuis, J.E., 2014a. Damaging biting behaviors in intensively kept rearing gilts: the effect of jute sacks and relations with production characteristics. Journal of Animal Science 92: 5193–5202. https://doi.org/10.2527/jas.2014-7918.

Ursinus, W.W., van Reenen, C.G., Reimert, I. and Bolhuis, J.E., 2014b. Tail biting in pigs: blood serotonin and fearfulness as pieces of the puzzle? PLoS ONE 9: e107040. https://doi.org/10.1371/journal.pone.0107040.

Valros, A. and Barber, C., 2019. Producer perceptions of the prevention of tail biting on UK farms: association to bedding use and tail removal proportion. Animals 9: 628. https://doi.org/10.3390/ani9090628.

Valros, A., Munsterhjelm, C., Hanninen, L., Kauppinen, T. and Heinonen, M., 2016. Managing undocked pigs – on-farm prevention of tail biting and attitudes towards tail biting and docking. Porcine Health Management 2: 2. https://doi.org/10.1186/s40813-016-0020-7.

Valros, A., Palander, P., Heinonen, M., Munsterhjelm, C., Brunberg, E., Keeling, L. and Piepponen, P., 2015. Evidence for a link between tail biting and central monoamine metabolism in pigs (Sus scrofa domestica). Physiology and Behaviour 143: 151–157. https://doi.org/10.1016/j.physbeh.2015.02.049.

Valros, A., Sali, V., Hälli, O., Saari,S. and Heinonen, M., 2021. Does weight matter? Exploring links between birth weight, growth and pig-directed manipulative behaviour in growing-finishing pigs. Applied Animal Behaviour Science 245: 105506. https://doi.org/10.1016/j.applanim.2021.105506.

van der Meer, Y., Gerrits, W.J.J., Jansman, A.J.M., Kemp, B. and Bolhuis, J.E., 2017. A link between damaging behavior in pigs, sanitary conditions, and dietary protein and amino acid supply. PloS ONE 12: e0174688. https://doi.org/10.1371/journal.pone.0174688.

Vargas Vargas, J., Craig, J.V. and Hines, R.H., 1987. Effects of feeding systems on social and feeding behavior and performance of finishing pigs. Journal of Animal Science 65: 463–474. https://doi.org/10.2527/jas1987.652463x.

Veit, C., Traulsen, I., Hasler, M., Tölle, K-H., Burfeind, O., große Beilage, E. and Krieter, J., 2016. Influence of raw material on the occurrence of tail-biting in undocked pigs. Livestock Science 191: 125–131. https://doi.org/10.1016/j.livsci.2016.07.009.

Verbeek, E., Keeling, L., Landberg, R., Lindberg, J.E. and Dicksved, J., 2021. The gut microbiota and microbial metabolites are associated with tail biting in pigs. Scientific Reports 11:20547. https://doi.org/10.1038/s41598-021-99741-8.

Wallgren, T., Westin, R. and Gunnarson, S., 2016. A survey of straw use and tail biting in Swedish pig farms rearing undocked pigs. Acta Veterinaria Scandinavica 58: 84. https://doi.org/10.1186/s13028-016-0266-8.

Watanabe, T., Noguchi, M., Numanoi, K., Haseyama, T. and Aoyama, M., 2015. Examination of managing methods to relieve tail biting in piglets-effects of chain/rope, sodium chloride, and diet form. Japanese Journal of Swine Science 52: 123–134. https://doi.org/10.5938/youton.52.3_123.

Wurtman, R.J. and Fernstrom, J.D., 1975. Control of brain monoamine synthesis by diet and plasma amino acids. American Journal of Clinical Nutrition 28: 638–647. https://doi.org/10.1093/ajcn/28.6.638.

Zonderland, J.J., Wolthuis-Fillerup, M., Van Reenen, C.G., Bracke, M.B.M., Kemp, B., Den Hartog, L.A. and Spoolder, H.A.M., 2008. Prevention and treatment of tail biting in weaned piglets. Applied Animal Behaviour Science 110: 269– 281. https://doi.org/10.1016/j.applanim.2007.04.005.

PART 3

Methods of addressing the problem

CHAPTER 10

Risk assessment

Sabine Dippel

Friedrich-Loeffler-Institut, Dörnbergstr. 25/27, 29223 Celle, Germany;
sabine.dippel@fli.de

Abstract

This chapter describes in which situations tail biting risk assessment is being used, i.e. daily care, farm planning and research. It outlines what should be considered for risk assessment in the different situations, which risk areas need to be covered and which parameters can be used. It also addresses how to use the respective risk assessment results in a practical way.

Keywords

parameters – daily care – farm planning – research

1 Introduction

Tail biting has been a decades-long problem in pig husbandry, which led to widespread tail docking. Rearing undocked pigs is more expensive because it needs more input (labour, material, space) and risk for tail injuries which are also costly, is much higher. Tail biting risk assessment is key for reducing tail injuries in undocked pigs, as well as in docked pigs.

1.1 *Meaning of tail biting in this chapter*

The term tail biting is used as an umbrella term in this chapter. It includes the behaviour (one pig biting the tail of another pig) as well as tail lesions (haemorrhaging, open wounds, swelling, loss of tail substance, necrosis etc.). Usually biting behaviour causes tail lesions, but the reverse is also possible when tail lesions from internal processes (chapter 14; Reiner *et al.*, 2019) or accidents attract investigation by the pen mates.

Tail biting is not the same as aggression. Aggression refers to fighting for social rank or resources and is usually directed at the front of the body (Prunier *et al.*, 2020). Generally, the tail is only injured in aggression if pigs have to queue for a resource ("sudden-forceful tail-biting"; Taylor *et al.*, 2010) or are bullied in rank fights (Turner *et al.*, 2006). Cannibalism may be a consequence of tail biting. Wounds have an interesting new smell and taste, so pen mates will investigate them with their snout and mouth, especially if there is not much else interesting in the pen.

2 What is risk?

Tail biting risk is the likelihood that tail biting will occur. Risk factors increase or decrease the risk. As tail biting is a multifactorial problem, a large number of risk factors are involved, and the strength of single factors may be amplified or weakened by other risk factors (interaction). The resulting situation is so complex that tail biting cannot be triggered reliably (van Putten, 1969; Ewbank, 1973). Grouping risk factors by area of influence (e.g. health or nutrition) is helpful for risk assessment because it provides structure. During analysis of risk, two or more factors may be confounded. This means their effects cannot be separated because they are strongly associated or only occur in certain combinations.

Everything that affects a pig contributes to tail biting risk. Imagine the pig as a bucket (Figure 10.1 A). The filling of the bucket is the total sum of the risk factors that affect the pig, i.e. the tail biting risk. In commercial pig farming there is a constant level of chronic risk originating from the basic environmental situation and

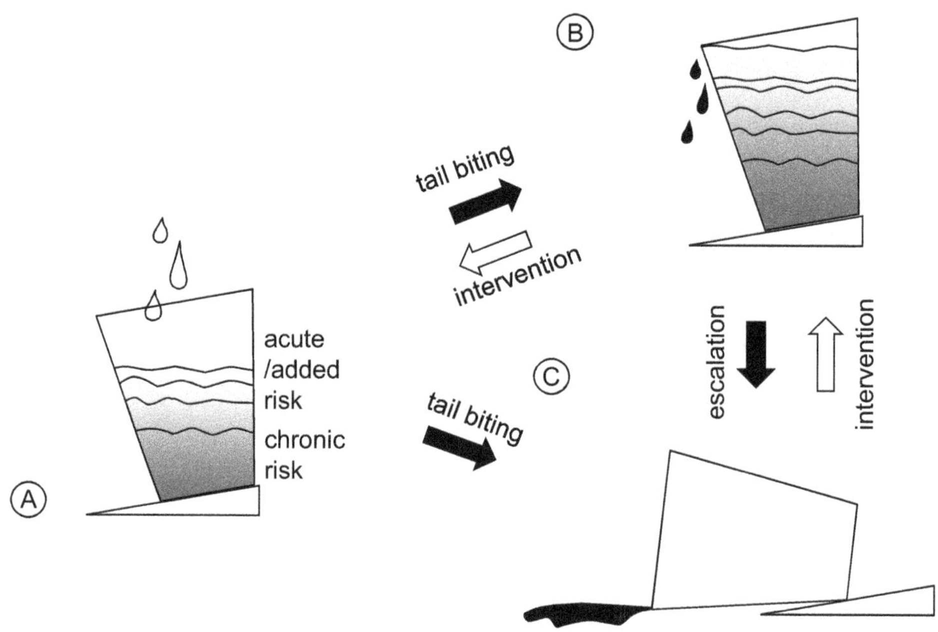

FIGURE 10.1 Tipping bucket model (modified after Bracke *et al.*, 2012; Valros, 2018)

pig characteristics. Any negative impact on a pig increases the filling of the bucket. A strong negative factor will raise the level of fill a lot, but many small negative factors may have the same effect. When too many negative factors have added to the bucket it will overflow; that is, tail biting will occur. This may range from 'trickling over the rim' (a slow increase in tail lesions; Figure 10.1 B), to 'tipping' (a sudden strong tail biting outbreak; Figure 10.1 C). Overflowing may also escalate into tipping. Intervention may stop overflowing, and may reduce tipping to overflowing, but it is virtually impossible to turn tipping into 'still room in the bucket' (A).

Good pig comfort translates into lower 'bucket filling'. Pig comfort means environments designed according to the pigs' needs and management that pays attention to detail.

2.1 *Why assess risk?*

Prevention of tail biting is key because once tail biting has started it is difficult to stop (Chou *et al.*, 2019). Intervention costs time and money, and tail biting is not only stressful and painful for the pigs but also unpleasant for their human caretakers. Tail biting may be a problem on any pig farm: in undocked pigs (Sell *et al.*, 2020), in docked pigs (Brocke *et al.*, 2018), in intensive housing (Courboulay *et al.*, 2009), or in some free range systems (Alban *et al.*, 2015) (see chapter 7). Each individual farm has its own mix of risk factors that determines its chronic tail biting risk (basic 'bucket filling'). Even farms that look the same, or pens in the same room, differ in details that together result in different risk levels. The different mix of risks is the reason that an intervention may work on one farm or in one pen but not on or in another; thus individual solutions are needed.

What is also different on each farm are the people. In this chapter, 'farmer' refers to persons involved in management decisions and 'caretaker' to those inspecting the pigs every day. This may be the same person or several persons, but all people on the farm need to be involved in tail biting reduction.

Successful prevention needs measurement of the risk situation, and monitoring of how it changes. This chapter uses 'assess' instead of 'measure' because the measurements range from subjective impressions of animal states, to categorising qualities, to measuring lengths of equipment, to recording sick animals over a year. The assessors may be people working on the farm or external persons. Keeping tail biting risk low requires good daily care combined with farm-specific long-term reduction of risks. This is supported by recommendation adjustments and deeper understanding based on research. Thus, there are three main times where risk assessment is generally useful:

- as part of daily animal care by the pig caretaker
- as part of farm management at intervals of several months (farm planning, with an external person)
- for research.

3 Risk assessment for daily care

Why: To detect fast-changing risks (what's 'dropping into the bucket' right now) and first signs of tail biting (check 'bucket filling') in order to act fast and prevent an outbreak. Risk assessment during daily care allows early detection and resolution of problems. This ensures that the bucket does not 'overflow' or 'tip' (i.e. tail biting does not start or escalate).

Where: Every pig farm.

Who: Pig caretakers.

When: At least once, but even better to be carried out several times per day.

Focus on: Pig signs and equipment checks.

Use results for: Keeping pigs comfortable, fast interventions, daily management decisions.

Key points:
- Must be feasible as part of daily routine.
- Usually informal. Some formalisation may be necessary, e.g. pen cards on larger farms.
- Fast translation of results into intervention actions is key for preventing (more severe) tail biting.

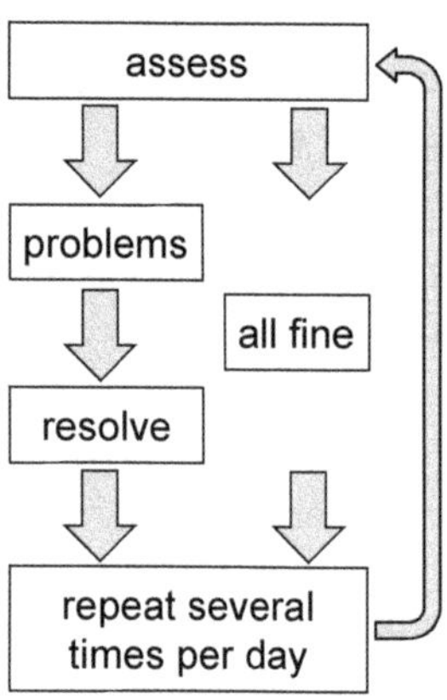

FIGURE 10.2
Flow chart for risk assessment as part of daily care

3.1 *Who?*

Each person taking care of pigs should do a tail biting risk assessment every day by checking that the pigs and husbandry are well. Combining it with fast intervention at the first signs of problems is key for preventing tail biting outbreaks. It is crucial in undocked pigs but also relevant for docked pigs.

Farmers and caretakers moving from docked to undocked pigs have to change their daily risk assessment and way of reacting to changes. Learning this 'new' way of animal care is required to reduce tail biting, especially severe lesions (Veit *et al.*, 2016). After a sufficiently long adaptation period, tail lesion prevalence in undocked pigs can be comparable to docked pigs (EFSA AHAW Panel, 2014). Caretakers, but also veterinarians and farm advisers, need training and support to adjust their mental priorities, recognise first signs in the pigs, and find good intervention strategies. Each farm should start with few groups of undocked pigs to allow time for adapting farm routines. Farm planning facilitates the transition.

3.2 *When?*

Risk assessment for daily care should be done several times per day (Figure 10.2). Being present when fresh feed is provided may help identify problems around feed access. Assessment during active times outside of feeding informs about the general situation.

3.3 *How?*

The assessment is done by inspecting all pigs and their husbandry. The guiding questions for every pen are: 'Do all pigs look good?', 'Do pigs behave differently to usual?' and 'Is everything provided as it should be?'. Refer to the iceberg indicators and recommended parameters in Table 10.2 to Table 10.9. Those marked 'x' in column D should be assessed daily and those marked '(x)' are things to keep in mind during assessment.

Assess the conditions from the perspective of a pig. Pig signs (e.g. lying behaviour, coughing, being agitated, tail posture) help and are important early problem indicators. They are indicated as iceberg indicators in Table 10.2 to Table 10.9. Regular positive interactions with the pigs (giving some enrichment, speaking calmly, petting them, if they like) can reduce tail biting risk (Büttner *et al.*, 2018). Giving some enrichment also makes it easier to inspect the pigs for lesions or observe tail posture. Hanging or tucked in tails may predict tail biting (Zonderland *et al.*, 2009; Lahrmann *et al.*, 2018). Tucked tails are a response to tail manipulation (Wilder *et al.*, 2020) while hanging tails may also suggest general discomfort or stress (Kleinbeck and McGlone, 1993).

Any alteration of the tail counts as a tail lesion for daily care (Table 10.10). Make sure you see every pig at least once per day. Once the first tail lesion has appeared in a pen, it will be more difficult to contain the situation. Thus, the aim is to recognise the first signs of problems in order to intervene before tail lesions become prevalent. Drinkers, enrichment etc. are easier to check and replenish, if they are in reach from the service alley. Consider moving them, unless this disturbs functional areas in the pen. Some formalisation of assessment may be helpful in certain situations, e.g. pen cards to help communication between different caretakers, or using a checklist for one week to train the eyes.

Some practical materials to help daily assessment and recognise pigs' signs have been developed and there may be more in other languages (AHDB WebHAT; FitForPigs; PigWatch; Scheepens *et al.*; Spoolder and PigWatch consortium, 2017; AHDB, 2018).

3.4 *What to do with results?*

Quickly react to any signs of problems discovered in daily risk assessment. Remove recognisable biters, provide new enrichment and then find and fix any potential causes (malfunctioning equipment, health problems etc.). Reacting fast is easier with an 'emergency kit' (LWK NRW, 2016). This refers to ready-to-use pens for separating biters and/or bitten pigs, interesting organic enrichment that the pigs aren't familiar with, and mineral blocks, extra drinking water, or a toxin binder (LWK NRW, 2016). These and other measures have been reported as useful by farmers in several countries (Bracke *et al.*, 2013; Valros *et al.*, 2016; Wallgren *et al.*, 2016; Haigh and O'Driscoll, 2019; Valros and Barber, 2019). The best options may differ between farms and situations, so trial and error is needed to determine what works best. If tail biting starts suddenly, consider recent changes as possible triggers. If it happens gradually or always in the same pen, check whether the resources and conditions in the pen (feed, water, temperature etc.) are as recommended. Tail biting risk assessment for farm planning may help at problematic times, but it is more useful, if it is established as a routine management measure.

4 Risk assessment for farm planning

Why: To get an overview of what increases risk on the farm (raises 'bucket filling') and determine measures to reduce chronic risk ('basic bucket filling').

Where: Every pig farm.

Who: Trained tail biting expert.

When: Routinely once per year or more often.

Focus on: Getting an overview of individual farm risk and identifying problem areas.

Use results for: Selecting medium- to long-term intervention measures.

Key points:

- All risk areas must be addressed.
- Level of detail may vary.
- Farm planning is a long-term activity.

Each farm has its own specific mix of risk factors (Taylor *et al.*, 2012). Farm planning aims to identify and improve problem areas and thus lower chronic risk, i.e. 'basic bucket filling'.

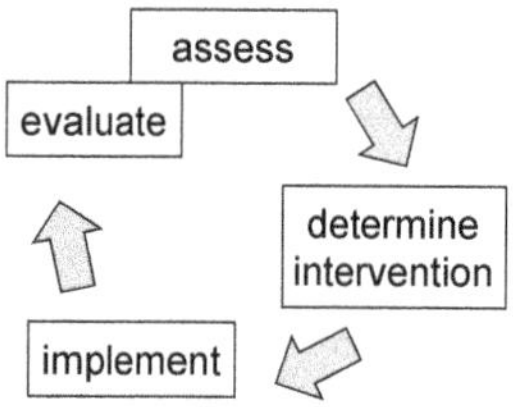

FIGURE 10.3
Flow chart for risk assessment as part of farm improvement planning

4.1 *Who?*

The assessment should be done by an external person who is trained for tail biting improvement. This avoids being blind to what you see every day, makes sure all areas are addressed equally and brings in expertise for discussing improvement measures.

4.2 *When?*

Farm planning is cyclic because the situation on a farm changes over time (Figure 10.3). Thus, the risk assessment should be repeated at regular intervals to monitor changes and adjust improvements. Interval length depends mainly on the production stage; possible intervals are six months for weaned piglets and 12 months for growing-finishing pigs. The intervals should not be too short that they don't allow time for improvements to be implemented and take effect. A focussed, supplementary assessment may be done in case of acute problems. The assessment is done on one day and thus focuses on risks that are present for longer periods of time or occur frequently.

4.3 *How?*

Make sure to address all risk areas listed below by using at least one iceberg indicator and one recommended parameter per area (Table 10.2 to Table 10.9). There is no mathematical tail biting model and thus no standard list of parameters. All parameters in Table 10.2 to Table 10.9 may be useful for farm planning and more have been suggested. The selected parameters should provide a good overview and identification of problem areas. The total number of parameters and level of detail should fit the assessor and farmer preferences to keep up motivation. Any assessment list needs to be supplemented with expertise and experience of the assessor, which is why the external person should have training in tail biting improvement. Once problem areas have been identified, expert consultation for those areas is recommended.

Several risk assessment systems with varying degrees of detail have been developed, the most extensive ones being the tail biting husbandry advisory tool (Taylor *et al.*, 2012, now an abridged online tool: AHDB WebHAT) and SchwIP (Madey, 2014; currently under revision). There are also additional national risk assessment systems in local languages (e.g. BRS; IZSLER; Stolk, 2019; overview in Dippel, 2022). However the scope of these risk assessments for complying with *Commission recommendation (EU) 2016/336* is often too narrow for supporting sustained farm improvement. Assessing the prevalence of tail lesions with and without loss of length (Table 10.10, level 2) reflects the general severity of the problem, but the number of pens is usually too small for calculating a true farm prevalence (Mullan *et al.*, 2009), and selecting problem pens gives a skewed result.

This type of risk assessment takes place in a sample of pens rather than in every pen on the farm. The focus is on the production stage (suckling piglets, weaned piglets, growing-finishing pigs) and husbandry system (combination of flooring, feeding, group size) where tail biting frequently or currently is a problem. It should include several rooms and several pens in each room. This is because conditions within a system and even in two pens in the same room may be different. The aim is to get an impression of the situation and how much it varies. The proportion of pens and pigs represented by the assessment can be calculated from the total numbers of pens and pigs.

Fewer pens can be assessed, if pigs are kept in groups of 50 or more. All resources (e.g. all drinkers) need to be checked in large pens, but clinical assessment (counting pigs with tail lesion or being lame) may be done on a representative sample. Make sure to include a cross-section of the group, e.g. by walking in a spiral and scoring each nth pig that is a meter ahead.

Tail biting risk assessment for farm planning needs to be kept separate from assessments for legal compliance or for farm assurance. Farm planning needs the farmer to trust the external person in order to be truly effective.

4.4 *What to do with results?*

The resources and conditions in the pen (feed, water, temperature, enrichment) as well as feed content information need to be compared against recommendations in order to know whether they are adequate, as inadequate provision will increase tail biting risk. As recommendations vary somewhat and might not fit the assessed system, pig signals should be used to help with interpretation. The external assessor presents the assessment results to and discusses them with the farmer. The farmer then decides which improvements he/she wants to implement. That way, improvements are going to be more likely to occur and be sustained. Improvement measures should be formulated as concrete actions, e.g. 'Give a handful of straw in each pen every day' rather than 'Improve enrichment'. Farmers with employees should include their caretakers in the knowledge gained from the assessment.

5 Risk assessment for research

Why: Investigate risk factors for tail biting (what goes into 'the bucket' and how much it adds to the 'filling'), underlying mechanisms and more.

Where: Specific farms and/or pigs.

Who, When, Focus: Depends on research question.

Use results for: Gaining knowledge for recommendations.

Key points:

- Many different set-ups with specific requirements.
- Assessor training and repeatability testing is required, also for single assessors.
- Assessment protocols should be comparable with other studies.

Tail biting risk assessment for research ranges from small scale experiments to large scale epidemiological studies. Experiments focus on finding out causal relationships

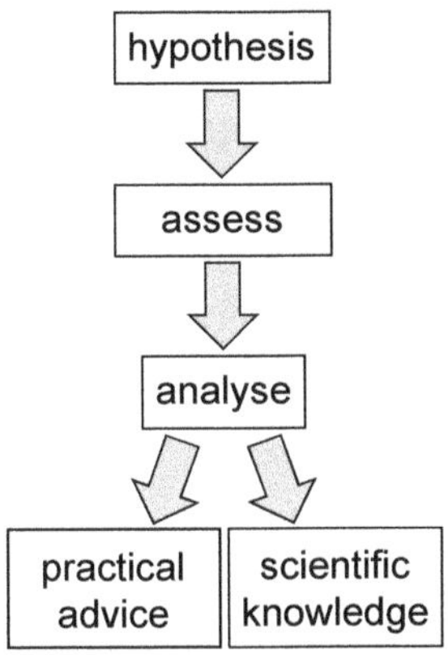

FIGURE 10.4
Flow chart for risk assessment for research

under controlled conditions, while epidemiological studies investigate multifactorial situations. Both are needed to better understand why pigs tail bite, and adapt and develop recommendations (Figure 10.4).

5.1 *Experiments*

Designed experiments test the causal effect of few selected factors, that is, how much 'filling' a certain factor contributes to the 'bucket'. They can compare the effect in otherwise identical situations. The clustering of pigs in replicates, rooms and especially pens should be accounted for in design and analysis (Bello and Renter, 2018; Seo *et al.*, 2018). Parameters that are usually used for correcting analyses (e.g. group size) are indicated in Table 10.2 to Table 10.9. Basic parameters from Table 10.2 to Table 10.9 can be used for documenting conditions to put results into context. Reporting confounding and publishing negative results is important as it contributes to the public knowledge base (Open Science Collaboration, 2015). The tail scoring key should at least distinguish loss of tail length from lesions without substance loss and possibly swellings from other lesions (Table 10.10 level 2 and part of 3). This allows the assessor to distinguish acute from chronic lesions. Assessors should be trained and their repeatability tested and reported. This applies to multiple assessors but also to a single assessor who may drift over time (Pfeifer *et al.*, 2019). All tail biting experiments have to cope with the sporadic nature of tail biting. Tail biting during the experiment may be too little or too much or too variable to draw firm conclusions (Veit *et al.*, 2016; Holinger *et al.*, 2018). This is why experiments should include several replicates over time.

5.2 *Epidemiological on-farm studies*

Epidemiological studies on tail biting investigate the strengths of risk factors in the multifactorial reality, i.e. how much 'filling' a certain mix of risks adds to the bucket. This is needed because interactions between risk factors may have a strong influence on whether pigs tail bite or not, and can be difficult to replicate in small scale experiments.

Epidemiological studies assess many farms, because the farm is the basic unit. Addressing the clustering of pigs within farm (production stages, systems, rooms, pens) and expected confounding in study designs improves the quality of results (Dohoo *et al.*, 2012).

Data for epidemiological studies can be collected in different ways (Table 10.1), each with its own specific methodology. The basic and recommended parameters as well as iceberg indicators from Table 10.2 to Table 10.9 may serve as a starting point for farm visits.

TABLE 10.1 Advantages and disadvantages of different epidemiological data collection methods

Method	Assess	Advantages	Disadvantages
one-day farm visit by trained assessors	T + R[a]	good data quality	limited number of farms, point-in-time data
repeated one-day visits by trained assessors	T + R	good data quality, information on variability	more limited number of farms
farm records	(T +) R	covers period of time, needed for most basic farm information	variable data quality, definitions may differ
farm caretaker observations	T (+ R)	covers period of time, frequent observations	assessor variation
phone survey (self-assessment)	T + R	many farms	very subjective (especially T)
farm planning	T + R	many farms	small, not representative sample of pens, assessor variation
farm assurance	T + R	many farms, good data quality (if trained assessors)	limited risk factors, assessor variation (if not trained)
abattoir: general	T	many farms	problems in earlier stages and pigs who died on farm not represented, point-in-time data, only some pigs of a farm may be delivered; see chapter 11
abattoir: trained assessors	T	good data quality	see above

a T = tail lesions, R = risk factors.

TABLE 10.1 Advantages and disadvantages of different epidemiological data collection (*cont.*)

Method	Assess	Advantages	Disadvantages
abattoir: meat inspection	T (+ R[b])	data are already being collected	assessor variation within and between abattoirs; only very severe tail lesions with inflammation

b Some health parameters.

Data collected by any method are difficult to analyse. Large numbers of factors, confounding, interactions and structural quality (skewed, missing values, different formats) are a statistical challenge. The results may be influenced by the analysis methods used (e.g. generalised mixed models, principal component analyses, decision trees) and how data are managed (e.g. outlier exclusion, categorisation). Thus, cooperation with analysis experts is advised.

Like for controlled experiments, the assessor(s) should be trained and tested and all results reported where possible. This also applies to resource-based parameters; it is important that the assessors know the definitions for data collected for other purposes. Defining parameters to be comparable with other studies helps interpretation of results and may allow combined analyses (EFSA AHAW Panel, 2014).

For farm visits, pen selection may follow different rules depending on the research question (random, case-control etc.). The proportion of pigs, pens, and rooms represented by the sample can be calculated from the total numbers in the assessed system. Estimating the true tail lesion farm prevalence requires a rather large sample of pens (Mullan *et al.*, 2009).

6 Risk assessment required by law

Commission recommendation (EU) 2016/336 makes tail biting risk assessment with certain criteria mandatory for EU member states. National implementations differ somewhat, and their scope is too narrow for the purposes addressed here.

7 Known impact strengths of risk factors

Scientific results only become scientific facts, if they have been replicated. This is difficult in tail biting research because of the multifactorial nature of biting. This means that the effect of risk factors differs depending on the presence of other risk factors, which makes it very challenging to determine the strength of the impact of each one.

However, we do know what generally influences tail biting risk, as research has provided information on relative factor strengths within the respective study populations. The next level is to combine this information into a mathematical model that can reliably predict tail biting risk for a specific farm or pen. This has not been achieved yet, and as such there is no 'true' risk assessment method for daily care or farm planning, only the rule to address all areas.

Existing research information has been combined into a decision support system (Bracke *et al.*, 2004), and used for meta-analyses (Averós *et al.*, 2010), toolboxes (EFSA AHAW Panel, 2014), or a relationship network (D'Eath *et al.*, 2014). However, combining existing research may be influenced by negative results not being published (Open Science Collaboration, 2015) and how frequently a factor is investigated; common or easy to investigate risk factors tend to be assessed more often. Expert consultation has also been used as a scientific tool (e.g. EFSA AHAW Panel, 2014). This can be very helpful but also needs careful consideration as expert opinion may be influenced by many external factors (Gustafson *et al.*, 2013) including public discussions (own unpublished data).

Epidemiological farm assessments provide information on the effect of factors in the multifactorial reality. Their results may be influenced by the study population and the parameters selected for inclusion in the assessment. Thus, providing information on confounding factors in the dataset (Moinard *et al.*, 2003; Grümpel *et al.*, 2018) is necessary for interpreting results because it reflects reality.

The number of epidemiological assessments is increasing (see below), with the largest epidemiological analysis so far having been carried out by the EFSA Panel on Animal Health and Welfare (2014). They performed a combined analysis of data assessed with the Welfare Quality® protocol on 242 farms in 5 countries. Even though they did identify risk factors, the explanatory power was too low for reliable strength estimates. Improving data comparability across epidemiological studies will allow further analyses with large sample sizes which will help us better understand the multifactorial relationships.

8 What to assess: risk factor areas and parameters

There are many ways to assess risk factors for tail biting. As a basic rule, the following risk areas should be addressed in each daily, farm planning and research risk assessment:
- physical infrastructure
- enrichment
- nutrition (feed and water)
- management

- health
- social environment
- individual pig related factors

How exactly each area is addressed is somewhat flexible. The list of factors and parameters in Table 10.2 to Table 10.9 is a suggestion. These can be rearranged, shortened, extended or be made more detailed. Some parameters may be more relevant in one situation than in others, and some are associated with two or more areas. More information for most areas can be found in other chapters of this book.

The list aims to include those parameters for which links have been found, those which may indicate problems that in turn increase tail biting risk, and those which may be helpful for putting the results into context. Iceberg indicators (ii) show the combined effect of several factors and thus point to problem areas. These are mostly animal-based parameters. Basic parameters (b) describe the general situation to help put risk assessment results into context. They stand for common risk factor combinations and often have been associated with tail biting. Some parameters are needed in order to adjust recommended threshold values (ar) for e.g. temperature, others to adjust analysis calculations (aa). Recommended parameters (rec) are known to indicate problems that increase tail biting risk or were associated with tail biting in at least two of the following studies: Moinard *et al.*, 2003; Cagienard *et al.*, 2005; Smulders *et al.*, 2008; Courboulay *et al.*, 2009; Mullan *et al.*, 2011; Taylor *et al.*, 2012; Temple *et al.*, 2012; EFSA AHAW Panel, 2014; Madey, 2014; Munsterhjelm *et al.*, 2015; Fohringer, 2016; Scollo *et al.*, 2016; Pandolfi *et al.*, 2017; Scollo *et al.*, 2017; Grümpel *et al.*, 2018; Kallio *et al.*, 2018; Leeb *et al.*, 2019; Reiner *et al.*, 2019; Sell *et al.*, 2020; Heinonen *et al.*, 2021 (epidemiological), Telkänranta *et al.*, 2014b; Telkänranta *et al.*, 2015 (experimental). Parameters in the studies may have been part of an index or been defined slightly differently. Parameters useful for daily care (D) are also indicated in the tables. Farm planners may want to use the entire list, while parameters used for research will depend on the objective. Both will need clear parameter definitions.

8.1 *Types of farming system* (chapter 7)

The farming system refers to consistent basic characteristics which provide context for farm planning and research: farm type, herd size, and production type (Table 10.2). The effects for all three of these parameters are inconclusive (EFSA AHAW Panel, 2014; Alban *et al.*, 2015; Kruse *et al.*, 2015; Munsterhjelm *et al.*, 2015; Kallio *et al.*, 2018), but they are important to record.

8.2 *Physical infrastructure* (chapter 7)

The physical infrastructure describes the direct environment of the pigs. While caretakers may focus on iceberg indicators, farm planners and researchers will also need to identify risk associated with resource-based parameters (Table 10.3).

TABLE 10.2 Suggested parameters for assessing farming system. Factors and parameters associated with several areas or factors, respectively, are listed only once and described in the text. ii = iceberg indicators, b = basic parameters, ar = for adjusting recommendations, aa = for adjusting analyses, rec = recommended parameters, D = assess for daily care

Factor	Parameter	ii	b	ar	aa	rec	D
farm type	production stages kept on farm						
herd size	number of pig places		x		x		
production type	conventional, label, organic		x				

TABLE 10.3 Suggested parameters for assessing physical infrastructure. Factors and parameters associated with several areas or factors, respectively, are listed only once and described in the text. ii = iceberg indicators, b = basic parameters, ar = for adjusting recommendations, aa = for adjusting analyses, rec = recommended parameters, D = assess for daily care

Factor	Parameter	ii	b	ar	aa	rec	D
husbandry system	proportion of slatted flooring		x	x		x	
	powered, or (controlled) natural ventilation		x				
	indoor, with outdoor run, outdoor		x			x	
bedding	none, some, deep litter		x	x		x	
space allowance	m^2 per pig, kilogram per m^2		x			x	
functional areas	clearly distinguished lying area		x	x		x	
	protected lying area (creep)					x	
thermal comfort	air temperature					x	
	relative humidity					x	
	air velocity					x	
	cooling possibilities						
	lying behaviour	x				x	x
	dirty pigs, wallowing in toilet area, defecating in intended lying area	x				x	x
	panting, shivering	x				x	x
air quality	noxious gases (NH_3, H_2S, CO_2)					x	
	dust						
	tear staining	x				x	x
	subjective air quality assessment	x				x	x

TABLE 10.3 Suggested parameters for assessing physical infrastructure (*cont.*)

Factor	Parameter	ii	b	ar	aa	rec	D
light	darker lying area						
	day-night-cycle						
	protection against direct sunlight						
other comfort	noise						x
	many flies						x
	stray currents						
outdoor pigs	shelters					x	
	cooling possibility						
	grass cover	x				x	x
	pigs per hectare						
	paddock rotation					x	

The physical infrastructure should provide good pig comfort. Suboptimal conditions increase chronic risk ('basic bucket filling') so that a negative event will be more likely to result in overflow, and consequently biting. The husbandry system sets the basis for this, especially flooring type and use of bedding (Courboulay *et al.*, 2009; Kallio *et al.*, 2018). Bedding improves lying and thermal comfort (except during hot periods) and serves as enrichment unless it's dirty. Overall, allthough there is no guarantee (Taylor *et al.*, 2012), tail biting risk is generally lower in systems that allow the pigs more control (e.g. temperature zones) and have better enrichment.

The space available per pig is a powerful parameter despite – or because of – its frequent confounding (Munsterhjelm *et al.*, 2015; Grümpel *et al.*, 2018). Specifying it in m^2/pig and kg/m^2 can be helpful. The pigs should be able to rest without disturbance from feeding or active pigs and at suitable temperatures. Thermal comfort is determined by the combined effect of temperature, humidity and air speed. The pigs' behaviour (lying, defecating, wallowing) is a useful problem detector (Jackson *et al.*, 2020). Panting, diarrhoea and tear staining may indicate thermal problems (but also have other causes). Temperature monitoring helps detect fluctuations, and regular climate checks are needed to adjust ventilation. All assessments need to be done where the pigs are, with focus on the lying area. Light, noise, flies and stray currents may be a problem on some farms. Preventing tail biting in outdoor pigs needs to refer to guidelines for good outdoor husbandry practices.

8.3 *Enrichment* (chapter 8)

Enrichment has a strong effect on tail biting because pigs are highly motivated to investigate their surroundings with their snout (root, chew). Caretakers should ensure every day that all pigs can access good enrichment. The importance of enrichment is supported by research results, and its provision should be monitored by farm planners.

Good enrichment is safe, fun to destroy, tastes good and smells new (Table 10.4). Bedding is more preventive, if it is frequently topped up, because this makes it more attractive to manipulate (Moinard *et al.*, 2003). A mixture of enrichment may be positive (Averós *et al.*, 2010), though some novel enrichment materials need to be kept back for outbreak intervention. A large proportion of pigs in the pens should be able to access enrichment at the same time in order to avoid competition. Enrichment should be at shoulder height or lower, fit into the mouth, and be clean. More pigs manipulating enrichment compared to pigs manipulating pen mates or the pen is a sign that enrichment provision is good (Weerd *et al.*, 2006).

TABLE 10.4 Suggested parameters for assessing enrichment. Factors and parameters associated with several areas or factors, respectively, are listed only once and described in the text. ii = iceberg indicators, b = basic parameters, ar = for adjusting recommendations, aa = for adjusting analyses, rec = recommended parameters, D = assess for daily care

Factor	Parameter	ii	b	ar	aa	rec	D
overall	ratio of pigs manipulating enrichment to manipulating other	x				x	x
quality	safe					x	
	manipulable, chewable, edible, rootable					x	
provision frequency	when broken, times per week, times per day		x			x	
access	proportion of pigs that can access it simultaneously	x				x	
	appropriate height						

8.4 *Nutrition* (chapter 9)

Water and feed are a physiological foundation. If they cannot be accessed or have insufficient quality, tail biting risk will considerably increase. While caretakers focus on access, farm planners and researchers also analyse quality. Pigs have been bred to grow fast, which requires good nutrition. Stress may make pigs more hungry and thirsty (Jankevicius and Widowski, 2003) which increases competition for water and feed and thus amplifies tail biting risk.

Pigs naturally drink by sucking, which is why bowls or troughs are more comfortable for them (Table 10.5). Nipple drinker height and flow rate must be suitable for the pigs' size to allow good drinking. A sufficient number of drinking places should be provided, preferably away from the feed. Pigs with liquid feeding (soup) also need drinking water (Vermeer *et al.*, 2009). Water quality can be reduced by the source, through dirty piping, or via soiled drinkers. Thus drinker flow and subjective water quality (soiling) should be checked daily.

Data about the effect of feeding systems are inconclusive (ad libitum / restricted, liquid / wet / dry, pelleted / granulated/ meal, on the floor / hopper / long trough). There seems to be some confounding with how feeding systems are implemented in different countries or other country-specific factors. Regardless, feed being delivered reliably is important, irrespective of system type (Scollo *et al.*, 2016). The number of pigs per feeding place in *ad libitum* systems is often confounded with space availability (EFSA AHAW Panel, 2014), and lesions from fighting may serve as problem indicator.

Feed should be free from contaminants. The nutrient content recommendations depend on production stage and gender. Higher dietary crude fibre combined with greater particle size (Millet *et al.*, 2010) but also roughage-like enrichment (Holinger *et al.*, 2018) may prevent stomach ulcers which are a source of discomfort. Expert advice and regular feed analyses are recommended.

TABLE 10.5 Suggested parameters for assessing nutrition. Factors and parameters associated with several areas or factors, respectively, are listed only once and described in the text. ii = iceberg indicators, b = basic parameters, ar = for adjusting recommendations, aa = for adjusting analyses, rec = recommended parameters, D = assess for daily care

Factor	Parameter	ii	b	ar	aa	rec	D
water	drinker type			x		x	
	drinker flow rate						x
	drinker height						
	pigs per drinking place					x	
	water quality					x	x
feed	*ad libitum* or restricted		x	x		x	
	liquid, wet or dry		x	x		x	
	pelleted, granulated or meal					x	
	feeder type					x	
	pigs per feeding place					x	
	reliability of feeding system					x	
	spoiled feed						x[a]
	provision of fibre					x	
	protein, lysin, tryptophan, NaCl content					x	

a Daily and farm planning: subjective. Farm planning and research: analyses.

8.5 *Management*

Management makes good pig comfort work, and it is particularly important at the start of a pigs' life. Caretakers and farm planners should bear this in mind, while researchers may want to investigate this more.

The people working on a farm are key for reducing tail biting because their knowledge, experience, motivation and attitude determine whether good pig comfort is achieved. Researchers have started to investigate the relationships (Bracke *et al.*, 2013; Valros *et al.*, 2016; Wallgren *et al.*, 2016; Haigh and O'Driscoll, 2019; Valros and Barber, 2019; Sell *et al.*, 2020), yet more research is needed to identify on-farm parameters that reliably reflect management quality. Suggestions are included in Table 10.6 (also see chapter 12).

More comfortable farrowing pens (Smulders *et al.*, 2008) with enrichment for suckling piglets can reduce tail biting risk even later in life (Telkänranta *et al.*, 2014b). Nowadays due to large litter sizes where the number of piglets exceeds the number of teats, many piglets are being cross-fostered or nursed artificially in order

TABLE 10.6 Suggested parameters for assessing management. Factors and parameters associated with several areas or factors, respectively, are listed only once and described in the text. ii = iceberg indicators, b = basic parameters, ar = for adjusting recommendations, aa = for adjusting analyses, rec = recommended parameters, D = assess for daily care

Factor	Parameter	ii	b	ar	aa	rec	D
general	number of pigs and or pens per caretaker		x			x	
	separation pens present						
	number of pig inspections per day						
	time until problems are addressed						
	tail biting emergency kit present						
	communication quality between caretakers						
	experience with and opinion on undocked pigs						
	caretaker tail lesion assessment differs from external assessment	x					
suckling piglets	farrowing system		x	x		x	
	litter size						
	cross-fostering						
	nursing						
	enrichment for suckling piglets					x	
	litters mingle before weaning						
	weight variation at weaning	x					
	suckling piglet mortality	x					

TABLE 10.6 Suggested parameters for assessing management (*cont.*)

Factor	Parameter	ii	b	ar	aa	rec	D
weaning	creep feeding of suckling piglets					x	
	weaning age					x	
	mixing at weaning					x	
	additional feed or water trough						
	reared in farrowing pen		x				
	diarrhoea	x				x	x
	runts	x				x	x
	weight variation at end of rearing	x				x	

to improve survival. These measures seem to increase tail biting risk, though data are limited and confounded with litter size and birth weights (Grümpel *et al.*, 2018).

Weaning is a high-impact event for pigs because many things change significantly at the same time; thus, anything that reduces the magnitude of these changes reduces tail biting risk. This includes using the same feed, enrichment and drinking system in farrowing and weaner pens, additional feed and water troughs after weaning, or letting litters mingle before weaning to reduce social stress (Lange *et al.*, 2021). More data are needed to estimate the effect of weaning age. This is difficult to study because most farms wean at a similar age or else have a different production system to conventional (label, organic), and there is often confounding with other suckling period parameters (Grümpel *et al.*, 2018). The effects of mixing litters at weaning (Scollo *et al.*, 2016; Grümpel *et al.*, 2018) and rearing in the farrowing pen (Lange *et al.*, 2021) are not clear, again, probably because they depend on the farm setting and way of mixing. Diarrhoea, presence of runts, pigs diverging in weight, and mortality are useful iceberg indicators for resource- or health-related problems that increase tail biting risk.

8.6 *Health* (chapter 5)

While nutrition is the physiological foundation, health is the physiological skyscraper: if it burns, things turn very bad very quickly. This is very important for caretakers and farm planners, and researchers are more and more investigating the internal processes that could trigger biting.

Overall, any form of health problem increases risk for tail biting (Boyle *et al.*, 2022). This is due to the stress they cause, but also to physiological processes related to immune system activation (chapter 5). Good hygiene and regular health checks help to maintain good health. All animal-based parameters in Table 10.7 are problem and often iceberg indicators. They include indicators for physical and mental wellbeing (being withdrawn, lethargic, depressed (sick); nervous or agitated) as well as pathogenic, physiological or physical infrastructure problems. Inflammation of

TABLE 10.7 Suggested parameters for assessing health. Factors and parameters associated with several areas or factors, respectively, are listed only once and described in the text. ii = iceberg indicators, b = basic parameters, ar = for adjusting recommendations, aa = for adjusting analyses, rec = recommended parameters, D = assess for daily care

Factor	Parameter	ii	b	ar	aa	rec	D
Diseases	herd health status					x	
	regular health checks						
	all-in-all-out		x				
Inflammation	mycotoxin analyses						
	inflammation of feet or teats	x					
health problems	sickness behaviour	x				x	x
	nervous or agitated	x				x	x
	respiratory problems	x				x	x
	diarrhoea	x				x	x
	rectal prolapse					x	
	lameness	x				x	x
	runts	x				x	x
	ear lesions	x				x	x
	circular wounds	x				x	x
	leg wounds	x					x
	mortality	x				x	
	carcase findings	x				x	

the coronary band, heel and teats has been associated with tail necrosis in suckling piglets (Reiner *et al.*, 2019). Respiratory problems (coughing, sneezing, red eyes, laboured breathing, nasal discharge) may be pathogenic or related to low air quality. A veterinarian should be consulted in case where there is a health problem indicated.

Presence of ear biting also indicates higher tail biting risk (Brunberg *et al.*, 2011; Telkänranta *et al.*, 2014a), especially in short-docked pigs (Goossens *et al.*, 2008). There is very little literature on the subject of ear biting, but it is thought that it differs from tail biting in being more linked to fighting (Prunier *et al.*, 2020) as well as having a pathogenic element (Pringle *et al.*, 2009).

8.7 *Social environment* (chapter 6)

Unbalanced social relationships in a group are stressful for the pigs and thus increase tail biting risk. Caretakers and farm planners should aim to create and enable stable, well balanced groups, while researchers investigate group dynamics.

TABLE 10.8 Suggested parameters for assessing social environment. Factors and parameters associated with several areas or factors, respectively, are listed only once and described in the text. ii = iceberg indicators, b = basic parameters, ar = for adjusting recommendations, aa = for adjusting analyses, rec = recommended parameters, D = assess for daily care

Factor	Parameter	ii	b	ar	aa	rec	D
group size	direct observation		x		x		
size variation	single pigs smaller or larger than group average	x				x	x
group stability	straight lesions on front of body	x				x	x
	frequency of mixing or splitting groups						

The effect of group size on tail biting is unclear, partly because of confounding with resource access. In addition, the mathematical likelihood for tail lesions increases with group size, which needs to be accounted for in calculations (Table 10.8). Large pens offer more possibilities for functional areas and thus pig comfort, but they are also more challenging for daily care. Assessors for farm planning and on-farm research should count the number of pigs in the group themselves, rather than depend upon numbers provided by the caretaker.

Smaller pigs have more problems in accessing resources which contributes to frustration, possible nutritional deficiencies, and thus tail biting risk. Size variation may also indicate health problems. Mixing and splitting groups disrupts the social order and leads to fighting which is stressful. Fighting later than a few days after mixing can indicate limited access to resources (Prunier *et al.*, 2020).

8.8 *Individual pig-related factors* (chapter 4)

The current 'bucket filling' of a pig is determined by its innate characteristics and its experiences. Caretakers should keep both in mind, farm planners should focus on improving experiences, and researchers are generally interested in both. Tail biting risk may gradually increase over the lifetime of the pig, or peak during specific stages. High risk periods should be identified for each farm during farm planning and be considered during daily care (Table 10.9). Things to consider include the weight of the pigs, which is needed to determine their resource needs (temperature, nutrition etc.), female pigs bite more than barrows, while entire and possibly immunocastrated males are intermediate. Nutritional recommendations including pigs per feeding space are also slightly different for the genders. The effects of breed and genetic line on tail biting are inconsistent and often confounded. Less backfat has been associated with higher tail biting risk (Moinard *et al.*, 2003). Note, that backfat recording techniques and thus numbers may differ between abattoirs.

TABLE 10.9 Suggested parameters for assessing individual pig-related factors. Factors and parameters associated with several areas or factors, respectively, are listed only once and described in the text. ii = iceberg indicators, b = basic parameters, ar = for adjusting recommendations, aa = for adjusting analyses, rec = recommended parameters, D = assess for daily care

Factor	Parameter	ii	b	ar	aa	rec	D
current weight or age	direct observation		x	x	x	x	(x)
gender	direct observation (+ interview)			x	x	x	(x)
genetics	breed						
	line					x	
	backfat thickness						
docking status	direct observation		x		x	x	(x)
pig history	abrupt change of drinking or feeding system					x	
	decrease of enrichment quality					x	
	previous problems health or tail / ear biting	x				x	
	mortality in previous stage	x				x	(x)
	pigs bought in		x				
	long transport (with mixing)						

Column D: (x) = consider only.

Strong negative events in the life of a pig (e.g. disease, drop in husbandry comfort) notably increase tail biting risk ('bucket filling') which often remains increased. Pigs with negative experiences need more daily attention. Previous production stages should be assessed where possible for farm planning and research. Farm planning should also address individual pig history where necessary. Risk is easier to manage, if pigs are raised on the same farm or bought in from one source only.

9 **Scoring tail lesions** (chapter 11)

There are many ways to score a tail. The system of choice should be reliable and easy to do. Tail scoring for daily care needs to identify the very beginnings of tail biting to allow early intervention. Prevalence of lesions with and without loss of the tail give an impression of escalation severity. Scoring for research depends on the objective and feasibility. Tail scoring training is obligatory for research, but farm

TABLE 10.10 Summary of first three levels of the German Pig Scoring System for tails (Bönisch *et al.*, 2017).

Level 1	Level 2	Level 3
tail intact	parameter absent[a]	parameter absent
tail not intact	lesion	skin lesion (scratches, scabs, wounds) necrosis (includes ring-like constriction) other (e.g. discolouration; optional) Swelling
	loss of length	partial [loss of length] complete [loss of length] (stump)
tail cannot be scored (e.g. too dirty)		

a Not visible from the given distance.

advisors, veterinarians and caretakers also profit from training, especially if they are only used to docked pigs.

Tail scores that are recorded should be as comparable with existing scoring systems as possible. The Welfare Quality® (2009) scoring system is widely used, but it is not very intuitive because it was intended for use after training. The German Pig Scoring Key for tail and ear lesions (Bönisch *et al.*, 2017) may be more useful for the purposes addressed here. Its level 1 (tail intact or not) is suitable for daily care and level 2 for farm planning (level 3 if needed).

Tail lesions recorded at the abattoir are a special case. They are traditionally recorded for food safety and underestimate the farm prevalence considerably (Brocke *et al.*, 2018; Kongsted *et al.*, 2020). Those records may nevertheless be useful for farm planning or research, because they cover long periods of time (chapter 11).

10 Conclusions

Tail biting is multifactorial: everything that impacts a pig has an influence on tail biting risk. The mix of risks differs between farms and within farms and changes over time, and the interactions between risk factors strongly influence whether tail biting happens or not.

No mathematical model can reliably predict tail biting, and therefore there is no standard risk assessment. It is clear though, that all risk areas need to be addressed in each risk assessment. Risk assessment also relies on subjective expertise, which requires experience and training.

11 Main hands-on/practical application for readers not involved in research

- Tail biting prevention means good pig comfort.
- All risk areas need to be addressed in each risk assessment.
- Daily risk detection and fast intervention is crucial for preventing tail biting outbreaks.
- Reduction of chronic risk requires regular improvement of farm-specific problems.
- Expert support and knowledge exchange help to navigate through the risk factor web.

Acknowledgement

I would like to thank Antonia Ruckli for her very helpful comments.

References

Agriculture and Horticulture Development Board (AHDB). AHDB Tail Biting WebHAT. https://webhat.ahdb.org.uk/. Accessed 3 February 2022.

Agriculture and Horticulture Development Board (AHDB), 2018. Identifying early signs of tail biting. https://ahdb.org.uk/knowledge-library/early-warning-signs-of-tail-biting-poster. Accessed 3 February 2022.

Alban, L., Petersen, J., Busch, M.E., 2015. A comparison between lesions found during meat inspection of finishing pigs raised under organic/free-range conditions and conventional, indoor conditions. Porcine Health Management 1, 4. https://doi.org/10.1186/2055-5660-1-4.

Averós, X., Brossard, L., Dourmad, J.-Y., Greef, K.H. de, Edge, H.L., Edwards, S.A., Meunier-Salaün, M.-C., 2010. A meta-analysis of the combined effect of housing and environmental enrichment characteristics on the behaviour and performance of pigs. Appl. Anim. Beh. Sci. 127, 73–85. https://doi.org/10.1016/j.applanim.2010.09.010.

Bello, N.M., Renter, D.G., 2018. Invited review: reproducible research from noisy data: revisiting key statistical principles for the animal sciences. J. Dairy Sci. 101, 5679–5701. https://doi.org/10.3168/jds.2017-13978.

Bönisch, K., Brocke, A. vom, Dippel, S., Grümpel, A., Hagemann, L., Jais, C., Lösel, D., Müller, A., Müller, S., Naya, A., Schrade, H., Späth, C., Veit, C., Wild, A., Lechner, M., 2017. Deutscher Schweine-Boniturschlüssel (DSBS; Version 30.05.2017) [German Pig-Scoring Key]. https://www.openagrar.de/servlets/MCRFileNodeServlet/openagrar_derivate_00046205/German-Pig-Scoring-Key_2017-06-30_en.pdf. Accessed 17 May 2024.

Boyle, L., Edwards, S., Bolhuis, J.E., Pol, F., Zupan Šemrov, M., Schütze, S., Nordgreen, J., Bozakova, N., Sossidou, E.N., Valros, A.E., 2022. The evidence for a causal link between disease and damaging behavior in pigs. Front. Vet. Sci. https://doi.org/10.3389/fvets.2021.771682.

Bracke, M.B.M., Hulsegge, B., Keeling, L., Blokhuis, H.J., 2004. Decision support system with semantic model to assess the risk of tail biting in pigs: 1. Modelling. Appl. Anim. Beh. Sci. 87, 31–44. https://doi.org/10.1016/j.applanim.2003.12.005.

Bracke, M.B.M., Lauwere, C. de, Wind, S.M.M., Zonderland, J.J., 2013. Attitudes of Dutch pig farmers towards tail biting and tail docking. Journal of Agricultural and Environmental Ethics 26, 847–868. https://doi.org/10.1007/s10806-012-9410-2.

Bracke, M.B.M., Vermeer, H.M., Bokma, M., van der Peet, C., Bolhuis, L., Leeijen, J., 2012. Checklist aanpak staartbijten bij (biologische) varkens. https://edepot.wur.nl/220045. Accessed 9 December 2021.

Brocke, A.L. vom, Karnholz, C., Madey-Rindermann, D., Gauly, M., Leeb, C., Winckler, C., Schrader, L., Dippel, S., 2018. Tail lesions in fattening pigs: relationships with postmortem meat inspection and influence of a tail biting management tool. Animal, 1–10. https://doi.org/10.1017/S1751731118002070.

Brunberg, E., Wallenbeck, A., Keeling, L.J., 2011. Tail biting in fattening pigs: Associations between frequency of tail biting and other abnormal behaviours. Appl. Anim. Beh. Sci. 133, 18–25. https://doi.org/10.1016/j.applanim.2011.04.019.

Bundesverband Rind und Schwein e.V. (BRS). Beratungshilfen, Checklisten und Downloads. https://www.ringelschwanz.info/weitere-infomationen/dokumente-links.html. Accessed 3 February 2022.

Büttner, K., Czycholl, I., Basler, H., Krieter, J., 2018. Effects of an intensified human-animal interaction on tail biting in pigs during the rearing period. Journal of Agricultural Science 156, 1039. https://doi.org/10.1017/S002185961800103X.

Cagienard, A., Regula, G., Danuser, J., 2005. The impact of different housing systems on health and welfare of grower and finisher pigs in Switzerland. Prev. Vet. Med. 68, 49–61. https://doi.org/10.1016/j.prevetmed.2005.01.004.

Chou, J.-Y., O'Driscoll, K., D'Eath, R.B., Sandercock, D.A., Camerlink, I., 2019. Multi-step tail biting outbreak intervention protocols for pigs housed on slatted floors. Animals 9. https://doi.org/10.3390/ani9080582.

Courboulay, V., Eugène, A., Delarue, E., 2009. Welfare assessment in 82 pig farms: effect of animal age and floor type on behaviour and injuries in fattening pigs. Animal Welfare 18, 515–521. https://doi.org/10.1017/S0962728600000932.

D'Eath, R.B., Arnott, G., Turner, S.P., Jensen, T., Lahrmann, H.P., Busch, M.E., Niemi, J.K., Lawrence, A.B., Sandøe, P., 2014. Injurious tail biting in pigs: how can it be controlled in existing systems without tail docking? Animal 8, 1479–1497. https://doi.org/10.1017/S1751731114001359.

Dippel, S. (Ed.), 2022. Inventory for tail biting risk assessment and pig observations. Friedrich-Loeffler-Institut, https://www.openagrar.de/receive/openagrar_mods_00079625. Accessed 23 May 2022.

Dohoo, I.R., Martin, W., Stryhn, H. (Eds.), 2012. Methods in epidemiologic research. VER Inc., Charlottetown, Prince Edward Island, Canada.

EFSA AHAW Panel, 2014. Scientific opinion concerning a multifactorial approach on the use of animal and non-animal-based measures to assess the welfare of pigs. EFSA Journal 5, 1–101. https://doi.org/10.2903/j.efsa.2014.3702.

Commission recommendation (EU) 2016/336. European Commission, 2016. Commission recommendation (EU) 2016/336 of 8 March 2016 on the application of Council Directive 2008/120/EC laying down minimum standards for the protection of pigs as regards measures to reduce the need for tail-docking.

Ewbank, R., 1973. Abnormal behaviour and pig nutrition. An unsuccessful attempt to induce tail biting by feeding a high energy, low fibre vegetable protein ration. The British Veterinary Journal 129, 366–369. https://doi.org/10.1016/S0007-1935(17)36440-0.

FitForPigs. FitForPigs (homepage with link to apps). https://www.fitforpigs.de/.

Fohringer, K., 2016. Prevalence of skin lesions and associated risk factors in organic pigs in six European countries. M.Sc., Vienna (AT).

Goossens, X., Sobry, L., Ödberg, F.O., Tuyttens, F.A.M., Maes, D., Smet, S. de, Nevens, F., Opsomer, G., Lommelen, F., Geers, R., 2008. A population-based on-farm evaluation protocol for comparing the welfare of pigs between farms. Animal Welfare 17, 35–41. https://doi.org/10.1017/S0962728600031961.

Grümpel, A., Krieter, J., Veit, C., Dippel, S., 2018. Factors influencing the risk for tail lesions in weaner pigs (Sus scrofa). Livest Sci 216, 219–226. https://doi.org/10.1016/j.livsci.2018.09.001.

Gustafson, L.L., Gustafson, D.H., Antognoli, M.C., Remmenga, M.D., 2013. Integrating expert judgment in veterinary epidemiology: Example guidance for disease freedom surveillance. Prev. Vet. Med. 109, 1–9. https://doi.org/10.1016/j.prevetmed.2012.11.019.

Haigh, A., O'Driscoll, K., 2019. Irish pig farmer's perceptions and experiences of tail and ear biting. Porcine Health Management 5, 30. https://doi.org/10.1186/s40813-019-0135-8.

Heinonen, M., Välimäki, E., Laakkonen, A.-M., Toppari, I., Vugts, J., Fàbrega, E., Valros, A.E., 2021. Evaluation of tail lesions of finishing pigs at the slaughterhouse: associations with herd-level observations. Front. Vet. Sci. 8, 650590(1–13). https://doi.org/10.3389/fvets.2021.650590.

Holinger, M., Früh, B., Stoll, P., Kreuzer, M., Hillmann, E., 2018. Grass silage for growing-finishing pigs in addition to straw bedding: Effects on behaviour and gastric health. Livest Sci 218, 50–57. https://doi.org/10.1016/j.livsci.2018.10.012.

Istituto Zooprofilattico Sperimentale Lombardia e Emilia-Romagna "B. Ubertini" (IZSLER). Classyfarm: Valutazione del Rischio Taglio Coda nel Suino da Ingrasso e Svezzamento [Tail docking risk assessment in fattening and weaning pigs]. Ministero della Salute. https://www.classyfarm.it/images/documents/VET-AZIENDALE_AGGIORNATO_06-23/SUINI-Benessere-valutazione-del-rischio-Taglio-coda-3.pdf. Accessed 10 May 2022.

Jackson, P., Nasirahmadi, A., Guy, J.H., Bull, S., Avery, P.J., Edwards, S.A., Sturm, B., 2020. Using CFD modelling to relate pig lying locations to environmental variability in finishing pens. Sustainability 12, 1928. https://doi.org/10.3390/su12051928.

Jankevicius, M.L., Widowski, T.M., 2003. Exogenous adrenocorticotrophic hormone does not elicit a salt appetite in growing pigs. Physiology & Behavior 78, 277–284. https://doi.org/10.1016/S0031-9384(02)00970-8.

Kallio, P.A., Janczak, A.M., Valros, A.E., Edwards, S.A., Heinonen, M., 2018. Case control study on environmental, nutritional and management-based risk factors for tail-biting in long-tailed pigs. Animal Welfare 27, 21–34. https://doi.org/10.7120/09627286.27.1.021.

Kleinbeck, S., McGlone, J.J., 1993. Pig tail posture: a measure of stress. Texas Tech University Agricultural Science Technical Report T-5-327, 47–48.

Kongsted, H., Foldager, L., Sørensen, J.T., 2020. Data from routine meat inspection is a poor indicator of the prevalence of tail lesions in undocked pigs. Porcine Health Management 6, 10. https://doi.org/10.1186/s40813-020-00149-z.

Kruse, A.B., Larsen, M.H., Skou, P.B., Alban, L., 2015. Assessment of human health risk associated with pyaemia in Danish finisher pigs when conducting visual-only inspection of the lungs. International Journal of Food Microbiology 196, 32–39. https://doi.org/10.1016/j.ijfoodmicro.2014.11.017.

Lahrmann, H.P., Hansen, C.F., D'Eath, R., Busch, M.E., Forkman, B., 2018. Tail posture predicts tail biting outbreaks at pen level in weaner pigs. Appl. Anim. Beh. Sci. 200, 29–35. https://doi.org/10.1016/j.applanim.2017.12.006.

Lange, A., Hahne, M., Lambertz, C., Gauly, M., Wendt, M., Janssen, H., Traulsen, I., 2021. Effects of different housing systems during suckling and rearing period on skin and tail lesions, tail losses and performance of growing and finishing pigs. Animals 11, 2184. https://doi.org/10.3390/ani11082184.

Leeb, C., Rudolph, G., Bochicchio, D., Edwards, S.A., Früh, B., Holinger, M., Holmes, D., Illmann, G., Knop, D., Prunier, A., Rousing, T., Winckler, C., Dippel, S., 2019. Effects of three husbandry systems on health, welfare and productivity of organic pigs. Animal, 1–9. https://doi.org/10.1017/S1751731119000041.

LWK NRW, 2016. Leitfaden für Hoftierärzte, Berater und Landwirte zur Haltung unkupierter Schweine im Rahmen der „Gemeinsamen NRW-Erklärung zum Verzicht auf das „routinemäßige“ Kürzen des Schwanzes bei Schweinen“ (Februar 2014). https://www.landwirtschaftskammer.de/landwirtschaft/tierproduktion/schweinehaltung/pdf/leitfaden-caudophagie.pdf. Accessed 27 January 2022.

Madey, D.P., 2014. Evaluation of a software-based intervention tool for the reduction of tail biting in German fattening pigs. Dr. med. vet., Hannover, Germany.

Millet, S., Meyns, T., Aluwé, M., Brabander, D. de, Ducatelle, R., 2010. Effect of grinding intensity and crude fibre content of the feed on growth performance and gastric mucosa integrity of growing–finishing pigs. Livest Sci 134, 152–154. https://doi.org/10.1016/j.livsci.2010.06.123.

Moinard, C., Mendl, M., Nicol, C.J., Green, L.E., 2003. A case control study of on-farm risk factors for tail biting in pigs. Appl. Anim. Beh. Sci. 81, 333–355. https://doi.org/10.1016/S0168-1591(02)00276-9.

Mullan, S., Browne, W.J., Edwards, S.A., Butterworth, A., Whay, H.R., Main, D.C.J., 2009. The effect of sampling strategy on the estimated prevalence of welfare outcome measures on finishing pig farms. Appl. Anim. Beh. Sci. 119, 39–48. https://doi.org/10.1016/j.applanim.2009.03.008.

Mullan, S., Edwards, S.A., Butterworth, A., Whay, H.R., Main, D.C.J., 2011. A pilot investigation of possible positive system descriptors in finishing pigs. Animal Welfare 20, 439–449. https://doi.org/10.1017/S0962728600003006.

Munsterhjelm, C., Heinonen, M., Valros, A.E., 2015. Application of the Welfare Quality® animal welfare assessment system in Finnish pig production, part II: associations between animal-based and environmental measures of welfare. Animal Welfare 24, 161–172. https://doi.org/10.7120/09627286.24.2.161.

Open Science Collaboration, 2015. Estimating the reproducibility of psychological science. Science (New York, N.Y.) 349, aac4716. https://doi.org/10.1126/science.aac4716.

Pandolfi, F., Kyriazakis, I., Stoddart, K., Wainwright, N., Edwards, S.A., 2017. The 'Real Welfare' scheme: identification of risk and protective factors for welfare outcomes in commercial pig farms in the UK. Prev. Vet. Med. 146, 34–43. https://doi.org/10.1016/j.prevetmed.2017.07.008.

Pfeifer, M., Eggemann, L., Kransmann, J., Schmitt, A.O., Hessel, E.F., 2019. Inter- and intra-observer reliability of animal welfare indicators for the on-farm self-assessment of fattening pigs. Animal 13, 1712–1720. https://doi.org/10.1017/S1751731118003701.

PigWatch. Marrit van Engen – Pig signals: the benefits of better observation [several languages]. FiBL. https://www.youtube.com/results?search_query=%22pigwatch+tutorial%22. Accessed 3 February 2022.

Pringle, M., Backhans, A., Otman, F., Sjölund, M., Fellström, C., 2009. Isolation of spirochetes of genus Treponema from pigs with ear necrosis. Veterinary Microbiology 139, 279–283. https://doi.org/10.1016/j.vetmic.2009.05.018.

Prunier, A., Averos, X., Dimitrov, I., Edwards, S.A., Hillmann, E., Holinger, M., Ilieski, V., Leming, R., Tallet, C., Turner, S.P., Zupan, M., Camerlink, I., 2020. Early life predisposing factors for biting in pigs. Animal 14, 570–587. https://doi.org/10.1017/S1751731119001940.

Reiner, G., Lechner, M., Eisenack, A., Kallenbach, K., Rau, K., Müller, S., Fink-Gremmels, J., 2019. Prevalence of an inflammation and necrosis syndrome in suckling piglets. Animal, 1–11. https://doi.org/10.1017/S1751731118003403.

Scheepens, K., Roozen, M., Hulsen, J., Vries, A. de, van Engen, M. Pig Signals (publisher page with all production stages, languages and formats). https://www.roodbont.nl/en/swine/pig-signals. Accessed 3 February 2022.

Scollo, A., Contiero, B., Gottardo, F., 2016. Frequency of tail lesions and risk factors for tail biting in heavy pig production from weaning to 170 kg live weight. Veterinary Journal 207, 92–98. https://doi.org/10.1016/j.tvjl.2015.10.056.

Scollo, A., Gottardo, F., Contiero, B., Edwards, S.A., 2017. A cross-sectional study for predicting tail biting risk in pig farms using classification and regression tree analysis. Prev. Vet. Med. 146, 114–120. https://doi.org/10.1016/j.prevetmed.2017.08.001.

Sell, A., Vidondo, B., Wechsler, B., Burla, J., Nathues, H., 2020. Risk factors for tail lesions in undocked fattening pigs reared on Swiss farms. Schweizer Archiv für Tierheilkunde 162, 683–695. https://doi.org/10.17236/sat00278.

Seo, S., Jeon, S., Ha, J.K., 2018. Editorial: guidelines for experimental design and statistical analyses in animal studies submitted for publication in the Asian-Australasian Journal of Animal Sciences. Asian Austral J Anim 31, 1381–1386. https://doi.org/10.5713/ajas.18.0468.

Smulders, D., Hautekiet, V., Verbeke, G., Geers, R., 2008. Tail and ear biting lesions in pigs: an epidemiological study. Animal Welfare 17, 61–69. https://doi.org/10.1017/S0962728600031997.

Spoolder, H., PigWatch consortium, 2017. Experiences with the protocol in 5 partner countries. PigWatch Task 1.1, February 2017. https://edepot.wur.nl/441485. Accessed 3 February 2022.

Stolk, K., 2019. Long tails in pigs: an on farm risk-assessment-tool. Master thesis, Utrecht (NL).

Taylor, N.R., Main, D.C., Mendl, M., Edwards, S.A., 2010. Tail-biting: a new perspective. The Veterinary Journal 186, 137. https://doi.org/10.1016/j.tvjl.2009.08.028.

Taylor, N.R., Parker, R.M.A., Mendl, M., Edwards, S.A., Main, D.C.J., 2012. Prevalence of risk factors for tail biting on commercial farms and intervention strategies. Veterinary Journal 194, 77–83. https://doi.org/10.1016/j.tvjl.2012.03.004.

Telkänranta, H., Bracke, M.B.M., Valros, A.E., 2014a. Fresh wood reduces tail and ear biting and increases exploratory behaviour in finishing pigs. Appl. Anim. Beh. Sci. 161, 51–59. https://doi.org/10.1016/j.applanim.2014.09.007.

Telkänranta, H., Marchant-Forde, J.N., Valros, A.E., 2015. Tear staining in pigs: a potential tool for welfare assessment on commercial farms. Animal. https://doi.org/10.1017/S175173111500172X.

Telkänranta, H., Swan, K., Hirvonen, H., Valros, A.E., 2014b. Chewable materials before weaning reduce tail biting in growing pigs. Appl. Anim. Beh. Sci. 157, 14–22. https://doi.org/10.1016/j.applanim.2014.01.004.

Temple, D., Courboulay, V., Velarde, A., Dalmau, A., Manteca, X., 2012. The welfare of growing pigs in five different production systems in France and Spain: assessment of health. Animal Welfare 21, 257–271. https://doi.org/10.7120/09627286.21.2.257.

Turner, S.P., Farnworth, M.J., White, I.M.S., Brotherstone, S., Mendl, M., Knap, P., Penny, P., Lawrence, A.B., 2006. The accumulation of skin lesions and their use as a predictor of individual aggressiveness in pigs. Appl. Anim. Beh. Sci. 96, 245–259. https://doi.org/10.1016/j.applanim.2005.06.009.

Valros, A.E., 2018. Tail biting. In: Špinka, M. (Ed.), Advances in pig welfare. Woodhead Publishing, pp. 137–166.

Valros, A.E., Barber, C., 2019. Producer perceptions of the prevention of tail biting on UK farms: association to bedding use and tail removal proportion. Animals 9. https://doi.org/10.3390/ani9090628.

Valros, A.E., Munsterhjelm, C., Hänninen, L., Kauppinen, T., Heinonen, M., 2016. Managing undocked pigs – on-farm prevention of tail biting and attitudes towards tail biting and docking. Porcine Health Management 2, 2. https://doi.org/10.1186/s40813-016-0020-7.

van Putten, G., 1969. An investigation into tail-biting among fattening pigs. The British Veterinary Journal 125, 511–517. https://doi.org/10.1016/S0007-1935(17)48710-0.

Veit, C., Traulsen, I., Hasler, M., Tölle, K.-H., Burfeind, O., große Beilage, E., Krieter, J., 2016. Influence of raw material on the occurrence of tail-biting in undocked pigs. Livest Sci 191, 125–131. https://doi.org/10.1016/j.livsci.2016.07.009.

Vermeer, H.M., Kuijken, N., Spoolder, H.A.M., 2009. Motivation for additional water use of growing-finishing pigs. Livest Sci 124, 112–118. https://doi.org/10.1016/j.livsci.2009.01.009.

Wallgren, T., Westin, R., Gunnarsson, S., 2016. A survey of straw use and tail biting in Swedish pig farms rearing undocked pigs. Acta Veterinaria Scandinavica 58, 1–11. https://doi.org/10.1186/s13028-016-0266-8.

Weerd, H.A. van de, Docking, C.M., Day, J.E.L., Breuer, K., Edwards, S.A., 2006. Effects of species-relevant environmental enrichment on the behaviour and productivity of finishing pigs. Appl. Anim. Beh. Sci. 99, 230–247. https://doi.org/10.1016/j.applanim.2005.10.014.

Welfare Quality®, 2009. Welfare Quality® assessment protocol for pigs. Welfare Quality® consortium, Lelystad, The Netherlands. ISBN 978-90-78240-05-1, 119 pp.

Wilder, T., Krieter, J., Kemper, N., Honeck, A., Büttner, K., 2020. Tail-directed behaviour in pigs – relation to tail posture and tail lesion. Appl. Anim. Beh. Sci. 233, 105151. https://doi.org/10.1016/j.applanim.2020.105151.

Zonderland, J.J., van Riel, J.W., Bracke, M.B.M., Kemp, B., den Hartog, L.A., Spoolder, H.A.M., 2009. Tail posture predicts tail damage among weaned piglets. Appl. Anim. Beh. Sci. 121, 165–170. https://doi.org/10.1016/j.applanim.2009.09.002.

CHAPTER 11

The evidence of tail biting: where, when and how to measure tail lesions

Anna Valros[1,*] *and Laura Boyle*[2]

[1]Research Centre for Animal Welfare, Faculty of Veterinary Medicine, Department of Production Animal Medicine, P.O. Box 57, 00014 University of Helsinki, Finland; [2]Teagasc Animal and Grassland Research and Innovation Centre, Pig Development Department, Moorepark, Fermoy, Co. Cork, Ireland; *Correspondence: anna.valros@helsinki.fi

Abstract

Measuring lesions caused by tail biting has many benefits: it can be used as an iceberg indicator for pig welfare (see Chapter 15) and as a benchmarking tool for producers to assess farm management both against themselves over time and compared to other pig producers. It is also a good indicator of risk for secondary infections and other health challenges. However, even though it may seem straightforward to evaluate tail biting-related lesions there is no uniform, generally accepted protocol available, and thus results from different sources are difficult to compare. This is further complicated by differences that arise in scoring tails on the live animal (on farm or in the abattoir) and scoring lesions on the dead pig's carcass. In addition, findings differ depending on the production stage when measuring on-farm and the site of measuring in the abattoir. This chapter discusses the pros and cons, as well as practical constraints of different sites for and methods of scoring tail damage. The outcome of tail lesion scoring is greatly affected by the chosen scoring system and scoring conditions. Thus, the scoring system should be selected according to the purpose: the preferred scoring method will differ if the aim is, for example to do scientific research on-farm for identifying risk factors or ways for intervention of outbreaks, or large-scale assessment for benchmarking purposes. Also results from different reports and studies can only be compared with great care, and considering the scoring system and conditions. Scoring tail lesions at the slaughterhouse is easier to standardise than on-farm, and abattoirs usually provide good conditions for systematic scoring. If all types of damage are included, including healed lesions, it also provides a good point to estimate lifetime pig welfare. Thus, measuring at the slaughterhouse can be recommended for use as an iceberg indicator of lifetime welfare of the pigs. It is also very useful for providing feedback on management success to farmers. To allow for benchmarking of tail lesion proportions, there is a need to develop a commonly adopted scoring system both for intact (enough) tails and for tail lesions.

DOI:10.3920/9789004694989_013

Keywords

lesion scoring – on-farm – abattoir – meat inspection – pig – damaging behaviour

1 General introduction

Even though tail biting-related lesions might appear straightforward to evaluate, there is no uniform, generally accepted definition available, and thus results from different sources are difficult to compare (Edwards and Valros, 2021, Honeck *et al.*, 2019). Numerous factors affect the outcome of an estimate of tail lesions. Firstly, the prevalence of tail damage is lower when the tails are docked (Hunter *et al.*, 1999, Grümpel *et al.*, 2018, Lahrmann *et al.*, 2017, Larsen *et al.*, 2018). Also the length of the remaining tail after docking affects the risk for tail biting, with Thodberg *et al.* (2018) showing a significantly decreased risk for damage only if tails were docked short (to 2.9 cm at docking). Secondly, the scoring outcome will differ depending on the conditions at the scoring site, and if it is done on live animals on-farm or post-mortem at the slaughter line, and where on the slaughter line (Carroll *et al.*, 2016, Valros *et al.*, 2020). Finally, the scoring method as well as the purpose of the scoring will influence the outcome: usually recordings for the purpose of scientific studies are more detailed, while those for more practical reasons, such as meat inspection or animal welfare labelling or assessment schemes are much less sensitive and detailed (Pandolfi *et al.*, 2017, Valros *et al.*, 2020, Welfare Quality®, 2009). This chapter will discuss the pros and cons, as well as practical constraints of different sites for and methods of scoring tail damage.

2 Recording tail damage on-farm

2.1 *Introduction*

Scoring tail damage on live animals on-farm will give the most representative picture of the current situation, and is especially important for producers to assess their own management practises and to identify potential risk factors or problematic periods during production.

Scoring systems for on-farm use include those used by researchers for tail biting assessment specifically (for examples, see Kritas and Morrison 2004, Wallgren *et al.*, 2019, Zonderland *et al.*, 2003) as well as systems which are part of holistic animal welfare assessment protocols (eg. Welfare Quality®, 2009 and SusPigSys, 2020); or used during more practical assessment of welfare in different animal welfare schemes (Real welfare (Pandolfi *et al.*, 2017), RSPCA assured (RSPCA, 2016)) or during herd health care visits by veterinarians (eg. the Finnish Sikava system (www.sikava.fi)). Scoring can be performed either on animal level (Welfare Quality®,

2009, Zonderland *et al.*, 2003), pen level (eg. Smulders *et al.*, 2008) or room or farm level (Sikava: www.sikava.fi).

Tail lesions are reported in all ages of growing pigs: in the nursing stage (Hakansson *et al.*, 2020, Vitali *et al.*, 2020, Ursinus *et al.*, 2014), the growing stage (Hakansson *et al.*, 2020, Valros *et al.*, 2021) and the finishing stage (Carroll *et al.*, 2018, Smulders *et al.*, 2008, Valros *et al.*, 2021). Some protocols, such as Welfare Quality® (2009) and SusPigSys (2020), are planned for scoring pigs in both the growing and finishing phase, while for example the Real welfare protocol (Pandolfi *et al.*, 2017) is for finishers only. However, the prevalence as well as the type and severity of lesions might differ throughout these age stages. Carroll *et al.* (2018) reported no severe lesions in pigs below 10 weeks of age, and an increasing prevalence of lesions with increasing age, while, controversially, Valros *et al.* (2021), reported less intact tails in the growing than in the finishing stage. Hakansson *et al.* (2020) developed a specific scoring scale for collecting data on piglets before weaning, due to the damage inflicted being less severe compared to older pigs.

2.2 *Scoring systems for use on-farm*

Scoring systems for tail lesions used in research projects are typically detailed and complicated, underlining the complexity of defining tail damage. Honeck *et al.* (2019) summarized and compared different protocols, and their paper is a good reference for a thorough reflection of different scoring keys, so only some examples will be given here. Honeck *et al.* (2019) divides scoring systems into those where tails are simply given one score each and those in which several dimensions are scored. Simple scoring systems usually combine different aspects of lesions, such as freshness and severity of the lesion. One such example is the Kritas and Morrison (2004) scale, which was used as such or modified in later studies, where score 0 represents no evidence of tail biting; score 1 healed or mild lesions, score 2 evidence of chewing or puncture wounds, but no swelling; score 3 evidence of chewing or puncture wounds with swelling and signs of infection; and score 4 partial or total loss of the tail. This single scoring scale thus includes several dimensions, with elements related to the timing of the lesion (healed), if the tail is infected or not, and if the tail length is reduced. On the other hand, it does not consider the size of the wound per se. An example of a system recording several parameters is the FareWellDock-scoring system (Wallgren *et al.*, 2019), which is based on Zonderland *et al.* (2003). It scores each tail for three features on separate scales: tail length, tail damage and lesion freshness. The German Pig Scoring Key proposed by Honeck *et al.* (2019) as a common scoring key to improve the comparability of studies is detailed, and contains 9 parameters for scoring, of which 5 are optional. The remaining, non-optional parameters are severity of lesion; necrosis type, ring-like constriction and loss of tail length. This scoring protocol is unique in that it has three levels all of which are transferrable between one another such that the most detailed version can be reduced to the simplest one and still be comparable.

In undocked pigs, a full length tail with no wounds is a good measure of an intact tail, while a tail with any type of lesion, or which is shortened, shows that the pig was bitten at some stage of its life. In Finland, where tail docking is banned, and thus all farms raise undocked pigs, the herd health scoring system Sikava (www.sikava.fi) scores percentage of intact tails on the farm as tails being of full length and curled, while tails that are shortened, injured or pressed down between the hind legs are not considered intact. Heinonen *et al.* (2021) showed that this assessment method was linked to what was seen in a detailed scoring of pigs from the same farms at the slaughter line. Using long hairs growing at the end of the tail as a sign of a full length tail, as in e.g. the SusPigSys-system (SusPigsSys, 2020), is a poor indicator of an intact tail. Instead, a fully intact tail can be identified by palpating the tail end: the last vertebrae of a pig tail is flattened and becomes narrower towards the tail tip. Thus, a tail with a thick end, or one with clear scar tissue, or lacking the flattened vertebrae shape is not intact.

The remaining length of a bitten tail indicates how serious and how prolonged was the biting, and probably also the degree to which the pigs welfare is impaired. Herskin *et al.* (2016) showed a larger behavioural reaction when tails were docked short and one can assume the same is also true for biting. Also, Valros *et al.* (2020) showed that tails bitten short are at greater risk of secondary infection than tails only missing a small part. We further suggest that if pigs tails are bitten short, this can be a potential indicator of poor management on the specific farm: tail biting outbreaks can often be spotted early on, and with proper intervention measures (Lahrmann *et al.*, 2018, 2019) can be stopped, or their severity at least reduced.

Surprisingly little information is available on the normal length of pig tails. As far as we know there are only two studies: Herskin *et al.* (2014) and Valros *et al.* (2020), both reporting that the length of the tail of a slaughter aged finisher pig is on average 30–32 cm. For pigs entering the finishing stage at about 30 kg, the length was about 21 cm (Munsterhjelm *et al.*, 2023). Tail length is often defined as the proportion of the tail which is missing. For example, Honeck *et al.* (2019) suggest that undocked tails should be scored as 1/3, 2/3 or more missing, but without specifying the length of a full tail. The SusPigSys protocol (SusPigsSys, 2020) scores tails as normal, half or 'stump' and defines half a tail to be over 10cm in finishers, and 5cm in growers, which is probably much less than half of the full length of a tail. Of course, using tail length as a measure of tail lesions is only useful for undocked pigs. If the pigs tails are docked, the tail length does not add value to lesion scoring, and does not confirm if the tail was bitten or not, unless the tail is more or less missing completely. Of course, from the pig´s point-of-view, it´s welfare was impaired by the docking procedure, the need for which is due to welfare challenges on the farm. Hence, measuring the percentage of intact, undocked tails acts as a good overall estimate of pig welfare (Pig welfare subgroup, 2019)

The damage type, or severity, is often estimated by size or appearance. For example, the FareWellDock system (Wallgren *et al.*, 2019) separates bite marks and bruises from open wounds, and defines the most serious cases as also being swollen. Swelling is an indicator of inflammation, and it is of course possible that for example a very recently bitten tail with a large wound is not swollen. The type and size of lesion is dependent on how big a piece of the tail is missing – as the tail becomes thicker towards the base, and as most tail biting lesions start from the tip of the tail, the closer to the base the wound is, the bigger it will be. Thus, large wounds commonly indicate the same welfare implications as a shortly bitten tail described above.

Tail lesion freshness is included in some scoring systems (eg. FareWellDock) to indicate how recently the lesion occurred. Fresh blood, or an open wound with freshly injured tissue indicates the biting occurred recently, while a wound fully or partly covered with a crust shows it occurred some days ago. If the skin is fully healed, but the tail is shortened, and/or there is visible scar tissue, the lesion is older, and probably occurred weeks ago. As far as we are aware, there is no literature on how quickly a crust forms on a tail biting wound, nor on how quickly a tail biting wound fully heals. Sandercock *et al.* (2016) however, showed that the skin of a docked tail heals superficially within 4 weeks. It can be speculated that it takes at least as long for a severely bitten tail to heal, probably even longer, as the tail might be injured several times, and probably has a higher risk of getting infected. The pigs in the study by Sandercock *et al.* (2016) were docked at three days of age. As wound healing is affected by the age of the pig (Yao *et al.*, 2001), the healing might take longer in older pigs. The freshness of tail damage thus only indicates a relative timeline, and does not allow for detailed timing of the occurrence of a lesion. As tails can be bitten at any stage of a grower-finisher pig's life, and probably heal within some weeks, including only fresh damage in a scoring used for single observations on a farm underestimates the prevalence of damaged tails.

Scoring a lot of tails in live pigs is very labour- and time consuming, and often limits the number of pigs that can be scored. In some large-scale studies (Scollo *et al.*, 2017) and when tail lesions are scored by farm staff (Sinisalo *et al.*, 2012), simple binary scoring systems are used. These provide a simplified outcome, and often do not consider (Welfare Quality®, 2009) or do not include very mild lesions (Scollo *et al.*, 2017). When choosing an appropriate scoring system for a study, it is important to consider the study aim. For example, if scoring is performed during a one-time visit to assess the general prevalence it might be enough to use a simpler method, without, for example, separating fresh blood and crust. On the other hand, if one wants to understand risk factors and prevention strategies, it is important to try to pinpoint at which time the tail biting occurred, as well as it's severity. If the aim is to study short-term intervention strategies for outbreaks, the freshness is an important parameter, while fully healed tails are not of interest. If choosing a detailed scoring system, including also mild lesions, it is important to be prepared for the observations. In our experience, additional people to help limit the pigs movements as well as having material to clean dirty tails and torches to improve visibility of lesions in poor lighting is crucial.

2.3 *Reliability of tail scoring*

Some studies attempted to validate the reliability of tail lesion scoring (Courboulay and Foubert, 2007, Pfeifer *et al.*, 2019). In general, as discussed above, it is usually easier to reliably score severe than mild tail lesions. This however, will make the results less reflective of the situation on the farm, and might underestimate some types of tail biting problems, such as the two-stage tail biting type (Taylor *et al.*, 2010). Also, as younger pigs have less moderate and severe wounds than older pigs (Carroll *et al.*, 2018, Hakansson *et al.*, 2020) such a crude method is unsuitable for all ages of pigs. Further, Pfeifer *et al.* (2019) reported that even with a rather crude binary scoring method, the inter- and intra-assessor reliability was not very good. Mullan *et al.* (2011) showed that differences between assessors might be caused by a difference in the scorers attitude to animal welfare: mild scores might be seen more as a sign of an underlying welfare problem in a pen, while severe lesions indicates that individual pigs are probably suffering from pain.

2.4 *Sampling strategies*

As tail lesion scoring is time-consuming, it is seldom feasible to score all pigs on a farm individually. It is, however, important to consider the sampling strategy carefully in order to achieve results that truly reflect the situation on the entire farm. Courboulay and Foubert (2007) suggested that by scoring 1/2 to 1/3 of the pigs in a room, one might get a valid reflection. If there is a high inter-pen variability in a measure – which can indeed be the case for tail lesions – this will increase the number of pens needing to be scored to give a good reflection of the farm as a whole (Mullan *et al.*, 2009a). Munsterhjelm at al. (*unpublished*) showed that specifically for tail lesion-related measures, scoring occurrence at pen level as a binary trait performs well in comparison to a farm prevalence measure.

Scoring systems developed for welfare assessment on farm-level are typically much simpler than those used for research purposes, and typically use a predefined selection of pens, typically 5–15 pens per farm (RSPCA, 2016, SusPigsSys, 2020, Welfare Quality®, 2009). For example, the RSPCA assessment (RSPCA, 2016) scores tails as having no lesions, mild lesions (a scratch of over 1 cm or a circular wound of over 0.5 cm in diameter) and severe (a part of the tail is missing, the tail is swollen, held oddly or has a scab covering the whole tip). Tails which are too dirty to be assessed should be recorded separately. The Welfare Quality protocol defines superficial lesions separately from lesions with fresh blood, swelling or signs of infection, but only the latter category is included in the final lesion score. The SusPigSys protocol (SusPigSys, 2020) scores tails as intact, mild (dry crust or fresh blood) and severe (blood or crust and swelling) as well as the tail length as normal, half or stump. The tail length does not differentiate between docked and undocked tails, which is recorded as a separate parameter. The Sikava herd health protocol used during farm visits by vets is an example of a very simplified scoring system: it only records a farm level prevalence of tail lesions as none or occasional (0%), a few (1–5%), several (6–10%) and many (>10%), without providing a definition of tail lesions (www.sikava.fi).

2.5 *Practical challenges*

Scoring tails in detail on live animals on-farm, comes with several practical challenges. To reliably score absence of lesions, the tails should in most cases be palpated. In other words, the observer needs to be in the pen and to touch the pigs. This is problematic as pigs are often not used to handling. Scared pigs move around more and are difficult to investigate and palpate properly. Some studies, where observations were performed from within in the pen do not report whether tails were palpated or not (eg. Lahrmann *et al.*, 2017, Scollo *et al.*, 2017), while Wallgren *et al.* (2019) reported that tails were inspected by palpating and Diana *et al.* (2017) that pigs were restrained, but not if the tails were palpated or not. Valros *et al.* (2021) scored tails while the pigs were restrained on a scale. Grümpel *et al.* (2018) reports that lesions were scored if visible from a 1 m distance, which would imply no palpation.

Courboulay and Foubert (2007) reported that mild lesions cannot be scored reliably from the corridor, though this might be possible for more severe lesions. However, Smulders *et al.* (2008) reported good inter-observer reliability using a 4-level score on commercial farms with observations performed from the corridor. Based on a scan of a few studies, not all authors report if tails were scored from the corridor or from within the pen – even in cases where mild lesions were scored (Kritas and Morrison, 2004, Marques *et al.*, 2012). The more animals in a pen, the more difficult it becomes to identify single animals, and the higher the stocking density, the more difficult it becomes to see tail lesions (Veissier *et al.*, 2013).

Pig tails are often dirty, mainly from dung present in the pen. In order to be sure of the absence of mild lesions, or to distinguish crusts reliably from dirt, tails should be cleaned before inspection. This is seldom done, or at least not reported. Mullan *et al.* (2009b) showed a correlation between dirtiness and a reduced level of tail lesions, indicating that dirty tails indeed are a problem for a reliable assessment. Pandolfi *et al.* (2017) reported that over 6 % of tails were too dirty to be scored for mild lesions, with the recorded prevalence of mild lesions (1.3 %) being much lower than the percentage of dirty tails. This further indicates that the reliability of the scoring is not very good, as a large percentage of mild lesions might go unnoticed.

Some studies report results from both docked and undocked animals using the same scoring system (Grümpel *et al.*, 2018, Lahrmann *et al.*, 2017, Thodberg *et al.*, 2018) and in all these studies more lesions occur in the undocked pigs, which, of course is to be expected. However, there is no validation of how well different scoring methods are suited for comparing lesions in docked vs non-docked tails. For example, the definition of a fully intact tail by the flattened last vertebra is clear and valid, allowing for identifying healed damage even when only a small proportion of the tail is missing. However, there is probably no way of identifying if a docked tail was bitten at an earlier stage and is already healed. The scoring key suggested by Honeck *et al.* (2019) is, as far as we know, the only system defining one parameter, namely loss of tail length, differently for scoring of undocked and docked tails. For docked tails the loss is related to the original docking length. This, however, can be problematic, as tails can be docked to various lengths even on the same farm, and as such, minor losses cannot be identified with this definition.

3 Recording tail damage at the slaughterhouse

3.1 *Introduction*

As outlined in the previous sections, scoring tail damage in live animals on-farm gives the most representative and accurate picture of the temporal pattern, scale and intensity of a tail-biting problem on-farm. However, assessment of tails at the slaughterhouse avoids potential problems associated with having to assess animals in crowded or dirty and poorly lit conditions (Edwards *et al.*, 1997, Velarde *et al.*, 2005). Indeed it has enormous advantages associated with the ability to evaluate large numbers of pigs from a variety of herds at one location and without attendant risks to biosecurity (Harley *et al.*, 2012a). Furthermore, inclusion of routine measurements of carcass tail lesions in the standard meat inspection (MI) process would expand its role as a health and welfare monitoring tool, as recommended in article 21 of the Regulation EC 625/2017 (de Luca *et al.*, 2021). Relaying such information to pig farmers and their veterinarians could not only contribute to reducing carcass losses but could also lead to general improvements in pig welfare on farm (Harley *et al.*, 2012a).

Tail lesions recorded at slaughter are strongly associated with farm productivity parameters, and are less common on farms that avail of record keeping through a farm advisory service (van Staaveren *et al.*, 2017a) or that engage with tail biting risk assessment schemes (eg. Schwip, vom Brocke *et al.*, 2019). Tail lesions are also associated with lung pathologies at slaughter (Boyle *et al.*, 2022, Valros *et al.*, 2020). Thus, they are negatively associated with better standards of farm management in general (Teixeira *et al.*, 2016, van Staaveren *et al.*, 2016). As well as severe tail lesions in the finisher stage, carcass tail lesions may also help identify farms with problems with poor body condition and bursitis in weaner pigs (van Staaveren *et al.*, 2017b), and tail lesions acquired both early and late in the pigs life remain visible on the carcass (Carroll *et al.*, 2018). Furthermore, while processing may damage carcasses post-mortem (Valros *et al.*, 2020), pre-slaughter management and handling and slaughter itself do not disrupt the visibility or presence of tail lesions on the carcass (van Staaveren *et al.*, 2015). Taken together these findings support the usefulness of tail lesions measured at the slaughterhouse as iceberg indicators of pig health and welfare on-farm.

One important point to consider is that pigs inspected at slaughter plants might not be representative of the on-farm population. Some pig producers send unthrifty, low body weight pigs or pigs recovered from illness or injury that spent time in hospital pens for slaughter to small/local slaughterhouses. Another related consideration is that death or euthanasia of severely tail bitten pigs on–farm means that what is recorded at the slaughterhouse underrepresents the extent of the tail biting problem/welfare problems on a farm (Marques *et al.*, 2012). Indeed, this is a key disadvantage of slaughterhouse based tail inspections.

3.2 *Constraints associated with the meat inspection process*

Currently, MI determines the health of animals both prior to death (ante mortem) and after death (post-mortem) predominately to protect human health and food quality. Carcasses are trimmed or condemned, partially or entirely/fully, if there are pathologies (or contamination) that threaten meat quality and safety. Abscessation is one of the most common such pathologies (Harley *et al.*, 2012b, Harley *et al.*, 2014) and is responsible for considerable economic losses to the pig industry (Harley *et al.*, 2014). Carcass tail lesions are strongly associated with abscessation, though there is no experimental evidence of a causal relationship (Boyle *et al.*, 2022, Gomes *et al.*, 2022). Nevertheless, tail lesion severity is linked to an increasing systemic temperature likely reflecting an increasingly severe systemic inflammatory response (Heinonen *et al.*, 2010, Teixeira *et al.*, 2020a). Such associations with carcass quality/safety is why tail lesions are already routinely recorded, albeit crudely, in the slaughterhouse as part of the MI process (EFSA, 2022). This recent exercise conducted by EFSA revealed however, that generally only severe lesions are recorded which is probably adequate for identifying farms with major tail biting problems. However, this explains why the results for tail lesions recorded at MI are usually much lower compared to animals inspected on farm (Busch *et al.*, 2004) or to figures generated from assessments by researchers (Keeling *et al.*, 2012, Valros *et al.*, 2020, vom Brocke *et al.*, 2019). Assessment of severe meat inspection findings shows better validity and repeatability than assessment of minor findings (Stärk *et al.*, 2014). However, vom Brocke *et al.* (2019) found a lower prevalence in results from meat inspectors compared to researchers even if only comparing the most severe category of tail lesion i.e. 'necrosis' indicating that even severe tail lesions are missed by meat inspectors. Further, healed tail damage, even when a large proportion of the tail is missing, is often not recorded as part of MI. At least in non-docked animals, also healed lesions, and a shortened tail length leads to an increased risk for condemnations (Valros *et al.*, 2020). Hence, recordings of tail damage based on data routinely currently collected at MI are unlikely to reflect the animals longer term welfare on the farm (Heinonen *et al.*, 2021).

In spite of the inadequacies of only recording severe tail lesions in the MI process at the slaughterhouse, linking ante and post mortem inspections of affected pigs is a potential source of valuable information (Ghidini *et al.*, 2021, Teixeira *et al.*,

2018, 2023ab). Ideally, inspectors identify pigs with severe tail lesions on arrival at the slaughterhouse at unloading and record them ante-mortem. Such pigs are detained/separated from the rest of the batch or distinguished in some other way (e.g. by tagging or tattooing) for closer inspection post-mortem (or they are euthanised if deemed too sick or too badly injured). The outcome of the post-mortem inspection is then linked to the ante-mortem tail lesion finding. However, research for the PIGWELFIND project (Boyle, 2018) identified several associated, suboptimal scenarios that occur in practice at the slaughterhouse. All further reduce the reliability of data on tail lesions recorded at MI. For example, pigs detected with severe tail lesions on arrival for slaughter (ante-mortem) are often recorded as cases of 'abscessation' with no mention of there being an associated tail lesion. As this condition can only be diagnosed post-mortem it suggests that inspectors utilise the term 'abscessation' synonymously with 'tail lesions'. This can also happen the other way around i.e. where abscessation detected post-mortem is recorded as 'tail lesion' in the absence of an actual inspection of the tail.

Teixeira *et al.* (2016) found considerable variation between meat inspectors in the way in which they classified and recorded MI findings. Variation in the terminology used extends to between slaughterhouses and countries making comparisons challenging (Alban *et al.*, 2022). Indeed, Nielsen *et al.* (2011) deemed it impossible to correct for the large differences in the practice of recording indicators relevant to pig health and welfare between Danish slaughterhouses. Alban *et al.* (2022) suggest the need for harmonised terminology and protocols for recording MI findings enabling comparison of data between meat inspectors, abattoirs, regions and countries. Specifically this should be accompanied by better training of meat inspectors in the identification and scoring of tail lesions of differing severities.

As mentioned above severe tail lesions are generally recorded, albeit crudely, when pigs are unloaded at the slaughterhouse (i.e. ante-mortem) (EFSA, 2022). Nevertheless, the probability of detecting less severe anomalies during routine ante-mortem inspection in abattoirs is relatively low (Jackowiak *et al.*, 2006, Schemann *et al.*, 2010, Teixeira *et al.*, 2020b). This is because many of the constraints which apply to on-farm assessments (discussed in section 2.5 of this chapter) of pigs tails also impede detailed inspection of large numbers of pigs on arrival at the slaughterhouse (i.e. pigs are moving and their tails are often dirty). Furthermore, the speed of unloading is often very fast and issues with crowding and time constraints are likely even worse in the lairage than on a farm (Teixeira *et al.*, 2020b).

3.2.1 Scoring carcass tail damage post-mortem

The most obvious benefits of slaughterhouse based versus farm based assessments lie in the ability to inspect the carcasses of pigs with tail lesions post-mortem (Stärk *et al.*, 2014). In contrast to the situation on-farm (Mullan *et al.*, 2009b, Pandolfi *et al.*, 2017), one of the major advantages of scoring carcass tail lesions is that lighting is good in the slaughterhouse and pig carcasses are clean and therefore lesions are more visible (Carroll *et al.*, 2016, Stärk *et al.*, 2014, Valros *et al.*, 2020). However, there is no standardised classifying and scoring system for tails routinely used post-mortem either across slaughterhouses in the EU or indeed in research. There are a few examples of pig health schemes that provide producers and veterinarians with reports on pathological, including tail, lesions, recorded at MI and/or reasons for condemnations and which allow benchmarking to various degrees (e.g. against previous batches, against average figures, against bottom/top producers) (e.g. Sanchez-Vazquez *et al.*, 2011, Willeberg *et al.*, 1984). In Italy, "CLASSYFARM" collects animal based measures on the farm and in the slaughterhouse, thereby enabling monitoring of health and welfare indicators along the entire production chain (https://www.classyfarm.it/). In some countries there are also growing attempts to link tail biting management/risk assessment tools to outcomes of post-mortem tail lesion assessments but these are still predominately at a research stage (e.g. Germany "Schwanzbei-Intervention program" or "SchwIP" – vom Brocke *et al.*, 2019; and the Irish Pig Health Check Programme – https://animalhealthireland.ie/programmes/pig-healthcheck/pig-healthcheck-database/). In Finland, most of the abattoirs provide farms with feedback about tail damage percentage in each batch, as based on post-mortem meat inspection findings. Since 2024 this is also used as a threshold for recieving a specific animal welfare subisidy related to improving the conditions on finishing farms (https://www.ruokavirasto.fi/tuet/maatalous/elaintuet/elainten-hyvinvointikorvaus/ehk-siat/EHK-siat-2024/).

There is an EU wide move towards systematic assessment of tail lesions at the slaughterhouse (EFSA, 2022a) This is not only seen as an important way to enhance the enforcement of the EU ban on tail docking (Pig welfare subgroup, 2019) but also as a means of monitoring general pig welfare on-farm (EFSA, 2022b). Recently, the European Food Safety Authority (EFSA), acting on an EU mandate under the Farm to Fork Strategy defined animal based measures, including tail lesions, collected and assessed in slaughterhouses to monitor the level of pig welfare on farms (EFSA, 2022b). The premise for this being the relevance of carcass tail lesions as an iceberg indicator of pig welfare on farm (see chapter 15 of this book, FAWAC, 2009). In addition to compliance (with legislation or quality assurance schemes) feedback of data on the different type and severity of tail lesions collected at the slaughterhouse could also act as an advisory system to pig producers to inform their herd health and welfare management plans (Harley *et al.*, 2012a, also cited in Valros *et al.*, 2020).

3.3 *Practicalities of assessing tails on the slaughterline*

The key consideration when inspecting carcass tail lesions post-mortem is at which stage in carcass processing and dressing, the carcasses should be observed. In previous research, tail and other skin lesions were assessed post-sticking/exsanguination (Carroll *et al.*, 2016, Valros *et al.*, 2020), post scalding (and before de-hairing – Bottacini *et al.*, 2018), post de-hairing (which includes singeing and whipping – Carroll *et al.*, 2016, Valros *et al.*, 2020) or post meat/veterinary inspection (Vitali *et al.*, 2021). Meanwhile Keeling *et al.* (2012) inspected tails post-sticking/exsanguination and after the first wash but before the carcasses were thoroughly washed and scalded. Valros *et al.* (2020) observed that before scalding tails are not only covered with hair and dirt but also sometimes with scabs which makes it difficult to assess the size or severity of potential underlying lesions. Accordingly, Carroll *et al.* (2016) showed that the percentage of pigs with severe skin lesions increased numerically after scalding and dehairing compared to post-sticking indicating that the lesions were concealed by hair and dirt. Similarly, both Valros *et al.* (2020) and Carroll *et al.* (2016) demonstrated improvements in the visibility of tail lesions following scalding and de-hairing (singeing and whipping).

Other authors expressed concern that slaughter processes may remove evidence of mild skin damage (Aaslyng *et al.*, 2013; Taylor *et al.*, 2010). An extreme example might be where carcasses are inspected at the dressing table which is post-veterinary/meat inspection (as per Vitali *et al.*, 2021). In some slaughterhouses meat inspectors remove badly damaged tails from the carcass, so the tail may be missing from the carcass at this point. Carroll *et al.* (2016) reported that evidence of milder skin lesions was removed when pigs passed through a singeing furnace followed by a scraping tunnel where rubber scrapers removed residual hair. The singeing process can also result in red-brownish marks on the tail but these are easily distinguishable from lesions caused by biting (Carroll *et al.*, 2016, Valros *et al.*, 2020). Nevertheless, if practically possible it is advisable to score carcass tail lesions on the line after scalding and before de-hairing as per Bottacini *et al.* (2018), thereby avoiding potential confounds associated with singeing.

Other aspects of carcass processing may leave marks which mimic tail lesions (Keeling *et al.*, 2012, Taylor *et al.*, 2010). Carroll *et al.* (2016) recounted informal observations that machinery-related damage to the carcass manifested as shredding and peeling of the skin. Meanwhile Valros *et al.* (2020) reported a slaughterhouse specific anomaly whereby the top 2cm of a high proportion of undocked tails were broken off during processing. This was clearly mechanically induced damage but points to the need to consider the degree and type of mechanical damage that occurs to the skin of the tail on an individual slaughterhouse basis. Mechanical damage is unlikely to obscure the relative findings from one factory but it might affect the absolute level of lesions across slaughterhouses.

Tail biting is generally not performed by pigs during the pre-slaughter period (loading on farm, transport, unloading and lairage) and as such there is little need to be able to discern tail lesions arising from tail biting on farm from those occurring in the pre-slaughter period (van Staaveren *et al.* 2015). Nevertheless, Valros *et al.* (2020) attributed marks on the tail base to damage occurring in the live pig in the pre-slaughter period caused by crowded conditions during transport and lairage and ante mortem handling. These authors disregarded such lesions when scoring tail lesions post-mortem.

Another practical consideration that depends on the slaughterhouse layout and design is how close the assessor can get to the tail which could influence the ability to palpate or manipulate it and to measure actual tail length. Some studies specifically positioned personnel so that they could touch and inspect the tails closely (e.g. Keeling *et al.*, 2012). However, D'Alessio *et al.* (2023) recently presented evidence of excellent agreement between visual only and visual plus handling scoring methods for bite marks and bruises (on docked tails). This indicates that manipulation of the tail might not be necessary to score lesions accurately. Valros *et al.* (2020) acknowledge the practical difficulties associated with their proposal to palpate the tail end to distinguish between fully intact and healed tails.

Slaughter line speed is a key consideration (van Staaveren *et al.*, 2019). In most studies, line speed facilitated tails being scored directly on the slaughter line. For example, Bottacini *et al.* (2018) scored carcass tail lesions on the line at a speed of 8 pigs per min giving 7–8 seconds to inspect each pig. Valros *et al.* (2020) involved two assessors to score and estimate tail length at a similar line speed. Meanwhile Vitali *et al.* (2021) working with a slightly faster line speed (6 seconds per pig) assessed tail, and other skin lesions, on the dressing line. However, some modern slaughter line speeds are too fast to enable in person carcass tail inspections. vom Brocke *et al.* (2019) used two cameras to automatically photograph tails after scalding because of a very fast line speed (up to 20 pigs/min or only 3 seconds to inspect a pig). Indeed, there are well-developed automated systems of detecting both tail lesions and tail length by camera in-line (i.e. TailCam developed by the PigWatch project – see Spoolder *et al.*, 2018 and Larsen *et al.*, 2019 and Blömke *et al.*, 2020). Unfortunately neither are operational at a commercial level. Even more novel developments involve classification of tail lesions and tail loss using artificial intelligence (Brünger *et al.*, 2019). However, it will be difficult to develop an automated system that can substitute humans to palpate the tail end and thereby differentiate between intact and healed undocked tails (Valros *et al.*, 2020).

In the absence of automated methods of tail lesion scoring, the optimal sampling strategy for research and for MI purposes must be decided. Meat inspectors expressed reluctance to score every pig on the line even if they were to be 'assisted' by a computerised recording system (van Staaveren *et al.*, 2019). Stakeholders suggested that only those pigs moved off-line for trimming or for closer inspection to a detention rail should be scored. However, such a sampling regime would unlikely yield reliable, representative results (see Pessoa *et al.*, 2023 for an example based on lung lesions). Sample size estimations must clearly include size of the study population, frequency of the lesion and size of the batch etc. Based on these parameters Carroll *et al.* (2016) determined to score every third pig on the line. Sanchez-Vazquez *et al.* (2011) deemed this adequate for detection of health and welfare issues post-mortem. However, if large batches are sent to slaughter plants with low capacity there is a risk that the sample of pigs inspected is dominated by animals from only a few farms (Keeling *et al.*, 2012). Carroll *et al.* (2016) increased their sample size by 15% to account for such potential clustering effects. Depending on the study objectives, small herds sending only occasional animals to slaughter might be excluded (e.g. Heinonen *et al.*, 2021).

3.4 *Factors influencing prevalence of tail lesions recording in studies*

Reports on the prevalence of tail lesions at the abattoir vary greatly, from below 10% (EFSA, 2007), over 20% to 30% (Carroll *et al.*, 2016; van Staaveren *et al.*, 2017a) up to 72.5% (Harley *et al.*, 2014). There may be actual differences in the prevalence of carcass tail lesions as they vary between farms (van Staaveren *et al.*, 2017a), abattoirs and regions (Harley *et al.*, 2012a). They could also vary depending on different data collection points on the slaughter line (Carroll *et al.*, 2016) or on the number of observers involved (Keeling *et al.*, 2012). However, the two main factors that influence carcass tail lesion prevalence is whether or not pigs are docked and the scoring system used. The majority of pigs in the EU are still docked (Nalon and De Briyne, 2019) and Valros and Heinonen (2015) estimated that this means a 2 to 4 fold reduction in the risk of tail biting. Furthermore, studies report on carcass tail lesions that are already healed by the time the pig reaches the slaughterhouse and that are represented as scars (Gomes *et al.*, 2022) while others report on freshness and lesions of different severity (or degree of healing) (Keeling *et al.*, 2012, Kritas and Morrison, 2007, Valros *et al.*, 2004).

Several studies involving mixed populations of docked and undocked pigs showed lower or almost no tail lesions on the carcasses of docked pigs (Hunter *et al.*, 1999, Gomes *et al.*, 2022, Penny and Hill, 1974). However, an Irish study where 99% of pigs were docked found a 72.5% prevalence of tail lesions with 2.5% of pigs affected by severe tail lesions (Harley *et al.*, 2014). In another study, severe tail lesions were seen in over 3% of the docked pigs (van Staaveren *et al.*, 2017a). Given that there is greater accuracy/less ambiguity in scoring severely injured tails it should be possible to directly compare the Irish figures with the prevalence of such lesions from studies in countries where tail docking is prohibited (e.g. Sweden and Finland). However, it must be remembered that scoring long tails is more complicated than scoring docked tails given that both injury level and tail length must be taken into account and lesions are generally classified differently. Keeling *et al.* (2012) reported figures for pigs with half or less of their tail of between 1.5 and 2% which were comparable to the slaughterhouses own routine recordings for severe lesions of 1.2–1.6%. Meanwhile, Valros *et al.* (2020) reported figures for 'major wound' of between 2.5–2.8% of undocked Finnish pigs. If these figures are roughly equivalent to the severe lesion category in the Irish studies it is clear that there is a similar or higher level of severe tail damage in some docked slaughter pig populations compared to the Scandinavian undocked population. Thus the lowered risk of tail biting associated with tail docking is likely not being realized in many countries. This is confirmed by a recent study by Vitali *et al.* (2021) who reported a figure of 4.44% severe lesions in docked Italian heavy pigs. Clearly, tail docking allows the keeping of pigs in conditions which would probably result in unacceptable levels of tail damage in undocked pigs.

Most studies do not define an intact tail any other way than that it lacks lesions (e.g. Kritas and Morrison, 2007, Valros *et al.*, 2004; vom Brocke *et al.*, 2018). Valros *et al.* (2020) suggested that tails with more than 75% of the average tail length remaining and with no visible fresh lesions could be considered 'intact enough' as associated carcasses had similar meat inspection findings to carcasses with fully intact tails. Gomes *et al.* (2022) reported pigs from three production systems as being undocked, docked mid-length or fully docked at slaughter. However, this was not applied based on the standards outlined for the three production systems because all three tail lengths were reported in all three systems making it difficult to interpret the results. Keeling *et al.* (2012) recommended that in situations where slaughterhouses kill docked and undocked pigs, information about the practice of tail docking at pig level, or at least herd level should be included in the analyses to facilitate easier comparison. They indicate that for docked pigs, recording the actual tail length does not add much to the information on the level of tail biting on a particular farm as it is impossible to know the precise length the tails were docked to and even then there is considerable variation between personnel doing the docking. They proposed that when tails are docked, only the part of the scoring scale relating to injury should be used and that both the injury and tail length scoring parts should be used on undocked pigs. Valros *et al.* (2020) recently validated such an approach for undocked pigs against pathological examinations.

The authors believe that complete tail loss that is thoroughly healed is relatively easy to discern on either a docked or undocked carcass. However, in docked populations healed partial tail loss on the carcass is almost impossible to determine though proportions of tail loss can still be expected (Keeling *et al.*, 2012). Indeed healed tail damage was not considered by Kritas and Morrison (2007) where the most severe category (i.e. score 4) is described as 'partial or total tail loss' but is depicted photographically as a severe, active/'open' lesion. This scoring system was widely employed in slaughterhouse based studies of docked pigs (e.g. Harley *et al.*, 2012b, 2014, Teixeira *et al.*, 2016, van Staaveren *et al.*, 2016). In contrast, vom Brocke *et al.* (2018) developed a 5-point scoring system similar to that of Kritas and Morrison (2007) where 'complete tail loss' (equivalent to score 4 in Kritas and Morrison, 2007) was represented by either healed or active injuries (described as 'perforated or healed skin surface'). Clearly healed lesions present as scars. Gomes *et al.* (2022) differentiated between scars caused by docking and those caused by biting. They considered mild scars as the remnants of docking while 'severe scars' were healed tail lesions caused by tail biting. They found that scarred lesions influenced within-batch probability for condemnations but did not differentiate mild from severe scars. With undocked Finnish pigs, Valros *et al.* (2020) showed similar findings where healed lesions, in combination with bite marks or bruises, were associated with partial carcass condemnations and abscesses. However, in this case, scars (healed lesions), were clearly the result of tail biting. This makes sense given the time taken for infections to establish within the body – so the scarred tails were at one stage bitten – infection became established elsewhere in the body and thereafter the injured tail healed (Hunter *et al.*, 1999). Arguably infection can also establish following docking which could explain Gomes *et al.* (2022) finding.

An earlier study by Valros *et al.* (2004) conducted prior to the ban on tail docking in Finland employed a recording method modified from Hunter *et al.* (1999). The authors used the terms 'healed' and 'fresh' to differentiate degrees of tail damage with 'fresh' tails having broken skin with infections and swelling. 'Fresh' implies recent tail damage but swollen and infected tails are chronic conditions. Hence, 'fresh' lesions did not necessarily represent recent tail biting. Indeed given that the purpose of lesion scoring at the slaughterhouse is to indicate evidence of longer term poor welfare/tail biting on–farm rather than recent outbreaks, the relevance of recording fresh tail lesions is dubious. In any case, the prevalence of truly 'fresh' or very recent tail damage could only be determined by the presence of fresh blood on the carcass. This would necessitate assessment before the pig is washed (i.e. directly after bleeding) a position which generally compromises the visibility of tail lesions (Carroll *et al.*, 2016). Finally, both vom Brocke *et al.* (2018) and Valros *et al.* (2020) excluded swelling of the tail from their scoring systems because they could not get sufficient agreement between observers.

4 Discussion and conclusions

As outlined in this chapter, scoring tail lesions on farm or at the slaughterhouse is not a simple or straightforward procedure. In the absence of automated methods for either it is laborious and time consuming. Thus, when comparing and discussing prevalence from different sources and countries, one should always take great care to understand the implications of the selected scoring system and conditions under which the data were collected. It is not possible, or reasonable, to use the same scoring system for different purposes (e.g. on-farm welfare assessment schemes, research, meat inspection), but the system should be chosen based on the aim. Honeck *et al.* (2019) proposed a constructive scoring system including different levels of detail that facilitates the transfer of results.

There are pros and cons with scoring tail damage both on-farm and at the slaughterhouse. Scoring on-farm makes it possible to get a full picture of the current situation, as it allows for a more detailed scoring on aspects such as freshness of the lesion. Furthermore, all pigs on the farm are potentially available for scoring. On the other hand, scoring live animals under typical farm conditions is challenging and laborious. Furthermore, quality of the lighting and dirtiness of pigs varies considerably between farms which also affects the scoring outcome. Scoring at the slaughterhouse, on the other hand, allows for more controlled, clean and light conditions, and is easier to standardise. Challenges include mechanical damage of tails during the slaughter process and high line speeds. These can be counteracted by selecting the scoring point carefully and by potential automatized, camera-based scoring systems. In accordance with Carroll *et al.* (2016) it is likely that even if processing damage to the tail is considerable, it is relatively easy to discern from lesions with an animal etiology and severe tail lesions are impossible to confuse with mechanical damage. One of the downsides is that not all pigs born on a farm make it to slaughter as a proportion die or are euthanised. These are arguably pigs that are more likely to be affected by tail lesions given that badly affected animals should not be transported. Nevertheless, Heinonen *et al.* (2021) recently concluded that tail lesion scoring at the slaughterhouse is a more precise measure than that performed on-farm as part of a herd health assessment.

Heinonen *et al.* (2021) argued that they could not discern the various combinations of percentages of (undocked) tails that were intact, healed, with minor acute wounds, and with major acute wounds in an on-farm situation owing to the challenges outlined above. They suggest that each of these categories provide different information about the management situation on the farm in relation to the type of tail biting present (Taylor *et al.*, 2010). Such information is crucial to the assessment of pigs lifetime welfare. Healed, but shortened tails indicate either early tail damage or tail docking as a piglet, and both are thus indicative of a reduced welfare level during an earlier stage of the pig's life (Valros *et al.*, 2020). Of course this relates specifically to undocked pig populations and though the EU pig industry is gradually moving towards a situation where docking will no longer be acceptable, the majority of pigs are still docked. Measurements of tail length are largely irrelevant in docked populations (Keeling *et al.*, 2012) but information on different types of tail lesions as opposed to just severe lesions as per meat inspection can also yield important information on pigs lifetime welfare (Carroll *et al.*, 2018). In accordance with Heinonen *et al.* (2021) van Staaveren *et al.* (2017a) reflected on the greater variation between docked herds in the prevalence of moderate lesions compared to severe lesions and therefore of their likely greater relevance to farm management practices. And thereby of the importance of feeding back information on lesions of a variety of severities to pig producers. As the EU pig industry gradually moves towards a majority undocked pig population it is crucial that any scoring system evaluates both tail damage and tail length, and thus also includes healed damage.

Clearly, tail docking allows the keeping of pigs in conditions which would result in unacceptable levels of tail damage in undocked pigs. In addition to being a potential benchmarking outcome, scoring tail lesions at the slaughterhouse, possibly as a part of meat inspection, provides a means for objective feedback to the farmers – allowing them to improve their on-farm conditions and management. This also opens the potential for reward systems, which could help motivate farmers to work against tail biting. The fact that healed tail lesions can stem from any phase of the pigs life poses some challenges. In production systems, where weaner pigs are sold to different producers, sometimes in a different country, for finishing, it is difficult to influence the quality of pigs received. Thus, an objective way of scoring weaners on arrival to finisher farms is required. This would avoid the problem that the finisher farm is punished for tail lesions (or docking) that occurred on the farm where the pigs originated.

It is important that scoring systems are comparable and fair to facilitate benchmarking between slaughterhouses within and between countries. Systems should be complicated enough to allow for a full picture of the lifetime welfare of the pig, if this measure is to be used as an iceberg indicator (see chapter 15). However, Keeling *et al.* (2012) discussed that despite training of observers, more detailed scoring systems for carcass tail damage were not reliable and emphasised the importance of more simple scoring systems for routine recording during meat inspection at the slaughterhouse. The Pig welfare subgroup (2019) and Valros *et al.* (2020) suggested methods that include evaluation of tail length (undocked pigs only) and size of fresh lesions separately. This is simple enough to work on large numbers of animals, with potential for automation, while still allowing both for benchmarking of intact (enough) tail proportions, and feedback with more detailed information to farmers. Nevertheless, while technology offers exciting prospects for online automated tail lesion and length measurement, the ability to discern the flattened vertebrae at the tail tip as suggested by Valros *et al.* (2020) is only achievable if the tail can be palpated and likely will not lend itself to automation. Thus, a definition of an intact enough tail is necessary for comparable results.

5 Take home messages

- The outcome of tail lesion scoring will be greatly affected by the chosen scoring system and scoring conditions. Thus the scoring system should be selected according to the purpose, and results should be compared to other studies or silmilar data sources with great care.
- Scoring systems need to be developed and validated to produce comparable results in both docked and undocked populations.
- Scoring tail lesions at the slaughterhouse is recommended for use as iceberg indicator of lifetime welfare of the pigs and as a feedback on success of attempts to address tail biting on-farm for farmers.
- To allow for benchmarking of tail lesion proportions, there is a need to develop a commonly adopted scoring system both for assessing intact (enough) tails and for tail lesions.

References

Aaslyng, M.D., Brandt, P., Blaabjerg, L. and Støier, S., 2013. Assessment and incidence of skin damage in slaughter pigs. In: Proceedings of the 59th International Congress of Meat Science and Technology, 18–23rd August, Izmir, Turkey pp. 13–23. Available at: https://www.dti.dk/_/media/53877_Skin%20damage%20ICOMST%202013-12042013.pdf.

Alban, L., Vieira-Pinto, M., Meemken, D., Maurer, P., Ghidini, S., Santos, S., Laguna, J.G., Laukkanen-Ninios, R., Alvseike, O. and Langkabel, N., 2022. Differences in code terminology and frequency of findings in meat inspection of finishing pigs in seven European countries. Food Control 132:108394. https://doi.org/10.1016/j.foodcont.2021.108394.

Blömke, L., Volkmann, N. and Kemper, N., 2020. Evaluation of an automated assessment system for ear and tail lesions as animal welfare indicators in pigs at slaughter. Meat Science 159:107934. https://doi.org/10.1016/j.meatsci.2019.107934.

Bottacini, M., Scollo, A., Edwards, S.A., Contiero, B., Veloci, M., Pace, V. and Gottardo, F., 2018. Skin lesion monitoring at slaughter on heavy pigs (170 kg): Welfare indicators and ham defects. PLoS One 13:e0207115. https://doi.org/10.1371/journal.pone.0207115.

Boyle, L.A., 2018. Development of ante and post mortem meat inspection of pigs as a welfare diagnostic tool (PIGWELFIND). Technology update. Teagasc publications. https://www.teagasc.ie/media/website/publications/2018/6210_TechnologyUpdate_Laura Boyle_PIGWELFIND.pdf.

Boyle, L.A., Edwards, S.A., Bolhuis, J.E., Pol, F., Šemrov, M.Z., Schütze, S., Nordgreen, J., Bozakova, N., Sossidou, E.N. and Valros, A., 2022. The evidence for a causal link between disease and damaging behavior in pigs. Frontiers in Veterinary Science 8. https://doi.org/10.3389/fvets.2021.771682.

Busch, M.E., Wachmann, H., Nielsen, E.O., Petersen, H.H. and Nielsen, J.P., 2004. Tail biting – can routine meat inspection data be used for classification of herds. In: Proceedings of the 18th International Pig Veterinary Society Congress, Hamburg, Germany, 27 June–1 July.

Brünger, J., S. Dippel, Koch, R., and C. Veit. 2019. 'Tailception': using neural networks for assessing tail lesions on pictures of pig carcasses. Animal 13: 1030–1036. https://doi.org/10.1017/S1751731118003038.

Carroll, G.A., Boyle, L.A., Teixeira, D.L., Van Staaveren, N., Hanlon, A. and O'Connell, N.E., 2016. Effects of scalding and dehairing of pig carcasses at abattoirs on the visibility of welfare-related lesions. Animal 10: 460–467. https://doi.org/10.1017/S1751731115002037.

Carroll, G.A., Boyle, L.A., Hanlon, A., Collins, L., Griffin, K., Friel, M., Armstrong, D. and O'Connell, N.E., 2018. What can carcass-based assessments tell us about the lifetime welfare status of pigs? Livestock Science 214: 98–105. https://doi.org/10.1016/j.livsci.2018.04.020.

ClassyFarm – Istituto Zooprofilattico Sperimentale della Lombardia e dell'Emilia Romagna. Available online: http://www.classyfarm.it/ (accessed on 20th May 2022).

Courboulay, V. and Foubert, C., 2007. Testing different methods to evaluate pig welfare on farm. Cambridge University Press: Animal Welfare 16: 93–196. https://doi.org/10.1017/S0962728600031316.

D'Alessio, R.M., McAloon, C.G., Boyle, L.A., Hanlon, A. and O'Driscoll, K., 2023. Comparison between two scoring methods to assess tail damage of docked pig carcasses during post-mortem inspection in Ireland. Veterinary Record Open e66. https://doi.org/10.1002/vro2.66.

De Luca, S., Zanardi, E., Alborali, G.L., Ianieri, A. and Ghidini, S., 2021. Abattoir-based measures to assess swine welfare: analysis of the methods adopted in European slaughterhouses. Animals 11: 226. https://doi.org/10.3390/ani11010226.

Diana, A., Manzanilla, E.G., Calderón Díaz, J.A., Leonard, F.C., Boyle, L.A., 2017. Do weaner pigs need in-feed antibiotics to ensure good health and welfare? PLoS ONE 12, e0185622. https://doi.org/10.1371/journal.pone.0189434.

Edwards, D.S., Johnston, A.M., Mead, G.C., 1997. Meat inspection: an overview of present practices and future trends. The Veterinary Journal 154: 135–147. https://doi.org/10.1016/S1090-0233(97)80051-2.

Edwards, S. and Valros, A. 2021. Understanding and preventing tail biting in pigs. In Edwards, S. (ed.), Understanding the behaviour and improving the welfare of pigs, Burleigh Dodds Science Publishing, Cambridge, UK (ISBN: 978 1 78676 443 0).

EFSA (European Food Safety Authority), 2007. Scientific opinion of the Panel on Animal Health and Welfare on a request from commission on the risks associated with tail biting in pigs and possible means to reduce the need for tail docking considering the different housing and husbandry systems. The EFSA Journal 611: 1–13 https://efsa.onlinelibrary.wiley.com/doi/pdf/10.2903/j.efsa.2007.611.

EFSA (European Food Safety Authority), 2022a. The use of animal-based measures at slaughter for assessing the welfare of pigs on farm: EFSA's AHAW Network exercise. EFSA Supporting publication 2021: EN-7028. 29 pp. https://doi.org/10.2903/sp.efsa.2021.EN-7028.

EFSA (European Food Safety Authority), 2022b. Scientific opinion on the welfare of pigs on farm. AHAW Panel (EFSA Panel on Animal Health and Welfare), EFSA Journal 20:7421, 315 pp. https://doi.org/10.2903/j.efsa.2022.7421.

Ghidini, S., Alborali, G.L., De Luca, S., Maisano, A.M., Guadagno, F., Conter, M., Ianieri, A., and Zanardi, E., 2021. Predictivity of antemortem findings on postmortem inspection in Italian heavy pigs slaughterhouses. Animals 11: 2470. https://doi.org/10.3390/ani11082470.

Gomes, A., Romeo, C., Ghidini, S. and Vieira-Pinto, M., 2022. The relationship between carcass condemnations and tail lesion in swine considering different production systems and tail lengths. Animals 12: 949. https://doi.org/10.3390/ani12080949.

Grümpel, A., Krieter, J., Veit, C. and Dippel, S., 2018. Factors influencing the risk for tail lesions in weaner pigs (Sus scrofa). Livestock Science 216: 219–226. https://doi.org/10.1016/j.livsci.2018.09.001.

Hakansson, F. and Houe, H., 2020. Risk factors associated with tail damage in conventional non-docked pigs throughout the lactation and rearing period. Preventative Veterinary Medicine 184: e105160. https://doi:10.1016/j.prevetmed.2020.105160.

Harley, S., More, S.J., Boyle, L.A., O'Connell, N.E. and Hanlon, A., 2012a. Good animal welfare makes economic sense: potential of pig abattoir meat inspection as a welfare surveillance tool. Irish Veterinary Journal 65: 1–12. https://doi.org/10.1186/2046-0481-65-11.

Harley, S., More, S.J., O'Connell, N.E., Hanlon, A., Teixeira, D.L. and Boyle, L.A., 2012b. Evaluating the prevalence of tail biting and carcase condemnations in slaughter pigs in the Republic and Northern Ireland, and the potential of abattoir meat inspection as a welfare surveillance tool. Veterinary Record 171:621. https://doi.org/10.1136/vr.100986.

Harley, S., Boyle, L.A., O'Connell, N.E., More, S.J., Teixeira, D.L. and Hanlon, A., 2014. Docking the value of pigmeat? Prevalence and financial implications of welfare lesions in Irish slaughter pigs. Animal Welfare 23: 275–285. https://doi.org/10.7120/09627286.23.3.275.

Heinonen, M., Orro, T., Kokkonen, T., Munsterhjelm, C., Peltoniemi, O. and Valros, A., 2010. Tail biting induces a strong acute phase response and tail-end inflammation in finishing pigs. The Veterinary Journal 184: 303–307. https://doi.org/10.1016/j.tvjl.2009.02.021.

Heinonen, M., Välimäki, E., Laakkonen, A-M., Toppari, I., Vugts, J., Fàbrega, E. and Valros, A., 2021. Evaluation of tail lesions of finishing pigs at the slaughterhouse: Associations with herd-level observations. Frontiers in Veterinary Science 8:650590. https://doi.org/10.3389/fvets.2021.650590.

Herskin, M.S., Thodberg, K. and Jensen, H.E., 2014. Effects of tail docking and docking length on neuroanatomical changes in healed tail tips of pigs. Animal 9:677–681. https://doi.org/10.1017/S1751731114002857.

Herskin, M.S., Di Giminiani, P. and Thodberg, K., 2016. Effects of administration of a local anaesthetic and/or an NSAID and of docking length on the behaviour of piglets during 5 h after tail docking. Research in Veterinary Science 108, pp. 60–67. https://doi.org/10.1016/j.rvsc.2016.08.001.

Honeck, A., Gertz, M., grosse Beilage, E. and Krieter, J., 2019. Comparison of different scoring keys for tail-biting in pigs to evaluate the importance of one common scoring key to improve the comparability of studies – a review. Applied Animal Behaviour Science 221:104873. https://doi.org/10.1016/j.applanim.2019.104873.

Hunter, E.J., Jones, T.A., Guise, H.J., Penny, R.H.C. and Hoste, S., 1999. Tail biting in pigs. I: the prevalence at six UK abattoirs and the relationship of tail biting with docking, sex and other carcass damage. Pig Journal 43: 18–32. Available at: https://www.cabdirect.org/cabdirect/abstract/20123070801.

Jackowiak, J., Kiermeier, A., Kolega, V., Missen, G., Reiser, D. and Pointon, A., 2006. Assessment of producer conducted antemortem inspection of market pigs in Australia. Australian Veterinary Journal 84:351–7. https://doi.org/10.1111/j.1751-0813.2006.tb12798.x.

Keeling, L.J., Wallenbeck, A., Larsen, A. and Holmgren, N., 2012. Scoring tail damage in pigs: an evaluation based on recordings at Swedish slaughterhouses. Acta veterinaria scandinavica 54: 1–6. https://doi.org/10.1186/1751-0147-54-3.

Kritas, S.K., and Morrison, R.B., 2004. An observational study on tail biting in commercial grower-finisher barns. Journal of Swine Health and Production 12: 17–22. Available at: https://www.aasv.org/jshap/issues/v12n1/v12n1p17.html.

Kritas, S.K. and Morrison, R.B., 2007. Relationships between tail biting in pigs and disease lesions and condemnations at slaughter. Veterinary Record 160: 149–152. https://doi.org/10.1136/vr.160.5.149.

Lahrmann, H.P., Busch, M.E., D'Eath, R., Forkman, B. and Hansen, C.F., 2017. More tail lesions among undocked than tail docked pigs in a conventional herd. Animal 11:1825–1831. https://doi.org/10.1017/S1751731117000490.

Lahrmann, H.P., Hansen, C.F., D'Eath, R.B., Busch, M.E., Nielsen, J.P. and Forkman, B., 2018. Early intervention with enrichment can prevent tail biting outbreaks in weaner pigs. Livestock Science 214: 272–277. https://doi.org/10.1016/j.livsci.2018.06.010.

Lahrmann, H.P., Faustrup, J.F., Hansen, C.F., D'Eath, R.B., Nielsen, J.P. and Forkman, B., 2019. The effect of straw, rope, and bite-rite treatment in weaner pens with a tail biting outbreak. Animals, 9, 365. https://doi.org/10.3390/ani9060365.

Larsen, H.D., Nielsen, G.G.B., Black, P., Vorup, P. and Hass, M.A., 2019. Test report. Description and functionality of a prototype system to measure and document tail length and tail lesions at the slaughterhouse. Danish Technological Institute https://www.dti.dk/_/media/74221_Final%20-Pigwatch%20test%20report%20D21.pdf.

Larsen, M.L.V., Andersen, H.M.L. and Pedersen, L.J., 2018. Which is the most preventive measure against tail damage in finisher pigs: tail docking, straw provision or lowered stocking density? Animal 12:1260–1267. https://doi.org/10.1017/S175173111700249X.

Marques, B.M.F., Bernardi, M.L., Coelho, C.F., Almeida, M., Morales, O.E., Mores, T.J., Borowski, S.M. and Barcellos, D.E., 2012. Influence of tail biting on weight gain, lesions and condemnations at slaughter of finishing pigs. Pesquisa Veterinaria Brasileira 32:967–974. https://doi.org/10.1590/S0100-736X2012001000003.

Mullan, S., Browne, W.J., Edwards, S.A., Butterworth, A., Whay, H.R., Main, D.C., 2009a. The effect of sampling strategy on the estimated prevalence of welfare outcome measures on finishing pig farms. Applied Animal Behaviour Science 119: 39–48. https://doi.org/10.1016/j.applanim.2009.03.008.

Mullan, S.M., Edwards, S.A., Butterworth, A., Whay, H.R. and Main, D.C.J., 2009b. Interdependence of welfare outcome measures and potential confounding factors on finishing pig farms. Applied Animal Behaviour Science 121:25–31. https://doi.org/10.1016/j.applanim.2009.07.002.

Mullan, S., Edwards, S.A., Butterworth, A., Whay, H.R., Main, D.C.J., 2011. A pilot investigation of Farm Assurance assessors' attitude to farm animal welfare as a confounding factor to training in pig welfare outcome measures. Animal Welfare 20(3): 413–421. https://doi.org/10.1017/S0962728600002979.

Munsterhjelm, C., Tuominen-Brinkas, M.M.J., Koskikallio, H.L., Heinonen, M., Ahlqvist, K.H.S., Valros, A., 2023. Pick your company: pen-level tail health affects tail lesion development in the individual pig during the finishing phase. 56th Congress of the International Society for Applied Ethology, Estonia 1–5. August 2023, p. 242.

Nalon, E. and De Briyne, N., 2019. Efforts to ban the routine tail docking of pigs and to give pigs enrichment materials via EU law: where do we stand a quarter of a century on? Animals 9: 132. https://doi.org/10.3390/ani9040132.

Nielsen, A.C., 2011. Data warehouse for assessing animal health, welfare, risk management and – communication. Acta Veterinaria Scandinavica 53: 1–4 https://doi.org/10.1186/1751-0147-53-S1-S3.

Pandolfi, F., Stoddart, K., Wainwright, N., Kyriazakis, I. and Edwards, S.A., 2017. The 'Real Welfare' scheme: benchmarking welfare outcomes for commercially farmed pigs. Animal 11:1816–1824. https://doi.org/10.1017/S1751731117000246.

Penny, R.H.C. and Hill, F.W.G., 1974. Observations of some conditions in pigs at the abattoir with particular reference to tail-biting. Veterinary Record 2:174–80. https://doi.org/10.1136/vr.94.9.174.

Pessoa, J., Boyle, L.A., McAloon, C., Manzanilla, E.G., Norton, T. and da Costa, M.R., 2023. Informative value of simplified lung lesions scoring systems to inform future codes for routine meat inspection in pigs. Journal of Porcine Health Management 9:31. https://doi.org/10.1186/s40813-023-00324-y.

Pfeifer, M., Eggemann, L., Kransmann, J., Schmitt, A.O. and Hessel, E.F., 2019. Inter- and intra-observer reliability of animal welfare indicators for the on-farm self-assessment of fattening pigs. Animal 13(8):1712–1720. https://doi.org/10.1017/S1751731118003701.

Pig subgroup of the EU platform on animal welfare. 2019. Output of the Commission's Subgroups on Transport & Pigs. Available online at: https://ec.europa.eu/food/system/files/2019-11/aw_platform_20191007_pres-03.pdf (accessed June 27, 2023).

RSPCA, 2016. RSPCA welfare standards for pigs. https://science.rspca.org.uk/sciencegroup/farmanimals/standards/pigs.

Sanchez-Vazquez, M.J., Strachan, W.D., Armstrong, D., Nielen, M. and Gunn, G.J., 2011. The British pig health schemes: integrated systems for large-scale pig abattoir lesion monitoring. Veterinary Record 169: 413–413. https://doi.org/10.1136/vr.d4814.

Sandercock, D.A., Smith, S.H., Di Giminiani, P. and Edwards, S.A., 2016. Histopathological characterization of tail injury and traumatic neuroma development after tail docking in piglets, Journal of Comparative Pathology 155:40–49. https://doi.org/10.1016/j.jcpa.2016.05.003.

Schemann, A., Hernández-Jover, M., Hall, W., Holyoake, P. and Toribio, J., 2010. Assessment of current disease surveillance activities for pigs post-farmgate in New South Wales. Australian Veterinary Journal 88:75–83. https://doi.org/10.1111/j.1751-0813.2009.00543.x.

Scollo, A., Gottardo, F., Contiero, B. and Edwards, S.A., 2017. A cross-sectional study for predicting tail biting risk in pig farms using classification and regression tree analysis. Preventative Veterinary Medicine 146: 114–120. https://doi.org/10.1016/j.prevetmed.2017.08.001.

Sinisalo, A., Niemi, J., Heinonen, M. and Valros, A., 2012. Tail biting and production performance in fattening pigs. Livestock Science 143:142–150. https://doi.org/10.1016/j.livsci.2011.09.019.

Smulders, D., Hautekiet, V., Verbeke, G. and Geers, R., 2008. Tail and ear biting lesions in pigs: an epidemiological study. Animal Welfare 17: 61–69. https://doi.org/10.1017/S0962728600031997.

Spoolder, H., Zebunke, M., Labyt, E., Godin, C., Tallet, C., Prunier, A., Dippel, S., Früh, B., Aaslyng, M.D., Larsen, H.D. and Vermeer, H., 2018, August. PigWatch: early automated detection of tail biting and aggression. In 69. Annual Meeting of the European Federation of Animal Science (EAAP) (Vol. 24, p. np). Wageningen Academic Publishers. https://hal.inria.fr/hal-01904403/.

Stärk, K.D.C., Alonso, S., Dadios, N., Dupuy, C., Ellerbroek, L., Georgiev, M., Hardstaff, J., Huneau-Salaün, A., Laugier, C., Mateus, A. and Nigsch, A., 2014. Strengths and weaknesses of meat inspection as a contribution to animal health and welfare surveillance. Food Control 39: 154–162. https://doi.org/10.1016/j.foodcont.2013.11.009.

SusPigSys 2020. Condensed protocol from Era-Net SusAn project "Sustainable pig production systems" (SusPigSys). Available at: http://dx.doi.org/10.13140/RG.2.2.17828.09605.

Taylor, N.R., Main, D.C.J., Mendl, M., Edwards, S.A., 2010. Tail-biting: A new perspective. The Veterinary Journal 186:137–147 https://doi.org/10.1016/j.tvjl.2009.08.028.

Teixeira, D.L., Harley, S., Hanlon, A., O'Connell, N.E., More, S.J., Manzanilla, E.G. and Boyle, L.A., 2016. Study on the association between tail lesion score, cold carcass weight, and viscera condemnations in slaughter pigs. Frontiers in Veterinary Science, 3:24. https://doi.org/10.3389/fvets.2016.00024.

Teixeira, D.L., Boyle, L.A. and Enríquez-Hidalgo, D., 2020a. Skin temperature of slaughter pigs with tail lesions. Frontiers in Veterinary Science 7:198. https://doi.org/10.3389/fvets.2020.00198.

Teixeira, D.L., Salazar, L.C., Enriquez-Hidalgo, D. and Boyle, L.A., 2020b. Assessment of animal-based pig welfare outcomes on farm and at the abattoir: A case study. Frontiers in Veterinary Science 7:576942 https://doi.org/10.3389/fvets.2020.576942.

Teixeira, D.L., Gibbons, J., Hanlon, A. and Boyle, L.A., 2018. The relationship between the statuses of slaughter pigs detained ante-mortem and their meat inspection outcome. In: Proceedings of the 69th Annual Meeting of the European Federation of Animal Science, Dubrovnik, Croatia p. 157. https://doi.org/10.3920/978-90-8686-871-1.

Teixeira, D.L., Salazar, L.C., Larraín, R. and Boyle, L.A., 2023a. The capacity of inspection on farm and at the abattoir to predict post-mortem outcomes in slaughter pigs: Aa study at animal level. Animal Science Journal 94: e13798. https://doi.org/10.1111/asj.13798.

Teixeira, D.L., Salazar, L.C., and Boyle, L.A., 2023b. The capacity of ante-mortem inspection of pigs at the abattoir to predict post-mortem findings and associated financial implications: a study at batch level. Animal Science Journal 94: e13843. https://doi.org/10.1111/asj.13843.

Thodberg, K., Herskin, M.S., Jensen, T. and Jensen, K.H., 2018. The effect of docking length on the risk of tail biting, tail-directed behaviour, aggression and activity level of growing pigs kept under commercial conditions. Animal 2: 2609–2618. https://doi.org/10.1017/S1751731118000563.

Ursinus, W.W., Wijnen, H.J., Bartels, A.C., Dijvesteijn, N., van Reenen, C.G. and Bolhuis, J.E., 2014. Damaging biting behaviors in intensively kept rearing gilts: the effect of jute sacks and relations with production characteristics. Journal Animal Science 92:5193–5202. https://doi.org/10.2527/jas.2014-7918.

Van Staaveren, N., Teixeira, D.L., Hanlon, A. and Boyle, L.A., 2015. The effect of mixing entire male pigs prior to transport to slaughter on behaviour, welfare and carcass lesions. PloS ONE 10(4): p. e0122841. https://doi.org/10.1371/journal.pone.0122841.

Van Staaveren, N., Vale, A.P., Manzanilla, E.G., Teixeira, D.L., Leonard, F.C., Hanlon, A. and Boyle, L.A., 2016. Relationship between tail lesions and lung health in slaughter pigs. Preventative Veterinary Medicine 127: 21–26 https://doi.org/10.1016/j.prevetmed.2016.03.004.

Van Staaveren, N., Teixeira, D.L., Hanlon, A. and Boyle, L.A., 2017a. Pig carcass tail lesions: the influence of record keeping through an advisory service and the relationship with farm performance parameters. Animal 11:140–146. https://doi.org/10.1017/S1751731116001117.

Van Staaveren, N., Doyle, B., Manzanilla, E.G., Calderón Díaz, J.A., Hanlon, A. and Boyle, L.A., 2017b. Validation of carcass lesions as indicators for on-farm health and welfare of pigs. Journal of Animal Science 95:1528–1536. https://doi.org/10.2527/jas.2016.1180.

Van Staaveren, N., Doyle, B., Hanlon, A. and Boyle, L.A., 2019. Multi-stakeholder focus groups on potential for meat inspection data to inform management of pig health and welfare on farm. Agriculture 9:40 https://doi.org/10.3390/agriculture9020040.

Valros, A., Ahlstrom, S., Rintala, H., Hakkinen, T. and Saloniemi, H., 2004. The prevalence of tail damage in slaughter pigs in Finland and associations to carcass condemnations. Acta Agriculturae Scandinavica Section A. 54: 213–219. https://doi.org/10.1080/09064700510009234.

Valros, A., Välimäki, E., Nordgren, H., Vugts, J., Fàbrega, E. and Heinonen, M., 2020. Intact tails as a welfare indicator in finishing pigs? Scoring of tail lesions and defining intact tails in undocked pigs at the abattoir. Frontiers in Veterinary Science 7:405 https://doi.org/10.3389/fvets.2020.00405.

Valros, A., Sali, V., Hälli, O., Saari, S. and Heinonen, M., 2021. Does weight matter? Exploring links between birth weight, growth and pig-directed manipulative behaviour in growing-finishing pigs. Applied Animal Behaviour Science 245: 105506. https://doi.org/10.1016/j.applanim.2021.105506.

Valros, A., Heinonen, M. Save the pig tail. *Porc Health Manag* 1, 2 (2015). https://doi.org/10.1186/2055-5660-1-2.

Veissier, I., Winckler, C., Velarde, A., Butterworth, A., Dalmau, A. and Keeling, L., 2013. Development of welfare measures and protocols for the collection of data on farms or at slaughter. In: Blokhuis, H., Miele, M., Veissier, I. and Jones, B. (eds) Improving farm animal welfare. Science and society working together: the welfare quality approach. Wageningen Academic Publishers, Wageningen, the Netherlands. pp. 115–146.

Velarde, A., Dalmau, A., Fàbrega, E. and Manteca, X., 2005. Health and welfare management of pigs based on slaughterline records. In: 56th Annual Meeting of the European Association for Animal Production, Uppsala, Sweden. 5–8th June pg. 379.

Vitali, M., Santacroce, E., Correa, F., Salvarani, C., Maramotti, F.P., Padalino, B. and Trevisi, P., 2020. On-farm welfare assessment protocol for suckling piglets: a pilot study. Animals 10:1016. https://doi.org/10.3390/ani10061016.

Vitali, M., Luppi, A., Bonilauri, P., Spinelli, E., Santacroce, E. and Trevisi, P., 2021. Benchmarking of anatomopathological lesions assessed at slaughter and their association with tail lesions and carcass traits in heavy pigs. Italian Journal of Animal Science 20:1103–1113. https://doi.org/10.1080/1828051X.2021.1944339.

Vom Brocke, A.L., Karnholz, C., Madey-Rindermann, D., Gauly, M., Leeb, C., Winckler, C., Schrader, L. and Dippel, S., 2018. Tail lesions in fattening pigs: relationships with post-mortem meat inspection and influence of a tail biting management tool. Animal 13:835–844. https://doi/org/10.1017/S1751731118002070.

Vom Brocke, A.L., Karnholz, C., Madey-Rindermann, D., Gauly, M., Leeb, C., Winckler, C., Schrader, L. and Dippel, S., 2019. Tail lesions in fattening pigs: relationships with post-mortem meat inspection and influence of a tail biting management tool. Animal, 13(4):835–844. https://doi.org/10.1017/S1751731118002070.

Wallgren, T., Larsen, A. and Gunnarsson, S., 2019. Tail posture as an indicator of tail biting in undocked finishing pigs. Animals 9:18. https://doi.org/10.3390/ani9010018.

Welfare quality® 2009. Welfare quality® assessment protocol for pigs (sows and piglets, growing and finishing pigs). Welfare quality® Consortium, Lelystad, Netherlands. Available at: Available at: http://www.welfarequalitynetwork.net/media/1018/pig_protocol.pdf.

Willeberg, P., Gerbola, M.A., Petersen, B.K. and Andersen, J.B., 1984. The Danish pig health scheme: nation-wide computer-based abattoir surveillance and follow-up at the herd level. Preventive Veterinary Medicine 3: 79–91. https://doi.org/10.1016/0167-5877(84)90026-6.

Yao, F., Visovatti, S., Johnson, C.S., Chen, M., Slama, J., Wenger, A. and Eriksson, E., 2001. Age and growth factors in porcine full-thickness wound healing. Wound Repair Regeneration 9(5):371–7. https://doi.org/10.1046/j.1524-475x.2001.00371.x.

Zonderland, J.J., Fillerup, M., Reenen, C.G., Hopster, H., Spoolder, H.A.M., 2003. Preventie en Behandeling van Staartbiljten Bij Gespeende Biggen; Wageningen UR: Animal Sciences Group: Lelystad, The Netherlands, 2003.

CHAPTER 12

Changing stakeholder behaviour

Grace Carroll[1,*], *Alison Burrell*[2] *and Lisa Graham-Wisener*[3]

[1]Animal Behaviour Centre, School of Psychology, Queen's University Belfast, UK; [2]Animal Health Ireland, 2–5 The Archways, Carrick-on-Shannon, Co. Leitrim N41 WN27, Ireland; [3]Centre for Improving Health Related Quality of Life (CIHRQoL), School of Psychology, Queen's University Belfast, UK; *Correspondence: g.carroll@qub.ac.uk

Abstract

Humans play a key role in determining the welfare status of animals kept under their control. For example, stakeholders in the farming industry such as veterinarians, consumers, retailers and legislators all have responsibility in determining pig welfare standards. Today, Human Behaviour Change (HBC) is a well-established field that is used to address important human health issues such as smoking cessation, increasing physical activity, and improving disease detection. However, despite the wealth of knowledge and research on HBC, the majority of 'interventions' regarding farming practices come in the form of top-down public policy changes that tend to lack a theoretical underpinning. The Behaviour Change Wheel provides a comprehensive framework for understanding the key predictors of human behaviour and has a user-friendly structure, making it a practical tool for those interested in changing human behaviour to reduce tail biting in pigs. This chapter provides an intervention design guide for those working with pigs that may be new to intervention design or who may be unfamiliar with the psychological foundations of Human Behaviour Change. A guide is provided along with hypothetical and real-world examples to aid understanding of this sometimes complex field, from defining the behaviour to be changed, to assessing the effectiveness of the designed intervention.

Keywords

Human Behaviour Change – intervention – pig welfare – COM-B – tail biting

1 Introduction

Humans play a key role in determining the welfare status of animals kept under their control. This is particularly true for farm animals; as farming has intensified,

 | DOI:10.3920/9789004694989_014

so too has the extent to which humans control their environment. Stakeholders in the farming industry such as veterinarians, consumers, retailers and legislators all have responsibility in determining animal welfare standards. However, the role of the farmer is arguably of particular importance. Pig farmers have the ability to determine stocking density, food and ventilation quality, and use of environmental enrichment, all of which are known to influence the prevalence of tail biting (Sonoda *et al.*, 2013). Several useful methods of tackling tail biting have been identified, many of which are discussed in this book. However, the success of all animal welfare solutions hinge on human behaviour, i.e., how humans choose to implement or not implement specific actions that impact the animals under our control. While the role of the human may seem obvious, there has historically been a lack of research on human behaviour in animal-related fields. Human Behaviour Change research is centred on gaining an understanding of the underlying reasons for specific behaviours, such as attitudes, social norms and habits. This understanding can then be utilised to attempt to bring about a desired change in the behaviour (Michie *et al.* 2013). It is recognised by psychologists that changing human behaviour is inherently challenging. However, carefully designed, behavioural interventions have the potential to tackle important issues at a low cost (Michie *et al.* 2018). Previously, many behaviour change interventions have been developed according to a common-sense approach using the 'ISLAGIATT principle' – *It Seemed Like A Good Idea At The Time* (Michie *et al.* 2013). This involves an intervention being developed prior to gaining a robust understanding of the behaviour that needs to be changed. Similarly, interventions are often designed without first gaining an understanding of the constructs to be targeted which are most likely to influence behaviour change. If an intervention developed in this way is *effective*, it is unclear why this was, and so, it is difficult to replicate. Similarly, if the intervention is *ineffective*, it is unclear why this was, making it difficult to identify improvements that could be made. There is widespread acknowledgement that behaviour change interventions are enhanced if they are informed by theory and empirical evidence. Developing a behaviour change intervention which is theoretically informed and evidence-based can help to target constructs with more precision that are known to influence behaviour, potentially resulting in a stronger effect.

Today, Human Behaviour Change (HBC) is a well-established field, with behaviour change theory used to address important human health issues such as smoking cessation, increasing physical activity, and improving disease detection (Coupe *et al.* 2021; Michie *et al.* 2011a; Michie *et al.* 2011b; Ogden, 2016; Oliwa *et al.* 2020). Policy guidance documents are available from the British Psychological Society, with recommendations on how to address behaviours associated with various human health issues. Informed by evidence from the Human Behaviour Change literature, a particular policy guidance document will, for example, outline the key factors determining whether an individual engages in a particular behaviour. For instance, physical activity is facilitated by personal motivation, social support and an environment that aids exercise, such as the availability of cycle paths (Uphill, 2018). Behaviour change interventions have been designed based on this knowledge, with evidence suggesting that goal setting and self-monitoring may be effective interventions for increasing physical activity. In contrast to human health behaviours, there is little information available on ways to change human behaviour to address farm animal health and welfare problems like tail biting in pigs. While there is a wealth of information on the causes of this important animal welfare issue (as discussed in other chapters of this book), tail biting nevertheless remains a pervasive issue.

Indeed, despite the wealth of knowledge and research on HBC, most interventions regarding farming practices, and recommended changes to improve animal health and welfare, come in the form of top-down public policy changes. Examples of this would be the introduction of new legislation or restrictions at a governmental or EU level and the imposition of penalties for those who do not comply with these changes. This top-down approach to changing farm practices can be successful, and is often an important catalyst for change. For example, Morgans *et al.* (2021) outline the example of a successful top-down intervention where new regulations saw the removal of battery cage egg production within the EU, which greatly improved animal welfare. However, relying solely on regulatory governance rather than considering the personal, environmental and social factors that affect change, can result in a disconnect from good farm practice, increased mistrust of authorities, and other unintended consequences (Escobar and Demeritt, 2017). Indeed, animal welfare can be particularly difficult to address via policy and legislation due to the number of stakeholders involved, each of which hold various interests and perspectives (Fernandes *et al.* 2019). Farm animal welfare may be particularly difficult to address given the multitude of stakeholders involved from farmers and meat processers, to assurance scheme providers, consumers and retailers. Often, farming stakeholders hold preconceived notions about others and the role they should play in improving animal welfare. Legislation and guidelines are informed by scientific evidence, and while this is a sound and sensible approach, recommended or mandated practices are often difficult to implement. This is often due to a failure in considering barriers that farmers may face in practice (Fernanded *et al.* 2019; Albernaz-Goncalves *et al.* 2021). As a result, there is increasing interest in bottom-up approaches that involve collaboration between stakeholders, with a focus on practice-led approaches to change (van Dijk *et al.* 2019). For interventions to result in embedded, relevant behavioural change at farm level, it is important to examine behaviour change interventions from what is known as a social ecological approach (Eldredge *et al.*, 2016). Taking a social ecological approach means that we are considering not only the person whose behaviour we want to change, but how their relationships, their community, and the society in which they live, impact on decision making and behaviour. The social ecological model is used extensively in human health promotion and systematic health behaviour intervention development. It views health as a function of an individual, but also of the individual's environment, including family and social networks, relevant organisations, communities and wider society. Taking a similar approach to changing the behaviours of relevant stakeholders on-farm will help the industry to move away from a solely top-down approach. It will also help in gaining an understanding of social norms within the pig industry which relate to tail biting, tail docking and other important issues. This increases the chances of achieving sustainable, meaningful changes to farm practice.

Many of the examples used to illustrate points in this chapter focus on changing farmer behaviour. It is acknowledged that tail biting is a multi-actor issue, being influenced by several factors such as consumer purchasing habits, supermarket requirements and assurance scheme standards. However, as farmers possess the most day to day influence on pig welfare, it is this stakeholder group that is referred to most often.

1.1 *The use of theory in changing behaviour – past and present*

Behaviour change interventions should be developed based on scientific understanding of the factors posited to influence human behaviour. A number of behaviour change theories have been developed and tested. The Theory of Planned Behaviour (TBP; Ajzen, 1985) is a seminal theoretical model within the behavioural sciences and is likely the theory most frequently applied by animal welfare science researchers. The TPB proposes that several key constructs determine an individual's intention; attitudes (e.g. 'I care about animal welfare'), subjective norms (e.g. 'this is how everyone else does things') and perceived behavioural control (e.g. 'I can't do anything to reduce tail biting on my farm'). Intention is then said to translate into behaviour. The TBP is a parsimonious model which was developed to allow prediction and explanation of a range of different behaviours (McEachan *et al.*, 2011). However, it has received various criticisms in the last decade, including a lack of explanatory power and limited use in informing behaviour change interventions, given the small number of factors posited to influence human behaviour (Sniehotta *et al.*, 2014). Some authors have argued that these criticisms stem from an unrealistic expectation that a model designed by theorists to be generalisable should be able to explain complex behaviour within highly contexualised settings (Gollwitzer and Oettingen, 2015). Indeed, there is a predominant focus within the TBP on rational reasoning to the exclusion of unconscious influences, such as impulsivity or habit, on behaviour. For example, according to the Theory of Planned Behaviour, a farmer with a positive attitude towards tail-biting interventions, such as adding straw to pens, would have the intention to add enrichment. This intention would then translate into the desired behaviour, i.e., the farmer would add straw to their pig pens. However, it is now recognised that an intention-behaviour gap exists, i.e., farmers (like most people), do not always do what they intend to do (Sniehotta, 2005). Even when a specific model such as the TPB is chosen, this may be from personal experience of using the model, or from a cursory analysis.

Consequently, key constructs that influence a target behaviour in a specified context may be omitted (Michie *et al.*, 2011c). These deficits have resulted in many elaborated versions of the model. For example, Winkel *et al.* (2020) examined influences on German pig farmer's intention to construct new pig housing to improve animal welfare. Here, the TPB was extended to include direct and indirect experience of constructing pig housing. Similarly, Salari *et al.* (2021) examined Iranian livestock breeders' intentions and behaviour with regard to animal welfare. In this case, the TPB was extended to include moral norms (personal beliefs about what is right or wrong) and self-identity (how a person perceives one' self, Salari *et al.* 2021). However, extending the TPB to include disparate constructs makes comparisons between studies difficult. Furthermore, other important influences on behaviour may be missed. Indeed, there are a range of external or 'real world' factors not easily accounted for within the TPB. For example, Regan *et al.* (2021) assessed farmers' decisions to engage in grass measurement on dairy farms and identified individual and external social and physical factors that influence performance of the target behaviour. Regan *et al.* (2021) outline the role of self-efficacy (belief in one's own capabilities), perceived skill and attitudes towards the practice, but stress the need to move away from behavioural frameworks which only examine individual factors to those that illustrate the role of tangible social support and environment on the uptake of grass measurement practices. Similarly, Shortall *et al.* (2016) found that barriers to implementing appropriate biosecurity on dairy farms included lack of adequate facilities and infrastructure, financial barriers, and time constraints. This demonstrates the importance of considering a comprehensive list of both internal and external factors that may influence farmer behaviour (Carroll and Groarke, 2019).

Many behavioural intervention developers may choose not to use a theoretical model to inform their intervention as existing models do not meet their needs (Michie *et al.* 2011c) or due to lack of awareness of the existence of such models (Carroll, Groarke and Graham-Wisener, 2022). In recent times, there has been increasing recognition of the utility of more comprehensive theoretical frameworks in allowing researchers to systematically develop behaviour change interventions that are grounded in a robust theoretical understanding of the predictors of their chosen behaviour. It may be difficult for researchers outside of the discipline of Health Psychology to know where to begin in developing interventions aimed at changing human behaviour. One useful evidence-based framework that is simple yet comprehensive is the Behaviour Change Wheel (Michie *et al.* 2011c).

1.2 *The Behaviour Change Wheel*

The Behaviour Change Wheel (BCW) is a synthesis of 19 theoretical frameworks of behaviour change (Michie *et al.*, 2014). The BCW provides a comprehensive framework for understanding the key predictors of a target behaviour, but also provides a systematic approach of how to then use this theoretical understanding to inform intervention development. This contrasts to most theoretical frameworks which are developed to provide an understanding of the key predictors of behaviour, rather than to inform behaviour change intervention design per say. Indeed, there is significant variability in terms of how theory is used to develop interventions (Michie and Prestwich, 2010). As a result of being comprehensive and accessible for use, the BCW is increasingly being applied by those working with farm animals and other animal species. Examples of behaviour change interventions aimed at improving animal health and welfare include how to increase reporting of stray cats for neutering (McDonald *et al.* 2018) and how to encourage higher welfare food choices in meat consumers (Cornish *et al.* 2019). However, theory-driven research related to Human Behaviour Change in the animal industries is still in its infancy. Much greater uptake is expected in the coming years due to the increase in awareness of the Behaviour Change Wheel (Michie *et al.* 2011c) outside of Health Psychology.

The BCW consists of three layers that provide a three-step approach to developing a behaviour change intervention (Figure 12.1). The inner-most layer of the BCW (the 'hub') is the Capacity Opportunity Motivation-Behaviour model (COM-B; Michie *et al.*, 2011c). The COM-B model supports the first step in the process by providing a framework for undertaking a behavioural analysis to understand the key constructs to target for behaviour change to occur. The next layer supports the second step in the process by providing nine Intervention Functions to select from which address particular constructs identified using the COM-B. The outermost

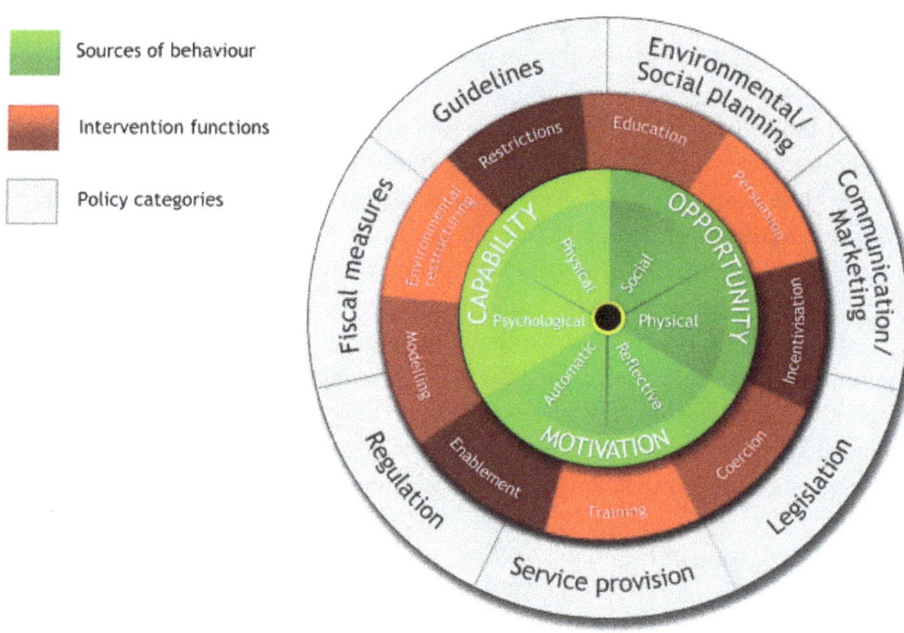

FIGURE 12.1 The Behaviour Change Wheel (BCW; Michie *et al.*, 2011c)

layer supports the third step in the process by providing seven Policy Categories to select from which can support the implementation of particular intervention functions.

This rest of this chapter provides a summary of each layer, alongside case studies relating to changing human behaviour towards addressing pig tail-biting. For a more detailed discussion of the BCW see Michie *et al.* (2011c; 2014).

2 Intervention design

2.1 *Understand the behaviour*

2.1.1 Define the behaviour

The first step in intervention design is to gain a clear picture of the behaviour that is to be changed. To define the problem of tail-biting in behavioural terms, it is important to be specific about who is involved in the behaviour (the individual, group, or population) and the behaviour itself. You are considering who needs to do what differently, when, where, and how. For example, is it the farmer who needs to change their behaviour? Or perhaps it is the veterinarian who provides the farmer with advice. Do you want to change the behaviour of just one farmer or a group of farmers? What specific behaviour related to tail biting do you wish to change? For example, you may want the farmer to add straw to pens or to reduce the stocking density.

2.1.2 Select a target behaviour

At this early stage in your behavioural analysis of the problem, it is necessary to be comprehensive in specifying all key behaviours that may influence the problem. It is important to consider the broader context of behaviour, considering that behaviours interact as a dynamic system and influence each other. Once you have a long list of specified behaviours, the next step is to select one target behaviour from this list for your intervention. When selecting the target behaviour, there are several criteria which may be considered (Michie *et al.* 2014). This includes;

2.1.2.1 *i) The potential impact of the change in behaviour*

Based on your knowledge and previous research, how likely is the chosen behaviour to reduce tail biting prevalence? Some tail biting solutions are known to be more impactful than others. For example, asking farmers to provide malleable enrichment to pens may be more effective than asking them to provide wooden blocks.

2.1.2.2 *ii) The likelihood of changing a specified behaviour (financial/human resources/acceptability/preference)*

Some behaviours may be relatively straightforward to change while others may be more challenging. For example, asking farmers to move from slatted flooring to solid flooring would be difficult to achieve due to the substantial financial cost of such a proposal. It is also likely that many farmers would find this idea unacceptable unless financial help was available.

2.1.2.3 *iii) The potential for the behaviour to spill over and influence other key behaviours (i.e. the centrality of the behaviour in the system)*

New behaviours have the potential to impact on other behaviours. These can have spill-over effects of varying magnitudes which may have desirable or unintended effects. For example, observing pigs to identify bitten or biter pigs may result in an increase in the detection of unrelated health and welfare issues which would be beneficial. Conversely, it might be the case that time spent adding straw to pens could take time away from daily health checks.

2.1.2.4 *iv) The ease of measuring if a behaviour has been changed*

To assess the effectiveness of any intervention, the behaviour of interest must be measured in some way. Intervention developers can directly measure the behaviour itself or infer its performance via more indirect means. For example, if aiming to change farmer behaviour, one option would be to observe the farmer performing the behaviour. However, this approach can sometimes be problematic as humans often change their behaviour when they know they are being observed. Instead, to assess use of environmental enrichment for example, invoices from enrichment providers could be obtained. Alternatively, the *outcome* of the behaviour can be assessed. For example, a reduction in tail lesion prevalence is a desired outcome for any Human Behaviour Change intervention related to tail biting. Therefore, recording tail lesion prevalence would enable the intervention developer to assess intervention effectiveness without the need to directly observe the farmer.

For a problem such as tail-biting where the causes are multi-factorial (Henry *et al.* 2021), these first few processes of defining and selecting a target behaviour may be challenging. It may be useful to undertake stakeholder engagement to support your behavioural mapping and decision-making (Carroll, Groarke and Graham-Wisener, 2021). This helps to support a bottom-up approach to intervention development by embedding insights and expertise from a range of relevant stakeholders. In the case of tail-biting, it will also be important to decide on the level at which the target behaviour will be specified. It may be that the behavioural target should be determined on a farm-by-farm basis. If this is not possible, or desired, group-level factors should be considered. For example, it is likely that the local context will be relevant in determining the success of a given intervention. If slatted floor systems are typically used in the country or region of interest, for instance, it may be difficult to encourage widespread use of malleable enrichment like loose straw (D'Eath *et al.* 2014; Haigh *et al.* 2019). In regions where solid flooring is commonplace, this target behaviour would be much more achievable.

2.1.3 Specify the target behaviour

Once you have selected your target behaviour, you need to specify the behaviour in significant detail. For example, 'add enrichment to pens', 'reduce stocking density' and 'add straw daily' are relevant target behaviours but lack detail. The TACTA principle of behavioural specification can be useful for specifying behaviours (Presseau *et al.* 2019); Target, Action, Context, Timing and Actors.

An example is provided in Table 12.1 of a target behaviour for reducing tail biting that is sufficiently detailed to allow for an assessment of intervention success further down the line. Furthermore, indicating details such as straw quantity and delivery schedule allows for cross-study comparisons to be made in the future.

TABLE 12.1 Specifying the target behaviour

Target behaviour	Add environmental enrichment to pens
Target (Who needs to perform the behaviour?)	The pig farmer and/or farm workers
Action (What do they need to do differently to achieve the desired change?)	Add 15 g of chopped straw per pig to each pen
Context (Where do they need to do it?)	In the growing and finishing units
Timing (When do they need to do it and how often?)	Once a day each morning
Actors (With whom do they need to do it?)	Can be done alone or with other farm workers

2.1.4 Identify what needs to change

Once you have specified the target behaviour, you then need to develop an understanding of what needs to change for the desired behaviour to occur. At this stage, researchers often use the COM-B to gain a detailed understanding of the sources of the target behaviour within their chosen context. This can be informed by existing evidence or by using the COM-B to conduct novel empirical research. The COM-B (Michie *et al.* 2011c) supports a behavioural analysis of the problem (pig tail-biting in this case) and helps to identify the sources of a specific human behaviour to be targeted via intervention. The COM-B proposes that Capability (*Psychological and Physical*), Opportunity (*Social and Physical*), and Motivation (*Automatic and Reflective*) need to be present for an individual to engage in a target behaviour. 'Capability' can be understood as an individual's physical and psychological ability to perform a behaviour. For example, does the farmer have the skills needed to identify tail biting? Does the farmer have enough knowledge of the causes of tail biting to deal with it effectively? 'Opportunity' is the physical and social environment that enables the behaviour. For example, do farmers have enough time or labour to add enrichment to pens? Is it common practice for other pig farmers in the local area to add enrichment? Lastly, 'Motivation' can be understood as the reflective (making plans, evaluating things that have already happened) and automatic (wants, needs, emotions, reflex responses) mechanisms that activate or inhibit the behaviour (Figure 12.2). For example, does the farmer intend to improve pig welfare? Is the farmer in the habit of doing daily welfare checks? The COM-B model can be used to develop an understanding of the sources of behaviour, allowing intervention developers to identify what needs to change for the desired behaviour to occur (i.e., what are the deficits in Capability, Opportunity and Motivation that prevent the desired behaviour from being performed?).

The COM-B supports a behavioural analysis of the 'problem' by recommending several key processes. This involves defining the problem in behavioural terms, selecting, and specifying a target behaviour, and then identifying what needs to change. For example, Farrell *et al.* (2023) used the COM-B model to develop a quantitative measure of behavioural patterns around antibiotic use on dairy farms. The COM-B model helped in identifying what needed to change by measuring farmers' Capability, Opportunity and Motivation to engage in prudent antibiotic use to address antimicrobial resistance. Similarly, Prosser *et al.* (2022) used the COM-B model to identify determinants of the use of control measures for bovine viral diarrhoea prevention. Likert scale questions, seeking level of agreement, were developed using the COM-B model as a framework. For example, 'I know how to control infectious disease in my cattle' assessed Psychological Capability while 'I do not have the time to control infectious disease in my cattle' assessed Physical

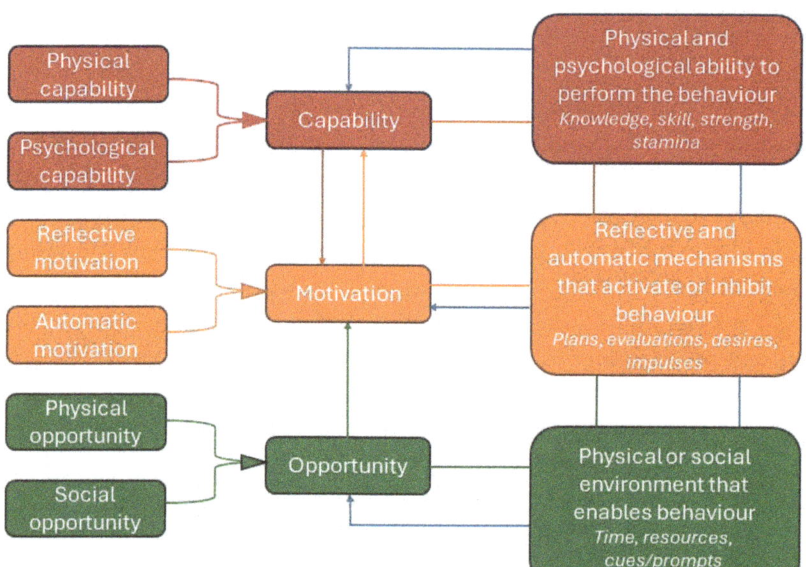

FIGURE 12.2 The Capability Opportunity Motivation-Behaviour model (COM-B; Michie *et al.* 2011c)

Opportunity (Prosser *et al.* 2022 supplementary materials). Prosser *et al.* (2022) found that appropriate action was taken when farmers had sufficient resources, an understanding of how to control infectious disease, and an understanding of why disease control matters. Information such as this is useful as it can inform interventions for farmers lacking in these areas. Qualitative research methods can also provide a rich understanding of target behaviours, including semi-structured interviews, focus groups or the more emergent method of qualitative surveys (Braun *et al.* 2020). For example, the COM-B was used to construct a semi-structured interview schedule by Walshe *et al.* (2023) to examine horse breeders' attitudes and behaviours towards parasite control practices and anthelmintic use on thoroughbred studs in Ireland. By using a qualitative approach guided by the COM-B, the study offered insights into the potential barriers for horse owners adopting sustainable parasite control practices and highlighted the need for end-user involvement in developing farm guidelines when addressing the issue of anthelmintic resistance in the thoroughbred horse industry.

If a more detailed analysis of barriers to the desired behaviour is required, researchers may also opt to use the Theoretical Domains Framework (TDF; Cane, O'Connor and Michie, 2012; McGowan *et al.* 2020). The TDF is a synthesis of 33 theories of behaviour, comprising of 14 domains of theoretical constructs such as social influences and knowledge. The TDF can be considered as an elaborated version of the COM-B, with the 14 domains each nesting within Capability, Opportunity or Motivation. For example, the TDF domains of *Knowledge* and *Memory, attention and decision processes* align to the COM-B component, 'Capability'. The TDF allows for systematic identification of barriers and facilitators of a desired behaviour (Michie *et al.* 2011; Carroll and Groarke, 2019). For example, similar to the COM-B, the TDF can be used as a template for creation of questionnaires, focus groups and interview schedules that address key factors known to influence human behaviour. The TDF has yet to be widely adopted by those carrying out animal health and welfare research. However, this is expected to change as an increasing number of researchers from various disciplines become aware of Human Behaviour Change theory. Definitions for each domain and sample interview questions can be found in Michie *et al.*'s (2011c) textbook (chapter 1, pp. 87–90).

3 Identify intervention options

Once the key barriers and facilitators of the desired behaviour have been established, the next step is to explore potential interventions. The analysis of Capability, Opportunity and Motivation will determine the Intervention Functions to be used (The second layer of the Behaviour Change Wheel; Figure 12.1). Michie *et al.* (2011c) recognise that a single behaviour change intervention can serve one or more functions. The term 'Intervention Function' is used in recognition of this. Michie *et al.* (2011c) list nine intervention functions, namely, *Education*, *Persuasion*,

Incentivisation, Coercion, Training, Restriction, Environmental restructuring, Modelling and Enablement. Intervention Functions likely to be useful in addressing each COM-B component have been agreed on via expert consensus (Michie *et al.* 2011c). For example, interviews with pig farmers could reveal that Physical Opportunity is the main barrier to removing the bitten pig from pens. Experts agree that barriers associated with Physical Opportunity are best addressed through the Intervention Functions of *Training, Restriction, Environmental restructuring* and *Enablement.* Intervention developers may therefore choose to focus the intervention design around one or more of these four functions. It is worth remembering however, that the Behaviour Change Wheel provides a *guide* to intervention developers. You still need to use your judgement in determining the most appropriate intervention for any given context (Michie *et al.* 2011c).

The APEASE criteria can be used to assess intervention ideas. APEASE stands for Affordability, Practicability, Effectiveness and cost-effectiveness, Acceptability, Side-effects/safety and Equity. For example, *Coercion* as an intervention function may not be successful as it is unlikely to be accepted by farmers while *Training* might be acceptable but not cost-effective to roll out on a mass scale. Any evaluation using the APEASE criteria will vary according to the given context, the actors involved, and the resources available. See Michie *et al.* (2011c, Table 1, page 23) for more information on this. The APEASE criteria can be used at this stage (identifying suitable Intervention Functions).

After suitable Intervention Functions have been identified, the next step in intervention design is to think about suitable Policy Categories that could support the delivery of the Intervention Functions. Policy Categories are actions that can be taken by those in authority to allow the intervention to be delivered (Michie *et al.* 2011c). An authority could be a research team, government legislators or assurance scheme provider. The seven Policy Categories outlined in the outer layer of the Behaviour Change Wheel (Figure 12.1) are as follows; *Communication/marketing, Guidelines, Fiscal, Regulation, Legislation, Environmental/social planning* and *Service provision.* Your ability to use any given Policy Category will depend on the level of influence you or your organisation has from a policy perspective. For example, an animal charity may conduct a mass media campaign (*Communication/marketing*) while a government body may ban a particular practice (*Legislation*). Veterinarians may provide a support service for farmers wishing to improve animal welfare (*Service provision*), while researchers at a university may disseminate recommendations for best practice based on a review of the available scientific evidence (*Guidelines*). Again, the APEASE criteria can be used to evaluate the proposed Policy Categories for a given context.

3.1 *Applying Human Behaviour Change Theory to the ban on routine tail docking*

For researchers new to Human Behaviour Change, and considering that most studies on HBC have traditionally been centred on human health (Carroll and Groarke,

2019), it may help to apply these principles to a problem relevant to pig farming. An example of a somewhat unsuccessful top-down attempt at human behaviour change comes from the EU-wide ban on tail docking. Tail docking is carried out to prevent tail biting in pigs (Sutherland and Tucker, 2011). However, tail docking does not address the root cause of tail biting. Rather, it simply masks the issue while animal welfare remains compromised (D'Eath *et al.* 2015). Furthermore, tail docking is painful for pigs (Valros *et al.* 2016) and is often ineffective (Harley *et al.* 2014). Consequently, tail docking should only be used as a last resort. Routine tail docking was banned in the EU in 2003. However, this ban is not currently enforced in as many as 20 EU Member States (Nalon and De Briyne, 2019). A cursory analysis of the intervention, i.e., the tail docking ban, is provided in Table 12.2, according to the APEASE criteria.

Likely barriers to effective implementation of the ban have been identified retrospectively. For example;

1. Tail biting is perceived as being worse than tail docking. Tail docking is therefore viewed as the better alternative of the two (European Commission, 2017).
2. Tail docking is socially accepted within certain farming communities (Albernaz-Goncalves *et al.* 2021)
3. There is a common view that enrichment materials are not compatible with many slurry systems in slatted housing. Therefore, farmers are often reluctant to add large quantities of enrichment materials such as straw as this can result in blockages in the slurry system (Wallgren *et al.* 2016)
4. There is a belief that changing the flooring system to allow for the provision of enrichment material would be a difficult task. Therefore, this option is seen as unrealistic for many intensive pig farmers (Buijs and Muns, 2019)

We can map these barriers to the Theoretical Domains Framework and COM-B model. For example, barrier 1, the idea that tail biting is worse than tail docking, maps to the TDF '*Belief about consequences*' which comes under the COM-B component Reflective Motivation. We can then consider suitable Intervention Functions. In this case, Reflective Motivation maps onto the Intervention Functions of *Education*, *Persuasion* and *Modelling* (Michie *et al.* 2011c, Table 2.2., p. 113). Barrier 2, that tail docking is socially accepted in certain farming circles, maps to the TDF '*Social influences*', which comes under the COM-B component, Social Opportunity. Social Opportunity, in turn, maps onto the Intervention Functions of *Restriction*, *Environmental restructuring*, *Modelling* and *Enablement*. These intervention Functions can then be explored along with their associated Policy Categories. The intervention developer should decide if they have the power to deliver an intervention via the selected route.

TABLE 12.2 Applying the APEASE criteria to the tail docking ban

	APEASE criterion					
Intervention	**Affordability**	**Practicability**	**Effectiveness/cost effectiveness**	**Acceptability**	**Side effects/ safety**	**Equity**
Banning the practice[a]	There is an implicit budget attached to the ban because it requires farmers to use alternative means of reducing tail biting risk, many of which may be unaffordable	The ban is not practical to deliver in practice without extensive resource allocation to support farmers in avoiding an increase in tail lesion prevalence	Unlikely to be effective as the ban does not deal with root cause of tail biting Unlikely to be cost-effective as the cost of addressing tail biting in other ways is too substantial for many farmers	Not acceptable to many farmers who may see tail docking as the most effective way of managing tail biting	A clear side-effect of implementation of the ban in existing farming systems would be an increase in tail lesion prevalence	A ban on routine tail docking impacts indoor pig farmers more than those using extensive systems

a tail docking is only to be carried out when other measures, such as proving appropriate enrichment, do not work

3.2 *Modelling as a promising intervention function*

Some Intervention Functions and Policy Categories may be more effective in changing pig farmer behaviour than others. Social aspects may influence farmer behaviour but are currently under-utilised. It is recognised that farming practices are learned within communities and via traditions (Nicolini, 2012; Prosser *et al.*, 2022). Furthermore, farmers with experience of a particular practice view it more positively (Albernaz-Goncalves *et al.*, 2021). For instance, Valros and Barber (2019) found that pig farmers that add bedding to their pens and do not dock tails rate adding bedding as being more important for reducing tail biting than other farmers. This may be because these farmers have experience in successfully dealing with tail biting without the need to tail dock. Therefore, modelling of husbandry practices that negate the need for tail docking may be one suitable method for addressing tail biting in pigs. Modelling appears to be a potentially useful intervention function when it comes to other farmer behaviours. For example, a *safe*food project report (Regan *et al.*, 2023) outlines suggested behaviour change interventions to reduce antibiotic use on farms. The report highlights the importance of peer-to-peer social support to model good farm practices, which could be adopted for the area of tail biting. According to Bandura's Social Cognitive Theory (1986), human behaviour is learned through observation of others. From observing others, farmers can form a picture of how to perform a behaviour and this can serve as a guide for when it is their turn to perform this behaviour in the future. Others modelling a behaviour can be a very effective method for behaviour change, particularly if the individual sees the 'model' receiving reinforcement for the behaviour (e.g., a financial reward, reduced tail biting, encouragement from others) within a supportive environment like farm discussion groups. This can improve a person's self-efficacy and behavioural capability. That is, they will feel more confident that they can perform the behaviour, learn practical skills from watching others and receive verbal persuasion that they too can do it. It is important to note that for observational learning to be effective, the 'model' farmers should be viewed as similar to the target group of farmers and should present a *coping* model rather than a *mastery* model. That is, the 'model' should outline the challenges to behaviour change as well as the successes, rather than simply presenting the successes or suggesting the change was simple and free from barriers. Modelling can be delivered via the policy categories of Communication/marketing or Service Provision. The methods of delivery of any related intervention will again depend on several factors which should be explored using the APEASE criteria.

Once the intervention functions and policy categories have been decided upon, the next step is to identify the intervention content.

4 Intervention content

4.1 *The Behaviour Change Technique Taxonomy*

The purpose of the Behaviour Change Technique Taxonomy (BCTT, Michie *et al.* 2013) is to aid the creation of intervention content. Behaviour Change Techniques (BCTs) are the 'active ingredients' of a behaviour change intervention, i.e. the underlying mechanisms that are proposed to change behaviour. These 'active ingredients' are both measurable and replicable (Michie *et al.*, 2013). For example, an intervention could be composed of several components, some or all of which may influence behaviour. Telling a farmer that they will be given a financial reward for reducing tail lesion prevalence, for instance, could be effective due to the material reward itself, but also the promise of a reward (Carroll and Groarke, 2019). Mapping interventions to the BCTT allows the researcher to understand what may or may not have influenced behaviour change. Detailed descriptions of the intervention should be provided alongside the behaviour change techniques that the intervention utilises. This mapping allows other researchers to replicate the research by insuring that all components of the original intervention are included. To create a common language to describe these 'active ingredients' and allow for precise reporting of interventions, Michie *et al.* (2013) created a taxonomy of 93 BCTs, based on an extensive review of behaviour change interventions, and grouped them based on expert consensus. This taxonomy is now used extensively in behavioural research to rigorously develop and evaluate intervention content and to systematically examine what 'active ingredients' are commonly used when conducting reviews of intervention literature. An example of a BCT is self-monitoring of behaviour. Self-monitoring is a very common BCT used to promote positive behavioural changes, such as using an activity tracker or smart watch to increase physical activity. Within human health research for example, self-monitoring has been illustrated as a BCT within a behaviour change intervention to successfully increase physical activity in patients with cardiovascular disease by counting daily steps via a pedometer (Kanejima, Kitamura and Izawa, 2019). The BCT Taxonomy provides a description of each of these to help in this mapping exercise.

BCTs should ideally be identified before implementation of any behaviour change intervention. However, as mentioned previously, intervention developers may be unaware of behaviour change theory or may choose not to use it. In this case, the BCT Taxonomy can also be applied retrospectively. This works best when there is a detailed description of the intervention available. Using this approach, McParland *et al.* (2018) retrospectively coded behaviour change techniques (BCTs) used to target public engagement with antimicrobial resistance using Michie *et al.*'s (2013) BCT Taxonomy. This was done so that the effectiveness of past interventions could be established, including the 'active ingredients' of the interventions that were successful or unsuccessful. Twenty studies were included in the final analysis. The intervention descriptions from each article were assessed to determine which BCTs were present. McParland *et al.* (2018) found that the most commonly used BCTs for this problem were '*Information about health consequences*', '*Credible source*' and '*Instruction on how to perform the behaviour*'. However, while less commonly used, the BCTs of '*Commitment*', '*Comparative imagining of future outcomes*', '*Monitoring of behaviour without feedback*', '*Information about antecedents*' and '*Feedback on behaviour*' were present in the most successful interventions. Interventions were coded as having between one and seven 'active ingredients', with a greater range of BCTs present in the more successful interventions. Indeed, interventions that involve groups of theoretically related BCTs are more effective than interventions that utilise only one BCT (Michie *et al.* 2018). The BCT Taxonomy is available to download in app form from iTunes and Google play app stores. Here, the taxonomy, BCT definitions and examples can be accessed in a user-friendly manner. Online training is also available (https://www.bct-taxonomy.com/).

4.2 *Mapping an intervention to the Behaviour Change Technique taxonomy*

At present, Animal Health Ireland (2022) provides a Targeted Advisory Service on Animal Health (TASAH) to commercial pig units in Ireland. This service, which is delivered by trained veterinary practitioners, involves reviewing factors associated with the risk of tail-biting with the aim of safely reducing the target behaviour: tail docking. TASAH aims to identify specific areas where the risk of tail-biting is particularly high and to provide practical recommendations to address these risks. BCTs are embedded in this consult. Table 12.3 describes some of the aspects of this tailored on farm intervention programme, corresponding BCTs and a description of these BCTs as provided by Michie *et al.* (2013).

For simplicity, only one BCT is provided per example in Table 12.3. However, as mentioned earlier, any one intervention can have several active ingredients. For example, an intervention involving a financial award would map to the active

TABLE 12.3 A real-world example of BCTs used during a structured veterinary practitioner consult

Intervention description	**Behaviour Change Technique**	**Definition of BCT from taxonomy**[a]
The TASAH-funded consult is delivered by a specialist-trained veterinary practitioner with a working knowledge and relationship with the farm	Credible source	Present verbal or visual communication from a credible source in favour of or against the behaviour

a Michie *et al.* (2013)

TABLE 12.3 A real-world example of BCTs used (*cont.*)

Intervention description	Behaviour Change Technique	Definition of BCT from taxonomy
Based on the risk assessment and consult, the farmer and approved veterinary practitioner agree on three changes to be made to farm routine to reduce the risk of tail biting on farm	Goal setting (behaviour); action planning	Set or agree on a goal defined in terms of the behaviour to be achieved; Prompt detailed planning of performance of the behaviour (must include at least one of context, frequency, durations and intensity)
Farmers are provided with data on their own risk assessment and anonymised data from other farms to act as a benchmark	Social comparison	Draw attention to others' performance to allow comparison with the person's own performance
Farmers are provided with information on the causes of and consequences of tail-biting in pigs to identify where changes to farm management need to be made.	Information about antecedents	Provide information about antecedents (e.g. Social and environmental situations and events) that reliably predict performance of the behaviour
Farmer and veterinary practitioner are asked to review the agreed actions and intended outcomes at their next consult, with changes made as necessary based on progress	Review behaviour goals; review outcome goals	Review behaviour or outcome goals jointly with the person and consider modifying goal(s) or behaviour change strategy in light of achievement.

ingredient of '*Material reward (behavior)*' (Michie *et al.* 2013) but may also contain other 'active ingredients' such as the promise of a reward, '*Material incentive (behavior)*' (Carroll and Groarke, 2019). Interventions that address behaviours related to reducing tail biting should be described in terms of these BCTs so that the successful features of the intervention that ignite change can be identified, measured and repeated in other similar situations.

4.3 *Promoting the use of BCTs amongst animal health professionals*

It is important to use strategies which have been evidenced to result in positive, meaningful behavioural changes and then use a common language to describe those strategies that energise behaviour change. Bull and Dale (2021) outline the value of training health and social care practitioners in embedding BCTs into their practice, and highlight the positive impact of this training to their confidence, perceived competence and intention to use BCTs. Regan *et al.* (2023) highlight the strong need for similar behaviour change training for veterinary practitioners, advisors and other animal health professionals when addressing changes to animal health management; this could certainly be applied to the complex issue of tail biting on farms. While it may seem difficult at first, embedding Human Behaviour Change theory into pig tail biting interventions has several benefits including increased replicability and the ability to identify what works and does not work when it comes to changing stakeholder behaviour. The BCW and related frameworks have been designed to be used by intervention developers from various backgrounds. However, consulting with a trained behaviour change specialist is always advised.

4.4 *Intervention dimensions*

Interventions should be reported in as much detail as possible. For example, the mode of delivery should be specified (Michie *et al.* 2013). The most suitable mode of delivery will depend on the target of the intervention (e.g. individual, group or population); for instance, an intervention involving identification of tail biting risk factors at farm-level could be delivered face-to-face, while interventions designed for populations (e.g. all pig farmers in a particular country) could be delivered via broadcast media such as TV or radio. The APEASE criteria can be useful here in evaluating the suitability of different modes of delivery; a smart phone app that helps identify tail biting risk factors may be cost effective, but may not be accepted by less tech-savvy end users. Other aspects of the intervention that should be described are outlined by Michie *et al.* (2013) and include specifying what was delivered, who delivered it, where it was delivered, to whom it was delivered, intensity (over how many contacts it was delivered), duration (over what period of time it was delivered) and fidelity (the extent to which it was delivered as intended, Michie *et al.* 2013).

4.5 *Intervention success*

An important but sometimes overlooked aspect of intervention design is measuring its success. This links back to the original target behaviour. The clearer the target behaviour is, the easier it will be to determine whether or not the behaviour has been achieved. The target behaviour can be measured directly, for example, measuring quantities of enrichment used by farmers. Target behaviour can also be assessed indirectly by focusing on the outcome of the behaviour, i.e. tail biting prevalence. For example, intervention designers may measure tail lesion frequency in experimental farms (those subject to the intervention) compared to control farms (those not subject to the intervention).

4.6 *Practical application of Human Behaviour Change theory to reduce tail biting in pigs*

- Be clear in the behaviour that you want to change. Do you want to change farmer behaviour or another stakeholder like veterinarians or abattoir workers? What behaviour do you want to see? Always be as specific as possible by thinking about who needs to do what, when, where, how often and with whom.
- Do not presume to know what needs to change. Use frameworks like the COM-B to systematically identify barriers and facilitators of change. This might involve using these frameworks to develop questionnaires or interview schedules, or, you might simply use the COM-B as an overall structure for grouping barriers to behaviour change that you identify from reviewing the literature.
- Decide how your intervention will be delivered. This will be informed by the barriers that you identified. For example, will your intervention educate, incentivise or coerce? Will the intervention be delivered via legislation or communication/marketing?
- Describe your proposed intervention in as much detail as possible and map these to the Behaviour Change Technique Taxonomy – what aspects of the intervention may be causing it to be successful or unsuccessful?
- Once you have rolled out your intervention, make sure to measure its success. Has the intervention brought about the desired result? The clearer the target behaviour, the easier it will be to assess intervention effectiveness.

References

Ajzen, I., 1985. From intentions to actions: a theory of planned behavior. In Action Control; Springer: Berlin, Germany. 11–39.

Albernaz-Gonçalves, R., Olmos, G. and Hötzel, M.J., 2021. My pigs are ok, why change? – animal welfare accounts of pig farmers. Animal 15(3): 100154. https://doi.org/10.1016/j.animal.2020.100154.

Animal Health Ireland (AHI), 2022. TASAH – Tail biting risk factor assessment (pigs) training. Available at: https://animalhealthireland.ie/training/tasah-tail-biting-risk-factor-assessment-training/.

Bandura, A., 1986. The explanatory and predictive scope of self-efficacy theory. Journal of Social and Clinical Psychology 4(3): 359–373. https://doi.org/10.1521/jscp.1986.4.3.359.

Braun, V., Clarke, V., Boulton, E., Davey, L., and McEvoy, C., 2021. The online survey as a qualitative research tool. International Journal of Social Research Methodology, 24:6, 641–654. https://doi.org/10.1080/13645579.2020.1805550.

Buijs, S. and Muns, R., 2019. A review of new knowledge on tail biting and tail docking. Agri-food and biosciences institute.

Bull, E.R., and Dale, H., 2021. Improving community health and social care practitioners' confidence, perceived competence and intention to use behaviour change techniques in health behaviour change conversations. Health and Social Care in the Community 29(1): 270–283. https://doi.org/10.1111/hsc.13090.

Cane, J., O'Connor, D., and Michie, S., 2012. Validation of the theoretical domains framework for use in behaviour change and implementation research. Implementation Science 7(1): 1–17. https://doi.org/10.1186/1748-5908-7-37.

Carroll, G.A. and Groarke, J.M., 2019. The importance of the social sciences in reducing tail biting prevalence in pigs. Animals 9(9): 591. https://doi.org/10.3390/ani9090591.

Carroll, G.A., Groarke, J.M. and Graham-Wisener, L., 2021. In: Human behaviour change models for improving animal welfare. Camerlink, I (ed.). Bridging research disciplines to advance animal welfare science: a practical guide, 273 pp.

Cornish, A., Jamieson, J., Raubenheimer, D. and McGreevy, P., 2019. Applying the behavioural change wheel to encourage higher welfare food choices. Animals 9(8): https://doi.org/524.10.3390/ani9080524.

Coupe, N., Cotterill, S., and Peters, S., 2021. Enhancing community weight loss groups in a low socioeconomic status area: application of the COM-B model and Behaviour Change Wheel. Health Expectations. 25(5): 2043–2055 https://doi.org/10.1111/hex.13325.

D'Eath, R.B., Arnott, G., Turner, S.P., Jensen, T., Lahrmann, H.P., Busch, M.E., Niemi, J.K., Lawrence, A.B. and Sandøe, P., 2014. Injurious tail biting in pigs: how can it be controlled in existing systems without tail docking? Animal 8(9): 1479–1497. https://doi.org/10.1017/S1751731114001359.

D'Eath, R.B., Niemi, J.K., Ahmadi, B.V., Rutherford, K.M.D., Ison, S.H., Turner, S.P., Anker, H.T., Jensen, T., Busch, M.E., Jensen, K.K. and Lawrence, A.B., 2016. Why are most EU pigs tail docked? Economic and ethical analysis of four pig housing and management scenarios in the light of EU legislation and animal welfare outcomes. Animal 10(4): 687–699. https://doi.org/10.1017/S1751731115002098.

Eldredge, L.K.B., Markham, C.M., Ruiter, R.A., Fernández, M.E., Kok, G., and Parcel, G.S., 2016. Planning health promotion programs: an intervention mapping approach. John Wiley and Sons, 704 pp.

Escobar, M.P., and Demeritt, D., 2017. Paperwork and the decoupling of audit and animal welfare: the challenges of materiality for better regulation. Environment and Planning C: Politics and Space 35(1): 169–190. https://doi.org/10.1177/0263774X16646771.

European Commission. 2017. Directorate-general for health and food safety. Directorate for health and food audits and analysis. Study visits on rearing pigs with intact tails: overview report, Publications Office. https://data.europa.eu/doi/10.2772/486504.

Farrell, S., Benson, T., McKernan, C., Regan, Á., Burrell, A.M.G. and Dean, M., 2023. Factors influencing dairy farmers' antibiotic use: an application of the COM-Model. Journal of Dairy Science 106:4059–4071 https://doi.org/10.3168/jds.2022-22263.

Fernandes, J., Blache, D., Maloney, S.K., Martin, G.B., Venus, B., Walker, F.R., Head, B. and Tilbrook, A., 2019. Addressing animal welfare through collaborative stakeholder networks. Agriculture 9(6): 132. https://doi.org/10.3390/agriculture9060132.

Golding, S.E., Ogden, J., and Higgins, H.M., 2019. Shared goals, different barriers: a qualitative study of UK veterinarians' and farmers' beliefs about antimicrobial resistance and stewardship. Frontiers in Veterinary Science, 6: 132. https://doi.org/10.3389/fvets.2019.00132.

Gollwitzer, P.M., and Oettingen, G., 2015. From studying the determinants of action to analysing its regulation: a commentary on Sniehotta, Presseau and Araújo-Soares. Health Psychology Review 9(2): 146–150. https://doi.org/10.1080/17437199.2014.919511.

Haigh, A., Yun-Chou, J. and O'Driscoll, K., 2019. An investigation into the effectiveness of compressed straw blocks in reducing abnormal behaviour in growing pigs. Animal 13(11): 2576–2585. https://doi.org/10.1017/S1751731119000715.

Harley, S., Boyle, L.A., O'Connell, N.E., More, S.J., Teixeira, D.L. and Hanlon, A., 2014. Docking the value of pigmeat? Prevalence and financial implications of welfare lesions in Irish slaughter pigs. Animal Welfare 23(3): 275–85. https://doi.org/10.7120/09627286.23.3.275.

Henry, M., Jansen, H., Amezcua, M.D.R., O'Sullivan, T.L., Niel, L., Shoveller, A.K. and Friendship, R.M., 2021. Tail-biting in pigs: a scoping review. Animals 11(7): 2002. https://doi.org/10.3390/ani11072002.

Kanejima, Y., Kitamura, M., and Izawa, K.P., 2019. Self-monitoring to increase physical activity in patients with cardiovascular disease: a systematic review and meta-analysis. Aging Clinical and Experimental Research 31(2): 163–173. https://doi.org/10.1007/s40520-018-0960-7.

McDonald, J.L., Farnworth, M.J. and Clements, J., 2018. Integrating trap-neuter-return campaigns into a social framework: developing long-term positive behavior change toward unowned cats in urban areas. Frontiers in Veterinary Science, 5:258. https://doi.org/10.3389/fvets.2018.00258.

McEachan, R.R.C., Conner, M., Taylor, N., and Lawton, R.J., 2011. Prospective prediction of health-related behaviors with the theory of planned behavior: A meta-analysis. Health Psychology Review 5: 97–144. https://doi.org/10.1080/17437199.2010.521684.

McGowan, L.J., Powell, R., and French, D.P., 2020. How can use of the theoretical domains framework be optimized in qualitative research? A rapid systematic review. British Journal of Health Psychology 25(3): 677–694. https://doi.org/10.1111/bjhp.12437.

McParland, J.L., Williams, L., Gozdzielewska, L., Young, M., Smith, F., MacDonald, J., Langdridge, D., Davis, M., Price, L. and Flowers, P., 2018. What are the 'active ingredients' of interventions targeting the public's engagement with antimicrobial resistance and how might they work? British Journal of Health Psychology 23(4): 804–819. https://doi.org/10.1111/bjhp.12317.

Michie, S., and Prestwich, A., 2010. Are interventions theory-based? Development of a theory coding scheme. Health psychology 29(1): 1. http://doi.org/10.1037/a0016939.

Michie, S., Ashford, S., Sniehotta, F.F., Dombrowski, S.U., Bishop, A., and French, D.P., 2011b. A refined taxonomy of behaviour change techniques to help people change their physical activity and healthy eating behaviours: the CALO-RE taxonomy. Psychology and health 26(11): 1479–1498. https://doi.org/10.1080/08870446.2010.540664.

Michie, S., Atkins, L., and West, R., 2014. The behaviour change wheel. A guide to designing interventions. 1st ed. Great Britain: Silverback Publishing, 329 pp.

Michie, S., Carey, R.N., Johnston, M., Rothman, A.J., De Bruin, M., Kelly, M.P. and Connell, L.E., 2018. From theory-inspired to theory-based interventions: a protocol for developing and testing a methodology for linking behaviour change techniques to theoretical mechanisms of action. Annals of Behavioral Medicine 52(6): 501–512. https://doi.org/10.1007/s12160-016-9816-6.

Michie, S., Hyder, N., Walia, A., and West, R., 2011a. Development of a taxonomy of behaviour change techniques used in individual behavioural support for smoking cessation. Addictive Behaviors 36(4): 315–319. https://doi.org/10.1016/j.addbeh.2010.11.016.

Michie, S., Richardson, M., Johnston, M., Abraham, C., Francis, J., Hardeman, W., Eccles, M.P., Cane, J. and Wood, C.E., 2013. The behavior change technique taxonomy (v1) of 93 hierarchically clustered techniques: building an international consensus for the reporting of behavior change interventions. Annals of Behavioral Medicine 46(1): 81–95. https://doi.org/10.1007/s12160-013-9486-6.

Michie, S., Van Stralen, M.M., West, R., 2011c. The behaviour change wheel: a new method for characterising and designing behaviour change interventions. Implementation Science 6: 42. http://doi.org/10.1186/1748-5908-6-42.

Morgans, L.C., Bolt, S., Bruno-McClung, E., van Dijk, L., Escobar, M.P., Buller, H.J., Main, D.C. and Reyher, K.K., 2021. A participatory, farmer-led approach to changing practices around antimicrobial use on UK farms. Journal of Dairy Science 104(2): 2212–2230. https://doi.org/10.3168/jds.2020-18874.

Nalon, E. and De Briyne, N., 2019. Efforts to ban the routine tail docking of pigs and to give pigs enrichment materials via EU law: where do we stand a quarter of a century on? Animals, 9(4), 132. https://doi.org/10.3390/ani9040132.

Nicolini, D., 2012. Practice theory, work, and organization: an introduction. Oxford University Press.

Ogden, J., 2016. Celebrating variability and a call to limit systematisation: the example of the Behaviour Change Technique Taxonomy and the Behaviour Change Wheel. Health Psychology Review 10(3): 245–250. https://doi.org/10.1080/17437199.2016.1190291.

Oliwa, J.N., Nzinga, J., Masini, E., van Hensbroek, M.B., Jones, C., English, M., and van't Hoog, A., 2020. Improving case detection of tuberculosis in hospitalised Kenyan children – employing the behaviour change wheel to aid intervention design and implementation. Implementation Science 15(1): 1–16. https://doi.org/10.1186/s13012-020-01061-4.

Presseau, J., McCleary, N., Lorencatto, F., Patey, A.M., Grimshaw, J.M. and Francis, J.J., 2019. Action, actor, context, target, time (AACTT): a framework for specifying behaviour. Implementation Science 14(1): 1–13. https://doi.org/10.1186/s13012-019-0951-x.

Regan, Á., Burrell, A., McKernan, C., Martin, H., Benson, T., McAloon, C., ... & Dean, M. (2023). Behaviour change interventions for responsible antimicrobial use on farms. Irish Veterinary Journal, 76(1), 8.

Regan, Á., Douglas, J., Maher, J., and O'Dwyer, T., 2021. Exploring farmers' decisions to engage in grass measurement on dairy farms in Ireland. The Journal of Agricultural Education and Extension 27(3): 355–380. https://doi.org/10.1080/1389224X.2020.1858892.

Salari, F., Yazdanpanah, M., Yaghoubi, J. and Forouzani, M., 2019. Understanding Iranian livestock breeders' intentions and behavior regarding nonhuman animal welfare. Society and Animals 29(3): 246–267. https://doi.org/10.1163/15685306-12341592.

Shortall, O., Ruston, A., Green, M., Brennan, M., Wapenaar, W., Kaler, J., 2016. Broken biosecurity? Veterinarians' framing of biosecurity on dairy farms in England. Preventative Veterinary Medicine 132: 20–31. https://doi.org/10.1016/j.prevetmed.2016.06.001.

Sniehotta, F.F., Presseau, J., and Araújo-Soares, V., 2014. Time to retire the theory of planned behaviour. Health Psychology Review 8(1): 1–7. https://doi.org/10.1080/17437199.2013.869710.

Sniehotta, F.F., Scholz, U., Schwarzer, R., 2005. Bridging the intention-behaviour gap: planning, self-efficacy, and action control in the adoption and maintenance of physical exercise. Psychology and Health 20: 143–160. https://doi.org/10.1080/08870440512331317670.

Sonoda, L.T., Fels, M., Oczak, M., Vranken, E., Ismayilova, G., Guarino, M., Viazzi, S., Bahr, C., Berckmans, D. and Hartung, J., 2013. Tail biting in pigs – causes and management intervention strategies to reduce the behavioural disorder. Berl Munch Tierarztl Wochenschr 126(3–4): 104–12.

Sutherland, M.A. and Tucker, C.B., 2011. The long and short of it: a review of tail docking in farm animals. Applied Animal Behaviour Science, 135(3): 179–191. https://doi.org/10.1016/j.applanim.2011.10.015.

Uphill, M., 2014. Behaviour change: physical (in) activity. The British Psychological Society. Retrieved from: https://www.bps.org.uk/sites/www.bps.org.uk/files/Policy/Policy%20-%20Files/Changing%20behaviour%20-%20physical%20%28in%29activity.pdf.

Valros, A. and Barber, C., 2019. Producer perceptions of the prevention of tail biting on UK farms: association to bedding use and tail removal proportion. Animals 9(9): 628. https://dx.doi.org/10.3390%2Fani9090628.

Valros, A., Munsterhjelm, C., Hänninen, L., Kauppinen, T. and Heinonen, M., 2016. Managing undocked pigs – on-farm prevention of tail biting and attitudes towards tail biting and docking. Porcine Health Management 2(1): 1–11. https://doi.org/10.1186/s40813-016-0020-7.

van Dijk, L., Buller, H.J., Blokhuis, H.J., van Niekerk, T., Voslarova, E., Manteca, X., Weeks, C.A. and Main, D.C., 2019. HENNOVATION: Learnings from promoting practice-led multi-actor innovation networks to address complex animal welfare challenges within the laying hen industry. Animals 9(1): 24. https://doi.org/10.3390/ani9010024.

Wallgren, T., Westin, R. and Gunnarsson, S., 2016. A survey of straw use and tail biting in Swedish pig farms rearing undocked pigs. Acta Veterinaria Scandinavica, 58(1): 1–11. https://doi.org/10.1186/s13028-016-0266-8.

Walshe, N., Burrell, A., Kenny, U., Mulcahy, G., Duggan, V., and Regan, A. 2023. A qualitative study of perceived barriers and facilitators to sustainable parasite control on thoroughbred studs in Ireland. Veterinary Parasitology, 317: 109904. https://doi.org/10.1016/j.vetpar.2023.109904.

Winkel, C., von Meyer-Höfer, M. and Heise, H., 2020. Understanding german pig farmers' intentions to design and construct pig housing for the improvement of animal welfare. Animals 10(10): 1760. https://doi.org/10.3390/ani10101760.

CHAPTER 13

Precision livestock farming

Mona Lilian Vestbjerg Larsen[1,2,*], *Lene Juul Pedersen*[2] *and Tomas Norton*[1]

[1]M3 BIORES – Measure, Model and Manage Bioresponses, Division Animal and Human Health Engineering, Department of Biosystems, KU Leuven, Kasteelpark Arenberg 30, B-3001 Leuven, Belgium; [2]Department of Animal and Veterinary Sciences, Aarhus University, Blichers Allé 20, 8830 Tjele, Denmark; *Correspondence: mona@anivet.au.dk

Abstract

Precision livestock farming (PLF) concerns the use of sensor technology to monitor the lives of livestock animals continuously and in real-time, with the purpose to manage livestock welfare, health, productivity as well as environmental impact of the livestock sector. The current development of PLF technologies provides a new and innovative opportunity and may be essential in addressing the tail biting problem due to its complex aetiology. PLF technologies have future potential to assist the farmer in the prevention of tail biting through automatic detection of tail biting behaviour and through prediction of tail damage. While the latter has received the greatest attention in the literature, results so far are unsatisfying, probably due to the multifactorial nature of tail biting. Based on the existing literature, a decrease in individual daily voluntary feed intake and an increase in pigs with lowered tails are the best-established indicators as predictors of tail damage, although they are probably not specific to tail damage. Prediction models developed so far are limited to the use of already available sensor technology including water usage measured by water-flow meters and pen temperature measured by thermometers; again, predictors shown not to be specific to tail damage. PLF technology also have future potential to assist the farmer in the prevention of tail biting by automatic risk factor assessment resulting in advice for the farmer on the most appropriate intervention strategy. These above-described potentials highly depend on the development of automatic monitoring methods for pig behaviour, physiology, and environmental parameters. Thus, the use of PLF technology to address the tail biting problem is currently limited by accuracy and reliability of such methods; a limitation that will hopefully diminish in the future with developments in technology and modelling techniques.

Keywords

sensor – technology – detection – prediction – modelling

1 Introduction

Precision livestock farming (**PLF**) concerns the use of sensor technology to monitor the lives of livestock animals continuously and in real-time, with the purpose to manage livestock welfare, health, productivity as well as environmental impact of the livestock sector (Berckmans, 2017). Recent reviews of the literature describe PLF as a young research field within the pig production, where focus is still mainly on development of technologies and less on implementation (Gómez *et al.*, 2021; Larsen *et al.*, 2021; Schillings *et al.*, 2021). Thus, the true value of PLF to animal welfare is still unclear, as well is the possible negative side-effects on animal welfare such as the risk of a dissociated human-animal relationship (Dawkins, 2021; Guzhva *et al.*, 2021, Tuyttens *et al.*, 2022). However, PLF technologies have the potential to assess livestock welfare and ensure good animal well-being by building digital representations of the animals, providing the farmer with a deeper insight into the needs of the animals through remote monitoring of animal behaviour, physiology, and their surrounding environment (Norton *et al.*, 2019). This chapter aims to describe the potential of PLF as a method to address the tail biting problem and to summarise what PLF technologies have currently been investigated for this purpose. It will also address what challenges are faced when developing PLF technologies for tail biting prevention and will end with suggestions for future research directions needed to be able to implement PLF technologies for tail biting prevention.

2 Potential of precision livestock farming

As clearly seen in previous chapters, tail biting is a complex welfare issue that occurs sporadically in time, with many associated risk factors and thus, no single solution exists to prevent tail biting. For several decades, research has been conducted to increase our understanding of tail biting in pigs including its causation and development. Despite this, tail biting remains one of the major welfare problems in pig production and tail docking is still widely implemented. The current development of PLF technologies provides a new opportunity to monitor factors associated with tail-biting and may even become key to prevent tail biting due to its complexity.

Currently, two strategies to use PLF technologies in the prevention of tail biting have been investigated: (1) detection of tail biting behaviour, (2) prediction of risk of tail damage. The latter has received the greatest attention and will be discussed in the next section of this chapter. Detection of tail biting behaviour is a relatively novel approach made possible with the current development in technology. Tail biting behaviour can be detected by the use of video recordings from 2D cameras, combined with computer vision and deep learning model techniques; so far obtaining an accuracy of almost 0.90 in the classification of automatically detected pair-wise pig interactions into tail-biting and no tail-biting interactions (Liu *et al.*, 2020). It is known to the authors that this method is currently being further developed and validated in a real-life situation in conditions unknown to the developed algorithm including new pigs and new housing facilities. Also, the method should be tested on tail-docked pigs as well, as a short tail could potentially make the detection more or less accurate and demand other parameters in the detection model. Automatic detection of tail biting behaviour may make it possible to use this behaviour as a predictor of its own consequence: tail damage. However, automatic detection of tail biting also have potential to ensure that not merely the damaging consequence is prevented, but also that the underlying welfare issue, including stress, frustration and boredom leading to tail biting, is addressed, and minimised as early as possible; a possible drawback of the strategy to predict the risk of tail damage. Beside these two strategies, the use of sensor technology (2D cameras) has also been investigated at the abattoir for automatic detection of tail lesions and tail losses (Blömke *et al.*, 2020; Brünger *et al.*, 2019). While only achieving a 74% agreement between automatic and manual human observations of tail lesions, this was similar to the agreement between different human observers with the main reason for disagreements being picture quality issues (Brünger *et al.*, 2019). Detection of tail lesions and losses at the abattoir can per definition not be considered a PLF technology

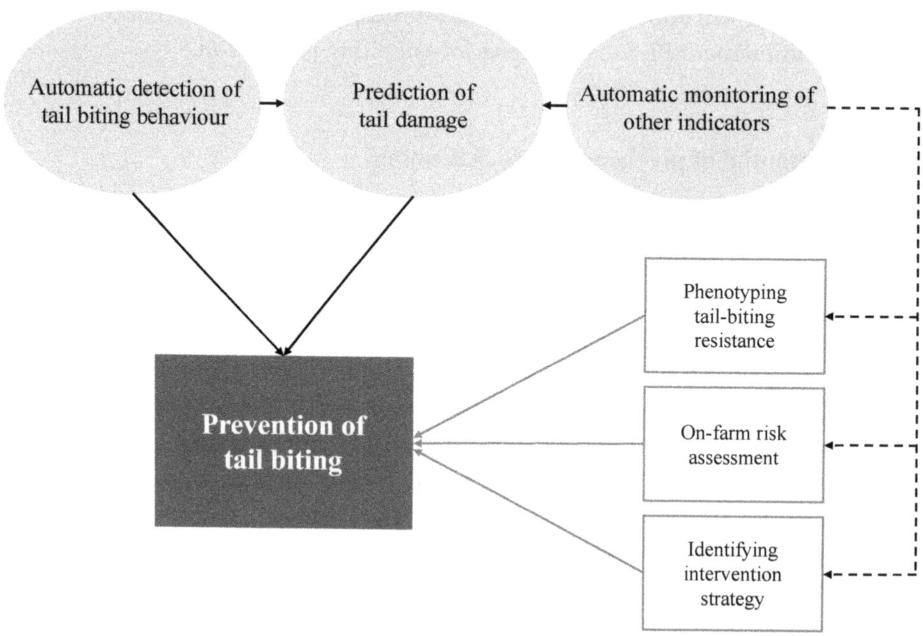

FIGURE 13.1 Illustration of the potential of PLF technologies to assist in the prevention of tail biting in pigs, through direct (––––) or indirect (– – –) routes. Automatic monitoring of other indicators includes pig behaviour, physiology, and the surrounding environment.

due to the lack of real-time monitoring with the potential to improve the lives of the animals being monitored. However, such use of sensor technology can still prove valuable by providing the farmer with benchmarkable feedback on the measures taken to prevent tail biting, and for the industry by creating data on tail biting prevalence for e.g., welfare labelling, also motivating the farmer with a higher economic return and providing documentation to consumers on animal welfare of purchased meat.

PLF technologies do not only have potential to inform the farmer of the timing and location of a present or future tail biting problem. It also has the potential to assist other methods addressing the tail biting problem (see Figure 13.1). This includes assistance in the development of tail-biting resistant phenotypes in breeding programmes, assistance in on-farm risk assessment (see Chapter 10), assistance in identification and choice of intervention strategy against tail biting, and assistance in the use of the pig tail as an iceberg indicator (see Chapter 14). Assistance in risk assessment and in choice of intervention strategy will be elucidated further in the 'Future directions' section of this chapter.

3 Prediction of tail damage

Prediction of tail damage, also termed early detection or early warning of tail damage, has received the greatest attention as a PLF strategy to prevent tail biting. This strategy aims to automatically monitor deviations in pig behaviour, physiology or in environmental parameters that occur prior to the development of serious tail damage (outbreaks). It is likely that several deviating indicators (predictors) will be needed due to the high number of risk factors associated with tail biting. The ultimate result of this strategy will be an alarm warning the farmer of high risk of tail damage in a specific location, making it possible for the farmer to prevent serious tail damage in time. To achieve this result, several steps have to be investigated and completed. First, indicators changing or deviating from their normal pattern prior to tail damage should be identified. Next, methods for automatic monitoring of these indicators should be developed. At last, prediction models can be developed, validated, and implemented based on these automatically monitored indicators. Thus, the last step is highly dependent on the success of, and progress made in the first and second step. Furthermore, the first and second step often demands interdisciplinary collaboration between animal scientists, engineers, and data scientists as it can be difficult to identify deviating indicators without being able to monitor them automatically, such as for indicators of pig behaviour and physiology where manual observations demands high workload and disturbance of the pigs. In the following, the research results obtained within each of these three steps will be summarised and an overview can be seen in Figure 13.2.

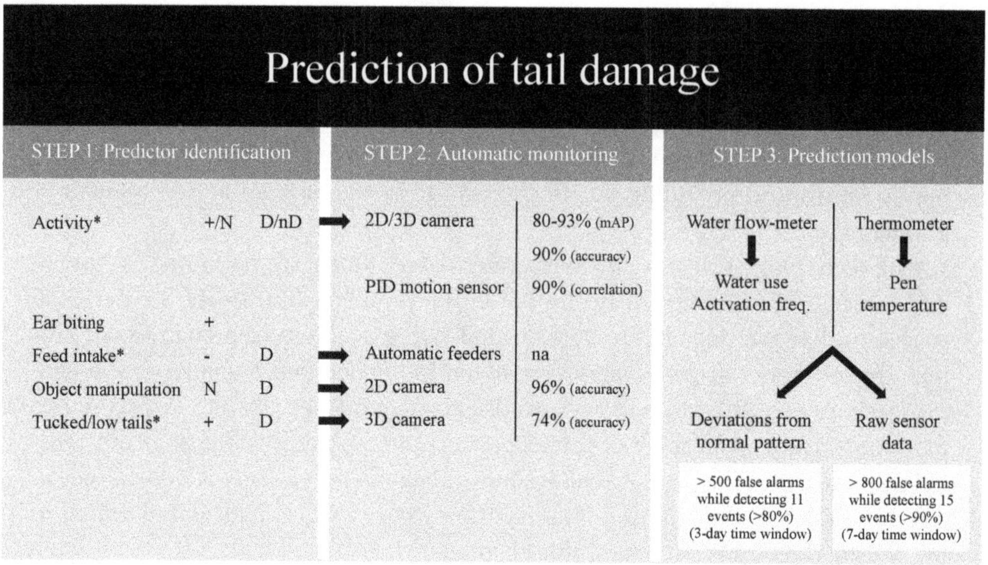

FIGURE 13.2 Simplified overview of research results obtained within each of the three steps to tail damage prediction. + the indicator increased prior to tail damage; – the indicator decreased prior to tail damage; N: the indicator did not change prior to tail damage; D: the indicator differed between pens with and without tail damage; nD: the indicator did not differ between pens with and without tail damage; * the indicator was investigated using automatic monitoring methods.

3.1 *Indicators changing prior to tail damage*

The indicator that seems most obvious to deviate from its normal pattern prior to tail damage is the tail posture of pigs, due to the pain that tail biting may cause to the pig tail. As a result, the pig may hide its tail (tuck the tail between the legs) to avoid further biting, or the pig may have a hanging motionless tail due to boredom/depression, a risk factor of tail biting. Tail posture in pigs has been shown to be affected by e.g., sex, enrichment, and stocking density when no tail damage is present (Larsen *et al.*, 2018; Wallgren *et al.*, 2019) and mixing (D'Eath *et al.*, 2018), although the latter seemed temporally confounded with the development of a tail damage outbreak. All of these are considered risk factors of tail biting and thus, the effects may have been found due to underlying frustration and perhaps even tail biting behaviour being present in the pen. Further, a recent study using remotely monitored tail posture of pigs on several UK commercial herds showed that low tails were associated with high prevalence of lameness and injuries due to aggression (indicated by ear injuries) (D'Eath *et al.*, 2021). Camerlink and Ursinus (2020) clarified in a recent review of the literature that the tail posture of pigs is a complex behaviour and caution is needed when interpreting this behaviour, especially in the case of the hanging tail. Also, tail posture may not be relevant in tail-docked pigs as their tail has a limited mobility. Despite this, tail posture is the most studied indicator for tail damage prediction. When comparing tail damage outbreak pens to pens without tail damage, Wedin *et al.* (2018) found a higher percentage of uncurled tails and tucked tails the day before tail damage was observed. Furthermore, they found an increase in the percentage of tucked tails the last seven days and a numerical higher percentage of uncurled tails was seen these seven days between tail damage and no tail damage pens. Larsen *et al.* (2018) found similar results with a higher percentage of lowered tails (hanging and tucked tails combined) in the tail damage pens the last three days prior to first tail damage, but with no increase in percentage of lowered tails across these three days. Lahrmann *et al.* (2018) also found a higher percentage of lowered tails the last three days prior to tail damage as well as an increase in this percentage from the second last to the last day; although with more pigs affected with tail lesions on the day of the tail damage outbreak. A similar definition of a tail damage outbreak was used by D'Eath *et al.* (2018) observing tail posture remotely from depth images from 3D cameras. They found that the percentage of low tails increased with an increasing rate over time prior to tail damage outbreaks, and with an increase from two weeks to one week prior to the outbreak. Further, they also found a difference between pens with and without a tail damage outbreak the last week prior to the outbreak. The remote monitoring method was used on several UK commercial pig herds with biweekly scoring of tail injuries. The study showed deviations in tail posture related to tail damage. When tails were scored with tail injury, with worsening tail injury and with tail loss, more pigs were detected with low tails and fewer with high tails (D'Eath *et al.*, 2021). All these results suggest that tail posture is a promising indicator for tail damage prediction, but also that tail posture in pigs may deviate from its normal pattern earlier than seven days prior to tail damage, arguing for the importance of automatic monitoring of this indicator. Automatic monitoring will also enable observation of tail posture during undisturbed periods, thereby removing the effect that arousal caused by human presence may have on tail posture of pigs.

Several studies evaluated tail posture as an immediate detector of tail damage for the individual pig. The odds of being scored with a tail wound was almost sixfold higher if the pig was scored with a tucked tail on the same day, even though only 28% of the pigs with a tucked tail were also scored with a tail wound (Larsen *et al.*, 2018). When tail posture was observed at feeding, pigs scored with a wound or an inflamed wound were 4 and 14 times more likely to also have a lowered tail, respectively, but only 55% of wounded tails were detected by a lowered tail (Wallgren *et al.*, 2019). In weaners, the observation of a tucked tail increased the risk of being scored with a tail wound 2–3 days later and the risk was increased even further if the pig was observed with a tucked tail two days in a row; nevertheless only 24% of the wounded tails were predicted and 30% were detected (Zonderland *et al.*, 2009). Furthermore, Wilder *et al.* (2020) found that received tail biting, but not received tail-in-mouth behaviour, increased the probability of a lowered tail; an effect that lasted for at least 80 minutes. These results show that a relation exists between tail posture and tail biting/damage, but also that the use of tail posture as sole indicator will result in many false positive (as well as false negative) tail wound predictions and detections. Thus, tail posture should be combined with other indicators in the case of tail damage prediction.

Other indicators investigated include pig activity/body posture, object manipulation, ear biting and feed intake; all of these are indicators that could be relevant for both tail-docked and undocked pigs, although they have primarily only been investigated in pigs with undocked tails. Larsen *et al.* (2019) found higher activity over the last 5 days prior to tail damage occurring (at least one pig in a pen scored with tail damage), but this difference was seen due to a decrease in activity in the pens without tail damage, making it difficult to use such a deviation for prediction of tail damage. Statham *et al.* (2009) found more pigs standing and less pigs lying inactive and sitting (no difference in lying active or locomotion) the last 4 days prior to tail damage in pens with tail damage compared to pens with no tail damage. Zonderland *et al.* (2011) found an increase in individual pigs' number of posture changes the last 6 days prior to the same individual being tail damaged (3 focal weaner pigs per pen; each pig as its own control), as well as a decrease in time spent lying ventrally and an increase in sitting/kneeling. Wedin *et al.* (2018) found similar results for sitting and lying ventrally the last 7 days prior to tail damage as well as an increase in lying laterally. However, they found no differences in these body postures between pens with and without tail damage, which was also the case in studies by Lahrmann *et al.* (2018) and by Larsen *et al.* (2020) using mean optical flow as a measure of pen level activity. However, mean optical flow increased the last three days prior to tail damage, not only in pens with tail damage but also in control pens without tail damage. A likely interpretation of these somewhat divergent results could be that increased level of activity indicates the presence of a risk factor that potentially may result in tail damage in the most sensitive pens, rather than being a unique predictor of tail damage as such. For manipulative behaviour, Larsen *et al.* (2019) found a general lower level of object manipulation (vertical wooden beams) the last 7 days prior to tail damage in the tail damage pens compared to the pens without tail damage, but no increase or decrease in object manipulation during the 7 days investigated. Also, Zonderland *et al.* (2011) found no change in manipulation of a chain or rubber toy the last 6 days prior to tail damage but did find an increase in received and performed ear biting.

The relation between tail damage and pig activity and manipulative behaviour seem less established than for tail posture. However, the mentioned studies are limited by e.g. low sample sizes, use of focal pigs, low number of observations per day or observations being limited to a specific time period during the day due to the high workload of the manual observations. This again argues for automatic monitoring of these indicators to obtain data in a much larger time window to exploits deviations not only over days but also in the dynamic patterns within days caused by the circadian rhythm. Munsterhjelm *et al.* (2015) did investigate the feed intake of pigs from automatic single-spaced feeders. They found that daily voluntary feed intake of pigs with future tail lesions deviated from the feed intake of other pigs in the pen at least the last 6 days prior to the tail lesion being observed and even further back in older animals and for animals being culled/dying afterwards as a consequence of the tail lesion. A similar investigative approach is currently lacking for the other indicators mentioned above except for tail posture. Based on the currently existing literature, a decrease in individual daily voluntary feed intake and an increase in pigs with lowered tails are the indicators best established as predictors of tail damage, although they will probably not be specific to tail damage (D'Eath *et al.*, 2021; Munsterhjelm *et al.*, 2015).

3.2 *Automatic monitoring of indicators*

Examples of automatic monitoring of indicators have already been mentioned above with the use of 3D cameras and depth images for tail posture detection (D'Eath *et al.*, 2021, 2018), and the use of automatic feeders and ear tag transponders for feed intake monitoring (Munsterhjelm *et al.*, 2015). These studies present the clear advantage of investigating deviations in possible predictors of tail damage by automatic monitoring. Even though low and not-low tails were only classified correctly in 74% of the tail posture detections, they still found an increase in percentage of low tails prior to tail damage outbreaks; this increase was not seen in pens without tail damage outbreaks (D'Eath *et al.*, 2018). Thus, the algorithm for tail posture detection needs improvement, but still show the potential of automatic monitoring in prediction of tail damage, also in commercial conditions (D'Eath *et al.*, 2021).

One algorithm has been developed for automatic detection of object manipulation (Chen *et al.*, 2020), being able to detect three different objects and manipulation of these with an over 96% accuracy. The algorithm calculates the duration of object manipulation on the object level, thus not discriminating between whether one or more pigs are manipulating the object at the same time. The only algorithm developed for monitoring of ear biting is automatic ear lesion detection at the abattoir, thus not monitoring in real-time (Blömke *et al.*, 2020). The algorithm was able to detect 77% of the ear lesions scored, while only 59% of the ears detected as having ear lesions actually had such lesions. However, the first step to develop an automatic detection algorithm for the actual ear biting behaviour has been taken by the development of an ethogram of both the biter and bitten pig including both non-vocal and vocal behaviours (Diana *et al.*, 2019).

Most algorithms developed for welfare monitoring in pigs focus on more general indicators of pig welfare such as pig activity and posture (Larsen *et al.*, 2021), which, as established earlier in this chapter, may also be predictors of tail damage. Thus, such algorithms also seem to be closer to implementation by being tested on several commercial farms and with different camera angles, pen designs, flooring types, lightning conditions, number of pigs and pig ages (Nasirahmadi *et al.*, 2019; Riekert *et al.*, 2020). Nasirahmadi *et al.* (2019) detected standing, sternal lying and ventral lying from 2D camera images of three commercial pig herds, recorded over a 1-year period, with a mean average precision of 93% across the three pig posture classes; indicating that the algorithm performed well in detecting each posture without too many false positive detections. Riekert *et al.* (2020) also used 2D camera images on different pen designs but varied the camera angles and lighting conditions as well based on the natural setup in the herd in different compartments. They detected lying and not-lying postures of pigs (and their position) with a mean average precision of 80%; however the performance decreased significantly when tested on pens that differed in almost every aspect from the pens used to develop the algorithm, and when tested during the night hours (Riekert *et al.*, 2021). Thus, obviously, the algorithm still needs improvement in generalisation to unseen conditions, but this is expected and is a necessary step in the development of such algorithms. Standing pigs can also be detected by 3D cameras and depth images by exploiting the obvious depth difference between standing and lying pigs (Matthews *et al.*, 2017). Pig overall activity can be monitored automatically by simply calculating the number of moving pixels between two consecutive images from camera sensors; this measure has been shown to have a strong positive correlation of 0.92 to activity observed manually by scan sampling (Ott *et al.*, 2014). A more correct representation of pig activity can be obtained with a relatively more complicated method, first locating each pig with an ellipse, then calculating the movement of these ellipses and lastly calculating the number of pigs in locomotion (accuracy = 90%, Kashiha *et al.* (2014)). Low-cost Passive Infrared Detector motion sensors have also been proven effective at measuring pig activity automatically, with a strong positive correlation to human observation of 0.90, although with a slight overestimation (Besteiro *et al.*, 2018b), and with the ability to monitor expected diurnal patterns in activity and variation with pig age, both at room and pen level (Besteiro *et al.*, 2018a; Ni *et al.*, 2017).

Overall, development of algorithms for automatic real-time monitoring has begun for most indicators identified as possible predictors of tail damage, except for ear biting. However, only tail posture, activity and feed intake has been investigated in relation to tail damage through automatic monitoring. Algorithms for automatic monitoring are also being developed for many other indicators as well including aggression, feeding behaviour, drinking behaviour, weight and vocalisations (Larsen *et al.*, 2021), which will hopefully make it possible in the near future to also investigate these as possible predictors of tail damage.

3.3 *Prediction models*

Most of the algorithms developed for automatic monitoring of pig welfare still lack performance evaluation in new conditions and thus, are still far from implementation in commercial pig production (Gómez *et al.*, 2021; Larsen *et al.*, 2021; Schillings *et al.*, 2021). This is probably also the reason why prediction models for tail damage so far are few, and have focused on already available sensor technology including water flow-meters and thermometers (Dominiak *et al.*, 2019a, 2018; Domun *et al.*, 2019; Larsen *et al.*, 2019b). Domun *et al.* (2019) used raw sensor data of water usage,

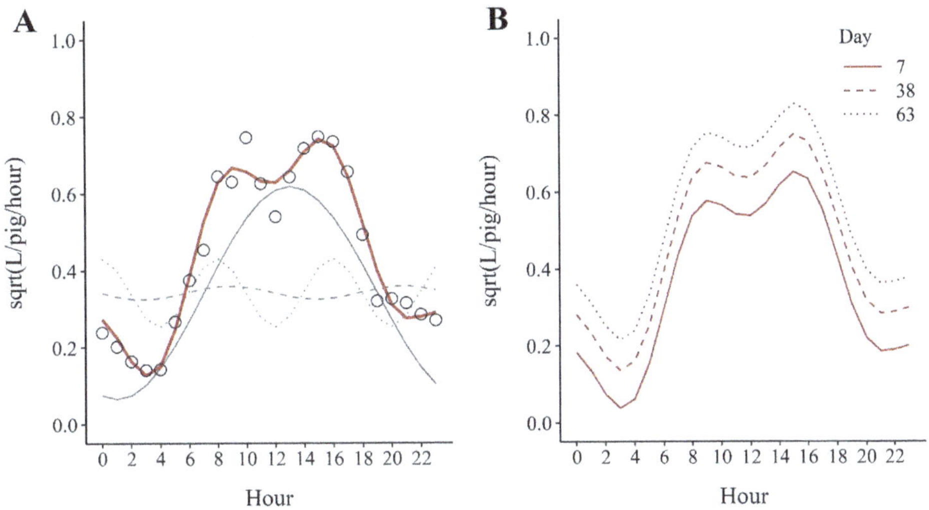

FIGURE 13.3 (A) Diurnal pattern of water usage averaged across all days of the 10 week study period in finisher pigs; circles: mean observation of the raw data for each hour; thin grey lines: the model-estimated three harmonic waves with 24-h (—), 12-h (- - -) and 8-h (·····) cycles; thick black line: the model-estimated sum of the three harmonic waves. (B) Diurnal pattern of water usage on three days during the study period illustrating the modelled growth trend starting from 30 kg pigs. Modified from Larsen *et al.* (2019b)

pen temperature (above solid and slatted flooring) and room climate parameters as input to a deep learning algorithm. The prediction model was able to predict over 90% of the 17 tail biting events with the longest prediction time window of 7 days, but also resulted in more than 800 false alarms. A similar challenge was identified by Larsen *et al.* (2019b) using the same study animals. They also included water usage, activation frequency of the drinking nipple and pen temperature above the solid and slatted flooring as data sources. However, they modelled the normal pattern of these data sources by a linear growth trend across the 10-week study period and the diurnal pattern by three harmonic waves. An example of the modelling approach is shown in Figure 13.3. They then used calculated deviations from this normal pattern, summarised to day level, as inputs to the prediction model. Again, the prediction model detected almost 80% of the 14 tail biting events with a prediction time window of 3 days but resulted in more than 500 false alarms. Dominiak *et al.* (2018) used a similar modelling approach and reported better prediction performance than either of the two models mentioned above, both when tested in known and unknown conditions. However, they did not report on number of detected tail biting events and false alarms and did include the day of the event in the prediction time window.

As indicated in the studies mentioned above, as well as in other studies using similar sensor data (e.g. Dominiak *et al.*, 2019b; Jensen *et al.*, 2017), the use of sensor data from water flow-meters and thermometers at pen level cannot be considered specific to tail biting events. These indicators may also deviate prior to and be predictive of other undesirable events such as pen fouling or diarrhoea, or others that have not yet been investigated. Thus, the non-optimal results with too many false alarms are not surprising. The used sensor data should be investigated in combination with other predictors of tail damage when these can be monitored automatically throughout the day and across the production period.

To summarise, research has been conducted within all three steps to tail damage prediction. However, it is evident from the available literature that the development of prediction models is highly dependent on the success in identifying appropriate predictors (indicators that deviate prior to tail damage) and in developing algorithms for automatic monitoring of these predictors. Such algorithms are still under development and thus, the knowledge of appropriate predictors and how they deviate from their normal pattern prior to tail damage is still deficient. Likely, we will see major progress within this research area in the future as the algorithms for automatic monitoring of many different behaviours in pigs become more accurate and validated in unknown conditions.

4 Challenges in using PLF to address the tail biting problem

Preventing tail biting through the use of PLF technology is not a straightforward task. As evident from the previous section, several predictors may be needed to predict tail damage, due to tail biting being a multifactorial challenge. Thus, multiple sensors may also be needed, decreasing the profitability of PLF as a solution. Focus should be on detecting as many indicators from as few different sensors as possible. However, the choice of sensors is limited as well. Accelerometers (e.g., neck collar, ear tag) have been used as sensors to monitor activity in sows (Chapa *et al.*, 2020), but may be less appropriate in the group-housed, more active, and exploratory growing pig, where the sensors risk destruction and where a sensor per animal may not be profitable. The same may apply to the use of RFID technology for individual recognition of pigs at a specific location in the pen; a technology that has been investigated in relation to e.g., feeding and drinking behaviour of growing pigs (e.g. Adrion *et al.*, 2018; Maselyne *et al.*, 2016). Furthermore, the sensors used need to be designed to be protected from both the pig and the indoor environment with high concentrations of dust and ammonia. Thus, sensors detached from the pigs such as cameras and microphones, or the already investigated thermometers and water flow-meters, need to be enclosed in equipment specifically designed for the purpose.

Another challenge in using PLF to address the tail biting problem is to choose an appropriate gold standard. The gold standard is the state-of-the-art method to determine whether tail biting is occurring and is used as the ground truth during training and testing of the PLF technology (Berckmans, 2013). If the gold standard is wrong, the output of the PLF technology will be wrong. Said in more common terms: 'rubbish in, rubbish out'. Thus, choosing the gold standard is one of the most critical steps in developing PLF technologies, and is a true challenge for tail biting because many different definitions are used in the literature. These range from tail-in-mouth behaviour, to tail damage outbreaks with several pigs scored with tail lesions. The appropriate gold standard for tail biting prevention will depend on the stakeholder in mind. If the aim is to assess the prevalence of tail damage on farms for documentation of on-farm animal welfare (e.g. for animal welfare label certification), the gold standard should measure an unacceptable level according to the specific label. If the stakeholder is the farmer aiming for preventing a degree of tail biting, the gold standard must align with the farmers threshold level for when to intervene, probably being serious tail damage. If the aim is to entirely prevent any tail damage, the first incidence of tail biting behaviour due to frustration responsible for performance of the behaviour would be chosen as the gold standard. Furthermore, it should be possible to measure the gold standard with both high validity and reliability. A valid gold standard measures the intended welfare issue, while a reliable gold standard measures each occurrence of the welfare issue, and the same degree of the welfare issue each time. To ensure consistency and avoid too much subjectivity in recording the gold standard for tail biting when using multiple observers, the authors have previously defined the gold standard as the occurrence of at least one bleeding tail wound combined with daily scoring (Larsen *et al.*, 2019b). However, even this definition can variate in severity as shown in Figure 13.4. Thus, choosing the appropriate gold standard of tail biting continuous to be an unresolved challenge.

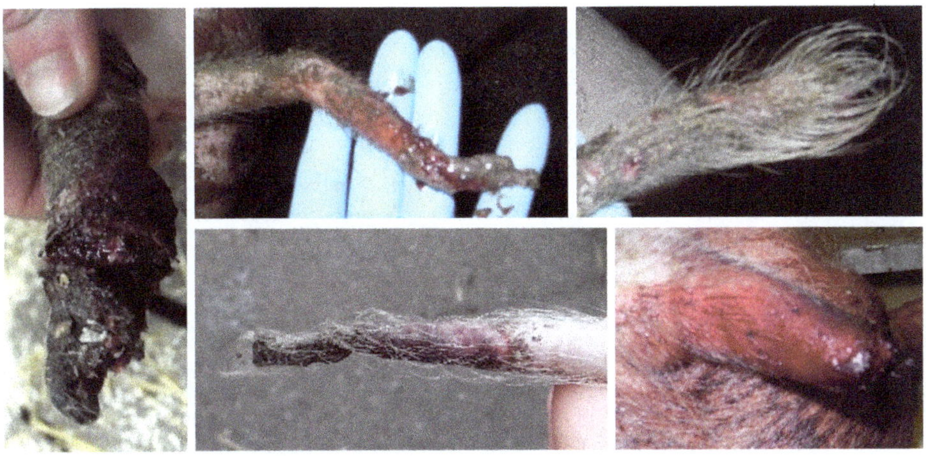

FIGURE 13.4 Examples of severity of the gold standard of tail biting defined as occurrence of at least one bleeding tail wound combined with daily scoring, as used previously by the authors (Larsen *et al.*, 2019b)

5 Future directions

So far, this chapter has focused on the use of PLF to detect tail biting behaviour or to predict tail damage in the near future. While being an important first step in prevention of tail biting and its detrimental consequences, it cannot stand alone. The next step involves the identification of the risk factor(s) responsible for the development of tail biting, and from that knowledge, the choice of the intervention strategy needed in the current case. Only when all these steps have been accounted for will tail biting be prevented from developing further, and possibly costly interventions will only be used when needed. As mentioned earlier, PLF also has potential to assist in this next step. For example, automatic monitoring of object manipulation can inform the farmer about the value assigned by the pigs to the provided objects over time. If manipulation of a provided object decreases at the same time as the detection of tail biting behaviour increases, it may indicate that the risk factor responsible is lack of proper enrichment and the appropriate intervention strategy could be to add the object in greater quantity, to clean the object or to replace the object to regain a high value associated with the objects provided. In the same manner, automatic detection of pig position and posture may indicate

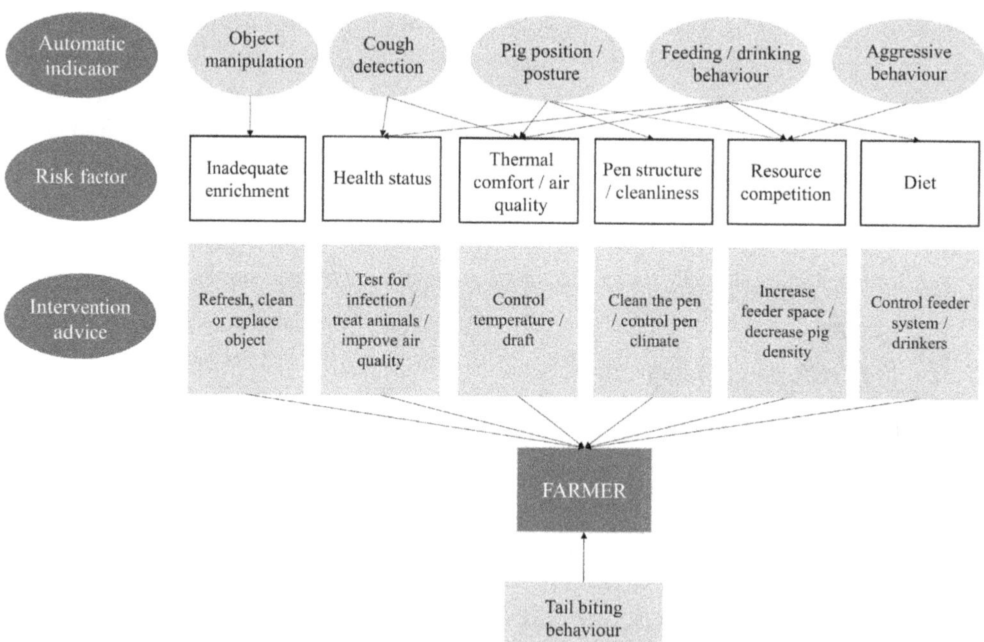

FIGURE 13.5 Example of the design of a decision support system for tail biting prevention providing the farmer with information on the presence of tail biting behaviour as well as advice on intervention strategy based on risk factor analysis conducted from automatically monitored indicators. Shown indicators and intervention advices are only a sample of possible choices to include in such a decision support system

the pigs' perception of the pen climate by the pigs e.g., changing from lying on solid to slatted flooring, changing from lying in sternal to ventral posture, or avoidance of the use of specific pen areas. Thus, changes in pig position and posture at the same time as an increase in tail biting behaviour could identify pen climate as the responsible risk factor; the appropriate intervention strategy could be to adjust/control the temperature, reduce number of pigs in the pen, or take precaution to avoid draft in certain pens. More of such examples can be seen in Figure 13.5. When combining both steps, as illustrated in Figure 13.5, a decision support system based on PLF technologies for tail biting prevention can be developed. Such a decision support system does not exist yet and the development will demand research into the relation between the automatically monitored indicators and the risk factors of tail biting; this research area is limited however by when algorithms for automatic monitoring of the specific indicators have all been developed.

6 Conclusion

The future potential for PLF to assist in prevention of tail biting has been presented, as well as challenges identified in using PLF as a method to address the tail biting problem. PLF can assist by automatic detection of tail biting behaviour or by prediction of tail damage, of which the latter has received the greatest attention. Results on the latter are so far unsatisfying, probably due to the multifactorial nature of tail biting. Furthermore, PLF has future potential to assist in automatic risk factor assessment resulting in advice for the farmer on the most appropriate intervention strategy. However, these potentials highly depend on the development of automatic monitoring methods for pig behaviour, physiology, and environmental parameters. Thus, the use of PLF to address the tail biting problem is currently limited by accuracy and reliability of such methods, a limitation that will hopefully diminish in the future with developments in technology and modelling techniques.

7 Highlights

- PLF could be essential to address multifactorial welfare problems including tail biting.
- PLF can assist through automatic detection of tail biting behaviour and prediction of tail damage from sensor data.
- PLF can also assist through automatic risk factor assessment resulting in advice about most appropriate intervention strategy.
- The field is currently limited by the correctness of the developed automatic methods to monitor pig behaviour and physiology – a limitation that will hopefully diminish with future developments in technology.

References

Adrion, F., Kapun, A., Eckert, F., Holland, E.M., Staiger, M., Götz, S., Gallmann, E., 2018. Monitoring trough visits of growing-finishing pigs with UHF-RFID. Computers and Electronics in Agriculture 144, 144–153. https://doi.org/10.1016/J.COMPAG.2017.11.036.

Berckmans, D., 2017. General introduction to precision livestock farming. Animal Frontiers 7, 6–11. https://doi.org/10.2527/AF.2017.0102.

Berckmans, D., 2013. Basic principles of PLF: gold standard, labelling and field data, in: Precision livestock farming 2013 – papers presented at the 6th European conference on precision livestock farming, ECPLF 2013. pp. 21–29.

Besteiro, R., Arango, T., Rodríguez, M.R., Fernández, M.D., Velo, R., 2018a. Estimation of patterns in weaned piglets' activity using spectral analysis. Biosystems Engineering 173, 85–92. https://doi.org/10.1016/J.BIOSYSTEMSENG.2017.06.014.

Besteiro, R., Rodríguez, M.R., Fernández, M.D., Ortega, J.A., Velo, R., 2018b. Agreement between passive infrared detector measurements and human observations of animal activity. Livestock Science 214, 219–224. https://doi.org/10.1016/J.LIVSCI.2018.06.008.

Blömke, L., Volkmann, N., Kemper, N., 2020. Evaluation of an automated assessment system for ear and tail lesions as animal welfare indicators in pigs at slaughter. Meat Science 159, 107934. https://doi.org/10.1016/J.MEATSCI.2019.107934.

Brünger, J., Dippel, S., Koch, R., Veit, C., 2019. 'Tailception': using neural networks for assessing tail lesions on pictures of pig carcasses. Animal 13, 1030–1036. https://doi.org/10.1017/S1751731118003038.

Camerlink, I., Ursinus, W.W., 2020. Tail postures and tail motion in pigs: a review. Applied Animal Behaviour Science 230, 105079. https://doi.org/10.1016/J.APPLANIM.2020.105079.

Chapa, J.M., Maschat, K., Iwersen, M., Baumgartner, J., Drillich, M., 2020. Accelerometer systems as tools for health and welfare assessment in cattle and pigs – A review. Behavioural Processes 181, 104262. https://doi.org/10.1016/j.beproc.2020.104262.

Chen, C., Zhu, W., Oczak, M., Maschat, K., Baumgartner, J., Larsen, M.L.V., Norton, T., 2020. A computer vision approach for recognition of the engagement of pigs with different enrichment objects. Computers and Electronics in Agriculture 175, 105580. https://doi.org/10.1016/J.COMPAG.2020.105580.

D'Eath, R.B., Foister, S., Jack, M., Bowers, N., Zhu, Q., Barclay, D., Baxter, E.M., 2021. Changes in tail posture detected by a 3D machine vision system are associated with injury from damaging behaviours and ill health on commercial pig farms. PLoS One 16, e0258895. https://doi.org/10.1371/JOURNAL.PONE.0258895.

D'Eath, R.B., Jack, M., Futro, A., Talbot, D., Zhu, Q., Barclay, D., Baxter, E.M., 2018. Automatic early warning of tail biting in pigs: 3D cameras can detect lowered tail posture before an outbreak. PLoS One 13, e0194524. https://doi.org/10.1371/JOURNAL.PONE.0194524.

Dawkins, M.S., 2021. Does smart farming improve or damage animal welfare? Technology and what animals want. Frontiers in Animal Science 2, 736536. https://doi.org/10.3389/FANIM.2021.736536.

Diana, A., Carpentier, L., Piette, D., Boyle, L.A., Berckmans, D., Norton, T., 2019. An ethogram of biter and bitten pigs during an ear biting event: first step in the development of a precision livestock farming tool. Applied Animal Behaviour Science 215, 26–36. https://doi.org/10.1016/J.APPLANIM.2019.03.011.

Dominiak, K.N., Hindsborg, J., Kristensen, A.R., 2019a. Predicting tail biting and diarrhea amongst growing pigs from drinking patterns: an evaluation of the predictive performances of volume and drinking, in: 9th European Conference on Precision Livestock Farming Cork, Ireland, 26–29 August 2019. p. 511.

Dominiak, K.N., Hindsborg, J., Pedersen, L.J., Kristensen, A.R., 2019b. Spatial modeling of pigs' drinking patterns as an alarm reducing method II. Application of a multivariate dynamic linear model. Computers and Electronics in Agriculture 161, 92–103. https://doi.org/10.1016/J.COMPAG.2018.10.037.

Dominiak, K.N., Hindsborg, J., Pedersen, L.J., Kristensen, A.R., 2018. Area-specific predictions of unwanted events using multivariate modeling of water data, in: 2nd International Conference on Agro Big Data and Decision Support Systems in Agriculture. pp. 27–30.

Domun, Y., Pedersen, L.J., White, D., Adeyemi, O., Norton, T., 2019. Learning patterns from time-series data to discriminate predictions of tail-biting, fouling and diarrhoea in pigs. Computers and Electronics in Agriculture 163, 104878. https://doi.org/10.1016/J.COMPAG.2019.104878.

Gómez, Y., Stygar, A.H., Boumans, I.J.M.M., Bokkers, E.A.M., Pedersen, L.J., Niemi, J.K., Pastell, M., Manteca, X., Llonch, P., 2021. A systematic review on validated precision livestock farming technologies for pig production and its potential to assess animal welfare. Frontiers in Veterinary Science 8, 660565. https://doi.org/10.3389/FVETS.2021.660565.

Guzhva, O., Siegford, J.M., Kolstrup, C.L., 2021. The hitchhiker's guide to integration of social and ethical awareness in precision livestock farming research monitoring individual animals using sensors, vision and plf view project behavior and welfare of laying hens in alternative housing systems view. Frontiers in Animal Science 2, 725710. https://doi.org/10.3389/fanim.2021.725710.

Jensen, D.B., Toft, N., Kristensen, A.R., 2017. A multivariate dynamic linear model for early warnings of diarrhea and pen fouling in slaughter pigs. Computers and Electronics in Agriculture 135, 51–62. https://doi.org/10.1016/J.COMPAG.2016.12.018.

Kashiha, M.A., Bahr, C., Ott, S., Moons, C.P.H., Niewold, T.A., Tuyttens, F., Berckmans, D., 2014. Automatic monitoring of pig locomotion using image analysis. Livestock Science 159, 141–148. https://doi.org/10.1016/J.LIVSCI.2013.11.007.

Lahrmann, H.P., Hansen, C.F., D'Eath, R., Busch, M.E., Forkman, B., 2018. Tail posture predicts tail biting outbreaks at pen level in weaner pigs. Applied Animal Behaviour Science 200, 29–35. https://doi.org/10.1016/J.APPLANIM.2017.12.006.

Larsen, M.L.V., Andersen, H.M.L., Pedersen, L.J., 2018. Tail posture as a detector of tail damage and an early detector of tail biting in finishing pigs. Applied Animal Behaviour Science 209, 30–35. https://doi.org/10.1016/J.APPLANIM.2018.08.016.

Larsen, M.L.V., Pedersen, L.J., Edwards, S., Albanie, S., Dawkins, M.S., 2020. Movement change detected by optical flow precedes, but does not predict, tail-biting in pigs. Livestock Science 240, 104136. https://doi.org/10.1016/J.LIVSCI.2020.104136.

Larsen, M.L.V., Andersen, H.M.-L., Pedersen, L.J., 2019a. Changes in activity and object manipulation before tail damage in finisher pigs as an early detector of tail biting. animal 13, 1037–1044. https://doi.org/10.1017/S1751731118002689.

Larsen, M.L.V., Pedersen, L.J., Jensen, D.B., 2019b. Prediction of tail biting events in finisher pigs from automatically recorded sensor data. Animals 9, 458. https://doi.org/10.3390/ANI9070458.

Larsen, M.L.V., Wang, M., Norton, T., 2021. Information technologies for welfare monitoring in pigs and their relation to Welfare Quality®. Sustainability 2021 13, 692. https://doi.org/10.3390/SU13020692.

Liu, D., Oczak, M., Maschat, K., Baumgartner, J., Pletzer, B., He, D., Norton, T., 2020. A computer vision-based method for spatial-temporal action recognition of tail-biting behaviour in group-housed pigs. Biosystems Engineering 195, 27–41. https://doi.org/10.1016/J.BIOSYSTEMSENG.2020.04.007.

Maselyne, J., Adriaens, I., Huybrechts, T., De Ketelaere, B., Millet, S., Vangeyte, J., Van Nuffel, A., Saeys, W., 2016. Measuring the drinking behaviour of individual pigs housed in group using radio frequency identification (RFID). animal 10, 1557–1566. https://doi.org/10.1017/S1751731115000774.

Matthews, S.G., Miller, A.L., PlÖtz, T., Kyriazakis, I., 2017. Automated tracking to measure behavioural changes in pigs for health and welfare monitoring. Scientific Reports 2017 71 7, 1–12. https://doi.org/10.1038/S41598-017-17451-6.

Munsterhjelm, C., Heinonen, M., Valros, A., 2015. Effects of clinical lameness and tail biting lesions on voluntary feed intake in growing pigs. Livestock Science 181, 210–219. https://doi.org/10.1016/J.LIVSCI.2015.09.003.

Nasirahmadi, A., Sturm, B., Edwards, S., Jeppsson, K.-H., Olsson, A.-C., Müller, S., Hensel, O., 2019. Deep learning and machine vision approaches for posture detection of individual pigs. Sensors 19, 3738. https://doi.org/10.3390/S19173738.

Ni, J.Q., Liu, S., Radcliffe, J.S., Vonderohe, C., 2017. Evaluation and characterisation of Passive Infrared Detectors to monitor pig activities in an environmental research building. Biosystems Engineering 158, 86–94. https://doi.org/10.1016/J.BIOSYSTEMSENG.2017.03.014.

Norton, T., Chen, C., Larsen, M.L.V., Berckmans, D., 2019. Review: Precision livestock farming: building 'digital representations' to bring the animals closer to the farmer. animal 13, 3009–3017. https://doi.org/10.1017/S175173111900199X.

Ott, S., Moons, C.P.H., Kashiha, M.A., Bahr, C., Tuyttens, F.A.M., Berckmans, D., Niewold, T.A., 2014. Automated video analysis of pig activity at pen level highly correlates to human observations of behavioural activities. Livestock Science 160, 132–137. https://doi.org/10.1016/J.LIVSCI.2013.12.011.

Riekert, M., Klein, A., Adrion, F., Hoffmann, C., Gallmann, E., 2020. Automatically detecting pig position and posture by 2D camera imaging and deep learning. Computers and Electronics in Agriculture 174, 105391. https://doi.org/10.1016/J.COMPAG.2020.105391.

Riekert, M., Opderbeck, S., Wild, A., Gallmann, E., 2021. Model selection for 24/7 pig position and posture detection by 2D camera imaging and deep learning. Computers and Electronics in Agriculture 187, 106213. https://doi.org/10.1016/J.COMPAG.2021.106213.

Schillings, J., Bennett, R., Rose, D.C., 2021. Exploring the potential of precision livestock farming technologies to help address farm animal welfare. Frontiers in Animal Science. 2, 639678. https://doi.org/10.3389/FANIM.2021.639678.

Statham, P., Green, L., Bichard, M., Mendl, M., 2009. Predicting tail-biting from behaviour of pigs prior to outbreaks. Applied Animal Behaviour Science 121, 157–164. https://doi.org/10.1016/J.APPLANIM.2009.09.011.

Tuyttens, F.A., Molento, C.F., & Benaissa, S. (2022). Twelve threats of precision livestock farming (PLF) for animal welfare. Frontiers in Veterinary Science 9, 889623. https://doi.org/10.3389/fvets.2022.889623.

Wallgren, T., Larsen, A., Gunnarsson, S., 2019. Tail posture as an indicator of tail biting in undocked finishing pigs. Animals 9, 18. https://doi.org/10.3390/ANI9010018.

Wedin, M., Baxter, E.M., Jack, M., Futro, A., D'Eath, R.B., 2018. Early indicators of tail biting outbreaks in pigs. Applied Animal Behaviour Science 208, 7–13. https://doi.org/10.1016/J.APPLANIM.2018.08.008.

Wilder, T., Krieter, J., Kemper, N., Honeck, A., Büttner, K., 2020. Tail-directed behaviour in pigs – relation to tail posture and tail lesion. Applied Animal Behaviour Science 233, 105151. https://doi.org/10.1016/J.APPLANIM.2020.105151.

Zonderland, J.J., Schepers, F., Bracke, M.B.M., Hartog, L.A. den, Kemp, B., Spoolder, H.A.M., 2011. Characteristics of biter and victim piglets apparent before a tail-biting outbreak. animal 5, 767–775. https://doi.org/10.1017/S1751731110002326.

Zonderland, J.J., van Riel, J.W., Bracke, M.B.M., Kemp, B., den Hartog, L.A., Spoolder, H.A.M., 2009. Tail posture predicts tail damage among weaned piglets. Applied Animal Behaviour Science 121, 165–170. https://doi.org/10.1016/J.APPLANIM.2009.09.002.

PART 4

The bigger picture

CHAPTER 14

The tail as an iceberg indicator: interrelationships with welfare problems

Manja Zupan Šemrov[1,*] *and Antonia Patt*[2]

[1]University of Ljubljana, Biotechnical faculty, Department of Animal Science, Groblje 3, 1230 Domžale, Slovenia; [2]Friedrich-Loeffler-Institut, Institute of Animal Welfare and Animal Husbandry, Dörnbergstr. 25/27, 29223 Celle, Germany; *Correspondence: manja.zupansemrov@bf.uni-lj.si

Abstract

This chapter addresses the need for animal welfare indicators that can be easily and quickly implemented. Using the theoretical concept of an iceberg indicator, this chapter develops a model for the assessment of domestic pig welfare that focuses on rearing and fattening pigs. The model assumes that the tail serves as a proxy for tail biting behaviour, which is associated with several aspects of sustained problems in pig production (i.e., underlying risk factors). The proposed complementary tail-based indicators are tail posture, tail lesions, and tail length. They provide an initial qualitative assessment of overall animal welfare status. These indicators are important because there is an urgent need for rapid and reliable protocols to assess pig welfare. Tail-based indicators help identify the extent of underlying welfare problems so that they can be quickly addressed. Although these indicators do not yet allow us to draw conclusions about positive welfare, a curled, intact tail is desirable as it provides an animal-based measure of better/improved welfare relative to other tail conditions and postures. While it may be difficult to fully utilise the model at this time given the high prevalence of pigs with docked tails, it has the potential to be implemented in a variety of pig production systems and across countries to assess and subsequently improve the overall welfare status of commercially housed pigs.

Keywords

pig welfare – welfare assessment – tail biting – tail docking – animal-based indicator

1 What is an iceberg indicator?

Animal welfare indicators are scientific measurements which need to adequately/reliably reflect the welfare status of an animal. They are categorised as animal-based and non-animal-based (i.e. resource-based or management-based) indicators (EFSA, 2012; Presi and Reist, 2011). Animal-based indicators, or outcome measures, show directly how the animal experiences the way it is kept and managed, and cover all indicators that are measured directly on animals, such as changes in behaviour, appearance, health, and physiology. Non-animal-based indicators, or input measures, are variables that are not measured on the animals but in their environment, and demonstrate what the animals are provided with to ensure good welfare and protection. Input measures are relatively easy to measure and provide information on housing systems and management (e.g., space availability, floor type, feeding space, weaning age, or the use of legally permitted mutilations such as castration or tail docking). However whereas input measures are directly linked to the production system the animals are housed in, output measures (i.e. animal-based measures) are less dependent on the production system and therefore are more easily applied across countries and across production systems. Consequently, animal-based indicators are generally considered more informative of animals' experiences than non-animal-based indicators and are the preferred method for assessing welfare.

The ability to summarise many different measures of welfare could be a useful tool for developing strategies to address welfare problems and find solutions. Given the difficulty of implementing more complex protocols for assessing animal welfare, one such tool is associated with the theoretical concept of animal-based "iceberg indicators" (FAWC, 2009). The theory is that a single indicator may reflect or 'capture' or summarise a number of welfare consequences/issues. If the proposed single, or limited number of indicators, provides reliable information on the overall state of animal welfare, assessments could be limited to these specific measures and the time required for assessment thereby reduced accordingly (Heath *et al.*, 2014). Iceberg indicators summarise or are correlated with multiple indicators, i.e., reflect the overall welfare status in an integrative manner, so that variance in animal welfare assessment is preserved. In addition, according to Spoolder *et al.* (2014), an iceberg indicator should enable a quick and comparatively inexpensive welfare assessment without sacrificing accuracy, and therefore be more practical compared to a detailed animal welfare assessment protocol (e.g., Welfare Quality, 2009, ProPIG, 2015, AssureWel, 2020). If the iceberg indicator points to welfare problems, a more detailed welfare assessment scheme could be applied to identify specific challenges.

Welfare Quality® developed welfare assessment protocols for multiple farm animal species relying almost entirely on animal-based measures. Due to its wide use in research and it being a reference point for the continuing development in welfare assessment, we refer to it frequently in this chapter. Animal-based measures are associated with considerable effort in terms of labour (e.g., implementing the Welfare Quality® protocols requires well trained personnel) and time (a complete assessment in growing pigs using the Welfare Quality® protocol takes around 6 hours (Veissier *et al.*, 2013)) such that the range of measures suitable for on-farm use in the context of self-assessment, quality assurance programs, or inspection is much narrower than when assessing under experimental conditions. However, the requirements of welfare assessment schemes vary between stakeholder groups, i.e., the implementation of a detailed assessment protocol might not be necessary in all contexts, for example when assessing compliance with legislation during an official inspection. In light of this, animal welfare assessment methods/protocols adapted to meet specific stakeholder needs emerged in recent years, e.g., Classyfarm in Italy (Alborali and Bertocchi, 2021), Real Welfare in the United Kingdom (Pandolfi *et al.*, 2017), Sikava in Finland (Sikava, 2022) for official inspections in pigs, or BEEP and KTBL guidelines for farmer self-assessment in France (Courboulay, 2020) and Germany (Rauterberg *et al.*, 2021; Zapf *et al.*, 2017). However, these protocols are still very time consuming. This shows the need for evaluation criteria or indicators that are easy and quick to implement for both scientists and stakeholders (e.g., farmers, veterinarians, farm advisors, official or private inspectors, NGOs, slaughterhouses), and that can be used irrespective of the type of production system (e.g., intensive, extensive, organic).

2 Attempts to derive iceberg indicators in pigs and other farm animal species

To date, there have been several attempts to identify potential iceberg indicators in various species. Using the Welfare Quality® protocol, Heath *et al.* (2014) investigated the suitability of individual indicators or subsets thereof in dairy cows. Surprisingly, in this study, Qualitative Behaviour Assessment (QBA, fixed term format included in the Welfare Quality® protocols), an integrative animal-based indicator of positive welfare developed by Wemelsfelder *et al.* (2001), did not facilitate classification of farms as well as, for example, 'absence of prolonged thirst'. Indicators derived from the updated AWIN (Animal Welfare Indicators) protocol for sheep, including QBA, body condition score, fleece cleanliness, a visual indicator assessing anaemia (FAMACHA© score), and flight distance in meters, were proposed as iceberg indicators and were useful in predicting overall welfare status in small extensive sheep farms (Hernandez *et al.*, 2020). Recently, Keeling *et al.* (2021), suggested five potentially suitable indicators for inclusion in a positive welfare protocol for cattle (ear position, play, allogrooming, brush use, and QBA).

In pigs, tail and skin lesions are among the most frequently cited animal-based indicators of welfare, and expert panels recommend their use in fattening pigs (EFSA, 2012). van Staaveren *et al.* (2017) provided some evidence that skin and tail lesions on the carcass are potential iceberg indicators of pig health and welfare on farms. However, they pointed out issues that must be considered before carcass lesions can be used as indicators of pig welfare in a commercial setting. These were to assess the cost of recording carcass lesions during meat inspection, the cost of not identifying problem farms, and the cost of erroneously visiting or penalising non-problem farms. In a recent study by Friedrich *et al.* (2020), iceberg indicators were derived from the Welfare Quality® protocol for sows and piglets. These included panting and shoulder sores for the principle of good housing, mortality and metritis for the principle of good health, and positive behaviour, stereotypies and frothy saliva (a newly proposed indicator for assessing stereotypies), for the principle of appropriate behaviour. In piglets, carpal joint lesions and undersized animals (good feeding), mortality, and sneezing (good health) were identified as indicators of animal welfare. However, the significance of these results is limited given the small sample size of only 13 farms. The aforementioned studies are similar in that they attempt to extract appropriate iceberg indicators from existing assessment schemes, often ending with a set of indicators reflecting different dimensions of welfare. Few studies investigated indicators that are not part of existing protocols, e.g., frothy saliva in sows mentioned above, or tear stains in rearing and fattening pigs (Telkänranta *et al.*, 2016). However, these new indicators are not well established, and therefore the current findings highlight the need for further research to find valid and reliable iceberg indicators for pigs.

3 The tail as an indicator of tail biting and animal welfare status

Although pig welfare has been regulated by EU legislation for 30 years, numerous intractable welfare problems remain. This is reflected in the persistence of the highly prevalent practice of tail docking in many European countries, despite the EU-wide legal ban on routine tail docking in 1994. Tail biting is identified as one of the most serious welfare problems in modern pig production. It is a multifactorial welfare problem affecting a large number of farms (30–70%; EFSA, 2007) and pigs (Nannoni *et al.*, 2014). Tail biting is associated with a variety of internal (e.g., genetic and health) and environmental (including feeding, management, housing conditions) risk factors (e.g., Taylor *et al.*, 2012; for a review, see Chapter 10) though for most, the underlying processes are poorly understood. Nevertheless, potential mechanisms underlying associations between environmental (D'Eath *et al.*, 2014), health (Boyle *et al.*, 2022) and early life (Prunier *et al.*, 2020b) risk factors are proposed. The common denominator is an inability to cope with a situation, or consequences of health limitations, indicating reduced welfare. For the pigs that are bitten (= victims), tail biting is likely to be painful and might result in negative affective states such as fear (Zupan *et al.*, 2012). Other negative impacts of tail biting on production costs are reduced growth (e.g., Sinisalo *et al.*, 2012), increased findings of health and meat quality issues at meat inspection (e.g., vom Brocke *et al.*, 2019), and carcass condemnation (e.g., Harley *et al.*, 2014). Consequently, tail biting indicates reduced welfare in the biter as well as the victim. Because of the large number and different categories of risk factors that vary from farm to farm, tail biting can be considered a promising iceberg indicator that allows conclusions to be drawn about the welfare status of pigs on farm. However, tail biting can be a behaviour of short duration and it is not exhibited regularly during the day (Wilder *et al.*, 2021). Hence, it is easily overlooked by human observers and difficult to assess reliably. An iceberg indicator of pig welfare should be readily observable at all times. Therefore, an intact tail may be the most appropriate iceberg indicator in domestic pigs (*Sus scrofa domesticus*) (FAWC, 2009; EFSA, 2012; Spoolder *et al.*, 2014).

It is due to the high prevalence of tail biting in pig husbandry that tail docking is still common practice in the EU (Nalon and De Briyne, 2019; Nannoni *et al.*, 2014). The demand for an end to docking is mainly a social demand in recent years. Eliminating tail docking would be a step forward for the pig sector in becoming more sustainable. As the tail is innervated, tail docking is painful (Prunier *et al.*, 2020a) and may cause chronic pain (Herskin *et al.*, 2015; Sandercock *et al.*, 2016), making this routinely performed procedure an animal welfare issue in itself. The fact that tail docking is 'likely to cause immediate pain and some prolonged pain' is even acknowledged in the European Directive 2008/120 laying down minimum standards for the protection of pigs (European Union, 2009). In addition, pigs with docked tails are still exposed to the suboptimal conditions (or different categories of risk factors, e.g., Taylor *et al.*, 2012; for review see Chapter 10) which contribute to tail biting (Valros and Heinonen, 2015). Tail biting is extremely challenging for producers/farmers because the high number, type and combination of risk factors vary between farms, making it difficult to determine the best/most effective preventive measures.

Where tail biting occurs, the definition of an 'outbreak' of tail biting varies from study to study, depending on the number of pigs affected, freshness, and severity of lesions (reviewed in Lahrmann, 2018). When lesions and (partial) tail losses, i.e., evidence/proxies of tail biting, are observed, intervention measures can be somewhat limited, although there are some attempts to find scientifically tested solutions (cited below) to reduce or stop outbreaks. Thus, by applying the proposed iceberg indicator model, we ultimately hope that the occurrence of severe tail lesions can be prevented and/or reduced because welfare problems are detected before they reach this stage.

When investigating the effectiveness of tail biting intervention measures once tail biting occurred, the most commonly studied enrichment materials that were tested are straw bedding/rooting substrates, with varying rates of success (reviewed in Henry *et al.*, 2021). Examining a multi-step intervention protocol under commercial housing conditions, Chou *et al.* (2019) found that the success of interventions (removal of biters, removal of victims, provision of three additional ropes) depended on the proportion of biters and victims per pen. The general problem with interventions is that the success of the method used depends on the type of tail biting outbreak, because the cause of a single outbreak can vary from farm to farm and outbreak to outbreak. Furthermore, there are different classifications of tail biting based on differences in biting behaviour (including temporal development) and aetiology: two-stage, sudden and violent, compulsive, or epidemic (Taylor *et al.*, 2010; Valros, 2018), which further complicates the choice of intervention. Therefore, each outbreak requires individual consideration and treatment, i.e., there is discussion on how removal of biters might be more effective for compulsive behaviour compared to other intervention methods (Chou *et al.*, 2019). Given the ongoing difficulties in reliably preventing tail biting (even in docked pigs), research is increasingly complemented by approaches aimed at early detection of tail biting on the farm to enable more effective implementation of intervention measures. Some of these also target the tail (Drexl *et al.*, 2022; Lahrmann *et al.*, 2018; Larsen *et al.*, 2018; Statham *et al.*, 2009; Wedin *et al.*, 2018; Zonderland *et al.*, 2009), with recent use of automatic detection (D'Eath *et al.*, 2018; D'Eath *et al.*, 2021).

4 An iceberg model for domestic pig welfare

In line with aforementioned earlier suggestions (FAWC, 2009; EFSA 2012; Spoolder *et al.*, 2014), we propose an iceberg model for domestic pig welfare in which the tail serves as the indicator that can reflect the animals' overall welfare status. Being associated with a high number and different categories of risk factors, the tail of a pig can be considered a realistic iceberg indicator that reflects various welfare problems (Figure 14.1). When talking about sustainable livestock production, aspects such as animal health, environmental protection, productivity, food safety, food quality, but also cost efficiency of production are interrelated (Buller *et al.*, 2018; Pethick *et al.*, 2011; Velarde *et al.*, 2015), and this model considers all of these aspects either directly or indirectly. This science-based animal welfare model can play an important role in objectively assessing pig welfare, but could also be of interest from the perspective of sustainable pig production.

We base this proposal on previous attempts to quantify the strength and interrelationships of associations between known risk factors, the pig's tail and its welfare (D'Eath *et al.*, 2014, Prunier *et al.*, 2020b, Boyle *et al.*, 2022). The model (Figure 14.1) focuses on rearing and fattening pigs on farm but is independent of the type of production system. Following the EU Common Agricultural Policy – direct payments to farms – and to improve the response of EU agriculture to societal demands for food and health, including animal welfare, this model can help to promote the welfare of pigs on farms by keeping their tails intact. It can also help structure an objective EU system for animal welfare compensation. Our model is also crucial in the context of One Welfare (Pinillos *et al.*, 2016) and the One Health concepts (e.g., Lerner and Berg, 2015), due to the lack of a unified model for both small and large (pig) farms. In general, introducing an animal-based indicator for use by multiple professionals in a variety of different contexts will help to put the pig at the centre of attention.

Animal welfare is multidimensional (Fraser 2008), which is reflected in protocols developed to assess animal welfare, e.g., Welfare Quality® (Keeling *et al.*, 2013), as they use multi-criteria assessment systems to obtain/aggregate the animals' overall welfare status. In our model we propose that for an initial qualitative assessment the overall welfare status can be integrated using only one measure.

Our iceberg model for domestic pig welfare includes a three-stage approach to assessing the problem of tail biting (highlighted as numbers in Figure 14.1). The **tail** is the visible **tip of an iceberg** (event-observed situation) above the water surface, from which three complementary tail-based indicators provide information on the overall welfare status of the pig (first stage: What is the situation?): Tail posture, tail lesions, and tail length (in ascending order of severity). These effectively summarise many risk factors of animal welfare that influence tail biting behaviour, i.e., the iceberg mass below the water surface. Tail biting behaviour is often difficult to observe because the presence of an observer can influence animal behaviour (Martin and Bateson, 1993) and/or because tail biting is a sporadic behaviour of short duration that is easily missed when observation periods are too short (Wilder *et al.*, 2021). This makes it difficult to implement and perform behavioural observations during inspections or daily checks by farmers. Consequently, in this model the overall animal welfare status at a given time (= What is the situation?) can be assessed using the tail as a proxy for tail biting behaviour (second stage: Why it occurs?).

In the model, the combined risk factor load (= sum of all risk factors the pigs are exposed to) of a given situation is represented by icebergs, where the extent of the risk factor load is related to the size of the iceberg i.e., the higher the risk factor load, the bigger the iceberg. The extent to which the tip protrudes above the water surface depends on the overall size of the iceberg, i.e., the bigger the iceberg, the bigger the visible tip. Based on the model of a cumulative risk bucket (Bracke *et al.*, 2012, also cited in Valros, 2018), the presence of risk factors might result in tail biting which becomes observable in the tail-based indicators (posture, lesions, length), i.e., the tip of the iceberg protruding above the water surface. According to this model, observable tail-based indicators infer the current overall welfare status and point towards existing welfare problems, i.e., the invisible part of the iceberg below the water surface, that needs to be addressed to improve the welfare of pigs (third stage: How can it be addressed?). In the context of the model, severity differentiates the tail-based indicators (i.e., posture, lesions, length) in ascending order. Smaller icebergs, i.e., smaller risk factor loads are associated with less severe welfare consequences, but are also likely to result in subtler tail-based indicators, i.e., can more easily be missed if tail-based indicators are not frequently collected. On the other hand, severe tail-based indicators are more easily/readily detectable but are also associated with higher risk factor loads and therefore more severe welfare consequences. In the following sections, we show that there are several correlations between animal-based measures and the measurable tail-based indicators supporting the suitability of the pigs' tail as an iceberg indicator.

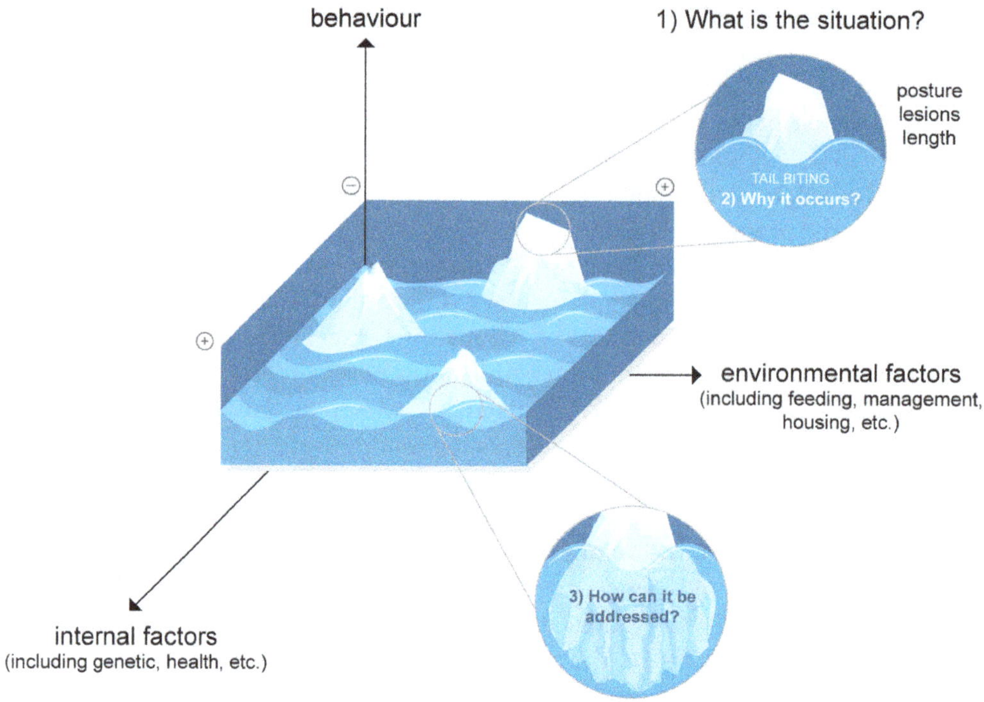

FIGURE 14.1 Iceberg model for domestic pig welfare. Given the multifactorial nature of tail biting, a tail can provide an overall assessment of welfare (first stage = What is the situation?) as observable tail-based indicator of the occurrence of tail biting (second stage = Why it occurs?), i.e., integrate the outcome of welfare when exposed to different combinations of environmental and internal risk factors at a given time. The indicators point towards existing welfare problems, i.e., the invisible part of the iceberg below the water surface that needs to be addressed to improve the welfare of pigs (third stage = How can it be addressed?). The size of the iceberg reflects the overall risk factor load the animals are exposed to at a given time. The specific combination of risk factors determines the icebergs position within the coordinate system. Risk factors have the potential to increase (= negative) or decrease (= positive) the risk of tail biting occurring.

Within the theoretical framework, the relationship between the occurrence of tail biting and the underlying risk factors are visualized by the icebergs' position within the theoretical **3D coordinate system.** The environmental factors (including feeding, management) are positioned on the x-axis ranging from environmental factors known to increase the risk for the occurrence tail biting (= negative) to environmental factors that help to decrease the risk for the occurrence of tail biting (= positive). The internal factors represented on the z-axis, cover/subsume a wide variety of factors, ranging from genetics to health. They are also positioned on the z-axis according to their potential to increase (= negative) or decrease (= positive) the risk for the occurrence of tail biting. The combined risk factor load/specific combination of risk factors at a given time determines the position of the iceberg within the coordinate system, i.e., the interaction between internal and environmental factors. Because different risk factors have different sensitivities to change over time, the time interval over which the size and position of an iceberg can change depends on the specific combination of risk factors. For example, while pig breed, pen design (e.g., access to feeders and drinkers), and litter size from which the animal is sourced may change only slightly over time, health status and feed composition are more likely to change, and access to enrichment materials is even more susceptible to change.

5 Evaluation of the three tail-based indicators

Even when using the tail as a proxy for tail biting, i.e., tail biting is not observed directly, there is variation in the visibility of indicators observed on the tail that reflect tail biting. Especially in the early stages, it may be difficult to detect the tail-based indicators of tail biting. When evaluating tail characteristics as iceberg indicators, we propose three different sub-indicators: tail posture, tail lesions, and tail length/loss. We argue below why they reflect different levels of severity and help to improve welfare when implemented as iceberg indicators.

5.1 *Tail posture*

Tail posture seems to play a dual role in the context of tail biting. Lowered tail posture is observable concurrent with fresh tail lesions (D'Eath *et al.*, 2021). However, changes in tail posture seem to be measurable before tail lesions or tail loss is observed, i.e., it serves as an early indicator of tail biting (D'Eath *et al.*, 2018; D'Eath *et al.*, 2021; Drexl *et al.* 2022; Lahrmann *et al.*, 2018; Larsen *et al.*, 2018; Statham *et al.*, 2009, Wedin *et al.*, 2018; Zonderland *et al.*, 2009). This enables efficient and rapid implementation of intervention measures. Consequently, tail posture is the subtlest of the three proposed tail-based indicators, indicating the lowest severity. Nevertheless, tail posture is likely more easily observed from outside the pen (Honeck *et al.*, 2019), than tail injury. A detailed description of the importance of the different tail postures and the characteristics of the tail muscles is in Chapter 2.

In addition to tail posture, movement of the tail is a potentially interesting indicator of pig welfare. However, Honeck *et al.* (2019) showed large inconsistency in the assessment methodology, making the measurement unreliable. This is also attributed to the fact that most studies examined tail-docked pigs, while no validation was yet performed to verify that the scoring method is applied equally to intact tails and tail-docked tails. In a short tail, it is only possible to score muscle movement at the base of the tail, i.e. tail flicking can still be observed in piglets whose tails were only 25% of their original length (Herskin *et al.*, 2016). The results of Paoli *et al.* (2016) suggest that docking to half the original length does not affect tail posture. Overall, however, docking reduces both the variability and visibility of tail posture and motion.

In research on tail biting predominantly in undocked pigs, the hanging and tucked tail (= low tail posture) is commonly studied. A low tail posture was more frequently observed in aversive housing conditions (no straw, high stocking density: Larsen *et al.*, 2018) and when an aversive event was anticipated (social isolation associated with negative, unpredictable interventions: Reimert *et al.*, 2013). In addition to tail injury and (partial) tail loss, low tail posture was also somewhat unexpectedly associated with other health parameters such as lameness, ear injury, or lesions due to social aggression (D'Eath *et al.*, 2021). Given these results, the authors concluded that it is plausible that low tail posture may be a general indicator of poor welfare. Thus, based on current evidence, tail posture appears to be a more reliable indicator of pig welfare than tail motion. Overall, it can be assumed that the welfare of pigs with a curled, intact tail is better than that of pigs with, for example a tucked tail and/or a tail with lesions or (partial) length loss. However, it is not possible to reliably infer a positive affective state. Considering the different outcomes of hanging tails, further studies are needed to reliably assess the different forms of hanging tails (e.g., relaxed versus tucked) (Camerlink and Ursinus, 2020) to determine the potential of a hanging tail as an indicator of a more positive welfare status.

5.2 *Tail lesions*

Tail lesions occur in all three different types of tail biting (Taylor *et al.*, 2010) and can be scored independently of tail length/loss. The variety of scoring methods is particularly wide when assessing tail lesions (for a review, see Honeck *et al.*, 2019). The level of detail, and thus severity, varies considerably depending on the scoring method used. Consistent with the arguments above, there are limitations with very detailed scoring system as they require trained personnel and are time-consuming, while a very simple scoring system may not detect a problem in time. For example, Chou *et al.* (2019) deliberately chose tail damage easily detected from outside the pen. Such an approach means animals are not disturbed by entering the pen and is likely better suited for assessments in an applied context (e.g., self-assessment, inspection). At the same time, scoring in this way also means that only fresh tail lesions of high severity are identified. In an attempt to work towards a more uniform assessment across EU Member States, the pig sub-group of the EU Platform on Animal Welfare proposed an assessment key (with a focus on assessments at the slaughterhouse) that differentiates between an intact tail, minor injuries and major injuries (EU Platform on Animal Welfare Subgroup Pig, 2019).

Tail damage, i.e., (partial) losses and lesions, is associated with secondary abscesses, bone and joint infections (reviewed by Boyle *et al.*, 2022), with the route of infection presumed to be via the cerebrospinal fluid and/or bloodstream (Huey, 1996). Although any skin penetration can be a portal of entry for bacteria, it is plausible to assume that (partial) tail losses are the most severe lesions, damaging not only connective tissue but also coccygeal vertebrae. In addition, studies reviewed by Boyle *et al.* (2022) show that there is an association between tail and respiratory lesions (pleurisy, pneumonia, and lung abscesses). These lesions, which can lead to partial or complete carcass condemnation at slaughter, are indicative of problems at the individual pig level as well as at the batch or farm level. Therefore, as suggested by the authors, tail lesions observed on the farm may indicate health problems that, if they heal, might not be detected until after the animal is slaughtered and processed. In this regard, the tail on the live animal is an iceberg indicator of current and past health problems.

5.3 *Tail length and tail loss*

When assessing tail length, it is important to consider whether or not the tails of the pigs were docked and, if docked, how short they were docked. When tails are docked (usually a few days after birth), the remaining tail length is reduced to two-thirds or one-half of normal tail length and one-quarter of normal length in finishing pigs (summarised by Sutherland and Tucker, 2011). The average length of intact tails in modern livestock breeds appears to be about 30 cm (Herskin *et al.*, 2015: 30.6 cm ± 0.6 cm; Valros *et al.*, 2020: 31.5 cm ± 2.5 cm; Villain and Ramonet, 2020: 29.7 cm) and is assumed proportional to their weight (Villain and Ramonet, 2020). When assessing tail length, it is difficult to distinguish between docked pigs and pigs with partial tail loss. Indeed, the proportion of the tail that is docked varies between farms, and can range from a few centimetres to more than 0.75% of the tail (Prunier *et al.*, 2020a). In both docked and bitten tails that heal, there is scarring of skin at the end of the tail and the shape of the tail tip is abnormal. However, in the absence of information on docking management, variable tail lengths indicate tail biting, whereas shortened but uniform tail lengths in pigs indicate docking (EU Platform on Animal Welfare Subgroup Pig, 2019). In addition, the presence of unusually shortened tails should be a warning sign of tail biting. Despite the high percentage of docked tails in many European countries, recording docking status is still important as it could indicate that housing conditions are suboptimal for pigs with intact tails. Based on the three different types of tail biting, (partial) tail losses tend to be associated with sudden, violent and compulsive tail biting (Taylor *et al.*, 2010). Consequently, monitoring this type of tail damage could help identify associated risk factors, i.e., potential underlying welfare problems.

6 Improve pig welfare by using the tail as an iceberg indicator

In applying the proposed iceberg model, the starting point is the assessment of the three tail-based indicators. Since the three indicators represent different levels of severity, their prevalence can be used to make an initial qualitative assessment of the overall welfare status of pigs. Any iceberg indicator should therefore be able to identify a threat to pig welfare at an early stage so it can be addressed quickly. Subsequently, for a more detailed understanding of the underlying welfare problems, risk assessments and more comprehensive welfare assessment protocols are applied to improve the situation and work towards a better welfare status. Assuming that the overall welfare status is similar in two farms based on the prevalence of the three tail-based indicators, this means that the size of the iceberg, i.e., the combined risk factor load is similar. However, its position, i.e., the number and combination of risk factors may vary. Before illustrating the application of the model with a theoretical example, it should be noted that the evidence on how some risk factors affect the occurrence of tail biting is sometimes still ambivalent (i.e., their position on the directed x- and z-axes in the model) and that the specific combination of and interaction between risk factors may vary and affect the occurrence of tail biting. Therefore, in our theoretical example, for illustrative purposes, we consider the relative importance of the risk factors derived from the analysis of international datasets performed by EFSA (2014) and from the expert opinion (EFSA 2007), but without claiming that the assumed combined risk factor load would add up as in the example (Figure 14.1). While in one farm the lack of suitable enrichment material is identified as a risk factor, in the other farm it is the provision of nipple drinkers as the only water source, the provision of wet feed, and dirty pens. Thus, icebergs of the same size are in very different positions within the coordinate system. If the model is applied to a farm with only a few risk factors of low relative importance and a farm with several risk factors of high relative importance, the overall assessment of animal welfare will diverge, i.e., they will differ in the prevalence and/or severity of tail-based indicators. Here, the icebergs are not only at different positions in the coordinate system, but are also of different sizes.

Depending on the risk factors identified after applying the iceberg model, there are several ways to apply the model in practise to first assess the welfare status of pigs and then work toward improving it. One example is to use the tail to focus on compliance with minimum legal requirements. Given the still high percentage of tail-docked pigs, focusing on the proposed tail-based indicators could help stakeholders gain the experience necessary to identify welfare problems and slowly transition to keeping pigs with intact, undocked tails. Another example is the use of tail-based indicators to provide incentives, such as through policy support programmes or labelling schemes. A review of case studies on pig labelling schemes as a tool to improve animal welfare found that while labelling often mandates intact tails, there is often a lack of implementation of animal-based indicators, which are more difficult to comply with than resource-based indicators (Sørensen and Schrader, 2019). On the other hand, multi-level labelling schemes and increasing animal welfare standards mean that farmers can continuously improve over time. In Finland, the percentage of intact tails measured during the official meat inspection will serve as the basis for granting an animal welfare subsidy to farmers called 'improving conditions for finishing pigs' from the beginning of 2023. The rationale is that improving housing conditions (closer monitoring of pig welfare, daily addition of manipulable material, space, adequate feeding and feed availability) will reduce tail biting and thus infections and the need to use antibiotics. In Finland, where tail docking is banned, intact tails are considered tails that are not injured/damaged by tail biting. In other words, the two indicators tail length and lesions are iceberg indicators of the overall welfare of finishing pigs (Ruokavirasto, 2022).

7 Conclusions

In this chapter, we propose an iceberg model for rearing and fattening pig welfare in which the tail serves as the iceberg indicator. This model makes two main assumptions: A pig's tail provides reliable information on overall animal welfare status and monitoring of its status mitigates or prevents the occurrence of tail biting by detecting problems related to internal (health, genetics, etc.) and environmental (nutrition, management, housing, etc.) factors in a timely manner. The size of the iceberg represents the cumulative load of underlying risk factors, while the specific combination of environmental and internal risk factors determines its position in the theoretical 3D coordinate system. When tail biting occurs, it can be indicated by three complementary tail-based indicators that reflect the tail's condition (length and/or lesions) or posture, reflecting different severity levels of tail biting. Given the lengthy protocols for assessing pig welfare and the difficulty of fully implementing the proposed model framework at this time, this forward-looking model could be useful as a descriptive, communicative, analytical, and visual tool that can be deployed in various pig production systems and used worldwide.

8 Short summary

- We propose an iceberg model for domestic pig welfare with the pig's tail as the iceberg indicator.
- Monitoring of the status of the pig tail provides an overall assessment of welfare and helps mitigate/prevent the occurrence of tail biting.
- Tail posture, lesions, and length are the proposed tail-based indicators.

Acknowledgement

We thank Françoise Pol of the Ploufragan-Plouzané-Niort Laboratory, EPISABE Unit, ANSES, B.P. 53, Ploufragan, France for her collaboration on the first draft of this chapter. We would like to thank her and Laura Boyle (Teagasc, Pig Development Department, Animal & Grassland Research and Innovation Centre, Moorepark, Ireland) for their comments on the text. Ivana Plohl is thanked for the graphical representation of the model.

References

Alborali, L.G. and Bertocchi, L., 2021. Benessere animale: Linee guida per la categorizzazione del rischio nell'allevamento suino dallo svezzamento all'ingrasso. Available at: http://www.classyfarm.it/.

AssureWel. 2020. AssureWel protocol for finishing pigs and dry sows. Available at: http://www.assurewel.org/pigs.html.

Boyle, L.A., Edwards, S.A., Bolhuis, J.E., Pol, F., Zupan Šemrov, M., Schütze, S., Nordgreen, J., Bozakova, N., Sossidou, E.N. and Valros, A., 2022. The evidence for a causal link between disease and damaging behavior in pigs. Frontiers in Veterinary Science 8: 771682. https://doi.org/10.3389/fvets.2021.771682.

Bracke, M., Vermeer, H., Bokma, M., van der Peet, C., Bolhuis, L. and Leeijen, J., 2012. Checklist aanpak staartbijten bij (biologische) varkens. Available at: https://edepot.wur.nl/220045.

Buller, H., Blokhuis, H., Jensen, P. and Keeling, L., 2018. Towards farm animal welfare and sustainability. Animals 8: 81. https://doi.org/10.3390/ani8060081.

Camerlink, I. and Ursinus, W.W., 2020. Tail postures and tail motion in pigs: A review. Applied Animal Behaviour Science 230: 105079. https://doi.org/10.1016/j.applanim.2020.105079.

Courboulay, V., Meunier-Salaün, M.C., Stankowiak, M. and Pol, F., 2020. BEEP: an advisory pig welfare assessment tool developed by farmers for farmers. Livestock Science, 240: 104107. https://doi.org/10.1016/j.livsci.2020.104107.

Chou, J.Y., O'Driscoll, K., D'Eath, R.B., Sandercock, D.A. and Camerlink, I., 2019. Multi-step tail biting outbreak intervention protocols for pigs housed on slatted floors. Animals 9: 582. https://doi.org/10.3390/ani9080582.

D'Eath, R.B., Foister, S., Jack, M., Bowers, N., Zhu, Q., Barclay, D. and Baxter, E.M., 2021. Changes in tail posture detected by a 3D machine vision system are associated with injury from damaging behaviours and ill health on commercial pig farms. PloS ONE 16: e0258895. https://doi.org/10.1371/journal.pone.0258895.

D'Eath, R.B., Jack, M., Futro, Al, Talbot, D., Zhu, Q., Barclay, D. and Baxter, E.M., 2018. Automatic early warning of tail biting in pigs: 3D cameras can detect lowered tail posture before an outbreak. PLoS ONE 13: e0194524. https://doi.org/10.1371/journal.pone.0194524.

D'Eath, R.B., Arnott, G., Turner, S.P., Jensen, T., Lahrmann, H.P., Busch, M.E., Niemi, J.K., Lawrence, A.B., Sandøe, P., 2014. Injurious tail biting in pigs: how can it be controlled in existing systems without tail docking? Animal 8: 1479–1497. https://doi.org/10.1017/S1751731114001359.

Drexl, V., Dittrich, I., Haase, A., Klingelhöller, H., Diers, S., Krieter, J., 2022. Tail posture as an early indicator of tail biting – a comparison of animal and pen level in weaner pigs. Applied Animal Behaviour Science 252: 105654. https://doi.org/10.1016/j.applanim.2022.105654.

EFSA Panel on Animal Health and Welfare, 2007. Scientific opinion of the Panel on Animal Health and Welfare on a request from commission on the risks associated with tail biting in pigs and possible means to reduce the need for tail docking considering the different housing and husbandry systems. The EFSA Journal 611: 1–13. https://doi.org/10.2903/j.efsa.2007.611.

EFSA Panel on Animal Health and Welfare, 2012. Scientific opinion on the use of animal-based measures to assess welfare in pigs EFSA Journal 10: 2512. https://doi.org/10.2903/j.efsa.2012.2512.

EFSA Panel on Animal Health and Welfare, 2014. Scientific opinion concerning a multifactorial approach on the use of animal and non-animal-based measures to assess the welfare of pigs. The EFSA Journal 12: 3702. https://doi.org/10.2903/j.efsa.2014.3702.

EU Platform on Animal Welfare Subgroup Pigs, 2019. Measuring tail damage at the slaughterhouse and setting common thresholds. Available at: https://edepot.wur.nl/553658.

European Union, 2009. Council Directive 2008/120/EC of 18 December 2008 laying down minimum standards for the protection of pigs (Codified version). Available at: https://eur-lex.europa.eu/legal-content/EN/ALL/?uri=CELEX%3A32008L0120.

FAWC (Farm Animal Welfare Council), 2009. Farm animal welfare in Great Britain: Past, present and future. Farm Animal Welfare Council, London, UK. Available at: https://assets.publishing.service.gov.uk/government/uploads/system/uploads/attachment_data/file/319292/Farm_Animal_Welfare_in_Great_Britain_-_Past__Present_and_Future.pdf.

Fraser, D. 2008. Understanding animal welfare. Acta Veterinaria Scandinavica 50 (Suppl 1):S1 https://doi.org/10.1186/1751-0147-50-S1-S1.

Friedrich, L., Krieter, J., Kemper, N. and Czycholl, I., 2020. Iceberg indicators for sow and piglet welfare. Sustainability 12: 8967. https://doi.org/10.3390/su12218967.

Harley, S., Boyle, L.A., O'Connell, N.E., More, S.J., Teixeira, D.L., Hanlon, A., 2014. Docking the value of pigmeat? Prevalence and financial implications of welfare lesions in Irish slaughter pigs. Animal Welfare 23: 275–285. https://doi.org/10.7120/09627286.23.3.275.

Heath, C.A.E., Browne, W.J., Mullan, S., Main, D.C.J., 2014. Navigating the Iceberg: reducing the number of parameters within the Welfare Quality® Assessment Protocol for dairy cows. Animal 8: 1978–1986. https://doi.org/10.1017/S1751731114002018.

Henry, M., Jansen, H., Amezcua, M.d.R., O'Sullivan, T.L., Niel, L., Shoveller, A.K. and Friendship, R.M., 2021. Tail-biting in pigs: a scoping review. Animals 11: 2022. https://doi.org/10.3390/ani11072002.

Hernandez, R.O., Sánchez, J.A. and Romero, M.H., 2020. Iceberg indicators for animal welfare in rural sheep farms using the five domains model approach. Animals 10: 2273. https://doi.org/10.3390/ani10122273.

Herskin, M.S., Di Giminiani, P. and Thodberg, K., 2016. Effects of administration of a local anaesthetic and/or an NSAID and of docking length on the behaviour of piglets during 5 h after tail docking. Research in Veterinary Science 108: 60–67. https://doi.org/10.1016/j.rvsc.2016.08.001.

Herskin, M.S., Thodberg, K. and Jensen, H.E., 2015. Effects of tail docking and docking length on neuroanatomical changes in healed tail tips of pigs. Animal 9: 677–681. https://doi.org/10.1017/S1751731114002857.

Honeck, A., Gertz, M., Grosse Beilage, E. and Krieter, J., 2019. Comparison of different scoring keys for tail biting in pigs to evaluate the importance of one common scoring key to improve the comparability of studies – a review. Applied Animal Behaviour Science 221: 104873. https://doi.org/10.1016/j.applanim.2019.104873.

Huey, R., 1996. Incidence, location and interrelationships between the sites of abscesses recorded in pigs at a bacon factory in Northern Ireland. Veterinary Record 138: 511–514. https://doi.org/10.1136/vr.138.21.511.

Keeling, L.J., Winckler, C., Hintze, S., Forkmann, B., 2021. Towards a positive welfare protocol for cattle: a critical review of indicators and suggestion of how we might proceed. Frontiers in Animal Science 2:753080. https://doi.org/10.3389/fanim.2021.753080.

Keeling, L., Evans, A., Forkman, B. and Kjaernes, U. 2013. The Welfare Quality® principles and criteria. In: Blokhuis, H.J., Jones, R.B, Veissier, I. and Miele, M. (eds.), Improving farm animal welfare. Science and society working together: the Welfare Quality approach. Wageningen Academic Publishers, Wageningen. 91–114.

Lahrmann, H.J.P., 2018. Tail biting outbreak in pigs: prevalence, early detection and targeted intervention. Ph.D. Thesis, University of Copenhagen, Kobenhavn, Denmark, 2018.

Lahrmann, H.P., Hansen, C.F., D'Eath, R., Busch, M.E. and Forkman, B., 2018. Tail posture predicts tail biting outbreaks at pen level in weaner pigs. Applied Animal Behaviour Science 200: 29–35. https://doi.org/10.1016/j.applanim.2017.12.006.

Larsen, M.L.V., Andersen, H.M.-L. and Pedersen L.J., 2018. Tail posture as a detector of tail damage and early detector of tail biting in finishing pigs. Applied Animal Behaviour Science 209: 30–35. https://doi.org/10.1016/j.applanim.2018.08.016.

Lerner, H. and Berg, C., 2015. The concept of health in One Health and some practical implications for research and education: what is One Health? Infection Ecology & Epidemiology 5: 25300. https://doi.org/10.3402/iee.v5.25300.

Martin, P. and Bateson, P.P.G., 1993. Research design. In: Measuring behaviour. An introductory guide (2nd ed.). Cambridge University Press, Cambridge, UK. 25–55.

Nalon, E. and De Briyne, N., 2019. Efforts to ban the routine tail docking of pigs and to give pigs enrichment materials via EU law: where do we stand a quarter of a century on? Animals 9: 132. https://doi.org/10.3390/ani9040132.

Nannoni, E., Valsami, T., Sardi, L. and Martelli, G., 2014. Tail docking in pigs: a Review on its short- and long-term consequences and effectiveness in preventing tail biting, Italian Journal of Animal Science 13: 1. https://doi.org/10.4081/ijas.2014.3095.

Pandolfi, F., Kyriazakis, I., Stoddart, K., Wainwright, N. and Edwards, S.A., 2017. The "Real Welfare" scheme: identification of risk and protective factors for welfare outcomes in commercial pig farms in the UK. Preventive Veterinary Medicine 146: 34–43. https://doi.org/10.1016/j.prevetmed.2017.07.008.

Paoli, M.A., Lahrmann, H.P., Jensen, T. and D'Eath, R.B., 2016. Behavioural differences between weaner pigs with intact and docked tails. Animal Welfare 25: 287–296. https://doi.org/10.7120/09627286.25.2.287.

Pethick, D.W., Ball, A.J., Banks, R.G. and Hocquette, J.F., 2011. Current and future issues facing red meat quality in a competitive market and how to manage continuous improvement. Animal Production Science 51: 13–18. https://doi.org/10.1071/an10041.

Pinillos, G.R., Appleby, M.C., Manteca, X., Scott-Park, F., Smith C. and Velarde A., 2016. One Welfare – a platform for improving and human and animal welfare. Veterinary Record 179: 412–413. https://doi.org/10.1136/vr.i5470.

Presi, P. and Reist, M., 2011. Review of methodologies applicable to the validation of animal based indicators of welfare. EFSA Supporting Publications 8: 171E. https://doi.org/10.2903/sp.efsa.2011.EN-171.

ProPIG, 2015. Available at: https://orgprints.org/id/eprint/28853/1/Handbook-ProPig.pdf.

Prunier, A., Devillers, N., Herskin, M.S., Sandercock, D.A., Sinclair, A.R.L., Tallet, C. and von Borell, E., 2020a. Husbandry interventions in suckling piglets, painful consequences and mitigation. In: Farmer, C. (ed.) The suckling and weaned piglet. Wageningen Academic Publishers, Wageningen, The Netherlands, pp. 107–138. https://doi.org/10.3920/978-90-8686-894-0.

Prunier, A., Averos, X., Dimitrov, I., Edwards, S.A., Hillmann, E., Holinger, M., Ilieski, V., Leming, R., Tallet, C., Turner, S.P., Zupan, M., Camerlink, I., 2020b. Early life predisposing factors for biting in pigs: review. Animal 14, 3: 570–587. https://doi.org/10.1017/S1751731119001940.

Rauterberg, S., Schrader, L., Zapf, R., Schultheiß, U. and Schubbert, A., 2021. Animal welfare indicators for pigs – guideline and tools for on-farm self-assessment by farmers. In: EAAP scientific committee (ed.) Book of abstracts of the 72nd annual meeting of the European Federation of Animal Science. August 30–September 3, 2021. Wageningen Academic Publishers, Wageningen, The Netherlands, p. 271. https://doi.org/10.3920/978-90-8686-918-3.

Reimert, I., Bolhuis, J.E., Kemp, B. and Rodenburg, T.B., 2013. Indicators of positive and negative emotions and emotional contagion in pigs. Physiology & Behavior 109: 42–50. https://doi.org/10.1016/j.physbeh.2012.11.002.

Ruokavirasto, 2022. Eläinten hyvinvointikorvaukset 2023–2027. Available at: https://www.ruokavirasto.fi/tuet/maatalous/elaintuet/elainten-hyvinvointikorvaus/ehk-siat/EHK-siat-2024/.

Sandercock, D.A., Smith, S.H., Di Giminiani, P. and Edwards, S.A., 2016. Histopathological characterization of tail injury and traumatic neuroma development after tail doking in piglets. Journal of Comparative Pathology 155: 40–49. https://doi.org/10.1016/j.jcpa.2016.05.003.

Sinisalo, A., Niemi, J.K., Heinonen, M. and Valros, A., 2012. Tail biting and production performance in fattening pigs. Livestock Science 143: 220–225. https://doi.org/10.1016/j.livsci.2011.09.019.

Sikava, 2022. Animal Health ETT (introduction in English). Available at: https://www.sikava.fi/PublicContent/IntroductionInEnglish.

Sørensen, J.T. and Schrader, L., 2019. Labelling as a tool for improving animal welfare – the pig case. Agriculture 9: 123. https://doi.org/10.3390/agriculture9060123.

Spoolder, H.A.M., de Jong, I.C., van Niekerk, T.G.C.M., van Reenen, C.G. and Vermeer, H.M., 2014. Iceberg indicators of animal welfare. In: Workshop Meeting of the Welfare Quality Network, Bristol, England, p. 13.

Statham, P., Green, L., Bichard, M. and Mendl, M., 2009. Predicting tail biting from behaviour of pigs prior to outbreaks. Applied Animal Behaviour Science 121: 157–164. https://doi.org/10.1016/j.applanim.2009.09.011.

Sutherland, M.A. and Tucker, C.B., 2011. The long and short of it: a review of tail docking in farm animals. Applied Animal Behaviour Science 135: 179–191. https://doi.org/10.1016/j.applanim.2011.10.015.

Taylor, N.R., Parker, R.M., Mendl, M., Edwards, S.A. and Main, D.C., 2012. Prevalence of risk factors for tail biting on commercial farms and intervention strategies. Veterinary Journal 194: 77–83. https://doi.org/10.1016/j.tvjl.2012.03.004.

Taylor, N.R., Main, D.C., Mendl, M., Edwards, S.A., 2010. Tail-biting: a new perspective. Veterinary Journal 186: 137–47. https://doi.org/10.1016/j.tvjl.2009.08.028.

Telkänranta, H., Marchant, J.N., Valros, A., 2016. Tear staining in pigs: a potential tool for welfare assessment on commercial farms. Animal, 10: 318–325. https://doi.org/10.1017/S175173111500172X.

Valros, A., Välimäki, E., Nordgren, H., Vugts, J., Fàbrega, E. and Heinonen, M., 2020. Intact tails as a welfare indicator in finishing pigs? Scoring of tail lesions and defining intact tails in undocked pigs at the abattoir. Frontiers in Veterinary Science 7: 405. https://doi.org/10.3389/fvets.2020.00405.

Valros, A., 2018. Tail biting. In: Spinka, M. (ed.) Advances in pig welfare. The Advances in farm animal welfare series. Woodhead Publishing, Sawston, UK, pp. 137–166.

Valros, A. and Heinonen, M., 2015. Save the pig tail. Porcine Health Management 1: 2. https://doi.org/10.1186/2055-5660-1-2.

van Staaveren, N., Doyle, B., Manzanilla, E., Calder´on Díaz, J., Hanlon, A., Boyle, L., 2017. Validation of carcass lesions as indicators for on-farm health and welfare of pigs. Journal of Animal Science 95: 1528–1536. https://doi.org/10.2527/jas.2016.1180.

Velarde, A., Fàbrega, E., Blanco-Penedo, I. and Dalmau, A., 2015. Animal welfare towards sustainability in pork meat production. Meat Science 109: 13–17. https://doi.org/10.1016/j.meatsci.2015.05.010.

Villain, N. and Ramonet, Y., 2020. Rearing long-tailed pigs housed on slatted floors. Tail condition and effect of housing enrichment (summary in English). Journées de la Recherche Porcine 52: 385–390.

Veissier, I., Winckler, C., Velarde, A., Butterworth, A., Dalmau, A. and Keeling, L.J., 2013. Development of welfare measures and protocols for the collection of data on farms or at slaughter. In Blokhuis, H.J., Miele, M., Veissier, I., Jones, B. (eds.) Improving farm animal welfare: science and society working together: The welfare quality approach. Wageningen Academic Publishers, Wageningen, The Netherlands.

vom Brocke, A.L., Karnholz, C., Madey-Rindermann, D., Gauly, M., Leeb, C., Winckler, C., Schrader, L., Dippel, S., 2019. Tail lesions in fattening pigs: relationships with postmortem meat inspection and influence of a tail biting management tool. Animal 13: 835–844. https://doi.org/10.1017/S1751731118002070.

Wedin, M., Baxter, E.M., Jack, M.C., Futro, A. and D'Eath, R.B., 2018. Early indicators of tail biting outbreaks in pigs. Applied Animal Behaviour Science 208: 7–13. https://doi.org/10.1016/j.applanim.2018.08.008.

Welfare Quality, 2009. Welfare quality assessment protocol for pigs. Welfare Quality Consortium, Lelystad, Netherlands, p. 120.

Wemelsfelder, F., Hunter, T.E.A., Mendl, M.T. and Lawrence, A.B., 2001. Assessing the 'whole animal': a free choice profiling approach. Animal Behaviour 62: 209–220. https://doi.org/10.1006/anbe.2001.1741.

Wilder, T., Krieter, J., Kemper, N. and Büttner, K., 2021. Observing brief and irregular behaviour of animals – the validity of short observation periods. Applied Sciences 11: 9770. https://doi.org/10.3390/app11209770.

Zapf, R., Schultheiß, U., Knieriem, U., Brinkmann, J. and Schrader, L., 2017. Assessing farm animal welfare – guidelines for on-farm self-assessment. Landtechnik 72: 214–220.

Zonderland, J.J., van Riel, J.W., Bracke, M.B.M., Kemp, B., den Hartog, L.A. and Spoolder, H.A.M., 2009. Tail posture predicts tail damage among weaned piglets. Applied Animal Behaviour Science 121: 165–170. https://doi.org/10.1016/j.applanim.2009.09.002.

Zupan, M., Janczak, A., Framstad, T. and Zanella, A.J., 2012. The effect of biting tails and having tails bitten in pigs. Physiology & Behavior 106: 638–644. https://doi.org/10.1016/j.physbeh.2012.04.025.

CHAPTER 15

The economic and legal aspects of tail biting and tail docking

Jarkko K. Niemi[1,*] *and Peter Stevenson*[2]

[1]Natural Resources Institute Finland (Luke), Bioeconomy and Environment, Kampusranta 9, FI-60320 Seinäjoki, Finland; [2]Chief Policy Advisor, Compassion in World Farming, River Court, Mill Lane, Godalming, Surrey GU7 1EZ, UK; *Correspondence: jarkko.niemi@luke.fi

Abstract

The costs of tail biting lesions across Europe are around €0.6 to €3.4 per finished pig. These costs are mainly caused by additional work, material and medication costs, carcass price discounts, impaired feed conversion efficiency and growth rate of pigs. Losses associated with tail biting lesions are influenced by the incidence and severity of tail biting on the farm. Additional work and inputs are needed also to control tail biting preventively. While low-cost measures can be economically profitable already at low levels of efficacy, high-cost measures require that they can mitigate substantial losses.

Widespread use of tail docking seems to be accepted because the alternative steps that producers are required to take before resorting to it are not specified in detail, and because it seems to be less risky and more profitable to produce docked than undocked pigs. Scientific research and a European Commission Recommendation indicate that when considering the legal obligation to try to prevent tail biting by changing inadequate environmental conditions or management systems, inspectors and pig producers should assess, and consider whether improvements are needed regarding thermal comfort and air quality; health status; diet; competition for space or feed; cleanliness; and the quality and quantity of enrichment material that is provided. If tail biting is still a problem on the farm, farmers should carefully consider what additional improvements are needed to prevent tail biting rather than assuming that no further action is necessary because certain levels, for example of enrichment, air quality, space allowance and health, have already been reached.

Keywords

economic impact – incentives – production costs – regulation

 | DOI:10.3920/9789004694989_017

1 Introduction

Tail biting is an important multifactorial animal welfare and economic problem in pig production. Various risk factors can lead to biting incidents and therefore multiple measures may be used to reduce the problem. To mitigate tail biting, pig producers can adjust farm management practices and the pigs' environment. Examples of such adjustments include the provision of additional pen space, access to point-source objects or long straw and other rooting materials, improvements in floor quality, pig herd health care, ventilation and feed and water supply, changes in genetics used on the farm as well as interventions in the pen once tail biting has been observed (e.g. EFSA, 2007; D'Eath *et al.*, 2014; Taylor *et al.*, 2020).

To limit the incidence of tail biting, most pig producers in Europe dock the tails of piglets. While tail docking reduces tail biting lesions, it does not eliminate tail biting or root causes of it. According to a survey conducted in 24 European countries (De Briyne *et al.*, 2018), 90–100% of pigs in most of these countries were tail-docked. Only in Finland, Norway, Sweden and Switzerland, less than 5% of the pigs were tail-docked, and in five other countries included in the survey the docking rate was below 90%. This is despite the fact that in the European Union (EU) tail docking may not be used routinely and it is allowed only as the last resort to control tail biting (Council Directive 2008/120/EC). In addition, some alternative forms of pig production do not allow tail docking at all.

Regulatory environment of pig farming and economic incentives provided to the farms influence pig producers' decision when they select the preferred approach to control the risk of tail biting. Although tail biting decreases the performance of pigs and causes additional costs, it may be rational for a profit-maximising pig producer to accept certain level of tail biting or to dock the tails rather than to eliminate the root causes of tail biting on the farm. Hence, when evaluating the cost and benefits of tail biting and tail docking, it is critical to take into account diverse impacts of tail biting, tail docking and measures to prevent tail biting.

An analysis of the regulatory environment and financial factors of pig farming can provide useful information which helps decision-makers to understand the choices of pig producers and how the producers can be motivated to enhance pig welfare. The change towards higher-welfare production practices can be incentivised by productivity gains obtained when adopting animal-friendly practices, by a price premium, or by a policy measure such as a subsidy which provides an additional incentive to the pig producer to revise his or her production practices (Niemi, 2020; Niemi *et al.*, 2020), or by imposing changes through regulations. Interventions to mitigate tail biting can be financially prohibitive if extra revenues and cost savings obtained from applying an intervention are smaller than the additional costs incurred because of the mitigation effort.

In this Chapter, the legal and economic considerations related to tail biting and tail docking in Europe are examined (Figure 15.1). Pig producer's decision whether to dock the tails of piglets is influenced by multiple factors, and each factor may

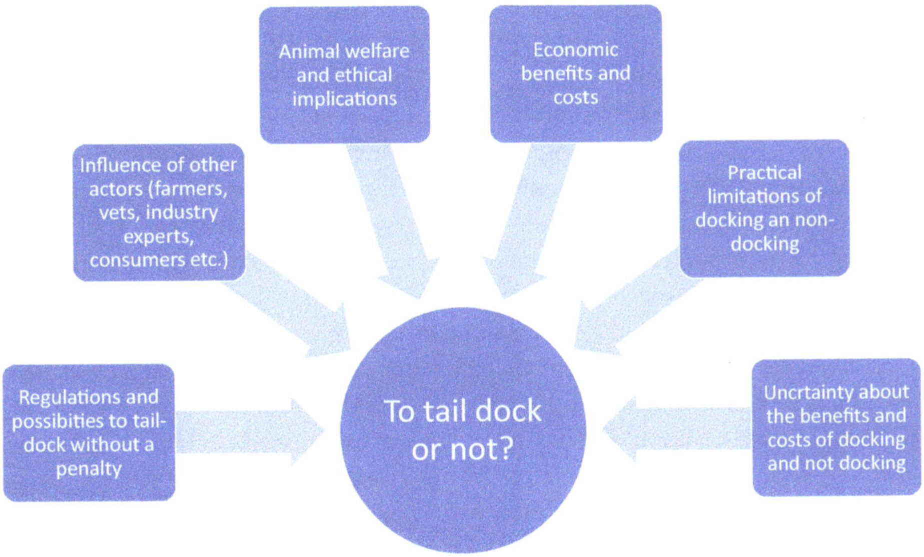

FIGURE 15.1 Overview of factors contributing to pig producer's decision as to whether to dock the tails.

have different weights in the decision-making of an individual producer. Legal regulations set the framework which limits producer's choice. While a farmer may face a prohibition on tail docking, legal provisions that allow docking as a last resort (when the farmer has tried to prevent tail biting by improving conditions on the farm but still has a tail biting problem) may permit tail docking without violating the regulations. Besides animal welfare and ethical concerns, other stakeholders may influence the attitudes and behaviour of farmer, and to how hard he or she is trying to manage without tail docking.

The economic viability of abandoning tail docking may be poor or uncertain which can reduce a producer's incentives to rear pigs with intact tails. In this context, it refers to that the expected return on pig production, when using the current production system and facilities, is lower when the tails are not docked, or that the return is more stable with than without the practice of tail docking (i.e. returns are more robust). In addition, practical limitations of current farming system may make it costly for farmer to abandon tail docking policy as docking may require costly changes in the pig house and production practices, such as lower stocking density or modification in housing.

2 Regulatory framework for tail docking in the EU

2.1 *Prohibition on the routine tail docking of pigs*

A prohibition on the routine tail docking of pigs was included in 1991 in Council Directive 91/630/EEC (European Council, 1991). The wording of the prohibition was amended to strengthen it by Commission Directive 2001/93/EC (European Commission, 2001). Then, in 2008, in order to bring the Directive and its amendments into one place, they were codified in Council Directive 2008/120/EC, hereafter referred to as 'the Pigs Directive' (European Council, 2008).

Annex I, Chapter I, point 8 of the Pigs Directive provides: 'Neither tail-docking nor reduction of corner teeth must be carried out routinely but only where there is evidence that injuries to sows' teats or to other pigs' ears or tails have occurred. Before carrying out these procedures, other measures shall be taken to prevent tail biting and other vices, taking into account environment and stocking densities. For this reason inadequate environmental conditions or management systems must be changed.'

Point 8 does not provide a total ban on tail docking. The prohibition is on routine docking and the wording of point 8 makes it clear that 'routine' docking is docking that is carried out without first trying to prevent tail biting by taking 'other measures' and in particular by improving conditions on the farm: to be specific by changing 'inadequate environmental conditions or management systems'. The effect of this is that farmers do not necessarily have to succeed in preventing tail biting without docking, but they must try very hard to do so.

The requirement to improve conditions before docking is a continuing obligation. If, despite having improved conditions on the farm, farmers still have a tail biting problem, they may tail dock. However, having docked, they must renew their attempts to improve the pigs' conditions, health and diet to get to the point where pigs no longer engage in tail biting and so tail docking is no longer needed. In a 2019 presentation the European Commission (hereafter referred to as 'the Commission') stated: 'As long as tails are being docked, farmers must engage in a continuous cycle of improvement measures to improve on farm conditions' (European Commission, 2019).

2.2 *Requirement to provide enrichment materials*

A requirement to provide enrichment was included in the 1991 Council Directive referred to above, with its wording being strengthened by the 2001 Commission Directive. This wording now appears in Annex I, Chapter I, point 4 of the Pigs Directive which provides: 'pigs must have permanent access to a sufficient quantity of material to enable proper investigation and manipulation activities, such as straw, hay, wood, sawdust, mushroom compost, peat or a mixture of such, which does not compromise the health of the animals'.

Over the years there has been considerable debate as to whether animal welfare legislation should focus on inputs and resources or on outcomes. Point 4 does both. It lays down examples of the materials to be used, but also specifies the required outcome i.e. the ability of pigs to be able to carry out proper investigation and manipulation activities. The word 'proper' guards against the provision of enrichment materials that are inadequate in quality or quantity; 'proper' indicates that pigs must genuinely be able to carry out their natural investigation and manipulation behaviours.

The words 'such as' in point 4 have been given much attention with some arguing that the reference to 'straw, hay, wood, sawdust, mushroom compost, peat' merely give examples and that very different materials can be used including metal chains or plastic toys or objects. This approach ignores the *eiusdem generis* principle of interpreting legislative language. *Eiusdem generis* means 'of the same kind or nature'. Under this approach when general words (e.g. 'such as') are accompanied by enumeration of a specific class of things, the general words must be interpreted to include only items of a similar nature or kind to those that are specifically mentioned.

Accordingly, where a material other than one of those specified in point 4 is used, the term 'such as' means that the material provided must be a natural material and/or must be as effective as those listed in fulfilling the objective of the legislation which is that pigs must be able to engage in 'proper investigation and manipulation activities'.

Points 4 and 8 are closely interwoven as point 4 focuses specifically on enrichment while, as indicated below, the provision of enrichment is also one of the key steps that must be taken under point 8 to try to prevent tail biting by changing 'inadequate environmental conditions or management systems'.

A key question is: what are the 'inadequate environmental conditions or management systems' that must be changed before (if tail biting persists) tail docking may be carried out? Scientific research has identified the factors that are most likely to cause tail biting. In 2007 the European Food Safety Authority (EFSA) produced a report that reviewed the scientific evidence in this area and, in an Opinion based on the report, concluded that 'the largest risk for being tail bitten is the lack of appropriate enrichment' (EFSA, 2007).

The Opinion goes on to state that the principal causal factors of tail biting are: 'the absence of straw, the presence of slatted floors and a barren environment. Absence of straw or a particulate, rootable substrate is an important hazard for tail biting. However, both the amount of straw (full bedding better than limited provision from a rack) and its form (long straw better than chopped) are also of importance. … there is little evidence that provision of toys such as chains, chewing sticks and balls can reduce the risk of tail biting.'

The EFSA Opinion also identified other measures that have been shown to lower the risk of tail biting including: ensuring that each pig has adequate feed intake and avoiding competition for feed; ensuring diet is adequate in salt and essential amino acids, and avoiding sudden change in diet composition, especially to a lower nutrient density; providing access to clean drinking water; avoiding heat or cold stress and high airspeed; ensuring pigs are in good health; avoiding mixing pigs; and removing tail biters and victims from the group.

Further guidance on what measures must be taken to meet the requirement to change 'inadequate environmental conditions or management systems' is provided by the EUWelNet's training tool for EU inspectors in respect of the provision of enrichment materials, tail biting and tail docking (EUWelNet, 2013). This points out that 'Tail biting is a stress-induced disorder. Though a lack of opportunity to explore and forage is a primary motivation for tail biting, other negative environmental and management factors can increase pigs' stress levels and the risk of tail biting occurring'.

2.3 *Commission recommendation*

The scientific and practical understanding of how to prevent tail biting and hence of what steps should be taken to comply with the prohibition of routine tail docking was formalised in 2016 by Commission Recommendation 2016/336 on the application of the Pigs Directive (European Commission, 2016a). This is a recommendation that was formally adopted by the Commission and published in the Official Journal. Though not legally binding, it carries weight in interpreting the requirements of the Pigs Directive on tail docking and the provision of enrichment.

Paragraph 1 of the Recommendation states that when applying the Pigs Directive's requirements on tail docking 'Member States should take account of the best practices guidance based on scientific knowledge set out in paragraphs 2 to 7'.

Paragraph 2 provides that Member States should:

a) ensure that farmers carry out a risk assessment of the incidence of tail-biting
b) establish compliance criteria with the requirements set out in the Pigs Directive.

The recommendation that Member States should establish compliance criteria is extremely helpful as such criteria will help inspectors and farmers know (i) what factors they must consider and (ii) what steps they must take as regards each of these factors in order to meet their obligation to change 'inadequate environmental conditions or management systems' before resorting to tail docking.

Paragraph 3 sets out the factors that should be checked when carrying out the risk assessment. These are the areas that should be considered when assessing what changes must be made under Annex I, Chapter I, point 8 of the Pigs Directive to try to prevent tail biting. Paragraph 3 of the Commission Recommendation states that the following factors should be considered:

- the provision of enrichment materials
- cleanliness
- thermal comfort and air quality
- health status
- competition for food and space
- diet.

Paragraph 3 adds: 'Based on the results of the risk assessment, appropriate management changes in farms should be considered, such as supplying appropriate enrichment materials, comfortable environmental conditions, assuring a good health status and/or providing a balanced diet for pigs'.

Paragraphs 4–6 focus on enrichment materials. Paragraph 4 sets out the characteristics enrichment materials should possess if they are to enable pigs 'to fulfil their essential needs'. It provides that enrichment materials should be:

a) Edible – so that pigs can eat or smell them, preferably with some nutritional benefits;
b) Chewable – so that pigs can bite them;
c) Investigable – so that pigs can investigate them;
d) Manipulable – so that pigs can change their location, appearance or structure.

Paragraph 5 adds that, in addition to possessing these core characteristics enrichment materials should be provided in such a way that they are:

a) of sustainable interest, that is to say, they should encourage the exploratory behaviour of pigs and be regularly replaced and replenished;
b) accessible for oral manipulation;
c) given in sufficient quantity;
d) clean and hygienic.

Paragraphs 4 and 5 take a rigorous approach to the characteristics and provision of enrichment. Clearly the approach taken by a proportion of farmers of simply providing metal chains and/or plastic toys with, in some cases, one or two pieces of wood to be shared by a large number of pigs is inconsistent with Paragraphs 4 and 5.

Paragraph 6 states: 'In order to fulfil pigs' essential needs enrichments material should meet all the characteristics listed in paragraphs 4 and 5'. The inclusion of the word 'all' is important. The characteristics set out in Paragraphs 4 and 5 are not suggestions from which a number may be selected; in the Commission's view all are necessary components for enrichment materials to fulfil the requirements of Points 4 and 8 of Annex I, Chapter I of the Pigs Directive.

Paragraph 6 continues by providing that enrichment materials should be categorised as:

a) optimal materials – materials possessing all the characteristics listed in paragraphs 4 and 5 and therefore such materials can be used alone;
b) suboptimal materials – materials possessing most of the characteristics listed in paragraphs 4 and 5 and therefore such materials should be used in combination with other materials;
c) materials of marginal interest – materials providing distraction for pigs which should not be considered as fulfilling their essential needs and therefore optimal or suboptimal materials should also be provided'.

Paragraph 7 mentions animal and non-animal indicators to monitor the welfare of the pigs.

The Commission Recommendation is accompanied by a Staff Working Document (European Commission, 2016b). This details how to assess enrichment materials and examines (both in its text and in Table I of Annex I) which materials qualify as optimal, which as suboptimal, and which are only of marginal interest. The Staff Working Document goes on to consider the other measures involved in preventing tail biting. In particular it looks at thermal comfort and air quality; health status; diet; and factors that may lead to competition e.g. high stocking density, and inadequate numbers of feeders or drinkers compared to the number of pigs in the group.

2.4 *Open norms*

The Commission Recommendation indicates which environmental conditions or management systems must be taken into account when assessing the changes which should be made to try to prevent tail biting. It states that farmers should consider enrichment, thermal comfort and air quality, competition for food and space, health, diet and cleanliness. However, neither the Recommendation, nor in most cases the Pigs Directive itself, provide specifics; they do not, for example, lay down maximum temperature or ammonia levels or minimum feeder space per pig. Hence, these are 'open norms' i.e. undefined requirements. The use of such open norms can make it difficult for inspectors and farmers to determine whether enough has been done to meet the requirement to try to prevent tail biting by changing environmental conditions or management.

However, the lack of specifics need not be a problem and indeed can be beneficial as it highlights the need for farmers and inspectors to carefully consider what must be done in light of the purpose of the legislation which is to ensure that genuine attempts are made to prevent tail biting by improving conditions on the farm.

For example, one legislative approach would be to provide that, in order to prevent tail biting, a certain minimum amount (cm) of feeder space must be provided per pig. The downside of this is that if the specified amount of feeder space is provided but, despite this, tail biting continues, producers may think that no increase in feeder space is required as they have provided the amount set out in the legislation. In contrast to this, if the required amount of feeder space is not specified and producer provides certain amount of feeder space, but still have a tail biting problem, one of the measures he or she must consider to prevent biting is increasing the amount of feeder space provided.

The Pigs Directive in a separate section from that which prohibits routine tail docking lays down minimum space allowances for pigs. The question arises whether producers who respect these minimum space allowances but still have a tail biting problem must increase the space provided to try to prevent tail biting before they may tail dock. Annex I, Chapter I, point 8 of the Pigs Directive provides that before carrying out docking 'other measures shall be taken to prevent tail-biting and other vices, ***taking into account*** environment and ***stocking densities***' (emphasis added).

The Commission has interpreted this as requiring space allowances to be increased beyond the minimum allowance laid down in the Directive if producers continue to have a tail biting problem. In 2019 the European Commission stated that improvement measures 'have to go beyond the otherwise minimum requirements if this is necessary to reduce tail docking. Improvement measures going beyond minimum standards will have to be taken to get to the point where tail docking is no longer routine' (European Commission, 2019).

EFSA (2007) identified the presence of fully slatted floors as a major risk for tail biting in large part because of the difficulty of providing effective enrichment on such floors. Despite this many farms use fully slatted floors and install such floors when old ones need to be replaced. In this respect, one may conclude that producers using fully slatted floors have not done enough to change inadequate environmental conditions or management systems. However, some countries have adopted regulations which prohibit fully slatted floors in pig houses (see e.g. D'Eath *et al.*, 2015).

3 Socio-economic framework of tail biting and tail docking

3.1 *Prevention of tail biting and tail docking as an economic problem*

In this Section we examine the economics of tail biting and tail docking. From the economic perspective, the question whether to tail dock can be considered as a decision problem where the pig producer selects the optimal measures to mitigate the risk and consequences of tail biting. Hence, tail docking is just one of the options to be considered. When producers are selecting their approach to control tail biting, several factors, such as legal, economic, social, ethical and animal welfare aspects may influence their choice. In economic terms, the choice can be regarded as an evaluation of which approach yields the highest utility to the producer. Because utility is typically measured by the financial result, the producer can be considered to compare the effort to prevent financial consequences of tail biting lesions (the costs) and the amount of consequences that can be prevented (the benefits). If the benefits are higher than the costs, then a rationally behaving producer would choose prevention instead of accepting a higher level of negative consequences. Moreover, the producer may choose to focus on reducing the consequences of tail biting events rather than mitigating tail biting itself.

The benefits are affected by several factors, such as how much the incidence of lesions can be reduced, how severe are the consequences which can be prevented and how uncertain it is that the expected benefits can be obtained. Uncertainty about the benefits may play an important role because the costs of preventive measures must be committed before the benefits are known. Preventive measures can be considered as interventions which reduce either the probability or the severity of tail lesions, and their financial consequences. Some examples of these measures include the provision of rooting materials, point source objects and other enrichments, selection of more robust animals, monitoring animals, provision of additional pen space and solid-floored pens and effort put to ensure good overall animal health in the pig herd, tail docking and a range of other measures (e.g. EFSA, 2007; D'Eath *et al.*, 2014; Taylor *et al.*, 2020). Although tail docking does not eliminate the root cause of tail biting, it may mitigate efficiently the financial consequences of tail biting and therefore it can be considered as a preventive measure.

When evaluating the profitability of preventive measures, the context of decision-making must be taken into account. For example, the provision of enrichments may be effective in one farm but not in another farm where other measures, or a combination of measures, are needed to reduce tail biting lesions. Moreover, measures can have different effects when applied in combinations rather than separately. Therefore, it should be considered whether inputs needed to mitigate tail biting lesions are substitutes. For example, when abandoning a measure makes it economically justified to adopt another measure, these measures can be considered as substitutes. Some inputs can be complements and it may be required that they are applied jointly. For example, applying plenty of long straw may require that pig house characteristics, such as the manure system, are adapted to tolerate long straw. Hence, applying long straw and adjustments in the manure system can be considered as complementary inputs in this context.

The discussion above implies that from an economic perspective, the costs and benefits of tail docking are evaluated against the costs and benefits of alternative ways to mitigate tail biting lesions. Hence, for tail docking not to be applied, there should be other measures which offer more cost-effective ways of mitigating the consequences of tail biting lesions than tail docking.

3.2 *Economic significance of tail biting and preventive measures*

Economically most important farm-level consequences of tail biting are summarised in Figure 15.2. Financial losses resulting from tail biting incidents include additional labour, feed, veterinary and other costs resulting from impaired production performance, treatment and care of bitten pigs and other pigs. These losses are influenced by the incidence and severity of tail biting, which depend on the context, such as which risk factors of tail biting are present on the farm. Revenues are also lost mainly because of reduced growth and increased mortality risk of bitten pigs. These effects lead to increased fixed costs per pig, although fixed costs per farm may not change. These effects likely also lead to impaired feed conversion ratio as slower growing pigs may use a higher proportion of their diet for body maintenance functions. In addition to tail biting itself, the measures used to prevent tail biting are costly.

The benefits of preventive measures are because of avoided cost burden of tail biting. Because tail biting is a multifactorial disorder and the efficacy of preventive measures may vary from case to case, these benefits may be uncertain. The costs of preventive measures include for example material and labour costs associated with the distribution, maintenance and cleaning of rooting materials, point source

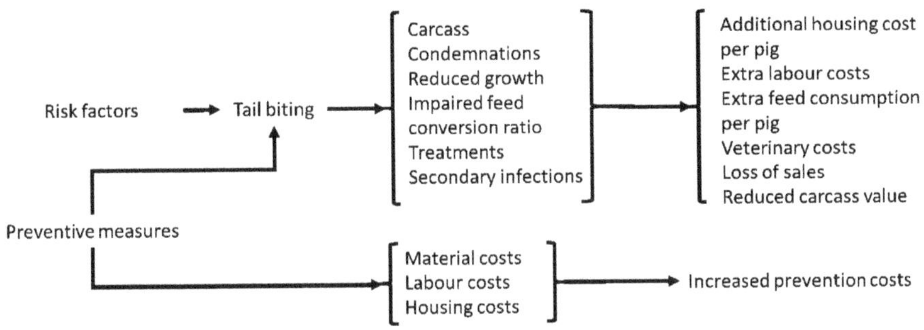

FIGURE 15.2 Economically most important farm-level consequences of tail biting and measures to prevent tail biting

objects and other enrichments, monitoring costs, and housing costs associated for example with the provision of additional pen space for pigs. Because tail docking is used to reduce losses caused by tail biting, the economic benefits of tail docking are 'avoided costs of tail biting'.

Previous studies suggest that the costs of tail biting lesions across Europe are around €2.0 (± €1.4) per finished pig, which corresponds to about 1–3% of the sale value of a pig when averaged across all pigs sold (Niemi *et al.*, 2021). Individual studies have reported a range of different estimates. A study from the UK reported tail biting lesions to cost the UK pig industry £3.51 million per year (Ref. Moinard *et al.*, 2003). A Dutch study estimated the costs of tail biting at about €2,400 per year for a typical finishing herd in the Netherlands, when 2% of pigs suffered from tail biting lesions (Zonderland *et al.*, 2011). D'Eath *et al.* (2016) reported €18.96 loss per victim of tail biting in Denmark. While Harley *et al.* (2014) found the economic costs of biting at €1.69 per produced pig, Niemi *et al.* (2021) reported a loss of up to €2.3 per slaughtered pig when the prevalence of tail biting lesions was on average 10%, and van Staaveren *et al.* (2020) found that the mean annual net profit decreased by €1.1 in high severe tail lesion Irish farms compared with low severe tail lesion farms.

Based on the studies mentioned above, economic losses due to tail biting are mainly due to extra work, material and medication costs and carcass price discounts associated with tail biting. In addition, impaired feed conversion ratio and growth rate of pigs can cause substantial economic losses, depending on the case. Besides the tail biting itself, it is important to recognize that additional work and inputs are needed also to control the problem preventatively.

3.3 *Factors contributing to the economic impacts of tail biting*

Several factors contribute to the losses of tail biting, and to the losses that could be avoided by applying preventive measures. The probability of occurrence of tail biting influences the benefits that can be obtained by tail docking and other preventive measures. The prevalence varies across countries and farms (e.g. Schrøder-Petersen and Simonsen, 2001; EFSA 2007; Hanley *et al.*, 2012; Sanchez-Vazquez *et al.* 2012;

vom Brocke *et al.* 2019; Valros *et al.*, 2020; Taylor *et al.* 2020). Official meat inspection records have typically reported lower prevalence levels than observational studies on farms or at slaughterlines. Meat inspection records often contain only cases which can compromise meat hygiene, whereas data collected for specific research can be more detailed and may also include less severe cases. However, less severe cases can also have economic consequences, for example, because of pigs must be treated or because they impact the production performance of pigs.

Studies provide information on the impacts of tail biting lesions on the production performance of growing pigs. A set of studies are summarized in Table 15.1. Tail biting lesions have been associated with up to 10% reduction in the average daily gain of growing pigs. However, the impact can be larger because production performance before the biting incidence can be considered unaffected by tail biting. For example, while Wallgren and Lindahl (1996) reported a 5% reduction in the average daily gain of tail bitten pigs, the reduction in the growth rate in the period between the biting incident and the time of slaughter was on average 11%. Moreover, part of the observed poor growth may be caused by lower genetic growth potential of the victims of tail biting. Sinisalo *et al.* (2012) suggested, that non-victims had 33.4 g/d greater average daily gain than victims when the difference was measured by using phenotypes, but only 10.8 g/d difference was measured when the genetic merit of each pig was taken into account.

While pigs with tail lesions are likely to consume less feed than pigs with intact tails (Wallenbeck *et al.*, 2013; Munsterhjelm *et al.* 2015), this may indicate a reduced feed conversion ratio. Van Staaveren *et al.* (2020) found that in a herd with more than 0.86% prevalence of severe tail lesions, 3.6% more weaner and 1.4% more finisher feed per year per pig was consumed and feed costs were €1.1 per pig higher than in a herd with less than 0.86% prevalence of severe tail lesions. Sinisalo *et al.* (2012) by contrast suggested that the impaired feed conversion ratio in tail bitten pigs may be associated also with the poorer genetics of bitten pigs rather than tail biting itself.

It is to some extent unclear how much tail biting influences pig mortality, because both no effect and positive impacts have been reported (Wallgren and Lindahl, 1996; England *et al.*, 1969; Nannoni 2014; McGlone *et al.*, 1990; see also Table 15.1). The severity of tail lesions may influence mortality, and the most severe lesions can be expected to raise mortality rates. Mortality caused by tail biting nevertheless has substantial economic consequences because the market value of the pig will be lost and in addition the carcass rendering will incur costs. Van Staaveren *et al.* (2021) considered that variable costs were increased due to a 2.4% increase in carcass disposal costs of dead animals.

Secondary diseases occurring after tail biting can have a major impact both the risk and the costs of tail biting. The costs of secondary infections can also be counted as costs related to tail biting. Tail bitten pigs are more likely to suffer from pleuritis and lung abscesses (Kritas and Morrison, 2007), and the additional costs of respiratory disorders in the victims of tail biting can cause an additional cost of several euros per finished pig (Niemi, 2021). Moreover, Niemi *et al.* (2012b) reported an odds ratio of 3.4 for the risk for locomotory disorders after a pig had been observed to have a bitten tail.

Studies suggest that pigs having tail lesions are subject to increased carcass condemnations at slaughter. The magnitude of effect however appears to vary from case to case as illustrated by Table 15.1, and only a few studies have indicated the economic value of elevated condemnations. Harley *et al.* (2014) found that tail lesions were significantly associated with carcass trimming, and that there was a negative effect of tail lesion severity on carcass weight. They estimated that on average €1.10 per slaughtered pig was lost because of carcass condemnations and trimmings, and an additional €0.59 per pig was lost because of lower carcass weight associated with tail lesions. In a Northern Ireland study, Harley *et al.* (2012) estimated, that

TABLE 15.1 Associations between tail biting and production parameters reported in selected studies

Parameter	Association	Sources
Average daily gain	1–3%	Sinisalo *et al.* (2012)
	5%	Wallgren & Lindahl (1996)
	5.5%	Camerlink *et al.* (2012)
	5%	van Staaveren *et al.* (2021)
	0–10%	Marques *et al.* (2012)
Feed conversion ratio	0%	Sinisalo *et al.* (2012)
Feed usage per pig	1.4%–3.6%	van Staaveren *et al.* (2021)
Chance in mortality percentage	+ 5.8%	Marques *et al.* (2012)
	3.9-fold[a]	Based on Moinard *et al.* (2003)
	~9–10%[a]	Calculated based on Heinonen *et al.* (2021)
Risk of locomotory disorders	3.4-fold	Niemi *et al.* (2012b)
	1.4 to 17.8-fold	Marques *et al.* (2012)
Risk of respiratory disorders or lung lesions	7.6-fold	Niemi *et al.* (2012b)
	2.8 to 11.2-fold	Marques *et al.* (2012)
Change in the percentage of condemned carcasses	26.4–41.5%	Marques *et al.* (2012)
	~7–20%	Calculated based on Heinonen *et al.* (2021)
	0.3–1.0%	Franco *et al.* (2021)
	5.4%	Valros *et al.* (2004)

a An association between mortality and tail biting indicators.

on average €32.12 was lost per condemned tail bitten carcass, which equalled to an average loss of €0.37 per pig slaughtered. Entire condemnations and abscesses constituted 77.3% and 24.6% of the losses, respectively, and 1.07% of all carcasses were condemned.

3.4 *Costs of preventive measures*

Because many risk factors for tail biting have been identified, also several measures to mitigate the risk of tail biting are known. Niemi *et al.* (2021) studied the costs and benefits of seven different measures to mitigate tail biting. They considered scenarios where the measure of straw provision to pigs cost €1.67–€4.17 per pig, improvements in pig house ventilation €0.35, genetic changes up to €2.7 and increased space allowance up to €3.72, provision of point-source objects up to €1.78, improvements in herd health €1.06–€3.41 and precision livestock farming based early warning systems cost €0.80–€6.00 per pig. Their results illustrated that the economic viability of preventive measures was dependent on the risk of tail biting without the preventive measure, the effect of the preventive measure on the occurrence of tail biting and the costs of implementing the preventive measure. Studies have reported levels of prevalence of tail biting lesions which are high enough, in some cases, to justify the more costly measures mentioned above. Measures which were considered the least expensive to apply, such as provision of point-source enrichment objects, or which provided wider production benefits, such as improvements in ventilation and herd health, became profitable at a lower level of efficacy than measures which were considered the most expensive to apply (e.g., straw provision, increased space allowance, automated early warning systems). Measures which were considered most efficient in reducing the risk of tail biting lesions, such as straw provision, can be cost-effective in preventing tail biting, especially when the risk of tail biting is high. At lower risk levels, the provision of point-source objects and other less costly but relatively effective measures could also play an important role. (Niemi *et al.*, 2021).

Taking into account risk factors present at the farm is critical when evaluating the profitability of a preventive measure, because preventive measure should address issues which are increasing the risk of tail biting at the farms. Studies have empahsized the importance of keeping the implementation costs of measures low when the measures are repeated daily (D'Eath *et al.*, 2016; Niemi *et al.*, 2021). Based on Niemi *et al.* (2021), improvements in ventilation, genetic change or provision of point-source enrichment objects can be financially viable, even when either the prevalence of tail biting lesions before implementing a measure or the efficacy of the measure is low, because these measures can be quite inexpensive to adopt. For farms struggling with more severe outbreaks of tail biting lesions, straw provision, improvement in herd health, increased space allowance by 0.2 m^2 or investment into an early warning system could be profitable if their costs were kept at a reasonable level and they were targeted to the problem that was causing elevated risk of tail biting lesions.

Taylor *et al.* (2016) suggested that tail biting can be expressed in three different descriptive behavioural types, namely, 'two-stage', 'sudden-forceful' and 'obsessive', each of which may have different motivational bases among pigs. Tail biting can emerge over time and the time between successive cases of tail biting in a pen is the shorter the more cases of tail biting have been observed in the pen (Niemi *et al.*, 2011; 2021). Therefore, effort taken to monitor the risk of biting and the immediacy of measures take to respond to the observed biting incidents may contribute to the costs of tail biting in pigs, although the time window of intervention may vary depending on the type of biting. As tail docking reduces the consequences of tail biting and automated monitoring systems may be costly to adopt, the immediacy of response to observations in the pen may have less relevance when rearing tail docked pigs compared to undocked pigs.

3.5 *Financial implications of terminating tail docking*

Tail docking itself incurs costs, because labour input and materials are consumed during the tail docking procedure. However, the cost of tail docking procedure itself is rather small (D'Eath *et al.*, 2015) and therefore it has only a minor direct economic contribution to the profit of pig rearing. The consequential effects of tail biting lesions have a more prominent contribution. Moreover, some studies have reported that tail docking and other mutilations may reduce the growth of piglets (Marchant-Forde *et al.*, 2009; Zhou *et al.*, 2013).

Tail biting lesions appears to be less common in docked pig populations than in undocked populations. This concerns the prevalence and severity of lesions, and economic consequences of tail biting. Studies (for example, McGlone *et al.*, 1990; Sutherland *et al.*, 2009) have found tail-docked pigs to have less tail lesions than intact tailed pigs. For example, Hunter *et al.* (2001) found that 2.4% of docked pigs and 8.5% of long-tailed (undocked and tipped pigs) in their sample showed evidence of tail biting. More recently, Lahrmann *et al.* (2017) found no clinical signs of injured tails within the tail docked group of pigs, whereas 23.0% of the undocked pigs had a tail lesion, and that the removal of pigs to a hospital pen was more likely in undocked pigs' pens (47.7% undocked pigs' pens and 22.9% docked pigs' pens).

Although tail docking does not eliminate the root causes of tail biting, it helps to control tail biting behavior among pigs. Tail-docked pigs can be reared in more barren conditions than undocked pigs. This may lead to a lower production costs per docked pig when compared to undocked pigs. For example, tail docking may facilitate both a lower space allowance and a lower usage of enrichment materials per pig and therefore lower fixed and enrichment costs per pig than not docking. The risk of tail biting and the space allowance are known to be nonlinearly associated to each other, so that providing a pig additional space at low space allowance levels reduces tail biting more effectively than at high space allowance levels. As pointed out by Haigh and O'Driscoll (2019), in countries where docking is completely prohibited (e.g. Finland, Sweden, Norway, Switzerland), maximum stocking densities for commercial pigs are lower than the EU minimum requirement. This could be one of the reasons why for example Finnish pig producers ranked stocking density lower among studied risk factors of tail biting than producers in other countries (Haigh and O'Driscoll, 2019).

D'Eath *et al.* (2016) compared four scenarios related to tail docking. 'Standard Docked' was a conventional housing scenario with tail docking meeting the recommendations for Danish production (space allowance 0.7 m^2/pig). 'Standard Undocked' was similar to 'Standard Docked' but with no tail docking. 'Efficient Undocked' and 'Enhanced Undocked' scenarios did not tail dock and had an increased floor area (0.9 and 1.0 m^2/pig, respectively) and provision of loose manipulable materials (100 and 200 g/straw per pig per day). They considered the likelihood of tail biting outbreak in a Standard Docked pen at 13.4%. In Standard Undocked pen the likelihood was 57% and an outbreak was expected to be larger in terms of the number of bitten pigs. In two other scenarios the likelihood of a tail biting outbreak was 27%. (D'Eath *et al.*, 2016)

Figure 15.3 illustrates the results of simulation scenarios RS1 of D'Eath *et al.* (2016). Given the setup of their scenarios, Standard docked practice was the most robust choice among the four scenarios. This implied that, under their assumptions, rearing docked pigs resulted in both the highest average return and the lowest variation of returns per pig. This implies that a profit-maximising, risk-averse pig producer would prefer tail docking over not tail docking.

The results of D'Eath *et al.* (2016) also show that rearing undocked pigs can be implemented in different ways which can lead to diverse financial outcomes. When discontinuing tail docking, it is critical to recognise, that small changes in production practices may be inadequate to successfully rear undocked pigs. For example, Larsen *et al.* (2018) pointed out that a combination of straw provision and lowered stocking density showed a similar hazard of tail damage as seen with only tail docking, which suggest adjusting simultaneously multiple features of the housing and management system may be required to achieve successful rearing results. Terminating tail docking must therefore be analysed as a system-level change.

An important reason for the use of tail docking is that it may allow the pig producer to eliminate the negative consequences of tail biting in intensive rearing conditions in an 'easy' way. Alternative solutions to mitigate tail biting may be considered complex, their benefits can be uncertain and the measures may be costly because of higher incidence of tail biting lesions and higher cost of adopting these measures.

Terminating the practice of tail docking likely requires that pigs are provided additional pen space and manipulable materials, point-source objects or other

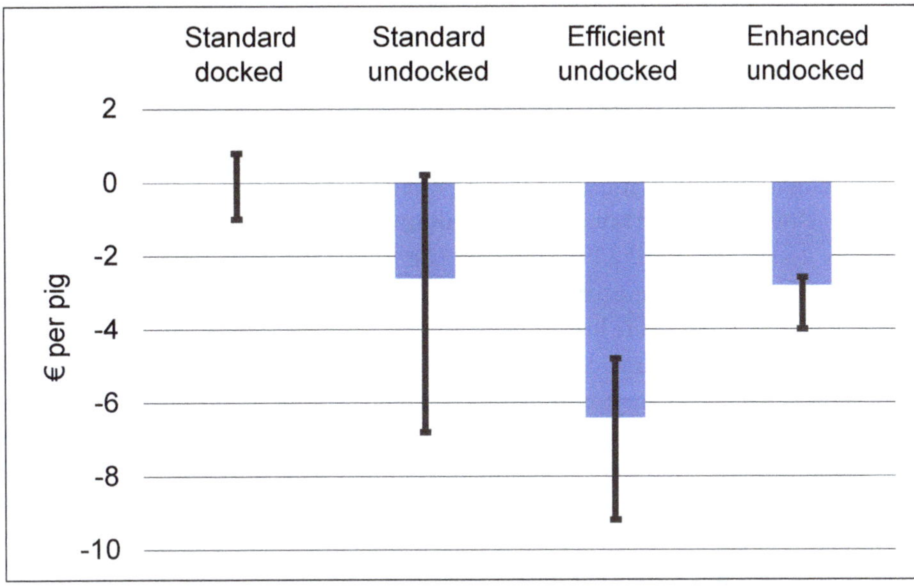

FIGURE 15.3 Impact of variation in the occurrence of tail biting and the adoption of three undocking policies on economic return per pig when compared to the Standard docked system (€0 = average return for the Standard docked system). Blue bars represent the average impact and thin black bars represent variation of simulated economic returns around the average impact.
SOURCE: CALCULATED BY USING DATA OF D'EATH *ET AL.* (2016, SIMULATION RS1)

enrichments are given in adequate amounts. However, terminating tail docking so that the risk of tail biting is kept under control may simultaneously reduce other costs, such as losses arising from respiratory and locomotory disorders and other animal health issues which are associated with the occurrence of tail biting. If such co-benefits can be obtained, then substantial benefits may be available. Additional space allowance may also facilitate higher slaughter weights. While the cost and revenue structure of rearing undocked pigs may be different from that of docked pigs, it does not necessarily imply that tail docking is economically the more viable option.

Besides production costs, terminating the practice of tail docking can be incentivised by public policies, such as animal welfare compensation paid for undocked pigs, or by market-based incentives such as possible price premium that may be available when supplying undocked pigs to the buyers. For example, some high animal welfare marketing systems pay attention to the absence of tail docking, to tail condition or to factors which can mitigate tail biting (e.g. Heinola *et al.*, 2020; Stygar *et al.*, 2022). However, the contribution of such systems to the profitability of pig farming is not well documented.

4 Conclusion

The costs of tail biting lesions across Europe are around €0.6 to €3.4 per finished pig. Economic losses due to tail biting are mainly caused by additional work, material and medication costs and carcass price discounts associated with tail biting, impaired feed conversion ratio and growth rate of pigs. Additional work and inputs are needed also to control tail biting preventatively.

Losses associated with tail biting lesions are influenced by the incidence and severity of tail biting. The economic viability of preventive measures depends on how much they can mitigate losses caused by tail biting lesions and the costs of preventive measures. While low-cost measures can be economically profitable already at low levels of efficacy, economic viability of high-cost measures require that they can mitigate substantial losses.

Scientific research and a Commission Recommendation indicate that when considering the legal prohibition on docking, unless the farmer has first tried to prevent tail biting by changing inadequate environmental conditions or management systems, inspectors and farmers should assess, and consider whether improvements are needed, regarding thermal comfort and air quality; health status; diet; competition for space or feed; cleanliness; and the quality and quantity of enrichment material that is provided.

Scientific research and a Commission Recommendation also show that to comply with Points 4 and 8 of Annex I, Chapter I of the Pigs Directive, enrichment materials must be edible, chewable, investigable and manipulable. In addition, they should be given in sufficient quantity and be of sustainable interest i.e. they should encourage the exploratory behaviour of pigs and be regularly replaced and replenished.

The Pigs Directive and the Commission Recommendation do not set specific parameters; they do not, for example lay down how much straw per pig should be provided if straw is used for enrichment or how much feeder space should be available per pig. Accordingly, in determining the quality and quantity of the various factors that must be considered when changing inadequate environmental conditions or management systems, inspectors and farmers should take account of the core objective of the legislation which is to prevent tail biting by improving conditions on the farm rather than by tail docking. Thus, if tail biting is still a problem on the farm, farmers should carefully consider what additional improvements are needed to prevent tail biting rather than assuming that no further action is necessary because certain levels, for example of enrichment, air quality, space allowance and health, have already been reached.

Widespread use of tail docking seems to persist on one hand because the alternative steps that producers are required to take before resorting to it are not specified in detail. On the other hand, an important reason to continue tail docking seems to be that, given the current housing facilities of farms, many pig producers expect it to be less risky and more profitable to produce docked than undocked pigs. This is related to the costs of tail biting and the uncertainty caused by tail biting. Preventing tail biting requires constant effort and this may be financially prohibitive when compared to masking the problem with tail docking. Hence, farming system changes, improved enforcement of the EU prohibition on routine tail docking or stronger legislation may be needed to encourage more producers to terminate the practice of tail docking. For the measures to prevent tail biting to be adopted and the practice of tail docking to be terminated in an economically sustainable way, changes must be implemented so that pig producers are better off after than before the change. This said, it must be noted that producers are obliged to comply with the EU ban on routine tail docking even if they find this incurs economic costs.

Changes in the pig carcass pricing and farm subsidies could also encourage producers to explore and implement new practices to reach this goal. Some high animal welfare marketing systems pay attention to the absence of tail docking, to tail condition or to factors which can mitigate tail biting. However, the contribution of such systems to the profitability of pig farming is not well documented. To gain market premiums for undocked pigs it is important to facilitate dialogue between pig producers and consumers. Moreover, technical specifications for international trade could also pay more attention to pig welfare because it would help to provide a level playing field for producers who give special attention to prevent tail biting without tail docking.

5 Summary

- The costs of tail biting lesions across Europe are around €0.6 to €3.4 per finished pig, and they are mainly caused by additional work, material and medication costs, carcass price discounts, and impaired feed conversion ratio and growth of pigs with tail lesions.
- In the European Union tail docking may not be used routinely and it is allowed only as the last resort to control tail biting (Council Directive 2008/120/EC).
- To comply with the Pigs Directive, producers must consider what improvements, such as additional space, enhanced air quality or enrichment that is new, edible, chewable, investigable and manipulable are needed to prevent tail biting.
- Tail docking seems to persist because the alternative steps required to take before resorting to it are not specific and because of economic benefits of tail docking.
- Changes in regulations, productivity gains, a price premium or a public support to enhance pig welfare can all encourage pig producers to mitigate tail biting.

References

Camerlink, I., Bijma, P., Kemp, B. and Bolhuis, J.E., 2012. Relationship between growth rate and oral manipulation, social nosing, and aggression in finishing pigs. Applied Animal Behaviour Science 142: 11–17. https://doi.org/10.1016/j.applanim.2012.09.004.

D'Eath, R.B., Arnott, G., Turner, S.P., Jensen, T., Lahrmann, H.P., Busch, M.E., Niemi, J.K., Lawrence, A.B. and Sandøe, P., 2014. Injurious tail biting in pigs: how can it be controlled in existing systems without tail docking? Animal 8: 1479–1497. https://doi.org/10.1017/S1751731114001359.

D'Eath, R.B., Niemi, J.K., Vosough Ahmadi, B., Rutherford, K.M.D., Ison, S.H, Turner, S.P., Anker, H.T., Jensen, T., Busch, B.E., Lawrence, A.B and Sandøe, P., 2016. Why are most EU pigs tail docked? Economic and ethical analysis of four pig housing and management scenarios in the light of EU legislation and animal welfare outcomes. Animal 10: 687–699. https://doi.org/10.1017/S1751731115002098.

De Briyne, N., Berg, C., Blaha, T., Palzer, A. and Temple, D., 2018. Phasing out pig tail docking in the EU – present state, challenges and possibilities. Porcine Health Management 4, 27. https://doi.org/10.1186/s40813-018-0103-8.

EFSA, 2007. The risks associated with tail biting in pigs and possible means to reduce the need for tail docking considering the different housing and husbandry systems – scientific opinion of the Panel on Animal Health and Welfare. The EFSA Journal 611: 1–98. https://doi.org/10.2903/j.efsa.2007.611.

England, D.C. and Spurr, D.T., 1969. Litter size of swine confined during gestation. Journal of Animal Science 28: 220–223. https://doi.org/10.2527/jas1969.282220x.

European Commission, 2001. Commission Directive 2001/93/EC of 9 November 2001 amending Directive 91/630/EEC laying down minimum standards for the protection of pigs. Annex, Chapter I, point 8. Official Journal of the European Union L316: 36–38.

European Commission, 2016a. Commission Regulation (EU) 2016/336 of 8 March 2016 on the application of douncil directive 2008/120/EC laying down minimum standards for the protection of pigs as regards measures to reduce the need for tail-docking. Official Journal of the European Union L62: 20–22.

European Commission, 2016b. Commission staff working document on best practices with a view to the prevention of routine tail-docking and the provision of enrichment materials to pigs. (SWD(2016) 49 final). European Commission, Brussels, Belgium, 15 p. https://ec.europa.eu/food/system/files/2016-12/aw_practice_farm_pigs_stfwrkdoc_en.pdf.

European Commission, 2019. Outcome of commission actions to reduce tail biting and prevent routine tail docking of pigs. Presentation to meeting of Civil Dialogue Group, 14 November 2019.

European Council, 1991. Council directive of 19 November 1991 laying down minimum standards for the protection of pigs (91/630/EEC). Annex, Chapter II, Section III, point 4. Official Journal of the European Communities L340: 33–38.

European Council, 2008. Council Directive of 18 December 2008 laying down minimum standards for the protection of pigs (2008/630/EEC). Annex I, Chapter I, point 8. Official Journal L 47, 18.02.2009 pp. 5–13.

European Food Safety Authority Opinion, 2007. Scientific opinion of the Panel on Animal Health and Welfare The risks associated with tail biting in pigs and possible means to reduce the need for tail docking considering the different housing and husbandry systems. The EFSA Journal (2007) 611, 1–13.

EUWelNet, 2013. Coordinated European animal welfare network. Available at: http://pigstraining.welfarequalitynetwork.net/.

Haigh, A. and O'Driscoll, K., 2019. Irish pig farmer's perceptions and experiences of tail and ear biting. Porcine Health Management 5: 30. https://doi.org/10.1186/s40813-019-0135-8.

Harley, S., More, S.J., O'Connell, N.E., Hanlon, A., Teixeira, D. and Boyle, L., 2012. Evaluating the prevalence of tail biting and carcase condemnations in slaughter pigs in the Republic and Northern Ireland, and the potential of abattoir meat inspection as a welfare surveillance tool. Veterinary Record 171: 621–627. https://doi.org/10.1136/vr.100986.

Harley, S., Boyle, L., O'Connell, N., More, S., Teixeira, D. and Hanlon, A., 2014. Docking the value of pigmeat? Prevalence and financial implications of welfare lesions in Irish slaughter pigs. Animal Welfare 23: 275–285.

Heinola, K., Kauppinen, T., Niemi, J.K., Wallenius, E. and Raussi, S., 2021. Comparison of 12 different animal welfare labeling schemes in the pig sector. Animals 11: 2430. https://doi.org/10.3390/ani11082430.

Heinonen, M., Välimäki, E., Laakkonen, A-M., Toppari, I., Vugts, J., Fàbrega, E. and Valros, A., 2021. Evaluation of tail lesions of finishing pigs at the slaughterhouse: Associations with herd-level observations. Frontiers in Veterinary Science 8: 650590. https://doi.org/10.3389/fvets.2021.650590.

Hunter, E.J., Jones, T.A., Guise, H.J., Penny, R.H.C. and Hoste, S., 2001. The relationship between tail biting in pigs, docking procedure and other management practices. The Veterinary Journal 161: 72–79. https://doi.org/10.1053/tvjl.2000.0520.

Kritas, S.K. and Morrison, R.B., 2007. Relationships between tail biting in pigs and disease lesions and condemnations at slaughter. Veterinary Record 160: 149–152. https://doi.org/10.1136/vr.160.5.149.

Lahrmann, H.P., Busch, M.E., D'Eath, R.B., Forkman, B. and Hansen, C.F., 2017. More tail lesions among undocked than tail docked pigs in a conventional herd. Animal 11: 1825–1831. https://doi.org/10.1017/S1751731117000490.

Larsen, M., Andersen, H. and Pedersen, L., 2018. Which is the most preventive measure against tail damage in finisher pigs: tail docking, straw provision or lowered stocking density? Animal 12: 1260–1267. https://doi.org/10.1017/S175173111700249X.

Marchant-Forde, J.N., Lay, D.C. Jr., Mc Munn, K.A., Cheng, H.W., Pajor, E.A. and Marchant-Forde, R.M., 2009. Postnatal piglet husbandry practices and well-being: the effects of alternative techniques delivered separately. Journal of Animal Science 87: 1479–1492. https://doi.org/10.2527/jas.2008-1080.

Marques, B.M.F.P.P., Bernardi, M.L., Coelho, C.F., Almeida, M., Morales, O.E., Mores, T.J., Borowski, S.M. and Barcellos, D.E.S.N., 2012. Influence of tail biting on weight gain, lesions and condemnations at slaughter of finishing pigs. Pesquisa Veterinária Brasileira 32: 967–974. https://doi.org/10.1590/S0100-736X2012001000003.

McGlone, J.J., Sells, J., Harris, S. and Hurst, S., 1990. Cannibalism in growing pigs: effects of tail docking and housing system in behavior, performance and immune function. Scientific Technical Report No T-5-283:69-71. Texas Tech University publications, Lubbock TX, USA. https://www.depts.ttu.edu/animalwelfare/research/documents/CANNIBALISMINGROWINGPIGS_McGlone1990.pdf.

Moinard, C., Mendl, M., Nicol, C.J. and Green, L.E., 2003. A case control study of on-farm risk factors for tail biting in pigs. Applied Animal Behaviour Science 81: 333–355. https://doi.org/10.1016/S0168-1591(02)00276-9.

Munsterhjelm, C., Heinonen, M. and Valros, A., 2015. Effects of clinical lameness and tail biting lesions on voluntary feed intake in growing pigs. Livestock Science 181: 210–219. https://doi.org/10.1016/j.livsci.2015.09.003.

Nannoni, E., Valsami, T., Sardi, L. and Martelli, G., 2014. Tail docking in pigs: a review on its short- and long-term consequences and effectiveness in preventing tail biting. Italian Journal of Animal Science 13: 3095. https://doi.org/10.4081/ijas.2014.3095.

Niemi, J.K., 2020. Animal welfare and farm economics: an analysis of costs and benefits. In: Ahmadi, B.V., Moran, D., D'Eath, R. (eds), 2020. Economics of farm animal welfare: Theory, evidence and policy. Wallingford, Cabi, UK, pp. 98–116. https://www.cabi.org/vetmedresource/ebook/20203308011.

Niemi, J.K., 2021. The economic cost of bacterial infections. In: Foster, N., Kyriazakis, I. and Barrow, P. (eds.), 2021. Advancements and technologies in pig and poultry bacterial disease control. Academic Press, pp. 1–23. https://doi.org/10.1016/B978-0-12-818030-3.00010-6.

Niemi, J., Bennett, R., Clark, B., Frewer, L., Jones, P., Rimmler, T. and Tranter, R., 2020. A value chain analysis of interventions to control production diseases in the intensive pig production sector. PLoS ONE 15: e0231338. https://doi.org/10.1371/journal.pone.0231338.

Niemi, J.K., Edwards, S.A., Papanastasiou, D.K., Piette, D., Stygar, A.H., Wallenbeck, A. and Valros, A., 2021. Cost-effectiveness analysis of seven measures to reduce tail biting lesions in fattening pigs. Frontiers in Veterinary Science 8: 682330. https://doi.org/10.3389/fvets.2021.682330.

Niemi, J.K., Sinisalo, A., Valros, A. and Heinonen, M., 2012a. Market and policy-oriented incentives to provide animal welfare: The case of tail biting. Paper presented at the European Association of Agricultural Economists (EAAE), 126th Seminar, June 27–29, 2012, Capri, Italy. http://purl.umn.edu/125957.

Niemi, J.K., Sinisalo, A., Valros, A. and Heinonen, M., 2012b. Hännänpurenta – syy vai seuraus? In: Schulman, N. and Kauppinen, H. (eds.), 2012. Maataloustieteen Päivät 2012 [online publication]. Suomen Maataloustieteellisen Seuran julkaisuja no 28. https://doi.org/10.33354/smst.75558.

Niemi, J.K., Sinisalo, A., Valros, A. and Heinonen, M., 2011. The timing and treatment of tail biting in fattening pigs. In: Hulgren, J., Persson, P., Nadeau, E. and Fogelberg, F. (eds.), 2011. 24th Book of abstracts; NJF congress food, feed, fuel and fun – Nordic light on future land use and rural development. Nordic Association of Agricultural Scientists, Uppsala, Sweden, 6 pp.

Sanchez-Vazquez, M.J., Nielen, M., Edwards, S.A., Gunn, G.J. and Lewis, F.I., 2012. Identifying associations between pig pathologies using a multi-dimensional machine learning methodology. BMC Veterinary Research 8: 151. https://doi.org/10.1186/1746-6148-8-151.

Schrøder-Petersen, D.L. and Simonsen, H.B., 2001. Tail biting in pigs. Veterinary Journal 162: 196–210. https://doi.org/10.1053/tvjl.2001.0605.

Sinisalo, A., Niemi, J.K., Valros, A. and Heinonen, M., 2012. Tail biting and production performance in fattening pigs. Livestock Science 143: 220–225. https://doi.org/10.1016/j.livsci.2011.09.019.

Stygar, A., Krampe, C., Llonch, P. and Niemi, J., 2022. How far are we from data-driven and animal-based welfare assessment? A critical analysis of European quality schemes. Frontiers in Animal Science 3: 874260. https://doi.org/10.3389/fanim.2022.874260.

Sutherland, A.M., Bryer, P., Krebs., N. and McGlone, J., 2009. The effect of method of tail docking on tail-biting behaviour and welfare of pigs. Animal Welfare 18: 561–570. https://doi.org/10.1017/S0962728600000993.

Taylor, N.R., Main, D.C.J., Mendl, M. and Edwards, S.A., 2020. Tail-biting: a new perspective. Veterinary Journal 186: 137–147. https://doi.org/10.1016/j.tvjl.2009.08.028.

Teixeira, D., Harley, S., Hanlon, A., O'Connell, N., More, S. and Boyle, L., 2014. Severity, prevalence, risk factors and economic implications of tail lesions in slaughter pigs. Available at: https://www.researchgate.net/profile/Dayane-Teixeira/publication/299802014_Severity_prevalence_risk_factors_and_economic_implications_of_tail_lesions_in_slaughter_pig/links/570562b408ae74a08e274abd/Severity-prevalence-risk-factors-and-economic-implications-of-tail-lesions-in-slaughter-pig.pdf.

Valros, A., Välimäki, E., Nordgren, H., Vugts, J., Fàbrega, E. and Heinonen M., 2020. Intact tails as a welfare indicator in finishing pigs? Scoring of tail lesions and defining intact tails in undocked pigs at the abattoir. Frontiers in Veterinary Science 7: 405. https://doi.org/10.3389/fvets.2020.00405.

van Staaveren, N., Boyle, L.A., García Manzanilla, E., O'Driscoll, K., Shalloo, L. and Calderón Díaz, J.A., 2021. Severe tail lesions in finisher pigs are associated with reduction in annual profit in farrow-to-finish pig farms. Veterinary Record 2021: e13. https://doi.org/10.1002/vetr.13.

Vom Brocke, A.L., Karnholz, C., Madey-Rindermann, D., Gauly, M., Leeb, C., Winckler, C., Schrader, L. and Dippel, S., 2019. Tail lesions in fattening pigs: relationships with postmortem meat inspection and influence of a tail biting management tool. Animal 13: 835–844. https://doi.org/10.1017/S1751731118002070.

Wallgren, P. and Lindahl, E., 1996. The influence of tail biting on performance of fattening pigs. Acta Veterinaria Scandinavica 37: 453–60. https://doi.org/10.1186/BF03548085.

Wallenbeck, A. and Keeling, L.J., 2013. Using data from electronic feeders on visit frequency and feed consumption to indicate tail biting outbreaks in commercial pig production. Journal of Animal Science 91: 2879–2884, https://doi.org/10.2527/jas.2012-5848.

Zhou, B., Yang, X.J., Zhao, R.Q., Huang, R.H., Wang, Y.H., Wang, S.T., Yin, C.P., Shen, Q., Wang, L.Y. and Schinckel, A.P., 2013. Effects of tail docking and teeth clipping on the physiological responses, wounds, behavior, growth, and backfat depth of pigs. Journal of Animal Science 91: 4908–4916. https://doi.org/10.2527/jas.2012-5996.

Zonderland, J.J., Bosma, B. and Hoste, R., 2011. Financiële consequenties van staartbijten bij varkens. Rapport 543. Wageningen UR Livestock Research, Lelystad, The Netherlands, 31 pp. https://edepot.wur.nl/188443.

CHAPTER 16

The ethical aspects of tail biting and tail docking

Bernice Bovenkerk[1,*], *Marc Bracke*[2,3] *and Anna Valros*[4]

[1]Wageningen University and Research, Hollandseweg 1, 6706 KN Wageningen; [2]Department of Animal Health and Welfare, Wageningen Livestock Research, Wageningen University & Research, 6708 WD Wageningen, The Netherlands; [3]Agricultural Biosystems Engineering, Wageningen University, 6700 AA, Wageningen, The Netherlands; [4]Research Centre for Animal Welfare, Department of Production Animal Medicine, Faculty of Veterinary Medicine, P.O.Box 57, University of Helsinki, Finland; *Correspondence: Bernice.bovenkerk@wur.nl

Abstract

In this chapter, the case of tail biting and docking in pigs is discussed from an ethical point of view. Utilitarian, welfarist, Kantian, care ethical, and virtue ethical theories are explained and applied to this case. A distinction is made between ideal and non-ideal ethical theories. Ideal theories provide an end-state towards which we should strive. According to some ideal theories, this end-state is the abolition of pig farming. Other ideal theories posit that it is morally allowed to farm pigs, as long as the pigs have good lives. However, ideal theories are not always well-equipped to deal with practical decision making in real world situations. Non-ideal theories therefore argue that we should focus on making piecemeal improvements to the current situation. We argue that we need both ideal theories to give meaning to our efforts and provide the 'dot on the horizon', and non-ideal theories to reflect on the steps towards that ideal. While there are clear differences of opinion between different theories there is convergence on a number of issues. First of all, when pigs have limited opportunities for carrying out their natural behaviours or are suffering from other health-and-welfare problems this may lead to psychopathological behaviour, such as tail biting. Therefore, we need to look at the underlying causes of tail biting and consider if alternatives are possible to tail docking; this is just the treatment of a symptom and does not solve the problem of reduced welfare that triggers the behaviour. Secondly, it is morally problematic to change the pigs by docking their tails rather than improving their environment (by providing enrichment etc.). Breeding pigs without tails, without biting impulses or pigs that are happy despite poor living conditions may also raise moral concerns, both from the viewpoint of welfare and from arguments beyond welfare, such as instrumentalisation, objectification, violation of integrity, and de-animalisation.

 | DOI:10.3920/9789004694989_018

Keywords

utilitarianism – animal rights – (non-)ideal theory – welfare

1 Introduction: Pig tails in Finland and the Netherlands

In this chapter, we will look at tail biting and tail docking from an ethical point of view. As we will show, different ethical theories may lead to different positions on whether, and if so how and how radically, we need to improve the current situation. We will start our discussion by making a short comparison between two EU countries, the Netherlands and Finland, as an illustration of how different positions can underpin different ways of dealing with the problem of tail biting.

According to the European Council Directive 2008/120/EC pigs' tails should not be docked routinely, but only as a therapeutic measure. While these EU-wide rules on pig welfare were already established in 1991 via Directive 91/630/EEC, in most European countries tail docking is still common practice. Currently in the Netherlands almost 100% of conventionally housed pigs' tails are still docked (Bracke *et al.*, in press). A number of relevant laws have been passed in the Netherlands. The 'Animal Law' has almost been amended, entailing that it would be no longer allowed to harm the health or welfare of an animal with the goal of housing animals in a particular way.[1] This amendment was supposed to reinforce the Animal Law, article 2.3, which states that pigs' tails can only be docked if the pig is no more than four days old, tissue damage due to biting has been observed in the group, and measures to adapt environmental factors (such as stocking density and enrichment) have not proven effective. The aim now is to completely phase out tail docking by 2040 in the Netherlands.

As tail biting is a multifactorial problem with several different possible causes, solving just one of the risk factors may not be sufficient to prevent it (EFSA, 2007). While research shows that changing housing conditions and providing enrichment can substantially reduce tail biting and the need for tail docking, there are economic obstacles to such changes (Niemi *et al.* 2021; Zonderland *et al.* 2008) (See chapter 14). As consumers will not pay a premium for meat from pigs with intact tails, it has been difficult for farmers to recover costs from efforts to keep tails intact. These factors have made it difficult for farmers to comply with the ban on tail docking. Perhaps related to this is the observation that conventional farmers in the Netherlands, who prefer preventive tail docking over the risk of tail biting, estimate the pain associated with tail docking lower, and have less faith in providing enrichment to prevent tail biting than organic pig farmers do (De Lauwere *et al.*

1 https://www.partijvoordedieren.nl/moties/amendement-vestering-over-het-niet-langer-aan passen-van-dieren-aan-het-systeem#:~:text=Dit%20amendement%20regelt%20dat%20 het,bepaalde%20manier%20te%20kunnen%20huisvesten.

2010). In the Netherlands, the Society for the Protection of Animals has developed a so-called Better Life Quality Mark, which provides labels (1 to 3 stars) for meat, eggs, and dairy that has been produced with presumably improved animal welfare (Ruis 2016). Forty percent of the pigs that are raised for the internal market has one of these labels (mostly one star). One-star pigs have a bit longer tails, though they are mostly still docked. Docking is banned in the 3 star version of Better Life, which includes organic pig farming.

In contrast, in Finland, the ban on tail docking, which is based on national legislation (Animal Welfare Decree 396/1996), is unconditional, and complied with. As far as the authors are aware, there have been no cases of illegal docking in Finland for many years. There are several features of Finnish pig production that make it more feasible to rear undocked pigs, such as a comparably larger space allowance, partly slatted flooring, use of manipulable materials, a good animal health status and meal feeding from long troughs (Valros, 2021). These include both legislative and voluntary measures, which go beyond the EU legislation, and might be, at least partly, a result of the tail docking ban. There are still considerable amounts of tail biting lesions reported for Finnish pigs (e.g. Valros *et al.* 2020), but on the other hand, the majority of the pigs go through life with a fully intact tail, or suffer only very minor lesions. Further, most Finnish producers are motivated to rear undocked pigs (Valros *et al.*, 2016), and a large between-farm variation in lesion levels indicate that it is indeed possible to successfully rear undocked pigs on commercial, intensive farms (Heinonen *et al.* 2021).

This brief illustration of the treatment of tail biting in the Netherlands and Finland shows that different positions can be taken on the prevention of tail biting and the need for tail docking. In this chapter, we will look at tail biting and tail docking from an ethical point of view. As we will show, different ethical theories may lead to different positions on whether, and if so how and how radically, we need to improve the current situation. According to some ethical views, it is sufficient to improve pig welfare, but others take an abolitionist approach to pig farming. There are also differences of opinion on the question of whether we should get to these end points abruptly (quickly) or piecemeal (gradually). In ethics, the latter is known as the difference between ideal and non-ideal theory approaches. What will become clear in our discussion is that there are different views and different moral frameworks, on what is meant by welfare in the first place, and on how to improve it. We will argue that *if* we decide to continue farming pigs, we need to start by changing their housing conditions and provide better enrichment, rather than change the pigs themselves, either by tail docking or by breeding pigs that are less likely to bite each other's tails. In our conclusion we will suggest that proponents of both ideal and non-ideal theory have their role to play in bringing about change and improving the interests of pigs. We firmly believe it is important to maintain an open attitude towards all different points of view, so that society can make informed decisions about which route and end-point to choose when it

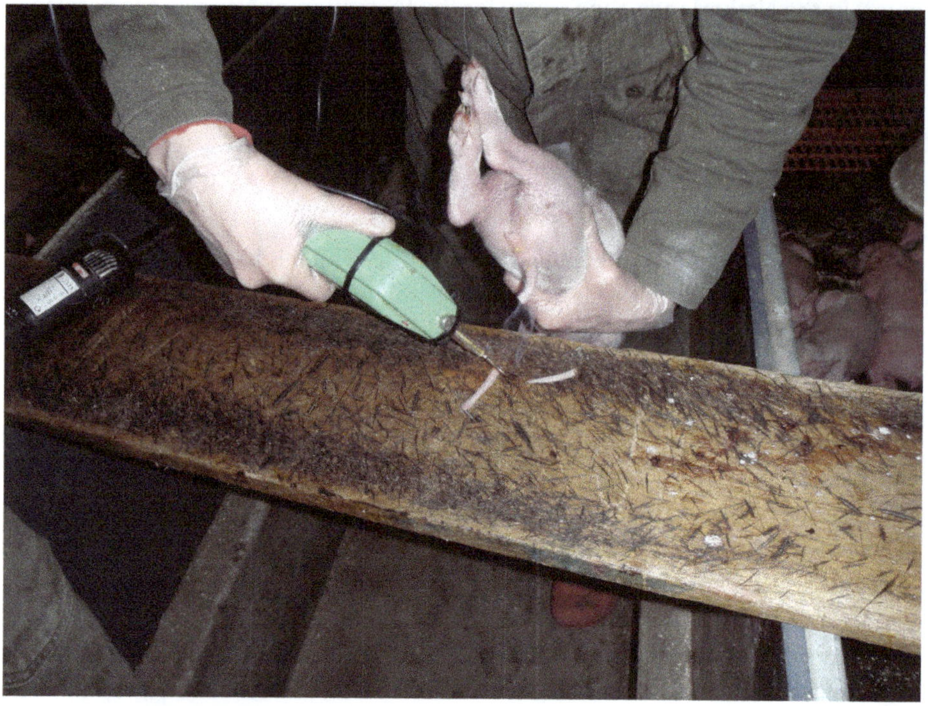

FIGURE 16.1 Tail docking of a piglet

FIGURE 16.2
Tail manipulation of an already seriously damaged tail

comes to pig farming, including solving the issues of tail docking, tail biting and providing proper enrichment for pigs (see also Bracke *et al.*, 2023).

2 Ethics

When we think about questions such as whether we should allow tail docking or whether and how we should improve pig welfare, we are dealing with ethics. These are questions not about how we *do* behave, but how we *should* behave; in other words, they are not simply questions about facts, but also about values. As a philosophical discipline ethics could be defined as a systematic reflection on morality. In turn, morality could be described as the totality of norms and values that a person or group regard as directive for action, and that are deemed very important by this person or group (Bolt 2003). Norms can be defined as action-guiding rules that express certain values, while values could be described as states of affairs, experiences, or qualities that are desirable in themselves. So for example, honesty is a value, and not misleading someone is a norm that follows from this value. Ethics thus is about right and wrong actions or attitudes; it does not ask empirical questions (i.e. questions about observations), but normative ones (i.e. questions about what ought to be the case). An answer to an empirical question (for example 'how many lesions does a pig have?') does not yet imply by itself what we ought to do (for example 'we need to start docking pig tails'). However, answers to empirical questions (i.e. descriptions) are often necessary as input to give better answers to ethical (i.e. prescriptive, or normative) questions as to how we ought to act. Docking tails as conducted in current pig farming, i.e. without providing anaesthesia or pain killers, hurts the pigs. This is a descriptive statement. It is a matter of fact; the statement that tail docking is painful, is at least either true of false. The fact that this statement is true in this case is not, by itself, enough to conclude that it is morally wrong to dock tails. When someone gets a vaccination, it hurts too, but that does not make it morally wrong to vaccinate someone. What is needed to conclude that the descriptive fact that tail docking is painful is morally wrong, is an underling value framework, which defines what is good and what is bad.

When dealing with animals it is important to use ethical reflection because using animals is not a neutral act; creatures are involved whose lives can go better or worse for them, depending on our actions. Moreover, there is increasing societal debate about animal use and those dealing with animals – be it as a farmer, veterinarian, or animal scientist – should be aware of and relate to the different positions in this debate. This means it is important to establish one's own position and to be able to justify this position with arguments. This also means being willing to regard one's own standpoint critically and not be influenced by one's own material interests. In other words, ethical theory should be applied with an open and honest attitude, and should not be used just to find arguments for the opinions one already has, or to justify one's economic or other personal interests. For example, some farmers have suggested that mutilations such as tail docking, teeth grinding and castration are not painful. In former days, scientists too conducted rather painful experiments on animals, so-called vivisection, under the assumption that animals cannot feel pain. However, by now it is generally agreed that many animals feel pain and that tail docking, tooth resection, and castration are painful to the pigs (Ison *et al.* 2014; Prunier *et al.* 2021). It is standardly thought that the infliction of pain is *prima facie* wrong (wrong at first sight) because pain is bad for the being that experiences it. As a result, the infliction of pain requires a strong justification (i.e. that the pain somehow leads to an improvement of other interests, such as health, in the case of a vaccination or that the interest in avoiding pain is outweighed by other interests).

Even though values are involved, this does not render ethical judgments subjective. Ethics is not just a matter of preference or taste; it is based on logical reasoning and it requires people to be transparent, consistent and unbiased in their viewpoints. One – so-called coherentist[2] – model of how we form moral judgements is offered by the theory of 'reflective equilibrium' (Rawls 1971; Van der Burg *et al.* 1998). In the original version of John Rawls, deliberation about principles of justice start from our considered convictions or intuitions; we need to discard those convictions that have been made in haste, with a personal bias, or that we are uncertain of (Rawls 1971, pp. 19–20). Then we move back and forth between our convictions, arguments, and principles, sometimes conforming our judgments to principle, sometimes changing the principles until we reach a (temporary) equilibrium (Rawls 1971, p. 20; see also Knight 2017 for a more in depth explanation of how the method works). Here we discuss one particular and simplified interpretation of the method of reflective equilibrium, namely that suggested by Rutgers (1998) because in our view it is particularly well applicable to practical cases such as the tail-biting and docking issue.[3] In this model moral judgements are based on three pillars: first, we are confronted with a situation that raises emotions or an intuition that something is going on that requires a moral justification (the pillar of moral intuition), then we look at the facts of the case (the pillar of facts) and try to square these with our first impression. Next, we look at the case in the light of ethical principles (the pillar of principles) and finally we try to achieve a balance (an equilibrium) between these three pillars. For example, in the case of tail docking a farmer may feel intuitively it is morally justified. However, faced with societal pressure, (s)he may look for a justification in the facts, arguing e.g. that if he would stop tail docking of young piglets, the pigs may later suffer from tail biting, and that the pain involved in the latter is a lot worse than the pain of tail docking. He or she may also have to square this with their moral principle that they would rather not cause pain to their pigs, but also that humans in need, e.g. poor people wanting to eat pork, are more important to him or her than the interests of the pigs, and that it is the farmer's primary duty to make sure people are provided with nutritious and affordable food.

In the schematic depiction of moral judgement formation in figure 16.3 there are double arrows between the three pillars, meaning that intuitions may have to be revised after consulting with the facts and principles and vice versa. So, for example, if a moral principle would lead to very counter-intuitive outcomes, this would be reason to refine the principles. Likewise, what are seen as morally relevant facts is dependent on one's ethical framework. Different people can even interpret the same factual information differently, because their underlying values differ. For example, in a study by Mullan *et al.* (2011) Farm Assurance assessors were asked to score pigs' welfare based on observations and images of specific pigs. A difference

2 Coherentism is a 'meta-ethical' view for determining whether a judgment is valid. If the judgement coheres with our other beliefs there is good reason to believe it is valid. It should be noted that there is discussion about whether this method is coherentist, foundational or something in between (see Knight. 2017). Most commentators hold it to be coherentist.

3 While the way we describe the method here is missing a bit of nuance, we think this suffices for our current purposes of explaining ethics in relation to tail docking.

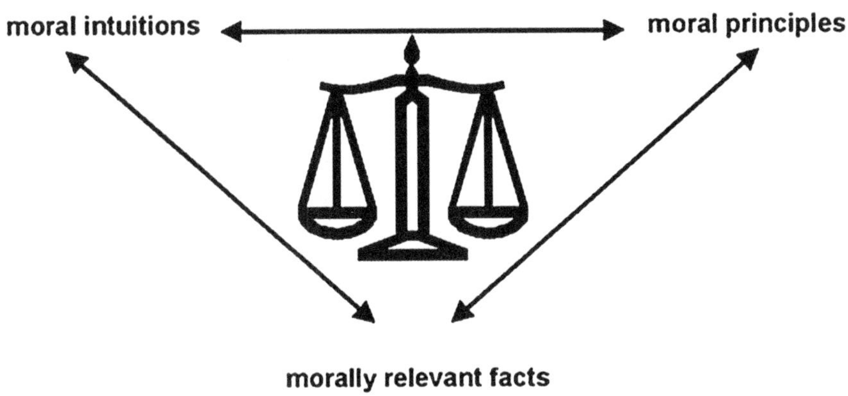

FIGURE 16.3 Reflective equilibrium
SOURCE: RUTGERS, B. 1998

was found between Farm Assurance assessors that held the view that pigs have the ability to experience positive or poor mental welfare; these more often recorded pigs as requiring hospitalization, compared to assessors who did not hold this view as strongly. Assessors who scored higher on the value 'people-centric, pigs as profit' recorded fewer pigs with severe tail lesions.

The following example further illustrates moral judgment formation according to Rutgers' reflective equilibrium method. On most dairy farms in the Netherlands the calf is separated from the mother soon after birth, and many citizens have the intuition that this is wrong. Looking at the facts, some welfare researchers argue that this is not problematic, because when the calf is taken away directly after birth no bond between mother and calf is formed and they therefore do not suffer (Hopster *et al.* 2016). This means that one would have to revise one's first intuition. This example also shows that a value judgment is involved here in establishing the facts: in the research by Hopster *et al.* (2016) suffering has been measured in terms of vocalisation and stress hormones. According to these measures, mother and calf do not suffer as long as they have not formed a bond. However, it is possible to question the validity and sensitivity of these measures in the case of immediate post-partum separation. This provides only one particular way of looking at suffering. When we look at ethical principles and how we may define welfare it becomes clear that animal welfare may comprise more than the direct discomfort experienced by the animals. It could also be argued that it is good for the cow's and calf's welfare if they are allowed to bond, e.g. because it's the natural thing to happen, even regardless of whether it is perceived as pleasant or not. According to some moral theories it could even be argued that they have a right to.[4]

---

4 Of course there are a lot more considerations around the issue of cow/calf separation, but here we just focus on suffering in order to give a simple illustration of how moral judgment formation according to the reflective equilibrium method works.

3 Ethical theories

Since animal ethicist Peter Singer wrote his influential book *Animal Liberation* in 1975, thereby drawing non-human animals (from here we follow the unscientific and morally problematic convention of referring to them as animals) into the moral community, most of the main ethical theories have been applied to animals. The two most influential contemporary theories in ethics are utilitarianism and (neo-) Kantianism. The former primarily looks at the consequences (utility) of our actions, while the latter focusses on our duties. Both are so-called action-guiding theories; their aim is not only to give a justification for why we should behave ethically and define what the ideal is that we should strive towards, but they also formulate general principles to inform moral decision-making. In that respect, they are universalist theories; the norms they formulate apply to all similarly situated individuals equally. Other major theories that have more recently been applied to animals either start from a criticism of this universalism (in particular care ethical theories) or do not focus on specific norms for action, but rather on character building (in particular virtue ethics). Below we have included a table with an overview of these four theories, as well as welfarism, which is another consequentialist theory that is often used by animal welfare scientists. In the next section we will discuss the different theories in a bit more depth when we show how they are applied to animals, and in the case of tail biting and docking.

Ethical theories should not be understood as simple decision-making models, where you put in all the relevant features of a case and the right answer comes rolling out. Rather, they present ways of reasoning and it is very well possible that, for example, one utilitarian will come to one conclusion and another utilitarian will come to a different conclusion. The same is true for Kantians, Care ethicists and Virtue ethicists: e.g. Kant himself rejected direct duties towards animals, while Tom Regan, an animal rights philosopher, holds that we have absolute duties towards animals (called rights). Moral reality is often complex and there may not always be one right answer. However, this does not mean that no *wrong* answers are possible; through the application of moral principles and logical reasoning, argumentative fallacies and inconsistencies in reasoning can be pointed out.

This means that we do not support moral relativism, which is the view that ethical viewpoints are subjective for one person or culture and in essence 'anything goes' in ethics. This would mean that no critical discussions on moral viewpoints are possible. Rather, here we argue from moral pluralism, which is the view that there is no one overarching theory, principle or value that overrides all others.[5] In practice, most people tend to be moral pluralists; they do not strictly follow Kantian

---

5 In fact one of the authors (MB) may not be a moral pluralist, in that he believes a moral theory like universal prescriptivism and a principle like human integrity (meaning that people should only do things that can truly be explained openly and honestly, Bracke *et al.* 2023), may well provide a (monistic) basis of other values and norms.

TABLE 16.1 Different ethical theories

Theory	Type of theory	Goal	How?	Who counts?
Utilitarianism	Consequentialist (only consequences count), universalist, action-guiding	To achieve happiness/the good	Maximize the good (happiness/ pleasure/utility) and minimize the bad	Each being that can experience happiness or pain & pleasure (sentient beings)
Welfarism	Consequentialist (only consequences count), universalist, action-guiding	To achieve good welfare	Maximize the good, defined as welfare	Each being that can experience happiness or pain & pleasure (sentient beings)
Kantianism	Deontological (focus on duties), universalist, action-guiding	To live according to reason/our rationality	By applying a rationality demand to each action (e.g. make sure what you do can be a universal law)	Autonomous, rational persons (if applied to animals: having preference autonomy)
Care ethics	Contextual (particularity of relations is fundamental)	The protection of vulnerable groups	Relations determine responsibilities	Those beings that we are entangled in relationships with
Virtue ethics	Not action-guiding; focus on cultivation of good character	The good/ virtuous life and good society	Building a virtuous character using practical wisdom	Not a relevant question to ask in virtue ethics

or utilitarian principles for example, but apply different principles to different cases. Farmers may, for example suggest that they should make their animals as happy as possible, e.g. by providing shelter and nutritious food, but also that they have a duty to feed the world. The former is more utilitarian in nature, the latter more of a Kantian requirement. Proponents of different theories in academia are constantly in discussion with each other, thereby refining their theories. It is very well possible that if you look at a problem from different theoretical frameworks you still arrive at the same answer. This is called convergence and may give us more confidence to adopt this viewpoint.

4 Different views on welfare

Before turning to animal ethics theories, we want to briefly turn our attention to the concept of animal welfare. Sometimes it is thought that when we are dealing with animals we need to think about their welfare and *also* look at ethical issues. However, animal welfare is an ethical issue in its own right and figures importantly in most animal ethics theories. Welfare is an evaluative term, combining both ethical and biological assumptions. The definition of animal welfare has changed over time; the focus of welfare scientists was, perhaps, first on how to measure it, so e.g. the animals' ability to cope with farming conditions. Animal ethicists and some welfare scientists, however, instead emphasised the subjective experiences of animals. As to the latter, most attention initially was directed at negative subjective experiences (poor welfare), while positive experiences are more recently being recognised as important (Bracke *et al.*, 2008b; Ohl *et al.* 2012), although they are more difficult to measure scientifically. The Farm Animal Welfare Council's Five Freedoms Approach, which was based on the Brambell report (1965) and has for a long time been the prevalent model to assess farm animal welfare, has been criticized for emphasising negative experiences, even though the 5th freedom was to express normal/natural behaviour (Korte *et al.* 2007). Arguably, these freedoms constitute a very practical way to assess welfare, but, depending on how they are applied in practice, the 5 freedoms model may neither fully guarantee good animal welfare (Harfeld 2016), nor line up well with the biology of the animals (e.g. animals have many more than 5 different welfare needs (Bracke *et al.*, 1999c)). Moreover, welfare needs do not come in 'absolutes' like 'freedoms' do; in fact the term 'freedom' is suggestive of a Kantian framework tending towards attributing absolute animal rights. Under such a framework livestock farming is often regarded as not acceptable at all and this seems contrary to the context for which the 5 Freedoms were proposed in the first place.

Broadly speaking, three conceptions on animal welfare can be distinguished: a function-based, a feeling-based, and a nature-based view (Fraser 2003). When we apply this to the case of tail biting, according to the function-based view we need to ask to what extent pigs are coping with farming conditions, e.g. in terms of stress and (pre-)pathological states indicating biological functioning ('fitness'). Tail biting could be regarded as a sign that certain pigs are not coping. According to the feeling-based view we need to ask how much pigs suffer from being tail bitten (and perhaps how happy pigs are who are doing the biting), and how this compares to the pain involved in tail docking. According to the nature-based view we should primarily ask how closely the pigs' managed environment aligns to their evolved environment and to what extent it allows highly motivated species-specific behaviours, such as rooting, wallowing, nest building and social behaviour, and, if not, to what extent this leads to abnormal, psychopathological tail biting behaviour. In addition, the question may be asked whether tail biting or cannibalism itself occurs naturally and whether tail docking inhibits the expression of natural, or species-specific, behaviour. For example, tails most probably play a role in communication between pigs, as tail movements and posture is related to the emotional state of the pig (Camerlink *et al.* 2020). Tail docking may therefore hamper the expression of emotional states/natural behaviour.

Of course, when assessing animal welfare it is important to recognise the potential relevance of all three conceptions of coping/biological functioning, feelings, and natural living/behaviour. Often they reinforce each other. However, it is also possible that they conflict with each other, and even reinforce conflict, leading to polarisation. For example, from a feeling-based view it may be better to dock the tails of all piglets when the probability of tail biting later in life is too high (e.g. above 25%; Bracke, 2017). From a nature-based view, however, docking may be problematic as it facilitates the keeping of pigs under conditions that limit their possibility to express natural behaviour, thus resulting in an increased risk for tail biting. Another illustration given by Fraser (2003) is telling in this context. He describes two animal welfare research groups, one from the EU and one from Australia, that each assessed the welfare of pigs in gestation stalls of the same set of pig farms and who came to opposite conclusions. The Australian group concluded from the function-based view focusing on growth, reproduction, and health that the pigs experienced good welfare. The EU group, in contrast, argued from the feelings- and nature-based views that the pigs showed fear, abnormal behaviour, frustration, and did not have the ability to live in groups, and to explore and root.

As described above, different (types of) people may have different personal or scientific interests and emphasize different aspects of welfare, and this may depend on one's ethical background. For example, a utilitarian who strives to maximize overall happiness is most likely to emphasize the feeling-based view of welfare, whereas an ecocentrist – who thinks decisions should be made on the basis of what is best for the collective, i.e. the ecosystem, species or population – would more likely lean towards the nature-based view of welfare, or perhaps not value farm animal welfare as something worthy of concern (as (s)he thinks suffering is a normal part of life). So even though many definitions of animal welfare contain elements of two or even all three of the welfare conceptions, each emphasizes one, based on the background normative framework and broader views about how animals should live and what a good human-animal relationship is (see also Bovenkerk *et al.* 2012). In practice, even experts such as veterinarians, ethologists and stress-physiologists may give different welfare scores to the same set of housing conditions (Bracke *et al.*, 2008a).

When assessing welfare we must take a number of factors into account: it makes a difference if we assess welfare at a particular point in time or over a longer period; it also matters if we take a short-term analytical or a more holistic view. For example, when tails are docked under general anaesthesia, this may solve the acute-pain aspect of tail docking, but it does not address other aspects of welfare. Firstly, only using anaesthesia does not take into account the long-term harm that tail docking often generates, including long-term pain (See Chapter 2). Secondly, it does not ask what the causes of tail biting are. As the causes of tail biting are related to stress or frustration (e.g. Valros *et al.* 2021) tail biting can be considered an ice-berg indicator of pig welfare, and this should not be treated by simply treating the symptoms (See Chapter 15). Thirdly, it does not consider bodily integrity (not modifying pigs to our needs). Finally, it may be important to realise that welfare assessment may be relative to the context (Bracke, 2001) and, related to that, to our own welfare state. For example, the more luxurious our own lives are, and the more we have been raised as compassionate citizens, the more likely we will be motivated to demand proper welfare conditions for animals too (Bracke *et al.*, 2023). In this sense welfare is a fuzzy concept, i.e. its meaning may be relative to the context in which the term is used. In the next section we discuss five different ethical theories: utilitarianism, reform welfarism, Kantianism, virtue ethics, and care ethics.

5 Animal ethics theories

5.1 *Utilitarianism*

Utilitarianism is a form of consequentialism. This means that when determining the right course of action in a particular moral question, we should only consider the potential consequences of different alternative courses of action, and choose the one with the best projected outcome (also called utility), for example the most overall happiness. According to utilitarianism, our aim should be to maximize good outcomes and minimize bad outcomes for all those involved in a moral issue/question. What is considered a 'good' or 'bad' outcome is dependent on what particular version of utilitarianism one holds. For example, hedonistic utilitarianism defines the good as happiness or pleasure and the bad as unhappiness or pain, while preference utilitarianism defines the good as fulfilled preferences (Singer 2011).[6] Hedonism is 'like-oriented', while preference utilitarianism is 'want-oriented', and these may not always coincide. For instance, a biter pig could be regarded as a kind of drug addict with a very strong motivation (wanting/preference) to frantically bite the tails of pen-mate pigs, but at the same time have a relatively low level of liking (hedonic pleasure) derived from the act of biting itself.

According to utilitarian animal ethics, capacities like language, reasoning, and moral agency are not by themselves relevant for moral consideration. All creatures that can minimally experience pain and pleasure – or in other words, that

6 Some hedonistic utilitarians, in particular Jeremy Bentham, treat happiness as more or less synonymous to pleasure, while others see happiness as something that is more encompassing than mere pleasure. John Stuart Mill, for example, thinks that some pleasures have a higher quality than others and will bring more happiness.

are sentient – have interests that should be taken into account in moral decision-making. The fact that people or pigs are relatively intelligent does not make them more entitled to moral consideration. It is their ability to be happy or unhappy that matters. However, being able to communicate what matters to you as a person, or to a pig by being vocal about it, may make moral agents more inclined to 'listen', as compared to, for example, the suffering of fish, who are less able to communicate their mental state to us. This, however, does not mean we should weigh the interests of the pigs or people higher than the interests of, for example, fish, as again, it is only (the amount of) happiness that matters. Belonging to a specific species thus does not in and of itself confer moral worth; in other words, human beings should not receive preferential treatment solely because they belong to the species 'human'. Otherwise, we would be committing 'speciesism', which – like sexism or racism – is discrimination on the basis of a morally irrelevant characteristic (Singer 2011).

Utilitarians weigh the interests (i.e. all pleasure & pain/preference satisfaction & frustration) of all those involved: they consider that we should treat equal interests equally and unequal interests unequally. Of course, what specific interests an individual has can be influenced by factors like age and species. Reason can also modulate or affect the type and amount (intensity, duration, and incidence) of pleasure and pain that is involved. For example, human beings but not pigs can suffer from being told they have an incurable disease. On the other hand, people might also sometimes suffer less than pigs, e.g. when they get vaccinated and understand this will protect them from severe illness. When weighing the interests of all sentient beings involved in our actions it helps to classify the types of stakeholders (e.g. farm animals, producers and consumers) and the types of interests that are involved. Donald VandeVeer (1979) has suggested that we distinguish basic, serious, and peripheral interests. Basic interests are, for example, access to food and water. Serious interests may restrict functioning severely, e.g. freedom of movement. Peripheral interests are luxuries that are not necessary for living a full life, such as chew toys provided in addition to natural bedding for pigs, or fur coats for people. VandeVeer argues that because most humans have more cognitive complexity their interests are often likely to weigh more than those of most non-human animals.[7] When the basic interests of humans and animals clash, human interests therefore generally trump those of animals in his view. However, peripheral or perhaps even serious interests of humans should not prevail over the basic interests of animals. For example, in order to produce veal, calves are taken directly from their mothers and put on an iron-deficient diet (EFSA 2012; EFSA 2022). They are not allowed to

7 Note that he tries to avoid speciecism by allowing that some human beings have lower cognitive complexity and their interests should weigh less than those with higher cognitive complexity. While this position is of course open to the criticism that this would mean that more intelligent people should get preferential treatment over less intelligent people, what is useful about his suggestion is that interests can be compared by classifying them as basic, serious, or peripheral.

graze and are slaughtered at a young age. This apparently serves a peripheral interest of humans, namely the preference for tender, white veal, and should not trump the more serious or even basic interests of the calves in living a long and healthy life and in the company of their mothers.

When we apply VandeVeer's reasoning to the case of tail biting and docking in pigs, we should first of all ask what type of interests are involved. On the side of humans, there is the interest in eating relatively cheap meat and on the side of the farmer in particular there is the interest in financial stability. Eating meat, and in particular eating cheap meat, could be considered a peripheral human interest, as people do not need meat to live a full life as there are many alternative and tasty sources of protein, and other nutrients found in pig meat (and red meat) may even pose a (minor) health risk to humans (WHO 2015). Financial stability of the farmer could be classified as a serious interest. The pigs have an interest in experiencing pleasure and avoiding pain. If pigs suffer from severe tail biting, which may lead to cannibalism, this indeed represents a basic interest. This could partly be compensated for by reducing the risk of biting by tail docking the pigs, although this also seems to classify as a serious interest, as a docked tail may lead to chronic (phantom) pain.[8]

Peter Singer (2011) argues that we should stop consuming meat, because the pleasure that people experience from eating meat does not outweigh the suffering that most animals endure under farming conditions. Interestingly, however, within the animal ethics community he is notorious for arguing that killing farm animals *as such* is not a moral harm. According to his so-called replaceability argument, if we could farm animals without detriment to their welfare, and kill them painlessly, we could simply replace one such 'happy animal' with a similarly happy animal. As long as the total amount of happiness or pleasure stays the same, Singer (2011) sees no problem in killing particular individual animals and replacing them with other individuals. This replaceability argument does not apply to most humans and to animals with high cognitive complexity, such as great apes and whales, as they have the capacity for self-reflexivity that is needed to have a 'preference for continued life'.

Of course, this is a theoretical argument. Singer is well aware that the usual processes of human-controlled reproduction, transport and slaughter in pig production are all well established as very stressful conditions. A more theoretical objection would be that the replaceability argument would also allow painlessly killing and replacing people who do not have a preference for continued life, such as infants or severely mentally-handicapped people (Pluhar 1995). It has furthermore been argued that a being does not need to have a preference for continued life in order for death to be harmful (Bovenkerk *et al.* 2016). According to the so-called 'foregone opportunities account' death is harmful not because it interferes with one's preference for

8 According to many animal ethicists, pigs in addition, have a basic interest in living, although not everyone agrees.

continued life, but because 'death forecloses the valuable opportunities that continued life would afford' (DeGrazia 2002, 61). In other words, life is instrumentally valuable for all sentient beings, and when one kills a sentient being one takes away all the potential future welfare that the being would have otherwise enjoyed. Of course, that also implies that if pigs are living in rather impoverished conditions, utilitarian reasoning may suggest that killing a pig is a morally good thing to do when it involves reducing the suffering of the pig (and enhancing the 'pleasure' of the consumer (taste) and the producer (income)).

What the foregoing discussion shows, is that within utilitarianism different positions can be taken. Some utilitarians think that rather than asking whether it is morally acceptable to dock pigs' tails, we should not keep pigs for meat production in the first place. If one adheres to the foregone opportunities account, the basic interest of pigs to have a continued life outweighs the serious or peripheral interests of humans to produce or consume pork. For utilitarians who adhere to the 'preference for continued life' account, it is not so much the death of the pigs itself that is considered harmful, but weighing the suffering of the pigs vis-a-vis the benefits to humans. In such a weighing process, we need to include the number of pigs that suffer, the number of human beings who benefit, and the degree of suffering of the pigs, as well as the degree of enjoyment of the human beings involved. As stated before, in Singer's view this calculation generally works out against farming most animals. However, if we would keep pigs in conditions that are very favourable to their welfare, farming pigs could be morally acceptable. Exactly how favourable their welfare would have to be to justify farming pigs would depend amongst others on how many pigs are farmed, how much improved welfare this would bring to how many people. Moreover, it would depend of course on how we define welfare; if we focus on the absence of pain we only have to weigh the severity of pain and the number of pigs involved; if we also focus on natural behaviour, there may be a presumption in favour of keeping tails intact. Another noteworthy corollary of the utilitarian calculus is that if a chicken and a pig have a similar level of happiness, then we should prefer to eat the pig rather than the chicken, because one pig can feed more people. Ideally then, utilitarian farmers may be looking for farming very large animal species, as the more human happiness can be generated from sacrificing one individual, the more likely the common utilitarian ethical principles will subscribe to it.

5.2 *Welfarism*

Another consequentialist theory, welfarism (also termed reform welfarism), starts from the premise that raising and killing animals for food production is not problematic as such, as long as the animals experience good welfare.[9] According to this view,

9 Welfarism is often taken to be a type of utilitarian theory, but strictly speaking this is not the case. Welfarism does not say that welfare should be maximised and it does not operate on a principle of equal and impartial consideration. Thanks to the anonymous reviewer for pointing this out.

which is adopted by many animal welfare scientists, animal liberation as proposed by the utilitarian philosopher Peter Singer, or animal rights as approached by the Kantian philosopher Tom Regan (Regan, 1983; see below), is too demanding. According to welfarism, all we need to do is to improve welfare step-by-step, e.g. by improving husbandry conditions (e.g. providing enrichment materials and more space), and if this is not possible, by taking measures affecting the animals, such as breeding animals that are more able to cope with farm conditions, or perhaps even changing the animals themselves (by for example tail docking in pigs or beak trimming in laying hens). Welfarism has been criticized for only arguing for limited reform, while providing a justification for the continued intensive use of animals (Haynes 2008; Schmidt 2011), hence leading to 'humane-washing'. Moreover, welfarism has been criticized for employing an overly narrow conception of animal welfare, only focussing on the absence of pain, hunger, illness, and discomfort (Haynes 2008).

Two issues are important to note here: firstly, as mentioned before, animal welfare as such plays a role in practically all animal ethics theories, and is not limited to utilitarianism or reform welfarism (see also Schmidt 2011). However for other theories, there are more concerns than welfare that matter morally; for Kantianism, for example, intrinsic value is also important. What is specific for reform welfarists is that they tend to have a hierarchical view of moral status, entailing that human interests almost always outweigh animal interests. This contrasts, in other words, with Singer's egalitarian view. Secondly, the reform welfarist position should be distinguished from approaches that argue that even though *ideally* we should thoroughly change current farming practices or end animal use completely, *in practice* animal welfare reforms are the best we can hope for under current circumstances (Garner 2013). This latter view is a so-called non-ideal theory. The reformist welfarist position, on the other hand, could be termed an ideal theory, as its ideal is to still use animals for human purposes, but to do so in a way that takes into account (a narrow version of) animal welfare. We will come back to this distinction between ideal and non-ideal theory later.

If we follow a welfarist approach, an important question is how the harm of tail biting compares to the harm of tail docking. In other words, how much tail biting should a farmer and veterinarian allow when deciding to stop tail docking? To be able to do such a weighing, we would need to estimate the amount of pain caused by docking and that caused by tail biting. Such a calculation procedure has been suggested by Bracke (2017) and Valros & Heinonen (2015). Bracke (2017) only looked at the welfare implications for the pigs, including e.g. the stress involved to induce biting in the biter pigs. As a rough estimate he suggested that the sum of pain experienced by a tail-docked pig population and that of an undocked pig population would be equal if 2 % of the docked pigs and 12 % of the undocked pigs are tail bitten. Valros & Heinonen (2015) further proposed that as pain is only one factor to be considered, the weighing should additionally consider aspects such as improved housing conditions if pigs are not docked, altered health status, and risk of infections due to docking or biting. Thus if we accept that a small number of undocked pigs in a pen will be bitten, the improved living and health standards for the rest of the group would mean that we sacrifice the interests of only a few pigs for the interests of the many. This is a type of reasoning where Kantianism may diverge from utilitarianism, as according to Kantians we may not sacrifice individuals for the greater good. On the other hand, if we dock all pigs to reduce a risk that will only be realized for a few, we sacrifice the many for the few, which is something that would be considered problematic both from a utilitarian (depending on the amount of pain experienced) and a Kantian perspective.

5.3 *Kantianism*

As mentioned above, welfare plays a role in practically all animal ethics theories. In Kantian theories, however, other issues besides welfare are deemed important, and in particular the view that individuals have inviolable rights (e.g. a right to life and a right to be free), and we have a duty to respect these rights. Furthermore, in contrast to utilitarianism, Kantian theories do not focus only on the consequences of our actions, but in the first place on the acceptability of actions themselves. This means that certain actions can be wrong in themselves even if they lead to the best overall consequences. In order to find out what is the right action we have to look for moral principles that are 'founded in practical reason, so that all rational beings have to accept this principle' (Kaldewaij 2013, p. 64). For Immanuel Kant this reason is the Categorical Imperative, i.e. the idea that one should only act in such a way that one's action could become a universal law. For example, if I wonder whether I should lie in a specific case, I need to think about what would happen if everyone were to lie in such a situation.[10]

For Kant only moral agents – those who can make moral judgments – have moral status – or in other words, have interests that need to be taken into account in our moral decision-making.[11] However, neo-Kantian animal ethicists have argued that in order to have moral status it is enough for an animal to be 'an agent in a much less demanding sense' (Kaldewaij 2013, p. 65). In Tom Regan's view, to have moral status animals need to be 'subjects-of-a-life', which can be taken to mean that they can experience their life subjectively. For this they need to have certain cognitive capacities: 'beliefs and desires; perception, memory and a sense of the future, including their own future; an emotional life together with feelings of pleasure and pain; preference – and welfare – interests; the ability to initiate action in pursuit of their desires and goals; a psychophysical identity over time; and an individual welfare in

10 This is one formulation of the Categorial Imperative, named the Universal Law formulation. As we will see, there is also another formulation that Kant uses, namely, the Humanity as an End in Itself formulation. In his view these are two formulations of the same principle. For a clear explanation see Timmons 2002.

11 For Kant humans should treat animals properly only in as far as improper treatment of animals was more likely to make us treat other humans improperly too.

the sense that their experiential life fares well or ill for them' (Regan 1983, p. 243). Other neo-Kantians argue that if an animal can have desires and can purposefully pursue them this is enough to attribute moral status to them (Kaldewaij 2013; Korsgaard 2011). Examples of animal desires and purposive action would be avoiding being tail bitten by tucking the tail between a pig's hind legs or by standing in the corner of a pen such that biter pigs cannot as easily bite its tail.

Kant formulated a principle that all rational beings have to accept. According to one formulation of this so-called Categorical Imperative, you should never use humanity (ourselves and others) solely as a means to an end, but always also as an end in itself (Timmons 2002). This could be taken to mean that we should never use others as mere instruments and this limits us acting out of our own selfish desires. From this principle a number of other duties can be derived, such as the duty of beneficence and non-maleficence (doing no harm). Note that Kantianism is also called a deontological ethical theory, a theory based on duties (rather than consequences in terms of utility as is the case in utilitarianism, which is therefore called a consequentialist theory). While Kant's formulation relates only to humans, neo-Kantian animal ethicists, such as Tom Regan and others, have applied this line of reasoning to animals as well, as they have argued that animals also have moral status (see above). They argue that animals should not be used solely as instruments for our use and that animals have certain rights which should not be violated. They hold that (most) animals have an intrinsic value in and of themselves that is not reducible to their instrumental value to us. In other words, they should not be 'instrumentalised' in the sense that they are not simply instruments for our purposes, but they have purposes in their own lives that are important to them. Animals moreover have dignity that can be violated when we use animals for certain purposes, such as when we train them to do tricks in a circus. Compared to utilitarianism, Kantian theories are rather absolute; if beings have moral rights, they can never be used solely for the purposes of others. Hence, these theories have generally led to an abolitionist stance on farming animals (and using them for experimentation or in circuses for example). Regarding the killing of animals, most neo-Kantian animal ethicists appear to hold the 'foregone opportunities account' of the harm of death. In the words of Regan (1984, p. 117): 'Death is the ultimate, the irreversible harm because death is the ultimate, the irreversible loss, foreclosing every opportunity to find satisfaction'.

Nevertheless, if we were to allow the use of animals for food production, we would at the very least have to take into account their own goals in life. This implies that we should not regard or treat an animal as an object (Brom 1997). If animals are altered to fit the environment we provide them, this may imply they are mere objects, changed to fit our needs. Treating an animal as an object is problematic because it leads to a denial of the animal's own interests and goals in life (see also Bovenkerk 2020). Thus, to solve problems with tail biting in pigs we should make changes to the pigs' environment rather than to the pigs themselves. Moreover, in this view making changes to animals, e.g. by docking tails, may be problematic even if this does not interfere with – or even improves – their welfare because it violates their integrity. Rutgers and Heeger (1999) describe integrity as 'the wholeness and intactness of the animal and its species-specific balance, as well as the capacity to sustain itself in an environment suitable to the species'. A pig whose tail has been docked is then no longer intact and this changes an important aspect of what it means to be a pig (see Bovenkerk 2020). Pigs whose tails have to be docked in order for them to produce pork economically are probably not living in an environment that is suitable to their species. Similar arguments have been made in feminist literature under the name of 'objectification'. We will turn to feminist care-ethical theories next.

5.4 *Care ethics*

In the relatively new field of care ethics or relational animal ethics it is argued that animal suffering is problematic not because it does not maximize happiness or violates rights, but because we care about animals and have relationships with them (Engster 2006). Feminist care ethicists criticize traditional animal rights theories for overlooking the interdependent, relational nature of both human and animal individuals (Donovan *et al.* 2007). Universalistic theories, such as utilitarianism and Kantianism do not have an eye for the relations of inequality and dependence that frequently exist within the family (between parents and children, and between husbands and wives for example) or on the farm (between farmers and 'their' animals). Care ethicists are focused on the value and quality of relationships and the conditions that are necessary to build and sustain those relationships. In relationships with animals it is important to acknowledge that they have feelings, and they point out that even though we often do not take notice, animals communicate these feelings. Unlike utilitarians and neo-Kantians they do not necessarily attribute *equal* moral status to animals (Engster 2006). Some care ethicists are in a sense speciesist, because they argue that every species is 'justified in caring first and foremost for its own kind since we all depend primarily on our own species for survival and development' (Engster 2006, p. 528). Other care ethicists think that it is not necessarily relationships with our own kind that matter most, but the quality of the relationships we are entangled in and whether these are reciprocal (Noddings 1984).

Care ethicists point out that because animals in the past have not been regarded as communicative beings, they have literally not been given a voice.[12] Farm animals

12 Like all feminist theorists, care ethicists focus on the implications of the way we use language. With language we can in- or exclude others and we can obscure or bring to light power imbalances. For example, many euphemisms are used when we refer to animals: we speak of pork instead of pig meat, we speak of euthanasia or culling instead of killing, (poor/reduced) animal welfare instead of animal suffering, enrichment instead of 'less impoverished'. We have words especially reserved for animals that aim to reinforce the divide between humans and other animals; animals have paws instead of legs and they are 'with young' instead of pregnant. The

have been made invisible – they are the 'absent referent' on our plates (Adams 1990); this means we are screened off from them, not only physically because we generally do not witness what happens on farms and in slaughterhouses, but also ideologically. Care ethicists take issue with the hierarchical nature of the relationship of owners and their animals, a hierarchy that is based on false dualisms in patriarchal societies; dualisms between, amongst others, man and woman, reason and emotion, mind and body, and humans and animals. They argue that through these dualisms mistreatment is rationalized and they criticize current systems in which hierarchy and power structures figure largely, e.g. the political bases for the mistreatment of animals (Donovan *et al.* 2007; Luke 1997). In general, care ethicists tend to call for a radical transformation of the social and political order, including industrial scale farming.

As part of their analysis of hierarchical power structures and false dualism, feminist thinkers draw parallels between the way women and animals are treated in patriarchal societies. Amongst others, they draw a parallel between the objectification of women and that of farmed animals. The concept of objectification refers to the previously discussed Kantian notion of instrumentalization. When a woman or an animal is objectified she is reduced to an object and adapted to the needs of those who objectify her. The system in which a woman or an animal lives can have this type of transformative effect. For example, some have argued that in highly intensive and technological production systems, animals are simply understood as production units and have become 'living parts of machinery' (Harfeld *et al.* 2016). According to Martha Nussbaum's (1995) conceptual analysis, objectification leads to a denial of a woman or animal's subjectivity and intrinsic value; it furthermore denies autonomy and treats the woman or animal as violable, replaceable, and a property that can be owned and dominated. This is regarded as leading down a 'slippery slope to abuse' and as being contrary to a relationship of care (see Bos *et al.* 2018 for a more thorough discussion of objectification). Adapting animals to the farming context rather than adapting the farm to the animal's needs, in this view is to objectify animals, partly even turning them into artefacts or commodities. If we apply this to the context of tail biting and docking it appears problematic to dock tails preventively, as this would mean adapting the pigs to the production process. On the other hand, it would be regarded as good care to provide ample enrichment to pigs when a tail biting outbreak starts, and to spend time and energy on solving the underlying problem and making sure it does not happen again.

---

word 'animal' itself has been problematized; it sets us humans apart from all the other animals, while in fact, biologically-speaking, we are undoubtedly animals too. While these seem semantic issues at first sight, they are tied up with values and in the end none of these terms is morally neutral. In fact, farmers can get rather upset when modern farming is referred to as factory farming or bioindustry, even though the latter may formally be a fairly accurate description of what this 'current type of farming' is mostly about.

According to care ethicists, we have a duty to care for those animals that cannot (adequately) care for themselves, and we should have special care for those who are most vulnerable. According to Palmer (2011) this duty is only present when we have somehow influenced the position animals are in, either by keeping them in captivity or by breeding them. In these cases, due to our past actions we have created a commitment to care for these animals, a commitment we do not have towards animals in the wild. An important question in care ethics is, logically, what is meant by proper care. According to Engster (2006, p. 522) 'the most general goal of caring is to help others to survive and function so that they take care of themselves and others, and pursue some conception of the good life'. What this entails more specifically differs between species and perhaps even between individuals, but care ethicists generally agree that raising animals in highly intensive systems where they are deprived of space, the ability to express important natural behaviours and the means to build meaningful social relationships with conspecifics does not entail proper care. There is disagreement among care ethicists, however, about the question of whether raising animals for consumption as such is morally allowed. According to some, 'raising animals for slaughter would not seem to be very conducive to treating them in a caring manner' (Engster 2006, p. 532). Others think that if animals were given the opportunity to express natural behaviour and develop social bonds and subsequently killed as painlessly as possible, raising animals for consumption is not in principle problematic. However, this would still call for quite substantial reforms to the ways in which animals are commonly kept.

If we apply this to the case of tail biting, we could argue that both the victims of tail biting and the biters should receive special care, because they are both victims of the way they have been raised. Zooming out to the social and political context of modern farming, we therefore need to ask whether pigs may be raised in environments where they cannot express natural behaviour in the first place. A final point which is related to virtue ethics (discussed next), is the fact that livestock farmers often strongly feel they are truly caring properly for their animals. Though some animal ethicists and animal activists may strongly object to this suggestion, our experience is that it is certainly how farmers feel. It should also not be underestimated what it is like to be a farmer, making sure thousands of pigs have sufficient food and water every day, a proper climate, shelter and health care, day in day out, and irrespective of periods when prices are really low. Considering this, farmers surely sometimes feel they are being exploited themselves.

5.5 *Virtue ethics*

Like care ethics, virtue ethics does not focus on universal principles that can be applied in all moral contexts. Virtue ethicists do not think that the question of whether animals should be attributed moral status is very helpful (Hursthouse 2019). As Lori Gruen (2010, np) explains, animals are simply members of our community that 'pull on us and it is in virtue of this indescribable pull that we recognize what is wrong with cruelty. Animals are individuals with whom we share a common life and this recognition allows us to see them as they are'. Virtue ethical theories do not focus on specific good or bad actions, but on general virtuous behaviour and having a good attitude. If we treat animals badly, we are displaying the wrong character traits. Virtuous character traits are, for example, respect, sensitivity, compassion, and moderation, and we do not cultivate these traits when we routinely harm animals. As Alvaro (2019, p. 634) puts it: 'Rather than arguing that animals have inherent rights or that they should be factored in a utilitarian calculus, we are better off realizing that raising animals for food, quite apart from rights and utility, is unvirtuous: e.g., callous, intemperate, self-indulgent, and base'. Virtuous people do not behave virtuously after reflection on the reasons to behave that way, but because they have cultivated a good disposition. They also do not act just on an emotional impulse or do what is right for self-interested reasons (Hursthouse 2019).

Jes Harfeld *et al.* (2016) advance a virtue ethical argument, claiming that experience of animals and their complex behaviour as it has evolved in their natural environment is necessary for people to develop the practical wisdom (*phronesis*) that will help us build moral character and make sound judgments regarding the treatment of animals. They argue that current conditions in livestock production are detrimental to our grasping the 'animalness' of animals and this makes us fail to give animals adequate ethical consideration. We need to experience the commonality between ourselves and animals both to acknowledge the moral considerability of animals and to enhance human self-understanding. The way animals are used, housed, and treated in their view leads to certain representations or perceptions of them, and these may lead to an attitude towards them as if they are mere things and their own experience as living and sensing individuals is overshadowed. They are in effect reduced to their functions within the system where there is room for few other capabilities and behaviours than 'giving off-spring, producing milk, and dying' (Harfeld 2016, p. 414). Housing animals in intensive systems which take the animals out of their environmental context means that they have very limited opportunities to show species-specific behaviour and they cannot clearly exhibit individual differences in personality. This may lead to an impoverished understanding of animals. These conditions keep us from building ethical relationships with animals. They also make us disregard the broader aspects of animal welfare, those having to do with the good life for an animal.

Virtue ethics can be applied to specific professions. A farmer should ask him (or her)self 'am I a good farmer'? An answer to this question could be based on specific virtues (which are mentioned in italics). For example, a *responsible* farmer will do research into what is the best and most *caring* way to keep his or her animals. Moreover, the farmer should refrain from being cruel to his or her animals not only because this will lead to unhealthy animals that will generate less income, but out of an attitude of respect and care for the animals. A conventional farmer who decides to take up the challenge to stop tail docking, may experience criticism from colleagues who see this as a risk that, if (s)he succeeds, they may later be forced to stop tail docking as well. It thus may take (quite) some *courage* to be a regular farmer and commit to providing proper enrichment and stop tail docking. If a farmer docks the tails of pigs due to problems with tail biting (s)he should do this out of an attitude of care for the pigs and should not do this routinely and preventatively.

6 Ideal versus non-ideal ethics

As has become clear in the foregoing discussion of animal ethics theories, many animal ethicists are critical of (intensive) animal husbandry in principle, and rather than discussing whether we are allowed to dock the tails of pigs, they argue that we should not keep pigs in intensive husbandry systems in the first place. In particular, those who argue that death is a moral harm for animals call for the abolition of livestock farming altogether. It might be easy to dismiss these ideas as too radical and unfeasible. After all, we can expect the demand for meat to increase at least for decades to come (OECD and FAO 2018). In this light, it seems naive to think that completely ending animal production is an achievable option in the short term. Still, we think it is important to take note of these views. First of all, when encountering a problem – such as tail biting – it makes sense to solve the root causes of the problem rather than solve the symptoms. Of course, differences of opinion exist about whether the root cause is the keeping of pigs as such, especially in intensive husbandry systems, or rather the way pigs are being kept and managed, for example with a lack of adequate housing and limited opportunities to carry out natural or species-specific behaviour. Secondly, if we were to accept the status quo – i.e. a majority of intensive and industrial scale pig farms – too easily, just because anything else may at the moment not seem feasible, we lack an ideal to work towards and this may actually hamper improvements to the current situation, because we fail to see different alternatives.

In ethics, a distinction is made between ideal theory and non-ideal theory approaches. As Robert Garner (2013, p. 10) explains, there is a debate in ethics which 'emerged as a result of increasing frustration by many at the discrepancy perceived between the abstract normative work of political philosophers, in which ideal political and moral principles are advocated, and the difficulty of applying such principles in the nonideal real world'. The word 'ideal' should here not be understood as utopian, implying that it could never be achieved; the ideal of ideal theories has to be in principle practically possible. However, as Zofia Stemplowska (2008, p. 320) explains, an ideal theory 'assumes that people, at least generally speaking, do what the theory demands of them. Such a theory is ideal in the sense that it assumes an idealized (untrue) version of reality: in the actual (nonideal) world, people do not always comply with what is required'. Of course, different people have different ideals; for some the ideal would be a world in which animals are no longer farmed for food, for others it might even be a world without domesticated animals (Francione 2010), for yet others it may be a world where livestock is kept but in a manner respectful of their species-specific needs.

Reflecting on the ideal situation, where everyone would comply with a theory's recommendation, is helpful because it forces us to consider what changes are necessary in order to come closer to an ideal and also to judge how far we actually are from that ideal at the moment (Stemplowski 2008). Because they have an image of the dot on the horizon, ideal theories can give us inspiration and reasons for change. They describe a kind of 'end-state' towards which we can strive. An ideal theory can therefore provide us with a long-term goal for (institutional) reform (Valentini 2012). However, 'the moral standards that we can derive from higher ethical theories are not always well-equipped to deal with practical decision making in real world situations' (van Herten 2021, p. 104). We therefore need non-ideal theory to give us clear recommendations. Certain ideal theorists, in particular abolitionists, whose only answer to the problem of tail biting and docking is saying that we should not farm pigs in the first place, may place themselves outside of further discussions on how to improve pigs' welfare ethically. However, this does not mean that their views are useless; they have an important role to play to show what the ultimate ideal could be. Yet this ideal theory needs to be supplemented by a non-ideal theory about what steps could be taken under non-ideal conditions.

In our view, therefore, we need both ideal and non-ideal perspectives; they need to reinforce each other. We need ideal theory to set the goals against which we can measure what actions are permissible; those actions are the steps necessary to eventually bring us to our ideal. We need this ideal 'to know we are on the right path and give meaning to our journey' (van Herten 2021, p. 104). But we also need the steps on this journey, and they are provided by non-ideal theories. What is important though, is that these steps as formulated by non-ideal theory do not take us further away from the ideal. For example, if our ideal would be the abolition of pig farming because we think pigs' integrity is regularly violated, an intermediate step may be to improve pig welfare. However, if we were to do this by breeding pigs without tails so that they can no longer be bitten, this may in fact violate the pigs' integrity even more and bring us further from our ideal. Improving welfare will generally mean that the motivation to end farming practices is reduced, moving away from the abolitionist ideal of some.[13] When acting as part of a non-ideal theory, it should be kept in mind that these actions are ultimately justified by the ideal theory. This implies that our measures to improve pig welfare are only justified if we at the same time

13 As Stemplowska (2008, p. 339) argues, therefore, 'this suggests that there is no real conflict between ideal and nonideal theory. Indeed, we should see all normative theorizing as part of a common project: the use of ideal theory techniques contributes to the use of nonideal theory techniques (and vice versa)'.

work towards our ideal situation, for example a situation where pigs are not farmed for food or where they are kept in circumstances where they can live a good life. Moreover, it should be clearly communicated what the final goal is.[14]

7 Changing the pig or changing the environment?

As has become clear so far, when we consider the case of tail biting – and docking – from an ethical perspective, one question is whether we should farm pigs for consumption in the first place. According to some ideal theories, the answer is a clear no. Other ideal theories posit that it is morally allowed to farm pigs, as long as the pigs have good lives, meaning that they experience good welfare, are able to carry out their species-specific behaviour, and are able to form social relationships with conspecifics (cf. Garner 2013). Certain non-ideal theories argue that in principle we should not farm pigs, but as this is not feasible in the short term, in practice we would be allowed to farm pigs under two conditions: 1) we undertake steps to improve their welfare (as long as this does not move us further away from our ideal), 2) we at the same time undertake steps to get closer to our ideal.

A second question that is raised is what we should do from a welfare point of view. According to utilitarian theories, we have to weigh different alternative options for action, and act in ways that maximize utility or happiness. This could mean that consumers should pay a bit more, and farmers should provide proper enrichment, care and housing conditions to allow rearing pigs with intact tails. As stated before, only considering the direct welfare implications for the pigs, a rough estimate is that about 12% tail biting in undocked pigs might be equivalent to 2% tail biting in a docked herd (Bracke, 2017). The weighing should additionally consider aspects such as changed housing conditions if pigs are not docked, altered health status and risk of infections due to docking or biting.

A third question is 'Do we think that all ethical issues have been solved whenever the welfare issues have been solved?'. For example, if we could breed pigs without teeth and without tails, or more sociable pigs that are less likely to bite or pigs that can deal better with the farming environment, would this satisfactorily solve the ethical problems? At first sight, this is a case where utilitarian theories conflict with Kantian or care ethical theories. After all, from a utilitarian point of view the goal is to maximize pleasure/happiness and to minimize suffering and these breeding interventions would alleviate a lot of suffering. According to Kantian theories, in

14 Note that in the context of ideal versus non-ideal theory two issues should not be intermingled: the difference between abolitionist and reformist theories on the one hand and the difference between piecemeal and radical change. Abolitionist theories tend to be ideal theories and advocate radical change. Reformist theories tend to have a different ideal than abolitionism (namely the improvement of welfare, which could also be considered an ideal) and they tend to advocate piecemeal change.

contrast, this may be problematic because it instrumentalises pigs. It may also violate the pigs' integrity, if we consider the tail or even biting behaviour as essential parts of the pig; we would then perhaps be starting to take away the 'pigness of a pig'. Similarly, for care ethics, this would be an example of the objectification of pigs, moulding them to farming conditions and thereby denying their subjectivity (Bos *et al.* 2018). Indeed even from a utilitarian point of view there could be objections to this approach. If we breed pigs without tails we take away a communication device. Moreover, biting behaviour (not necessarily directed towards tails, but as part of the pigs' normal behaviour repertoire while eating and foraging) might cause happiness in pigs, and breeding this out of pigs may mean we would be taking away possibilities for pigs to be happy. We could, perhaps, also breed pigs that release more endorphins and are in a continuous state of pleasure. What if we could find a (party) drug that brings pleasure in a more artificial way? If this drug were safe for consumers and would solve all the welfare problems associated with tail biting and tail docking, many people would probably still have fairly strong moral intuitions that such a way of solving welfare issues is not morally desirable, even though it may be a lot better than the current situation in global pig farming.[15]

At first sight, then, there is convergence between theories that it is better to change the farming environment than change the pig to fit the environment. After all, changing the pig to fit the environment seems to take us further away from the ideals of either abolition or creating circumstances where pigs can lead good lives. However, not all changes are problematic (e.g. just cutting a pig's hair is not a morally problematic change). Moreover, de-breeding bad side-effects of our previous breeding choices appears to be morally acceptable or even required. For example, if pigs have a higher likelihood of getting arthritis due to the way they have been bred, this is something we should turn back if we can. This relates to a discussion in animal ethics about so-called 'painless livestock', where it is argued that under non-ideal circumstances it is better to knock out the capacity to experience pain in livestock than to continue to raise them in ways detrimental to their welfare (Shriver 2009). However, it seems more realistic and morally appropriate to make breeding choices that target the causes of pain. For example, one source of pain in livestock is mastitis; wouldn't it be better to breed cows with smaller udders (i.e. de-breeding a side effect of a previous choice to breed for maximising milk yield) than to continue breeding cows with large udders and then knock out their capacity to feel the pain from mastitis? The upshot of this discussion is that intentions matter. If we de-breed unfavourable side effects of prior breeding choices, the intention is to undo past harms. If, however, we were to breed for example sociable swine in order to enable upscaling pig production facilities, this would be morally dubious.

---

15 An interesting follow-up idea would be whether we should use such breeding or drug solution to solve human welfare problems too (e.g. the need of some to eat 'real meat').

8 Conclusion

We have discussed several different ethical theories and applied these to the problems of tail biting and docking. While there are certainly disagreements between these different theories, and theorists, there is also in fact a high degree of convergence, at least within the field of animal ethics. First of all, we need to look at the underlying causes of tail biting and consider if alternatives are possible to tail docking, which in the end is just the treatment of a symptom that does not solve the problem of reduced welfare (stress) that seems to trigger the behaviour. When pigs are housed in suboptimal conditions they become stressed and frustrated, and this can lead to harmful social behaviour. When they have nothing to relieve this frustration and stress the biting can easily get worse.

Secondly, from the perspective of most animal ethicists and their diverse theoretical backgrounds it is problematic to change the pigs by docking their tails rather than change their environment (by providing enrichment etc.). Breeding pigs without tails, without biting impulses or pigs that are happy despite poor living conditions may raise moral concerns, both from the viewpoint of pig welfare and from arguments beyond welfare, such as arguments about instrumentalisation, objectification, violation of integrity, and de-animalisation. On the other hand, changing pigs to undo problems that were caused by past breeding choices appears less morally problematic.

Finally, we need both ideal theory, to show us the dot on the horizon, so to speak, as well as non-ideal theory, to give clear recommendations under non-ideal circumstances. Briefly coming back to our comparison of the Dutch and Finnish ways of dealing with tail biting, it could be argued that Finland shows that the ideal of stopping tail docking is attainable, if the underlying causes of tail biting are addressed. The Dutch system of certification with the different stars for improved welfare conditions is an interesting case to consider from the point of view of ideal versus non-ideal theory. When this system was implemented at first, it seemed to be a non-ideal theory recommendation which took a step in the right direction. However, for some ideal theorists, in particular abolitionists and those who favour much higher welfare standards, the certification system can actually be regarded as taking us further away from the ideal. After all, people can clear their conscience by buying this so-called 'Better Life' meat, and may not be prepared to invest in further welfare measures. According to some the certification system is now in fact hampering improvements as well as running counter to abolitionist ideals (Eskens 2021).[16] In any case, ethical theories agree that we need to create conditions that allow for informed moral decision making, and this requires an open mind towards various

16 In fact, it may even be questioned whether the worst-off Better-Life pigs have actually had a better life compared to the best-off conventional pigs (see also De Jonge *et al.*, 2023).

more and less ideal ethical positions (prescriptive viewpoints) and towards knowing (all) the 'facts' (descriptive states of affairs).

9 Take home messages

- Different ethical theories such as utilitarianism, welfarism, Kantianism, care ethics and virtue ethics have different bases for moral judgments.
- While there are disagreements between these theories about what is good or bad, there is convergence on a number of issues related to tail biting.
- We need to first solve the underlying causes of tail biting, such as having limited opportunities for carrying out natural behaviour, and health and welfare problems caused by housing conditions and management decisions.
- We need to consider if alternatives to tail docking – such as providing enrichment – are possible, as docking is not only painful but disrespectful to the bodily integrity of the animals. Tail docking is also just a symptomatic treatment that does not address the cause of this psychopathological tail-biting behaviour.
- Breeding pigs without tails, without biting impulses, or pigs that are happy despite poor living conditions also raise serious moral concerns.
- We need to consider both health and welfare arguments, and arguments beyond welfare, such as the views that tail docking may lead to instrumentalisation, objectification, and de-animalisation of the pigs, and may violate their integrity.
- It seems helpful to distinguish between the proximate and ultimate objectives of moral reasoning, i.e. between ideal and non-ideal moral theories. We need both an ideal to know what ultimate aim to strive for and a theory that reflects on the steps that can be taken under non-ideal conditions.

References

Adams, C., 1990. The sexual politics of meat: a feminist-vegetarian critical theory. New York: Continuum.

Alvaro, C., 2019. Is animal suffering really all that matters? The move from suffering to vegetarianism. Journal of Agricultural and Environmental Ethics 32, 633–645. https://doi.org/10.1007/s10806-019-09793-0.

Bolt, I., Verweij, M., Delden, J.J.M., 2003. Ethiek in praktijk. Assen: Koninklijke Van Gorcum.

Bos, J.M., Bovenkerk, B., Feindt, P.H. and van Dam, Y.K., 2018. The quantified animal: precision livestock farming and the ethical implications of objectification. Food Ethics 2, 77–92. https://doi.org/10.1007/s41055-018-00029-x.

Bovenkerk, B., 2020. Ethical perspectives on modifying animals: beyond welfare arguments. Animal Frontiers 10(1): 45–50. https://doi.org/10.1093/af/vfz055.

Bovenkerk, B. and Braithwaite, V., 2016. Beneath the surface: killing of fish as a moral problem, in F. Meijboom and E. Stassen (eds), The end of animal life: a start for ethical debate . Ethical and societal considerations on killing animals. Wageningen: Wageningen Academic Publishers, pp. 227–250.

Bovenkerk, B. and Meijboom, F.L.B., 2012. Fish welfare in aquaculture: explication the chain of interactions between science and ethics. Journal of Agricultural and Environmental Ethics 26, 41–61. https://doi.org/10.1007/s10806-012-9395-x.

Bracke, M.B.M., 2001. Modelling of animal welfare: the development of a decision support system to assess the welfare status of pregnant sows. PhD Thesis. Wageningen University, Wageningen, 150 pp.

Bracke, M.B.M., Spruijt, B.M. and Metz, J.H.M., 1999. Overall welfare reviewed. Part 3: welfare assessment based on needs and supported by expert opinion. Netherlands Journal of Agricultural Science 47: 307–322. Available at http://library.wur.nl/ojs/index.php/njas/article/view/468/184.

Bracke, M.B.M., 2017. Weighing tail biting against tail docking. March 27, 2017. FareWellDock.eu. Available at http://farewelldock.eu/weighing-tail-biting-tail-docking/. Accessed 23-2-2-20.

Bracke, M.B.M., S.A. Edwards, B. Engel, W.G. Buist and B. Algers, 2008a. Expert opinion as 'validation' of risk assessment applied to calf welfare. Acta Veterinaria Scandinavica 50: 29. Available at https://actavetscand.biomedcentral.com/articles/10.1186/1751-0147-50-29. Accessed 22-10-2019.

Bracke, M.B.M., S.A. Edwards, J.H.M. Metz, J.P.T.M. Noordhuizen and B. Algers, 2008b. Synthesis of semantic modelling and risk analysis methodology applied to animal welfare. Animal 2: 1061–1072. https://doi.org/10.1017/S1751731108002139.

Bracke, M.B.M. 2018. Chapter 6. Chains as proper enrichment for intensively-farmed pigs? In: Spinka, M. Advances in pig welfare. 2017, pp. 167–197. Elsevier. Available at http://www.sciencedirect.com/science/article/pii/B9780081010129000058 Accessed 27-8-2018. Author version available at http://edepot.wur.nl/416659. Book chapter with supplement and technical details available at http://farewelldock.eu/chain-as-enrichment-with-supplement/. Author version at Wageningen UR: https://www.wur.nl/en/Publication-details.htm?publicationId=publication-way-353232373832.

Bracke, M.B.M., K. de Greef, C. van der Peet-Schwering, M. Gerritzen and H. Vermeer. In press. Bouwstenen voor een risicobeoordeling dierenwelzijn in de varkensketen [Building blocks for a risk assessment of animal welfare in the pig chain]. Wageningen Livestock Research. https://doi.org/10.18174/541702.

Bracke, M.B.M., I.J.J.M. Boumans, H.J. Nijland, and E.A.M. Bokkers. 2023. Review: connecting circularity to animal welfare calls for a 'novel' conceptual framework based on integrity. Animal, 17: 100694. https://doi.org/10.1016/j.animal.2022.100694.

Brambell, R., 1965. Report of the technical committee to enquire into the welfare of animals kept under intensive livestock husbandry systems.

Brom, F.W.A., 1997. Onherstelbaar verbeterd (irrepairably improved). Assen: Van Gorcum.

Camerlink, I. and Ursinus, W.W., 2020. Tail postures and tail motion in pigs: a review. Applied Animal Behaviour Science 230, 105079. https://doi.org/https://doi.org/10.1016/j.applanim.2020.105079.

De Briyne, N., Berg, C., Blaha, T., Palzer, A. and Temple, D., 2018. Phasing out pig tail docking in the EU – present state, challenges and possibilities. Porcine Health Management 4 (27). https://doi.org/10.1186/s40813-018-0103-8.

DeGrazia, D., 2002. Animal rights: a very short introduction. New York: Oxford University Press.

De Jonge, Iris, Marc Bracke, Theun Vellinga, Wijnand Sukkel and Gerard Migchels, 2023. Het perspectief van biologische landbouw. Effecten van het vergroten van het areaal biologische akkerbouw en melkveehouderij op klimaat, natuur en dierenwelzijn. Rapport. Wageningen Livestock Research. 10.18174/629169.

De Lauwere, C., Zonderland, J. and Bracke, M., 2010. 'Stoppen met staarten couperen geen utopie'. V-focus: Varkens & Pluimvee.

Donovan, J. and Adams, C. (eds), 2007. The feminist care tradition in animal ethics. New York: Columbia University Press.

Edwards, A. and Valros, A., 2021. Understanding and preventing tail biting in pigs In S. Edwards (Ed.), Understanding the behaviour and improving the welfare of pigs [BDSP-BK2019-0338] (Burleigh Dodds Series in Agricultural Science). Burleigh Dodds Science Publishing.

EFSA (European Food Safety Authority), 2007. The risks associated with tail biting in pigs and possible means to reduce the need for tail docking considering the different housing and husbandry systems. Scientific Opinion of the Panel on Animal Health and Welfare. The EFSA Journal 611, 1–13.

Engster, D., 2006. Care ethics and animal welfare. Journal of Social Philosophy 37(4): 521–536.

EFSA (European Food Safety Authority), 2012. Scientific opinion on the welfare of cattle kept for beef production and the welfare in intensive calf farming systems. https://www.efsa.europa.eu/en/efsajournal/pub/2669.

EFSA (European Food Safety Authority), 2022. Welfare of calves. https://doi.org/10.2903/j.efsa.2023.7896.

Eskens, E. 2021. Comment: animals in 'non-ideal ethics' and 'no-deal ethics'. In: B. Bovenkerk and J. Keulartz (eds), Animals in our midst: the challenges of co-existing with animals in the anthropocene. Cham: Springer, 275–284.

Francione, G., 2010. Animal welfare and the moral value of nonhuman animals. Law, Culture and the Humanities 6(24). https://doi.org/10.1177/1743872109348989.

Fraser, D., 2003. Assessing animal welfare at the farm and group level: The interplay of science and values. Animal Welfare, 12, 433–443. https://doi.org/10.1017/S0962728600026038.

Garner, R., 2013. A theory of Justice for Animals. Oxford: Oxford University Press.

Harfeld, J.L., Cornou, C., Kornum, A., and Gjerris, M., 2016. Seeing the animal: on the ethical implications of de-animalization in intensive animal production systems. Journal of Agricultural and Environmental Ethics, 29, 407–423. https://doi.org/10.1007/s10806-016-9611-1.

Haynes, R.P., 2008. Animal welfare. Competing conceptions and their ethical implications. Dordrecht: Springer.

Held, V., 2006. The ethics of care: personal, political, and global. Oxford University Press.

Herskin, M.S., Di Giminiani, P. and Thodberg, K., 2016. Effects of administration of a local anaesthetic and/or an NSAID and of docking length on the behaviour of piglets during 5 h after tail docking Research in Veterinary Science 108, 60–67. https://doi.org/10.1016/j.rvsc.2016.08.001.

Hopster, H. and Bergsma, K., 2016. Vroegtijdig scheiden van melkkoe en kalf. Implicaties voor gedrag en gezondheid in het licht van de gangbare praktijk. Van Hall Larenstein, University for Applied Sciences.

Hursthouse, R., 2011. Virtue ethics and the treatment of animals. In: Beauchamp, T.L. and Frey, R.G., The Oxford handbook of animal ethics. https://doi.org/10.1093/oxfordhb/9780195371963.013.0005.

Ison, S.H. and Rutherford, K.M.D., 2014. Attitudes of farmers and veterinarians towards pain and the use of pain relief in pigs. The Veterinary Journal 202 (3): 622–7. https://doi.org/10.1016/j.tvjl.2014.10.003.

Kaldewaij, F., 2013. Does fish welfare matter? On the moral relevance of agency. Journal of Agricultural and Environmental Ethics, 26, 63–74. https://doi.org/10.1007/s10806-011-9372-9.

Knight, C., 2017. Reflective equilibrium. In: Blau, A. (ed.), Methods in analytical political theory. Cambridge: Cambridge University Press. https://doi.org/10.1017/9781316162576.005.

Korsgaard, C., 2011. Interacting with animals: a Kantian account. In T. Beauchamp & R.G. Frey (Eds.), The Oxford handbook of animal ethics. New York: Oxford University Press.

Korte, S.M., Olivier, B., and Koolhaas, J.M., 2007. A new animal welfare concept based on allostasis. Physiology & Behavior, 92(3), 422–428. https://doi.org/10.1016/j.physbeh.2006.10.018.

Mullan, S., Edwards, S.A., Butterworth, A., Whay, H.R. and Main, D.C.J., 2011. A pilot investigation of Farm Assurance assessors' attitude to farm animal welfare as a confounding factor to training in pig welfare outcome measures. Animal Welfare 20, 413–421 https://doi.org/10.1017/S0962728600002979.

Niemi, J.K., Edwards, S.A., Papanastasiou, D.K., Piette, D., Stygar, A.H., Wallenbeck, A. and Valrors, A., 2021. Cost-effectiveness analysis of seven measures to reduce tail biting lesions in fattening pigs. Frontiers of Veterinary Science. https://doi.org/10.3389/fvets.2021.682330.

Noddings, N., 1984. Caring: a feminine approach to ethics and moral education. Berkeley and Lost Angeles: University of California Press.

Ohl, F., and Van der Staay, F.J., 2012. Animal welfare: at the interface between science and society. The Veterinary Journal 19: 13–19. https://doi.org/10.1016/j.tvjl.2011.05.019.

Palmer, C., 2011. Animal ethics in context. New York: Columbia University Press.

Prunier, A., Tallet, C., and Sandercock, D.A., 2021. Evidence of pain in piglets subjected to invasive management procedures. In S. Edwards (Ed.), Understanding the behaviour and improving the welfare of pigs [BDSP-BK2019-0338] (Burleigh Dodds Series in Agricultural Science). Burleigh Dodds Science Publishing.

Rawls, J., 1971. A theory of justice, original edition. Cambridge, London: The Belknap Press of Harvard University Press.

Regan, T., 1983. The case for animal rights. University of California Press, Berkeley.

Ruis, M.A.W., 2016. Weet wat je eet: keurmerken voor een beter leven – Dierenwelzijn in beeld.

Rutgers, L.J.E., and Heeger, F.R., 1999. Inherent worth and respect for animal integrity. In: Dol, M. *et al.*, (ed), Recognizing the intrinsic value of animals: beyond animal welfare. Assen: Van Gorcum, pp. 41–51.

Rutgers, B., 1998. The use of the reflective equilibrium method in normative veterinary ethics. In: Van der Burg, W., van Willigenburg, T. (eds), Reflective equilibrium. Essays in honour of Robert Heeger. Kluwer Academic Publishers.

Schmidt, K., 2011. Concepts of animal welfare in relation to positions in animal ethics. Acta Biotheor 59: 153–171. https://doi.org/10.1007/s10441-011-9128-y.

Scollo, A., Contiero, B. and Gottardo, E., 2016. Frequency of tail lesions and risk factors for tail biting in heavy pig production from weaning to 170 kg live weight. The Veterinary Journal 207, 92–98 https://doi.org/10.1016/j.tvjl.2015.10.056.

Shriver, A., 2009. Knocking out pain in livestock: can technology succeed where morality has stalled? Neuroethics 2(3): 115–124. https://doi.org/10.1007/s12152-009-9048-6.

Singer, P., 2011. Practical ethics. Cambridge: Cambridge University Press.

Singer, P., 1975. Animal liberation. New York: New York Review.

Stemplowska, Z., 2008. What's ideal about ideal theory? Social Theory and Practice 34 (3), 319–340. https://doi.org/10.5840/soctheorpract200834320.

Timmons, M., 2013. Moral theory. An introduction (second edition). Lanham: Rowmann and Littlefield Publishers.

Thodberg, K., Herskin, M.S., Jensen, T. and Jensen, K.H., 2018. The effect of docking length on the risk of tail biting, tail-directed behaviour, aggression and activity level of growing pigs kept under commercial conditions. Animal 12:2609–2618. https://doi.org/10.1017/S1751731118000563.

VandeVeer, D., 1979. Interspecific justice. Inquiry: An Interdisciplinary Journal of Philosophy 22 (1–4): 55–79. https://doi.org/10.1080/00201747908601866.

Van der Burg, W., van Willigenburg, T. (eds), 1998. Reflective equilibrium. Essays in honour of Robert Heeger. Kluwer Academic Publishers.

Valentini, L. 2012. Ideal vs. non-ideal theory: a conceptual map: ideal vs non-ideal theory. Philosophy Compass 7 (9): 654–664. https://doi.org/10.1111/j.1747-9991.2012.00500.x.

Valros, A, 2021: Review: the tale of the Finnish pig tail – how to manage non-docked pigs? Animal, in press: https://doi.org/10.1016/j.animal.2021.100353.

Valros, A. and Heinonen, M., 2015. Save the pig tail. Porcine Health Management 1 (2). https://doi.org/10.1186/2055-5660-1-2.

Valros, A., Munsterhjelm, C., Hänninen, L., Kauppinen, T. and Heinonen, M., 2016. Managing undocked pigs – on-farm prevention of tail biting and attitudes towards tail biting and docking. Porcine Health Management 2:2, https://doi.org/10.1186/s40813-016-0020-7.

Valros, A., Välimäki, E., Nordgren, H., Vugts, J., Fàbrega, E. and Heinonen, M., 2020. Intact tails as a welfare indicator in finishing pigs? Scoring of tail lesions and defining intact tails in undocked pigs at the abattoir. Frontiers in Vet Sci 7:405. https://doi.org/10.3389/fvets.2020.00405.

Višak, T., 2011. Killing happy animals: explorations in utilitarian ethics.

WHO (World Health Organization), 2015. Cancer: carcinogenicity of the consumption of red meat and processd meat. https://www.who.int/news-room/questions-and-answers/item/cancer-carcinogenicity-of-the-consumption-of-red-meat-and-processed-meat.

Zonderland, J.J., Wolthuis-Fillerup, M., van Reenen, C.G., Bracke, M.B.M., Kemp, B., den Hartog, L.A. and Spoolder, H.A.M., 2008. Prevention and treatment of tail biting in weaned piglets. Applied Animal Behaviour Science 110 (3–4): 269–281. 10.1016/j.applanim.2007.04.005.

Conclusion

Keelin O'Driscoll and Anna Valros

This book serves as a testament to the volume of academic effort that has gone into tackling the problem of tail biting in pigs in recent decades. For several of the topics covered in the book the number of scientific papers and reports (e.g. those produced by the European Food Safety Authority, EFSA) that were available to draw upon was so great that it was challenging to condense the content into meaningfully concise chapters. Nevertheless, despite this enormous investment of time and resources, there has been very little change over time in most countries when it comes to the proportion of pigs' tails that are docked, and the amount of damage inflicted by biting that is visible on cleaned carcases on the slaughter-line. Along with this, the fact that EU legislation and other national governmental policies enacted to phase out the practice of docking is practically ignored in most countries means that research and policy alone are not sufficient to bring about systemic change.

We know from both research and from working directly with stakeholders in the pig industry that tail biting is almost a certainty at some stage when rearing pigs commercially. It is therefore understandable why producers, the majority of whom only have experience of the intensive indoor type system that is now most prevalent, are resistant to transitioning away from what they consider as a harmless procedure – docking – as a preventative measure. In fact, in systems where docking is routine, producers may consider it the most welfare friendly course of action available to prevent tail biting (Valros and Barber, 2021). However, even though producers often deem even low levels of biting to be problematic, a positive correlation between the perceived seriousness of biting and the amount that the producer sees as acceptable has been reported (Valros *et al.*, 2016, Haigh *et al.*, 2019; Kakanis *et al.*, 2023). It may be the case that in farms where tail biting regularly occurs, producers are so accustomed to its occurrence that the frequency or severity threshold for it to be a considered a serious adverse event is higher than on farms where it is more infrequently observed.

Specific risk factors that can cause sudden or extreme tail biting are often easily identifiable. These are the factors that are most often researched, with several discussed in depth in Chapters 3 through 9, with detail of how to practically assess their presence in Chapter 10. Producers in different countries often rank these most common risk factors in order of seriousness in a broadly similar order, indicating validity in the level of associated risk when it comes to practical animal management in typical systems. However, there is often a lack of awareness that tail biting can result from long term, accumulation of stressors. Chronic stress in pigs typically results from a mismatch between their highly driven innate motivations,

 | DOI:10.3920/9789004694989_019

determined by their evolutionary history (chapter 3), and their living conditions in commercial systems. Acute events that on the surface appear to be the sole triggers of outbreaks thus often simply act as tipping points, the straw that breaks their backs. This feeds into the reported difficulty in identification of specific causes of outbreaks, and in predicting and preventing them. The fact that tail docking is not a cure for biting, but simply a 'band aid' that enables management of pigs in systems that cause chronic stress (evidenced by tail biting) is not widely understood. In fact, Albernaz-Gonçalves *et al.* (2021) reported that pig producers felt that practices such as antibiotic use (and tail docking) that alleviate the risk of health issues are all that is needed to maintain good welfare. In that study, producers felt that abnormal behaviours '*are inherent to animal production and not a welfare concern*'.

Simplistically then, it could be suggested that improvements to the way that producers and stockpersons, at both individual and industry level, are informed and educated about animal welfare would result in a shift in how animals are managed. However, provision of information is clearly not enough, evidenced by the lack of impact of the ever growing, and outwardly communicated, scientific knowledge and understanding of the area. Research into the causes of and methods to alleviate the risk of tail biting has increased exponentially during the past 15 years, and it has become a consistent requirement of funding that comprehensive communication and dissemination plans are defined at proposal stage, with the aim of ensuring that scientific results have real world impact. As well as establishing the EU Reference Centre for Animal Welfare on Pig (described in Chapter 1), the EU funded a Cost action (GroupHouseNet, https://www.cost.eu/actions/CA15134/ 2016–2020) that aimed to '*provide the European livestock industry with innovations in breeding and management for pigs … that are needed … without necessitating painful tail docking …*', and in 2022 launched a thematic sub-group on pigs within the EU Platform on Animal Welfare to '*provide technical expertise and opinions to the Commission on the ways the Union's instruments can be adapted to improve the welfare of pigs at farm level*.' To date, however, none of these endeavours have resulted in meaningful change in the EU. Increasingly, scientists and policy makers are aware that to change human behaviour in any context, it is essential to understand better the underlying motivations and capabilities of the actors that are targeted. The information provided in Chapter 12 forms a hugely valuable introduction to this scientific area, which will likely be pivotal in accomplishing the aim to phase out tail docking, as top down policy and communication of research outputs has failed.

Nevertheless, a significant barrier to change, regardless of whether or not stakeholders in the industry are motivated themselves to stop docking, is that in most countries the pig industry operates in a challenging financial environment, with minimal physical or economic elasticity built in. Modern pig farms are enclosed systems, in the main constructed using concrete and other slowly degrading materials, with the lifespan of infrastructure often 25 years or more; thus once a system is in place, without external financial supports it is economically unfeasible to change it, potentially for decades. This issue is understood by law makers, evidenced by the long lead in time before enforcement of laws related to system changes such as the move towards group housing of sows, and in poultry the requirement to provide laying hens with enriched cages. To date there has been limited co-ordination even across the EU when it comes to putting financial supports or incentives in place which specifically work towards reducing the risk of tail biting. Nevertheless, both within and outside the EU, there is growing awareness in the commercial sphere of a potentially untapped market when it comes to higher welfare pig meat. Quality assured pig-meat, produced to standards that require primary producers to comply with higher welfare standards, and that reward them with a higher pig price, are emerging and growing in market share globally. In some instances, these have been developed by government agencies (e.g the Danish 3 Heart Welfare Scheme) and by NGO's (e.g. RSPCA assured), but often they are established by commercial entities such as retailer chains (e.g. Waitrose, UK). These assurance schemes have not necessarily emerged due to a sense of benevolence toward the pigs or producers, but due to commercial awareness of changing consumer demands, and the potential to develop and exploit untapped markets. The EU has already established a compulsory labelling standard when it comes to eggs; the development of a similar, easily understood system in relation to pig meat production would benefit not only consumers, but producers who have an interest in becoming early adopters of higher welfare systems.

Finally, although tail biting is hugely appreciated as being a welfare issue for pigs, it also impacts the welfare of both humans and the environment, so should really be considered a One Welfare issue. The concept of One Welfare is relatively new, building upon the One Health idea recognising the interconnections between animal welfare, human wellbeing and the environment (García Pinillos *et al.*, 2016). There is a strong link between animal welfare and human well-being, and the impact of routinely dealing with injured and sick animals due to tail biting can only have a negative impact on the humans who are responsible for immediately dealing with the affected animals. Beyond this, the direct and indirect financial consequences of tail biting outlined in Chapter 15 have a longer term negative impact on the viability of individual pig enterprises, affecting both owners and their employees. Moreover, treatment of wounds resulting from tail biting outbreaks often necessitates use of antimicrobials, potentially contributing to the risk of antimicrobial resistance, and which has impacts far beyond the farm gate. With regard to the environment, there is empirical evidence that tail biting has a negative impact on pig performance and efficiency, and thus on farms where tail biting is endemic a greater proportion of environmentally impactful resources such as feed and energy are used than in farms where it is not prevalent. Consideration of the negative impact that current pig production systems have not only on pig, but also on human and environmental well being can only accelerate the move towards systems that improve welfare standards for all.

The changes to the systems that are typically used to rear pigs that have occurred during the past century have resulted in a situation where it is almost unimaginable that tail biting will cease to be a routine, yet significant problem in most countries, if pigs are not docked. The optimisation of performance and efficiency at the expense of all else, which has grown from a complex relationship between economic necessity, government policy and scientific knowledge, has resulted a culture in the pig industry of considering increases in animal output, while maintaining or minimising inputs, to be termed as 'improvements' or 'advances' in pig production. The contents of this book form a convincing argument as to how this strategy and its resultant approach to pig management, while initially well intentioned, is not only outdated but hugely damaging. A century is but the blink of an eye not only in evolutionary but in historical terms, so in theory it should be a simple task to acknowledge and reverse the mistakes that have been made during the past 100 years. However, the unprecedented progress that has been made in producing healthy and affordable food makes this hugely challenging. Nevertheless, we have reached the point where it is essential, for the sake of animals, humans, and the planet, that swift and effective action is taken, so that tail biting in pigs becomes an issue of the past.

References

Albernaz-Gonçalves, R., Olmos, G., and Hötzel, M.J. 2021. My pigs are ok, why change? – animal welfare accounts of pig farmers. Animal 15:100154 https://doi.org/10.1016/j.animal.2020.100154.

Bracke, M.B.M., De Lauwere, C.C., Wind, S.M.M. and Zonerland, J.J. 2013. Attitudes of Dutch pig farmers towards tail biting and tail docking. J Agric Environ Ethics 26:847–868 https://doi.org/10.1007/s10806-012-9410-2.

Haigh, A. and O'Driscoll, K. 2019. Irish pig farmer's perceptions and experiences of tail and ear biting. Porcine Health Management 5:30 https://doi.org/10.1186/s40813-019-0135-8.

Kakanis, M., Marinou, K., and Sossidou, E.N. 2023. Greek pig farmers' perceptions and experiences of tail biting and tail docking. Animals 13:672. https://doi.org/10.3390/ani13040672.

García Pinillos, R., Appleby, M., Manteca, X., Scott-Park, F., Smith, C. and Velarde, A. 2016. One Welfare – a platform for improving human and animal welfare. Veterinary Record 413. https://doi.org/10.1136/vr.i5470.

Valros, A. and Barber, C. 2019. Producer perceptions of the prevention of tail biting on UK farms: association to bedding use and tail removal proportion. Animals 9:628 https://doi.org/10.3390/ani9090628.

Valros, A., Munsterhjelm, C., Hänninen, L., Kauppinen, T. and Heinonen, M. 2016. Managing undocked pigs – on-farm prevention of tail biting and attitudes towards tail biting and docking. Porcine Health Management 2:2 https://doi.org/10.1186/s40813-016-0020-7.

www.ingramcontent.com/pod-product-compliance
Lightning Source LLC
Chambersburg PA
CBHW080936240926
51849CB00004BA/469
9789004691346